AF849062

Caucasus and Central Asia Political, Social and Economic Issues

Biomes of the Caucasus

A Comprehensive Review

Caucasus and Central Asia Political, Social and Economic Issues

Additional books and e-books in this series can be found on Nova's website under the Series tab.

This book is dedicated to my teachers, colleagues, friends and internationally renowned biologists Prof. Dr. Alexander F. Emelyanov

and

(to the memory of) Prof. Dr. Izyaslav M. Kerzhner,

Zoological Institute of the Russian Academy of Sciences (RAS)

CAUCASUS AND CENTRAL ASIA POLITICAL, SOCIAL AND ECONOMIC ISSUES

BIOMES OF THE CAUCASUS

A COMPREHENSIVE REVIEW

ARNOLD GEGECHKORI

nova science publishers
New York

Copyright © 2020 by Nova Science Publishers, Inc.

All rights reserved. No part of this book may be reproduced, stored in a retrieval system or transmitted in any form or by any means: electronic, electrostatic, magnetic, tape, mechanical photocopying, recording or otherwise without the written permission of the Publisher.

We have partnered with Copyright Clearance Center to make it easy for you to obtain permissions to reuse content from this publication. Simply navigate to this publication's page on Nova's website and locate the "Get Permission" button below the title description. This button is linked directly to the title's permission page on copyright.com. Alternatively, you can visit copyright.com and search by title, ISBN, or ISSN.

For further questions about using the service on copyright.com, please contact:
Copyright Clearance Center
Phone: +1-(978) 750-8400 Fax: +1-(978) 750-4470 E-mail: info@copyright.com.

NOTICE TO THE READER

The Publisher has taken reasonable care in the preparation of this book, but makes no expressed or implied warranty of any kind and assumes no responsibility for any errors or omissions. No liability is assumed for incidental or consequential damages in connection with or arising out of information contained in this book. The Publisher shall not be liable for any special, consequential, or exemplary damages resulting, in whole or in part, from the readers' use of, or reliance upon, this material. Any parts of this book based on government reports are so indicated and copyright is claimed for those parts to the extent applicable to compilations of such works.

Independent verification should be sought for any data, advice or recommendations contained in this book. In addition, no responsibility is assumed by the Publisher for any injury and/or damage to persons or property arising from any methods, products, instructions, ideas or otherwise contained in this publication.

This publication is designed to provide accurate and authoritative information with regard to the subject matter covered herein. It is sold with the clear understanding that the Publisher is not engaged in rendering legal or any other professional services. If legal or any other expert assistance is required, the services of a competent person should be sought. FROM A DECLARATION OF PARTICIPANTS JOINTLY ADOPTED BY A COMMITTEE OF THE AMERICAN BAR ASSOCIATION AND A COMMITTEE OF PUBLISHERS.

Additional color graphics may be available in the e-book version of this book.

All pictures belong to the author (data of pictures are given in brackets)

Library of Congress Cataloging-in-Publication Data

Name: Gegečkori, Arnold, 1940- author.
Title: Biomes of the Caucasus: a comprehensive review / Arnold Gegechkori, Biomes of the Caucasus (BCs), Head of Biodiversity Division of Faculty of Exact and Natural Sciences, Tbilisi State University.
Identifiers: LCCN 2019006909 | ISBN 9781536152340 (hardcover) | ISBN 9781536152357 (ebook)
Subjects: LCSH: Biotic communities--Caucasus.
Classification: LCC QH540.83.C28 G44 2019 | DDC 577.8/2--dc23
LC record available at https://lccn.loc.gov/2019006909
ISBN: 978-1-53615-234-0

Published by Nova Science Publishers, Inc. † New York

CONTENTS

ACKNOWLEDGEMENTS

I am very grateful to Prof. Dr. I.D. Batiashvili, Head of the Chair of Entomology of the University of Agriculture, who has created opportunities for me to find a way to science in the Center of Zoological and Biogeographical Studies, in the Zoological Institute, Russian Academy of Sciences, St. Petersburg (former Leningrad).

I am deeply indepted to Dr. M.M. Loginova who get me started with the taxonomy of Psylloidea (Hemiptera). I would like to express here a special, great thanks to my colleague and good friend Prof. Dr. A.F. Emeljanov, who introduced me to the field of such an interdisciplinary branch of natural science namely historical biogeography. My warmest thanks are extended to outstanding entomologists – 'face of ZIN' Prof. Dr. I.M. Kerzhner for his encouragement while writing my doctoral dissertation, a checklist of the Psylloidea of the USSR, for his insightful advice and useful discussions on 'how to make science' and for creating a nice atmosphere at the Hemiptera Department (to the latter two persons I dedicate the present monograph). I would like to thank Prof. Dr. O.L. Krizhanovski, an outstanding zoogeographer, for examining my comprehensive monographs. I owe him a debt of gratitude for his encouragement and advice to make comparable field expeditions out of the Cacuasus in various regions of the former USSR and also for intellectual guidance.

I would like to thank very much the Director of the former State Museum of Georgia, Prof. Dr. L. Chilashvili, Deputy Director Prof. Dr. A. Djavakhishvili (an incomparable thinker) of the same Museum, for implementation of field trips across the Caucasus and, travel opportunities in all regions (Middle Asia, Siberia, Russian the Far East, etc.). To the General Director of the Georgian National Museum Prof. Dr. D. Lordkipanidze, a world renowned paleanthropologist, for kindly supporting my scientific activity until the present. I thank some foundations: the George Soros Foundation, WWF, UNESCO Scientific Programme 'Man and Biosphere' and the Academy of Science of Georgia which gave me grants ten times. Many thanks to Deputy Director Prof. Dr. V.F. Zaitsev, for arranging field expeditions in Middle (Central) Asia via the Zoological Institute (RAN) of Tadjikistan - Dr. V. Bajeva, in the Russian Far East - Dr. A. Konovalova, in northwestern Iran - Dr. M. Soufi and Dr. N. Eslami, and others.

Sometimes, the present monograph requires a comparison of its biomes with biomes and ecosystems of other countries, regions and continents. Therefore, it needs collaborations with other people and institutions. I would like to express my warmest thanks to Prof. Dr. E. Khmaladze (UNSW, Sydney) and his family for financial support enabling me to make field expeditions in Tasmania and New Zealand. I want to extent my thanks to Prof. Dr. C.

Sutherland, Prof. Dr. P. Greenaway, Prof. Dr. B. Fox, Prof. Dr. M.Fox, Prof. P. Adams, Dr. H. Martin, Dr. S. Mooney, Dr. P. Banks (UNSW, Sydney), Dr. J. Cassis (Institute of Biodiversity, Sydney), Prof. Dr. M. N. Clayton, Dr. A. Zaslavsky, Dr. F. Burstein (Monash University, Melbourne); Prof. Dr. L.W. Irungu, Prof. Dr. M.M. Gaguchi (University of Nairobi, Kenya), Mr. E.J. Gerenta (general director of National Parks of Tanzania); Dr. N. Penny, biologist J. Kinyon (Academy of Sciences, San Francisco, USA) and Dr. Y. Kakabadze, former Minister of the Environment of the Republic of Ecuador (1998-2000). All of the above mentioned persons supported me all the way up to the submission of my monographs (especially biogeography), including this present book.

I am always grateful to all the working group of Prof. Dr. H. Joosten (and his family) at the Greifswald University (Germany) for generously offering their time, for his good scientific advice and his input by reading some sections of a draft of this manuscript; biologist Mathias Krebs for his valuable support in many ways and to Prof. Dr. I. Abashidze for providing valuable information about the arboreal plants of the Caucasus and adjacent regions.

I really appreciate the advice and encouragement of Prof. Dr. A. Battimer, geographer and former president of IGU (2000-2004), Prof. Dr. M. Archer, director of the Australian Museum (1999-2004), and Prof. Dr. E.H. Blum Winfried, pedologist, Secretary-General of IUSS.

I also express thanks to the staff of the Department of Zoology of the State Museum of Georgia and the laboratories of the Chair of Biodiversity of the Ivane Javakhishvili Tbilisi State University, especially to Maia Chubinidze for her tireless support to finish the manuscript of the present book. I would like to thank Dr Trevor Cartledge, my Editor and proof reader for his efforts and dedication during the latter stages of the production of the manuscript. I would like to personally thank all my family members for their continued support throughout my endless field expeditions and scientific missions abroad. In conclusion, most certainly, my cordial thanks to the breathtaking nature of the Caucasus, significantly of Georgia, which inspired so many poets, writers, artists, scientists….; to which was dedicated my recent scientific-literature, color book-album 'This is Georgia' (in Georgian and English, 544pp.).

INTRODUCTION

The Caucasus is a geopolitical region which encompasses four independent countries. The North Caucasus includes the following republics of the Russian Federation: Adygeya, Karachayevo–Cherkessia, Kabardino–Balkaria, Northern Ossetia, Ingushetia, Chechnya, Dagestan, and parts of the Krasnodar and Stavropol Provinces. The area to the South of the Greater Caucasus Range, i.e., the South Caucasus, includes all of Armenia, Azerbaijan, and Georgia (Gvozdetsky 1963).

The area is situated between the Black Sea (west) and the Caspian Sea (east) in a triangle of three massive mountain peaks; the Elbrus (5,642 m a. s. l., the highest summit) in Kabardino-Balkaria, the Kazbek (5,033 m a.s.l.) in Georgia, and the Ararat (5,165 m a.s.l.) found in Turkey, close to the Armenian border (Gvozdetsky 1954, 1963).

The beauty of the Caucasus is much celebrated in Russian literature, most notably in Pushkin's poem 'Captive of the Caucasus', Lermontov's novel 'A Hero of Our Time' and Tolstoy's novels 'The Cossacks and Hadji Murad'

The Caucasus is not only the natural bridge between Europe and Asia. Throughout its history, the region was the permanent battle-ground between two continents of several dominant imperial powers, namely the Greeks, Romans, Persians, Turkish Ottomans and Mongolians, before the region, over the course of 200 years, was finally incorporated into the Russian Empire (Longuet-Marx (Ed.) 1998). This region is still under the threat of post-Soviet wars and interethnic conflicts at different levels (O'Loughlin *et al.* 2007; Coene 2010).

The region covers about 441,000 km^2 (Gvozdetsky 1954; Dumitrashko 1966, 1977). It should be stressed that the definitions of 'Caucasus Biodiversity Hotspot' and 'Caucasus Ecoregion' concepts also includes parts of northeastern Turkey (the Black Sea coastal area) and part of northwestern Iran (the Caspian Sea coastal region), covering a total area of 580.000 km^2 (Williams *et al.* (Eds.) 2006).

The word 'Kavkaz' is very old though the origin of the word has not been determined exactly. Pliny the Elder's *Natural History* (AD 77-79) derives the name of the Caucasus from the Scythian *kroy-khasis* ('ice-shining, white with snow') (*https://en.wikipedia.org/wiki/Caucasus*). According to the other version it may be connected with the Hittite *Kaz-kaz*, the name of the people who lived on the southern coast of the Black Sea. For the first time it was mentioned (as Káukanoç) in the Greek tragedy Prometheus is bound by Aeschylus, the ancient Greek dramatist (*c. 525 BC/524 BC – c. 456 BC/455 BC*) in which Prometheus is bound to a rock as punishment from Zeus for providing fire to the mortal humans (*encyclopedia2.thefreedictionary.com/Kavkaz*).

According to the 11th century Georgian chronicler Leonti Mroveli, Kavkas was a son of Togarmah and grandson of biblical Noah's third son Japheth. After the fall of the Tower of Babel and the division of humanity into different languages Togarmah settled with his sons Kartlos, Haik, Movakos, Lekos, Heros, Kavkas, and Egros between two inaccessible mountains, presumably Mount Ararat and Mount Elbrus (Qoranashvili 1996).

The Czech linguist and archaeologist Betrich Hrozny suggested that Kavkaz comes from the word 'Kash' or 'Kuoh', the name of the land of an early Indian civilization, which has not left any written documents (Hollubek & Krikva 2006).

Scientific names of vascular plants follow Czerepanov (2007), for mammals Wilson & Reeder (2005) and for birds Beaman (1994). The geological time scale of the Earth's history (terminology), sequence and time of geological periodization during the last 600 mya follow Eicher and McAlester (1980). Consulted dictionaries of biological and a more general scientific nature include Grebenshikov (1965); Reimers (1990); IUCN multilingual dictionary of conservation terms (1976); Afanasjeva *et al.* (1979); Menabde (1983); Makashvili (1995) and Maruashvili (1985).

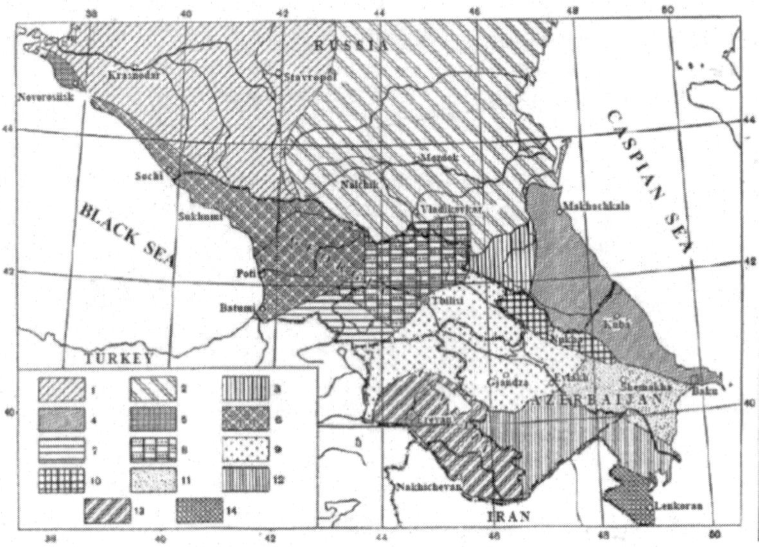

Figure 1. Schematic map of the natural-historical regions of the Caucasus. 1. Western part of the North Caucasus; 2. Central part of the North Caucasus; 3. Inner Dagestan; 4. Eastern part of the North Caucasus; 5. Novorossiiskaya Oblast (Region); 6. West Caucasus; 7. Meskhet-Javakheti; 8. Upper Kartalinia (Kartli); 9. Central Transcaucasia; 10. Alazani-Avtorani Region; 11. Shirvani Region; 12. Zangezur-Karabakh region; 13. Southern Transcaucasia (middle reaches of the Aras); 14. Talysh (Gulisashvili 1964).

Georgian terms (ancient, classical, modern, biological and other natural sciences) were translated into English using Rayfield (Ed-in-chief) (2006).

Globally threatened species are indicated using the IUCN criteria with the following abbreviations: EX-extinct, EW-extinct in the wild, CR-critically endangered, EN-endangered, VU-vulnerable, NT-near threatened, LC-least concern, DD-data deficient, NE-not evaluated.

Basically, only current publications that involved references to older literature are included because inclusion of all of the primary literature would have resulted in an excessively long bibliography.

Terms and concepts may differ between English and Russian. Examples include the usage of the terms Middle Asia and Central Asia, Arcto-Tertiary Flora and Turgai Flora, and finally soil classification. As far as is relevant these differences are explained in the text.

Psyllids (Hemiptera, Psylloidea) as a Key Indicator Group for Understanding of the Caucasian Biota

The overview of biomes is first and foremost the result of my long-term comprehensive (faunistic, floristic, biogeographical) studies of the wildlife of the Caucasus proper and Caucasus Ecoregion in general. In this study I focused first of all on the psyllids or jumping plant-lice (Insecta: Hemiptera, Psylloidea).

Psyllids are small (1-5mm) phloem feeding insects that are typically monophagous (feed on a single plant species) or narrowly oligophagous (feed on a few related species). In their nymphal stages in particular, they inhabit almost exclusively perennial dicotyledonous plants (Gegechkori 1968; Loginova 1968; Hodkinson 1974; 2009, Burckhardt 1989). There are about 4000 described species distributed world-wide in all biogeographical regions (Li 2011; Burckhardt, pers. comment). Together with aphids, scales and white-flies they form the monophyletic group, Sternorrhyncha (Homoptera), which was until recently considered basal within the true bugs (Hemiptera) (*Percy: www.psyllids.org/*). Psyllids have recently been considered as vectors of serious plant diseases, as economically important pests in agriculture and forestry and as potential control organisms of exotic invasive plants (Burckhardt & Ouvrard 2012). The current taxonomy of Psylloidea is provided by Burckhardt and Ouvrard (2012) based on a cladistic and phenic study of larval and adult morphological characters of the world fauna by White and Hodkinson (1985). The superfamily Psylloidea in the former Soviet Union contains about 520 species (Gegechkori & Loginova 1990) and in the Caucasus – about 210 species (Gegechkori 1984a) (Figure 2).

Figure 2. Gall-inducing psylloid *Trichochermes walkeri* feeds on *Rhamnus cathartica*.

The data were gathered during almost thirty years (1962-1990) of field work in all biomes and altitudinal zones of the 14 natural-historical regions of the Caucasus and in all vegetation seasons (early spring – late autumn). During the years 1990-2018 similar excursions were made but with a lower intensity. Field excursions involved all Strict Natural Reserves, National parks and Biosphere Reserves of the Caucasus. Data for interregional comparison were collected in the Caucasus Ecoregion (Turkish Caucasus, Iranian part of Hyrcanian refugium), Middle (Central) Asia, Siberia and the Russian Far East (1981-1991), in Asia Minor (Turkey, 1995-1998, 2006), Asia Interior (Iran, 2012) and in other parts of the world (1986, 1989, 2000-2003), involving all biomes of the world (Gegechkori 2008) and almost all 'Biodiversity hotspots' (CEPF 2000). The results are presented in various monographs (Gegechkori 1975a, 1978, 1984a, 1985, 2007, 2008; Gegechkori & Jibladze 1976; Gegechkori & Loginova 1990), and numerous articles.

Low mobilty of psyllids and, therefore, poor powers of dispersal allows them to be used successfully as a highly indicative (bio-indicator) group for environmental research (Hodkinson 1974, 2009; Loginova 1968; Gegechkori 1984a, b).

In the Caucasus and other mountainous countries psyllids display a well expressed altitudinal distribution with their host plants (Gegechkori 1984a; Conci *et al.* 1996).

Psyllids have an almost worldwide distribution (on all major land masses, except Antarctica). They occur in all terrestrial biomes from the tropics up to both of the subpolar regions, the Antarctic and tundra where suitable host plants are found (Hodkinson 1980, 2009).

All these and other ecological characteristics make Psylloidea a very useful group of invertebrate animals for studies of ecological biogeography in different biomes and habitats including the habitat-rich Caucasus region (Ecoregion) (e.g., *Some aspects of the ecology of Psylloidea of the Caucasus*) (Gegechkori 1984a: pp.70-132).

They are rather well fossilized (Loginova 1968). For example, in Dominican amber (Miocene) all extant families are represented except for Carsidaridae and Homotomidae (Burckhardt & Ouvrard 2012). From the Caucasus the single well-known fossil psyllid fauna (almost 32 species) is known from mid-Miocene deposits of the North Caucasus (the area of Stavropol and its surrounding). They belong mainly to species of two genera *Psylla* Geoffroy and *Trioza* Förster (*sensu lato*). The insect fauna looks rather modern, close to or congeneric with extant genera (Bekker-Migdisova 1964).

According to fossil materials and present distribution patterns of psylloids on the former Laurasian (Klimaszewski 1973; Hodkinson 1980, 1983, 1986, 1988; Gegechkori & Loginova 1990; Burckhardt & Basset 2000; Li, 2011) and Gondvanian continents (Hollis 1976, 1984, 2004; Hodkinson 1983, 1989; Burckhardt 1987, 1988a; Taylor 1990), they are obviously well affected by vicariant events. Psyllids are a very useful group in testing hypotheses about the reconstruction of the history of ancestral biotas in response to the changing geography of the Earth (e.g., *Distributional patterns of some psyllid taxa in southern countries in relation to plate tectonic theory* (Gegechkori 1985: pp.31-43) and *The origin and evolution of Psylloidea of different biomes of the Earth* (Gegechkori 1985: pp.46-233).

Psyllids, as mentioned, are generally highly host specific – monophagous or narrowly oligophagous, being restricted to just one or a few closely related host plants. Such close relationships between the hosts and consumers are hypothesized as a mechanism of a close corespondence between the phylogenies of the insects (psyllids) and their host plants (vascular, largely dicotyledonous) (Hodkinson 1984: 327pp.; Gegechkori 1985: pp. 187-194, pp. 251-275; Brown & Hodkinson 1988: pp. 280-284).

These and other materials help to determine the history of the formation of regional fauna, the biogeographic aspects, etc. (e.g., *The principal stages of formation of the Psylloidea of the Caucasus* (Gegechkori 1984a: pp. 269-278).

The present research and review have the aim to increase our knowledge about the set of some organisms inhabiting the Caucasus with particular reference to their habitats. The data presented here are meant to supply a database for nowadays, the presence and distribution of some species, particularly the keystone species. It is highly likely that these factors will be changed in species composition and their range and correlated with either climatic changes (e.g. global warming) or direct impacts (agriculture, tourism, pollution, poaching, etc.).

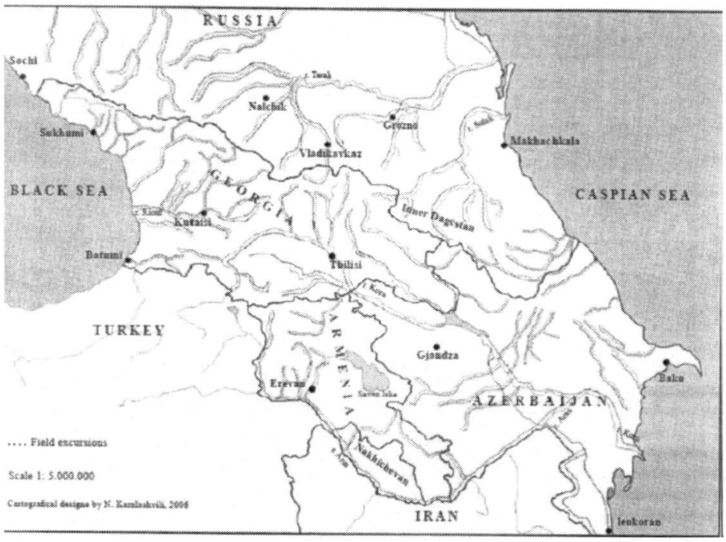

Figure 3. Field excursions of Dr. Arnold Gegechkori across the Caucasus (1962-2018).

Figure 4. The zoological expedition of a team of scholars of the Georgian National Museum, crossing the lower reaches of the river Aras (1972).

A Brief History of Botanical and Zoological Study in the Caucasus

The Caucasus has attracted the attention of geologists, geographers, archaeologists, and anthropologists for years. For 300 years, this region has already been of great interest to naturalists, especially botanists. From the beginning the initiative and guidance was in the hands of explorers, mainly from Western Europe, with Joseph de Tournefort (1656-1706) being the pioneer student of the Caucasus flora (1702-1703). Peter Simon Pallas (1741-1811), Samuel Gottlieb Gmelin (1744-1774), Johann Anton Güldenstädt (1745-1781), Friedrich August Marschall von Beberstein (1768-1826), Christian von Steven (1781-1863), Carl Anton von Meyer (1795-1855), Fridriech Hohenacker (1798-1874), Alexander Davidovič von Nordmann (1803-1866), Karl Koch (1809-1876), Ernest Rudolf Trautvetter (1809-1889), Pierre Edmond Boissier (1810-1885), Franz Ruprecht (1814-1870), Jakov Sergejevič Medvedev (1847-1923), Vladimir Ipolitovič Lipskij (1863-1937), Martin Rikli (1868-1951), Nikolaj Adolfovič Bush (1869-1941), and Eduard August Rübel (1876-1960) were among the first researchers. The first description of the vegetation zonation in the mountains of Georgia was provided in 1742 by the outstanding Georgian geographer Vakhushti Bagrationi (1696-1757).

An important impetus to the study of fauna and vegetation of the Caucasus was the establishment of the Caucasian Museum in Tbilisi[1] (1852), which, since 1866, has consisted of separate departments, for geology, botany and zoology. The Department of Zoology was headed by the incomparable naturalist Gustav Radde (1831-1903), who was also the director of the entire Museum (1866-1903).

Apart from the Caucasus, Radde organized field expeditions to adjacent countries and regions, including Iran, Turkey, Middle Asia, the Crimea and the Balkan Peninsula (1881-1901). Famous zoologists like Johan Karl Edward Ivanovich von Eichwald (1795-1876), Konstantin Ljudvigoviėc Bramson (1842-1909), Nikolai Yakovlevich Dinnik (1847-1917), Julius Lederer (1821-1870), Otto Staudinger (1830-1900), Hugo Feodorovič Khristof (1831-1894), Nicholas (Nicolay) Mikhailovič Romanov (1859-1919)[2], Ludwik Frantiszek Mlokosievich (1831-1909), Boris Petrovič Uvarov (1888-1970) and others contributed to the rich faunal collections of the Museum (Gegechkori & Joosten (Eds.) 2009).

N. M. Albov, F. Bayern, N. Kuznetsow, J. Medwedew, A. Schelkownikow, B. Shishkin, D. Sosnovsky, K. Steven, G. Woronow, and partially also the above-mentioned zoologists, especially Gustav Radde, participated in collecting the plants of the Caucasus (Gegechkori & Joosten (Eds.) 2009).

---

[1] Tbilisi (Georgian თბილისი) formerly (until 1936) known as Tiflis. Tblisi it is the capital of Georgia (in 1922-1936), the Imperial administration of the Caucausus during the Russian rule from 1801-1917, as well as the Transcaucasian Socialist Federative Soviet Republic from 1922 to 1936, etc. (*https://en.wikipedia.org/wiki/Tbilisi*)

[2] Grand duke Aleksander Mikhailovich of Russia (in Russian Александр Михайлович), was a branch of the Romanov dynasty. He was three years old when, in 1862, the family moved from Russia to Tiflis (Georgia) where his father, Grand Duke Michael Nikolaevich of Russia, was named Viceroy of the Caucasus. Nicholas spent his childhood and youth in Georgia, where his family (parents, brothers) lived for twenty years. From his youth in the Caucasus, Nicholas, a scholar and an eminent historian, developed an interest in wildlife (botany, entomology). He had rich collections of rare butterflies (part of his collections are deposited in the Georgian National Museum – A.G.). He published a ten-volume work entitled 'Discussions on the Lepidoptera' (*https://en.wikipedia.org/.../Grand_Duke_Nicholas_Mikhailovi...*).

As we see most of the eighteenth century investigations of natural history in the Caucasus were carried out by Western Europe's scholars (mainly Germans) via St. Petersburg's Academy of Sciences (founded in 1724 by order of Czar Peter the First. Researchers of European and then Russian origin prevailed in it for about 150 years.

According to Maharramova (2015), the first publication on the flora of the region was written by M. von Bieberstein in *Flora Taurico-Caucasica* (1808) who described about 2,000 Caucasian plant species. During the 19th century, Caucasian species were included in *Prodromus* of de Candolle (1824-1873), *Flora Rossiaca* of Ledebour (1841-1853, about 3,000 species), *Flora Orientalis* of Boissier (1867-1884) and *Flora of the Caucasus* of V.I. Lipskiy (1899, about 4,500 species). In 1901, N. Kuznetsov, N. Bush and A. Fomin began to publish *Flora Caucasica Critica*, but publication was interrupted during World War I and the Russian Revolution. A great contribution to the flora of the region was made by outstanding A.A. Grossheim, who published, among other titles, the comprehensive *Flora of the Caucasus* (1928-1967), *The analysis of the Caucasian Flora* (1936), *Manual of the Caucasian Plants* (1949), and the first *Flora of Azerbaijan (*1934-1936). The next major work describing Caucasian species was *Flora of the USSR* (V.L. Komarov (Ed.) 1934-1960). During the 20th century, many works on the flora of different regions of the Caucasus were published. By the 21st century, old publications were out-of-date and a new *Caucasian Flora Conspectus* was edited by the outstanding A.L. Takhtajan (2003-2008). First data on the endemic species of all the regions belonging to the Caucasus Ecoregion were given in the *Red list of the Endemic Plants of the Caucasus* (Solomon *et al.* 2014).

The first study of the fauna of the Caucasus was published by entomologist E. Ménétriés: *Catalogue raisonné des objets de zoologie recueillis dans une voyage au Caucase et jascu'aux frontiéres actualles de la Perse* (St. Petersbourg 1832). This work was published on the basis of a field expedition (1829-1830) in the eastern Caucasus. Further reports on the richness and diversity of the animal kingdom of the Caucasus can be found on the basis of travel notes during the 19th century by K. E. von Eichwald: *Fauna Caspio-Caucasica* (1841). The next publication *Meletemata entomologica* (1845-1846) by F.A. Kolenati contains extensive taxonomic studies in entomology (Coleoptera, Dermatoptera, Lepidoptera, Neuroptera, Aphaniptera, Anoplura), arachnology and, extensively, Chiropterology. In connection with a faunistical study one must refer with profound gratitude to the A. Nordmann – *Die in Gabiete der Fauna Taurico-Caucasica* (1851); as a result of expeditions to the Caucasus the species of butterflies have been listed by E.F. Eversmann (1851), and other entomologists, J. Lederer (1852, 1860, 1864) and O. Staudinger (1871), etc. A significant role in the development of Lepidoptera was played by N.M. Romanov (1884-1887) in *Les Lépidoptéres de la Transcaucasie* (in Romanoff, *Mem. Lépid*). He greatly expanded the list of species of butterflies of the South Caucasus (Didmanidze 2005).

As already mentioned, the most famous explorer and traveller during the late 19th and early 20th centuries was G. Radde – one of the most many-sided, encyclopedically educated scholars. He pioneered complex and multiple field expeditions across the Caucasus, excursion routes (over 35 years) covering hundreds of kilometers, included the most remote places of the region from the 'jungles' of Colchis and Hircan forests up to the towering snowcapped mountain of the Greater Caucasus. Scientific interests consisted of both natural (geography, botany, zoology) and cultural (ethnography) heritages of the North- and South Caucasus. These most complete and comprehensive investigations are cited in numerous monographs (volumes) and articles. For example, *Ornis Caucasica* (1884), *Die Sammlungen des Kaukasischen Museums* (alternative text in Russian: *Kollektsii Kavkazskogo Muzeia*). Actually, it was descriptions of the collections of the Caucasian Museum and his own biography in six volumes. The first volume was dedicated to a zoological study of the region (two volumes from six were not finished during his life). If we take into account the fact that he set up the Caucasus Museum in 1867 (founded in 1852) and Public Library (founded in 1846) in Tiflis, his contribution to the development of natural history in this region is priceless (Gegechkori 2012b) (Figure 5).

Figure 5. Gustav Radde (1831-1903).

The next list of major works dedicated to the Caucasian animal kingdom include: *Beasts of the Caucasus* by N.J. Dinnik (1910), *Mammals of the Caucasus region*, vol. 1-2 by K. A. Satunin (1915-1920) (Gegechkori 2012b).

Traditions of elaborated research on fauna were continued among the academic expeditions of the beginning of twentieth century. During the period following the political changes in Russia and the Caucasus, field expeditions and collecting of faunistical materials in this region increased significantly. This increase was mainly through the Zoological Institute of the Academy of Science of the USSR (Leningrad) and similar institutions within the Republics of the North - and South Caucasus. It should be pointed out a series of publications *Animal kingdom of the USSR* (in Russian) in five volumes of which vol. five is dedicated to various groups of animals of the Caucasus. A great contribution to the mammalian fauna of the Caucasus was made by N.K. Vereshchagin (*The mammals of the Caucasus; a history of the evolution of the fauna* (1959), translated from Russian into English 1967) (Gegechkori 2002).

The 20th century can be considered as the heyday of the study of Caucasian nature with the activities of two outstanding scientists A. A. Grossheim (1888-1948) and A. L. Takhtajan (1910-2009). Their direct and indirect pupils from Armenia, Azerbaijan and Georgia have also made a valuable contribution to botanical science. Worth mentioning are, among others, I. S. Abashidze, S. D. Agadjanov, A. G. Dolukhanov, R. I. Gagnidze, V. Z. Gulisashvili, L. M. Kemularia-Natadze, N. N. Ketskhoveli, A. L. Kharadze, A. A. Kolakovski, A. K. Magakjan, L. B. Makhatadze, G. S. Nakhutsrishvili, L. I. Prilipko, I. S. Safarov, M. E. Sakhokia, I. I. Tumadjanov and others (Maharramova 2015).

An intensive study of the fauna of the Caucasus was carried out by P.A. Zaitsev, a native scientist from Georgia, and the Russian naturalists G.Y. Bey-Bienko, N. S. Borchsenius, O.L.

Krizhanowski, L.A. Portenko, M. Ryabov, A.A. Stackelberg, M.E. Ter-Minasian, N.K. Vereshchagin, A.K. Zagulyaev and others (Gegechkori 2012b).

From 1970 onwards, the staff of the Department of Zoology of the Georgian National Museum (as the former Caucasian Museum was named since 1852), led by A.M. Gegechkori (1970-2005), resumed Radde's tradition to study biodiversity in all regions of the Caucasus and in adjacent countries. E.A. Didmanidze and I.A. Skhirtladze were the first women scientists who, for 20 years (1970-1989), collected invaluable entomological material from every natural zone of the North- and South Caucasus.

Today the Georgian National Museum holds about 160,000 herbarium specimens (including samples collected in the period 1805-1810 by Ch. Steven), about 480,000 animal specimens of which 295,000 are invertebrate and 33,000 vertebrates with many duplicates. It includes the unique specimens of butterflies presented to the Museum by the cousin of the Russian Tsar N. Romanov. Some collections belong to animal species that are today extinct in the Caucasus, e.g., the Turanian (Caspian) tiger and the Caucasian mountain bison. Many stuffed animals were prepared for the Museum by I. Chkhikvishvili with the assistance of the German naturalist K. Krell (1932) (Gegechkori 2002).

The flora and fauna of the Transcaucasian countries and Ciscaucasian republics of the Russian Federation are investigated by a large network of research institutions, mainly belonging to the National Academies of Sciences. Most institutions carry out research projects on the biodiversity of the Caucasus, as do the respective chairs of botany, zoology, ecology and forestry Faculties of Universities, and the scientific staff of botanical gardens, zoological parks, strict nature reserves (zapovedniks), and national parks.

The first ever national University in the Caucasus (*Alma Mater*) was opened on January 26 (new style February 8), 1918 in Tbilisi ('White Temple'), symbolically on the Memorial Day of King David Agmashenebeli (the Builder).

Recently, the 100th anniversary of Ivane Javakhishvili Tbilisi State University (TSU) was held during 2018 (with nationwide and international celebrations that took place throughout the year). Under the aegis of UNESCO the relevant decision was made at the 39th session of the UNESCO General Conference in Paris upon proposal by member states. The University laid the foundation for a European-type higher school in Georgia based on Georgian educational traditions. It should be remembered that one of the first higher educational institutions was the Fazisi Academy in Colchis (4th century) followed by the Gelati and Ikalto Academies in medieval Georgia (11th-12th). Now, TSU is the only University, not only in Georgia but also throughout the Caucasus region, which has been ranked among the top two percent of Universities in the world, according to U.S. News & World Report.

The chairs of Botany and Zoology of the State University of Georgia were founded in 1919 by the Russian botanist S. G. Navashin and zoologist (entomologist) B. P. Uvarov as the first chair-holders. In 1852 the Royal Garden of Tiflis was awarded the new status of 'Botanical Garden'. Consequently, with the foundations of the Caucasian Museum (1852) and the Botanical Gardens, Tiflis became the center of botanical investigations in the Caucasus.

The Caucasus possesses a very wide range of natural habitats and climatic regions. This has led to a similar richness in numbers and variety of plant and animal species (Gegechkori 2008).

The biodiversity of the Caucasus is exceptionally rich, conditioned by diverse climatic and topographic features and by a complex geological history. 'Almost all species of Europe and

Asia coming together in the Caucasus and live here side by side' (Anisimov 1916 cited in Gegechkori 2007).

Various global organizations have identified the Caucasus as one of the biodiversity hotspots of the world. Now this biodiversity is being lost at an alarming rate due to unsustainable logging, overgrazing, poaching, infrastructure development, and pollution. Conservation strategies in the Caucasus Ecoregion are supported by a range of scientific, governmental, and non-governmental organizations, including the Global Environmental Facility (GEF), the World Bank (WB), and the US Agency for International Development (USAID). Non-governmental, environmental organizations carrying out programs in the region include Greenpeace, IUCN, CNF and WWF. WWF has a Country Office in Georgia, Tbilisi (WWF – Caucasus), which has been active since 1990.

As well as these international and regional institutions, research centres in the four Caucasus countries are involved in biodiversity conservation. The Center for Biodiversity Conservation and Research (NACRES) was founded in 1989 (Tbilisi). Its conservation activities in the South Caucasus are based on scientific research (e.g., 'Conservation of *Capra* species in Georgia', 'Brown Bear Conservation in Georgia', the research programme also included the famous male leopard from Vashlovani National Park). Another non-governmental organization is the Georgian Centre for the Conservation of Wildlife (GCCW, Tbilisi 1994) with research into living nature and conservational and educational programmes as its main goals. It is also important to mention the Azerbaijan Ornithological Society (1986), the Armenian Society for the Protection of Birds (1999), and the Russian Bird Conservation Union (1993) (Gegechkori & Joosten (Eds.) 2008).

Box. 1. The Phenomenon of Blumenbach's 'Caucasian race'

The dawn and dusk of scientific anthropology focus on the Caucasian race. The systematization of race concept during the Enlightenment period did not receive much support. However, the idea received its influential theoretization in the writings of a colleague of Meiners from the University of Göttingen, Blumenbach. He proposed a new, polygonal and major hypothetical racial division.

Johann Friedrich Blumenbach (1752-1840), German physician, natural scientist, physiologist and comparative anatomist. He is often cited in the scholary literature as 'the father of anthropology'. In the historiography of physical anthropology, Blumenbach appears as a Founding Father of racial science and the classification of races (Painter 2003). His M.D. thesis – *De generis humani varietate nativa* (1775), later was translated in English ('On the natural variety of mankind').

Many historians of science believed that Blumenbach belongs with Buffon, Linnaeus, Kant, Lamark, and Cuvier as Enlightenment thinkers who all made significant 18th century contributions to the field of scientific anthropology. Blumenbach was the first to show the value of comparative anatomy in the study of man's history, and he is well known as being one of the first to propose a scientific classification of the races mankind (*http://www.blumenbach.info/-/Proposal_Pt_3.html.*).

Blumenbach's thesis was published in three editions. In 1775 the first edition discussed four geographically defined varieties of humans, while the second edition in 1781 outlined five geographically defined varieties. He developed this classification further in the third edition in 1795 (see below), which is the definitive volume, where he provided generic rather than geographical labels. He identified the major unique features of humans as: the large brain, speech, erect posture, two free hands, naked skin and the hymen in women (Blumenbach 1865; Painter 2003, 2011).

Blumenbach underlined the unity of humanity - that all humans are one species – *Homo sapiens*. However, he saw gradations among humans, but no distinct species or subspecies. He attributed differences between five human types, such as variations in stature, color and even the shape of the skull to environmental factors, largely to climate which exerts its full influence, and produces its most deteriorating effects in a savage of society. Another factor, according to Blumenbach, is pure biotic, i.e., contemporary society, the level of civilization in which men are educated and the mode of life which is attracted to their general appearance, the form of expression of their emotion and their whole aspect. Blumenbach viewed mankind as an object of natural history, an animal as of one group and superior to another one. He surveyed characteristics of human races, mapped their distribution and invented comparative physiological anatomy. Unlike Mainers, Blumenbach based his anthropological classification primarily on craniology after deciding that there was more to racial differences than skin pigmentation (Baum 2006 and references therein).

In addition to the index of skin color, he focused on other bodily measurements, especially of skulls. His most important anthropological work – 'On the Natural Variety of Mankind' – was based on a collection of sixty human crania. Among them was the skull of a young, Georgian woman sent in 1772 from Russia (St.Petersburg) by a famous scientist, Ash. Blumenbach reduced a diversity of skulls to five main varieties. Unlike Camper's idea that only the measurement of facial angles was important, Blumenbach measured skulls along several lines, placing skulls from around the world in a horizontal line on a table and studying them from the rear view. This approach would allow the identification of various parts of the skulls – forehead, the size and angle of the jawbone, the angle of the teeth, Camper's facial angle in profile and so on – that formed the national character (Painter 2003, 2011; Teo 2009).

The physical traits of Caucasoid crania are still recognized as distinct within modern forensic anthropology in contrast to the Mongoloid and Negroid races.

In the first edition of his work (MD thesis, 1775) Blumenbach pointed out that climate and culture determine the outward appearance of various kinds of people, portraying brown skin as less beautiful than white. Having initially built upon the Linnaean scheme, Blumenbach in the 1780s and 1790s departed from Linnaeus's four-way classification of humans. The second edition of his work (1781) added Malays and introduced the five-fold racial division: Caucasian (white race), Mongolian (yellow race), Malayan (brown race), Ethiopian (black race) and American (red race) with the Caucasian example in the center. He wrote: 'The middle one is a very symmetrical and beautiful one of a Georgian female . . .' (Painter 2003, 2011 and references therein; Baum 2006).

He stressed that five races were more consonant to nature than Linnaeus's four-way racial classification of *Homo sapiens*. In the 1795 third edition of his work, Blumenbach improved the description of his five varieties of humankind and specifically in this edition described one of his races as the **'Caucasian variety'**: color white, cheeks rosy; hair brown or chestnut-colored; head subglobular; face oval, straight, its parts moderately defined smooth, nose narrow, slightly hooked, mouth small. The primary teeth placed perpendicularly to each jaw; the lips (especially the lower one) moderately open, the chin full and rounded. In general, that kind of appearance which, according to our opinion of symmetry, we consider most handsome and becoming. To this first variety belong the inhabitants of Europe (except the Laps and the remaining descendants of the Finns) and those of Eastern Asia, as far as the river Obi, the Caspian Sea and the Ganges; and lastly, those of Northern Africa' (Blumenbach 1795, third edition).

With regard to the Caucasian race or *Varieties Caucasia*, the scientist named it after the Caucasian peoples from the southern Caucasus region, whom he considered to be the archetype for the grouping (Oxford English Dictionary: 'a name given by Blumenbach').

Another reason the Caucasus had such an attraction to Blumenbach was because of its proximity to Mount Ararat where according to the Biblical account Noah's ark eventually landed after the flood. The Caucasus was also the traditional site of Prometheus and the Golden Fleece – myths featured in the Caucasus (Schiebinger 2004 and references therein). According to Painter (2003) 'its very vagueness probably made it more attractive than the more precisely located 'Georgia, whose name lacked this Caucasus mythological and symbolic charm'.

Later anthropologists of the 19th and early 20th century (Pritchard, Broca, Morton, Dixon, etc.) came to recognize other Caucasian morphological features (Lieberman 2002).

Blumenbach was regarded by his contemporaries as one of Europe's greatest scientists; indeed, his scientific heritage has been largely influential in the thought of several generations of early anthropologists in Europe and the United States. A number of outstanding, rare theorists have taken up his racial theory making it the firm basis for scientific racism.

Despite the number of human races offered by racial taxonomists: three (Cuvier), four (Kant), four (Linnaeus), six (Buffon), eight (Agassit), twenty-two (Morton), etc., within their racialized systems, Blumenbach is generally regarded as being the first to develop a scientific concept of race as applied to humans.

The science of 'Craniology' was introduced by Blumenbach. The scientists use his method of studying skulls today. Those who are working in forensics still need to use his basic opinions to determine age, sex, race and other factors based on human skulls.

Though recognized as a significant figure during his life, unlike the founding 'fathers' in other scientific disciplines, Blumenbach's fame and influence waned in the latter half of the 19th century. He was largely forgotten by the turn of the 20th century, frequently appearing in studies on race and racism mainly in negative a sense (e.g., Gould 1994; Baum 2006; Rhopal 2007; Teo 2009; Painter 2011 and so on). On the other hand, some specialists take into account the fact that Blumenbach lived in an age when ideas of progress, and the cultural superiority of European ways, dominated political and social life. Despite all of this, he was 'largely, passively recording the social view of his time' and, convinced of the superficiality of racial variation, Blumenbach defended the mental and moral unity of all peoples (Gould 1994).

Blumenbach taught many students, such as the naturalist Alexander von Humboldt, Karl Ernst von Hoff. When he was appointed curator at the University of Göttingen, the collection of human skulls housed 85 specimens. When he died (1840), the collection had grown to 245 crania.

As we see, human diversity has long fascinated scientists, but unfortunately it also has led to discrimination. The dominant western cultural ethos, that whites were 'superior' and blacks 'inferior' (and even 'ape like'), was commonly reflected in science books published between 1880 and 1980. In the 1930s, most negative ideas of eugenics such as rankings of intelligence (IQ), racial segregation, racial hygiene (inferiority), sterilization, genocide, and contrary, racial superiority, thanks to proponents of eugenics among scientists (Ernst Rüdin, Eugen Fischer, Stuart Chamberlain) were used to correlate human physique with physiological traits (intelligence, criminality, etc.) as justification for the racial policies (pure 'Nordic' or 'Aryan') of Nazi Germany. This argument of pre-Nazi and Nazi ideology was strongly influenced by social Darwinism (with the appearance of Herbert Spencer's social Darwinism) (Hofstadter 1992).

What is race from today's viewpoint? George Gaylord Simpson stated that: 'Races of men have', or perhaps one should say 'had', exactly the same biological significance as the subspecies of other species of mammals' (Simpson 1966), i.e., biological variation in keeping with general zoological practice (Keita et al. 2004 and references therein).

As late as the 2000s, anthropologist Hewson Smith concluded that race is more than just a biological fact (Stockwell 2002). Mayr (2002) considering that since every local population of *Homo sapiens* has its own genetic pool and a mutational history that is unique to the local environment. Despite this, difference between individuals on the planet may have less to do with genetics, but more to do with culture and social geography. Therefore, geographical and cultural factors are more significant than racial differences based on skin color or genetic differences. Hence, there is no innate superiority of 'whites' vs inferiority of 'blacks' and other skin-colored people. For Mayr, a superiority of a culture of race is thus not due to a set of innate characteristics, but rather due to favorable climate, and to the accident of history, being in the right place, first, at the right time (Smith 2002).

'. . . Skin color or facial features, according to Bamshad and Olson (2003), – traits influenced by natural selection – are routinely used to divide people into races. But groups with similar physical characteristics as a result of selection can be quoted as different genetically. Individuals from sub-Saharan Africa and Australian aborigines might have similar skin pigmentation, but genetically they are quite dissimilar. In contrast, 'two groups that are genetically similar to each other might be exposed to different selective forces'.

Based on the relationship between the patterns of biological and sociocultural variation in extant humans, some researchers continue to conceptualize the term in widely different ways. Another explanation of race involves the following definition: 'all living humans belong to the same species, *Homo sapiens*, and the latter itself includes one subspecies, *Homo sapiens sapiens*' (Keita *et al.* 2004).

Although still used in general contexts, race has often been replaced by other words which are less ambiguous, such as populations, people(s), ethnic groups (a group of people sharing the same culture, history, language, etc.), or communities, depending on context (Keita *et al.* 2004.)

Therefore, biological evolution has stopped but, according to the same author, cultural evolution continues. Despite arousing controversy, both in the past and today, many scientists categorized humans into four distinct races (white or Caucasian, black or African, yellow or Asian, and red or Native American), and within racial groups, as it was previously mentioned, some phenotypical sub-races.

The term Caucasian was largely introduced by Blumenbach and is still used today. For example, it is used in anthropology (Howells 1997), forensic anthropology (Eckert 1997), psychology (Teo 2009). The terms either 'Caucasian' or 'Caucasoid' is used in race genetics in relation to human population origination including the popular notion of an 'African (Real) Eve', woman. She lived in East Africa about 300,000 years ago, and all of us are descendants of this female in the global scale migrations of early members of the genus *Homo* from early Pleistocene until Last Glacial Maximum; the paths by which Neolithic pastoralists spread from the Middle East toward Europe, West Asia, and North Africa (Cavalli-Sforza *et al.* 1994).

'Caucasian' is broadly recognized in the craniology, epidemiology, forensic archaeology, especially in the USA (http.//an.wikipedia.org(wiki) Caucasian race). Among other races the white race, under the name 'Caucasian' is still exhibited in many museums of the world.

Time has shown that some of Blumenbach's views were wrong (e.g., 'scientific romanticism', which means that maximal beauty of people comes from the Caucasus, and the probability that humans were first created in this area). Despite the changes made in Blumenbach's theories with respect to advances in biology when evolution and natural selection became widely accepted, he is – one of the least racist thinkers of his day - still highly credited for his works within anthropology and comparative anatomy.

Chapter 1

GENERAL CHARACTERISTICS OF THE CAUCASUS

1.1. OROGRAPHY

The Caucasus region (N 47^0 15′ to N 38^0 25′, E 36^0 30′ to 50^0 20′) is mainly mountainous and, as mentioned, covers approximately 441,000 km^2 of which about 15% is above the treeline. It occupies the area between the Black Sea, the Sea of Azov in the west and the Caspian Sea in the east and is therefore also known as the Caucasus Isthmus (Gvozdetsky 1954) (Figure 6).

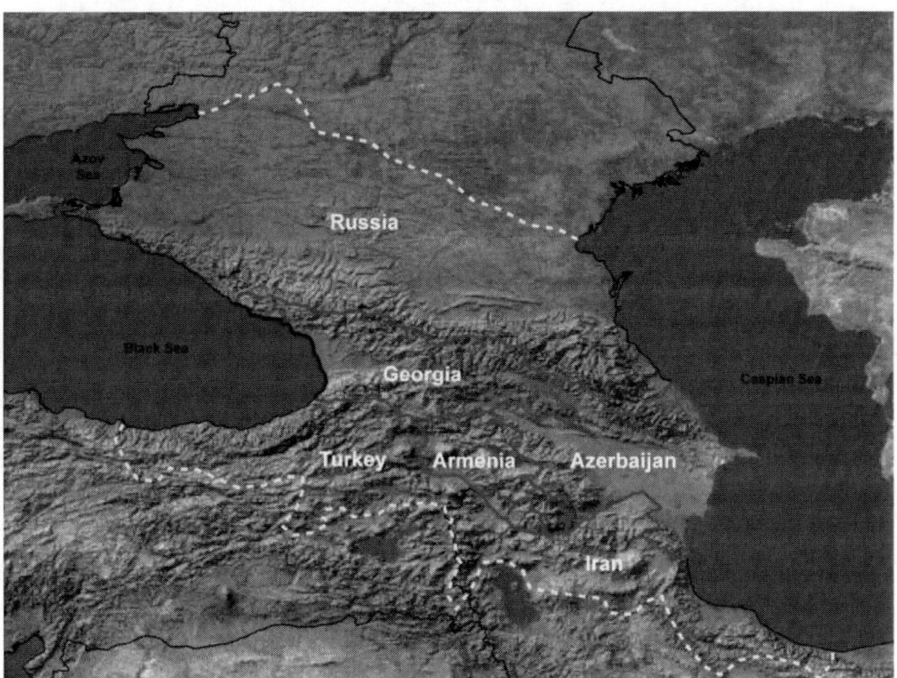

Figure 6. Caucasus Ecoregion (480,000 sq. km.) with the Caucasus proper (~441,000sq. km).

The northern border is formed by the Kuma-Manych depression, the southern border by Adjara (westward) and the Aras River in the South Caucasus (eastward); i.e., by the borders of

Turkey and Iran. The Kuma-Manych Depression is bounded by the Sea of Azov and Kerch Strait on the west, and the Caspian Sea on the east (Gvozdetski 1954; Prokhorov (Ed.-in-chief) 1973; Gabrielian (Ed.) 1986).

The Caucasus is a geographical hinge that connects Eastern Europe with the Asian Interior and has thus an extraordinary crossroad location between Europe and Asia. It is characterized by the complex and imposing mountain ranges (Gvozdetsky 1958).

Its position between two continents has long been and still is the source of controversy between geographers and politicians: does the region belong to Europe or to Asia? The opinion of the scientists is divided, some consider the Caucasus to be a part of Europe, whereas others suggest the region to be part of Asia. Similarly different political bodies have different opinions.

According to Coene (2010), the main viewpoints are the following:

- The Caucasus belongs entirely to Asia because the Europe-Asia border passes through the Kuma-Manych depression (see also *Coincide Atlas of the World, second edition* 2008).
- The Caucasus belongs entirely to Europe because the Europe-Asia border passes along the border between the South Caucasian countries, on the one side, and Turkey-Iran, on the other.
- The Northern Caucasus belongs to Europe and the Southern Caucasus to Asia, because the Europe-Asia border passes along the Greater Caucasus Range.
- The part of the Caucasus north of the rivers Rioni and Kura belongs to Europe and the part to the south to Asia, because these rivers form the Europe-Asia border.
- The main western part of the Caucasus belongs to Europe and a smaller part in the east (basically most of Azerbaijan and small parts of Georgia, Armenia and the Caspian Sea coast of the Russian Federation) to Asia, because the Europe-Asia border passes along the landscape borders. This last version - as Coene indicates, - is the most widely accepted one.

The Caucasus is often divided into two parts: the North Caucasus (Ciscaucasia) and the South Caucasus (Transcaucasia), separated by the Main Ridge of the Greater Caucasus. The North Caucasus thus comprises the northern macro slope of the Greater Caucasus Range and the adjacent foothills and lowlands, whereas the southern macro slope is part of Transcaucasia. The Greater Caucasus includes three provinces: the northern macro slope, the southern macro slope, and the central part. The area is further subdivided into western, central and eastern sectors with the borders between the sectors corresponding to the highest mountains, Mt Elbrus (5,642metres above sea level) and Mt Kazbek (5,033m) (Dumitrashko 1966, 1977; Khain & Koronovski 1998; Gabrielian (Ed.) 1986; Volodicheva 2003; Eppelbaum & Khesin 2012).

Geomorphologically the Caucasus contains the following main landforms (Table 1): (1) the North Caucasus lowlands; (2) the Greater (Great) Caucasus Range; (3) the Intermontane Depression (between the Greater Caucasus and the Lesser Caucasus) comprising: (3a) the Colchis Depression adjacent to the Black Sea in the west, and (3b) the Kura-Aras Depression adjacent to the Caspian Sea in the east; (4) the Transcaucasian Uplands consisting of three mountain systems: (4a) the Lesser Caucasus, (4b) the Talysh Mountains, and (4c) the South Caucasus (Armenian-Javakhetian) Volcanic Upland, which extends into Turkey and Iran.

There are three main lowlands in the South Caucasus: the Colchis, Kura, and Lenkoran (Gvozdetsky 1960; Mil'kov & Gvozdetsky 1976).

The Colchis Lowland (also known as Kolkhida or Rioni Lowland) and the Kura Lowland, divided by the transverse Surami Ridge, comprise the Rioni-Kura tectonic depression. The Colchis Lowland is a young accumulative plain formed during the Pliocene-Quaternary period when the whole region experienced uplift. The Miocene marine basin, which existed in its place, was filled by sediments, the thickness of which reaches 700m. However, a number of rivers, beginning in the Greater Caucasus Ridge and the mountains of Adjara and Imereti (West Georgia), drain the Colchis Lowland delivering and depositing a vast amount of alluvium at a rate exceeding 100 million m^3a^{-1} to the coast and beyond (Gvozdetsky 1960; Mil'kov & Gvozdetsky 1976).

The Colchis Lowland has a flat topography with altitudes gently increasing landwards. It comprises relict river deltas and river-mouth bars and ramps and also relict alluvial deposits. Originally, much of the lowland was swampy and, although extensive improvements have been made, swamps still occupy large areas. The warm and moist climate of the Colchis Lowland preconditions the development of lush vegetation containing many relict species which have survived from the Tertiary; the vegetation in the foothills is especially rich floristically while on the lowland itself, plant communities are less diverse because of the lowland's young age. Little of the climax vegetation survived on the Colchis Lowland with the exception of riparian forests because it is a major agricultural region growing tea, grapes, and citrus fruits. These and many other species were introduced in the 19th century and at present agricultural ecosystems dominate. Of introduced plants, eucalyptus, used in amelioration projects, is particularly widespread (Gulisashvili 1964; Gegechkori 2008).

The Kura (or Kura-Aras) Lowland extends between the Surami Ridge and the Caspian Sea. Absolute altitudes decline from the west (450m above sea level) to the east where the delta of the Kura is located 27m below mean sea level. The lowland is composed of marine sediments in the east, deposited during the Quaternary transgressions of the Caspian, and alluvium delivered by the Kura and the Aras. Protected by the Greater Caucasus from the north and the Likhi or Surami Range from the west, the Kura Lowland has a dry climate with 200-400mm precipitation per annum. Desert and semi-desert and sparse, arid woodlands represent the zonal vegetation. The ground water table is close to the surface and topographic depressions are often swampy. Wetland vegetation communities developing in such habitats are represented largely by *Phragmites australis* and *Bolboschoenus maritimus* while riparian forests develop along the Kura and the Aras Rivers (Gvozdetsky 1960).

Bordering on the Kura-Aras Lowland are southern Kobustan, the Absheron Peninsula with its hills and mud volcanoes, the Iori-Adzhinouri highland region and the Kura sloping plain (in the foothills of the Lesser Caucasus). Also belonging to this territory are the Shida Kartli (Gori) plain and the Alazani-Agrichai longitudinal valley (the Alazani-Avtoran intermountain area) (Grossheim 1932b; Gvozdetsky 1960).

The folded Talysh Mountains (with heights up to 2,477m) and the coastal Lenkoran Lowland are located south of the Kura Lowland, extending between the eastern foothills of the Talysh and the Caspian. This region has a very warm and moist climate with annual precipitation totals exceeding 1,200mm. Climax vegetation communities comprise many Tertiary species but natural forests have been largely destroyed and replaced by agricultural ecosystems. Natural ecosystems are preserved in coastal wetlands, which are widespread in Lenkoran, and in nature reserve (Hyrcan [or Hyrkan] National Park) (Grossheim 1926, 1948).

Table 1. Main Geomorphological Landforms of the Caucasus (Gvozdetsky 1960)

North Caucasus Lowlands	Western North Caucasus Lowland	
	Stavropol Elevation	
	Eastern North Caucasus Lowland	
Greater Caucasus Range		
Intermontane Depression	Colchis Depression	
	Kura-Aras Depression	
	Lesser Caucasus	Adjara-Trialeti mountain system
		Lokh-Karabakh mountain system
		North Armenian mountain system
	South Caucasus (Armenian-Javakhetian) Volcanic Upland	

The North Caucasus Lowlands are composed of three parts: the Western North Caucasus Lowland, the Stavropol Elevation and the Eastern North Caucasus Lowland.

The Caucasus region is dominated by the massive Caucasus mountain range. This range is sometimes called the Main, Major or Greater Caucasus to distinguish it from another range, for example further south is the orographic unit called the Lesser (or Minor) Caucasus.

The Greater Caucasus is Europe's largest and highest mountain system. It runs from west to east for approximately 1100 km extending from the vicinity of the Taman Peninsula on the Black Sea to the Absheron Peninsula in the Caspian Sea, on an axis oriented west-northwest to east-southeast. Its total area is about 145,000km^2. The Greater Caucasus (Figure 7) with its complex of branching and parallel ridges is generally about 100 km wide, but narrows to 60 km in the Terek River Valley and widens to 135 km in Dagestan. The northern slopes are generally rather gentle and full of terraces and inclined plateaus, whereas the southern slopes are generally steeper and partly fractured. The highest summit of the Greater Caucasus (and therefore of the entire Caucasus) is the already mentioned Mt Elbrus (Figure 8). The highest peak in the South Caucasus is Mt Shkhara (5,068 m) (Figure 9). The ridge between Mt Elbrus and Mt Kazbek is called the Central Caucasus, where most peaks reach heights of between 4,000 and 5,200m. Mt Shkhara and Mt Dykh Tau (5,203 m) in the middle of the Greater Caucasus are very close to each other and represent the highest point of the main watershed ridge of the Caucasus. This is the land of permanent snow and glaciers. The ridge west of Mt Elbrus is called the West Caucasus. The frontal ranges and ridges on the northern side of the western and central Caucasus have a cuesta character. In places, karst is strongly developed in limestone (Gvozdetsky 1958; Geresimov (Ed.) 1966; Gvozdetsky & Mil'kov 1976; Volodicheva 2003; Eppelbaum & Khesin 2012).

The Greater and the Lesser Caucasus are separated by the river valleys of the Rioni (Colchis, in the west) and the Kura (in the east) and connected by the 102 km long Likhi (Surami) Range (900-2,471 m a.s.l.) which has originated by submeridian tectonic faulting and divides Georgia into western and eastern parts. On the other hand, it separates the Colchian Lowlands from the Kura-Aras Basin. Both of these lowland areas are deep depressions partly filled by enormous deposits of Tertiary sediment (several kilometers thick) and Pliocene and Quaternary alluvia (Gvozdetsky & Mil'kov 1976).

Figure 7. The Greater Caucasus Range (GCR), geographical backbone of the Caucasus, the highest landform in Europe. Panoramic view from the Lagodekhi Reserve, E. Georgia. In the foreground: growth of *Acer trautvetteri* (1972).

Figure 8. Mount Elbrus, 5,642 m a.s.l. (1976).

Figure 9. Mount Shkhara, 5,068 m a.s.l. (1983).

The Lesser Caucasus 'with its arcuate N-convex shape is a most heterogenous structure' (Eppelbaum & Khesin 2012). It extends from the Black Sea to the Aras River Valley. The Lesser Caucasus in western Georgia links to the easternward to the Pontic Mountains of northeastern Turkey, and the Talysh Mountains of South-eastern Azerbaijan link to the Alborz (Elburz) mountain range of northern Iran.

The Lesser Caucasus consists of three highlands: 1) the Adjara-Trialeti, (Figure 10), 2) the Lokh-Karabakh, and 3) the North Armenian mountain systems. The highest peaks of Transcaucasia are Mt Giamish (3,724m) on the Murovdagh ridge (the Lesser Caucasus), Mt Gomorgoi (Kijmurkei, 2,477 m) in the Talysh Mountains, and Mt Aragats (4,090m) on the South Caucasus Volcanic Upland. Characteristic features of the Volcanic Upland (Armenian Highland) (Figure 11) are extinct volcanoes, lava plateaux and plains, uplifts formed by lava and tufa, and in the south folded ranges (Aiotsdzor) and large intrusive massifs, for example, the southern part of the Zangezur Range with Mt Kaputdzhukh (3,904m). A tectonic depression, the Middle Aras extends along the Azerbaijan and Armenia border. There are no glaciers in the Lesser Caucasus (Gvozdetski 1960; Maruashvili 1975; Khain & Koronovski 1997; Eppelbaum & Khesin 2012).

The mountains, particularly those of the Greater Caucasus, determine distinct climatic and biotic gradients, both vertical and horizontal, provide a large variety of habitats and form natural barriers that limit the migration of species (Volodicheva 2003).

Figure 10. Western Lesser Caucasus (WLC), striking area of relict and endemic arboreal plants found nowhere else. Part of the Lesser Caucasus, scenic view from Zemo (Upper) Adjara, W. Georgia (1970).

Figure 11. View encompasses Javakheti volcanic plateau. Surface of highland remnant of lava that covered Javakheti since prehistoric time (1998).

1.2. Geology

According to L.V. Eppelbaum & B. E. Khesin (2012), many outstanding geologists have studied the Caucasus. Here should be mentioned G.V. Abich, I.M. Gubkin, V.E. Khain, K.N. Paffenholtz and others.

Geologic terms and the length of geologic time follow Eicher and McAlester (1980).

The origin of the Caucasus Isthmus is considered on the basis of a modern understanding of Wegener's theory (e.g., Khain & Ryabukhin 2002 cited in Eppelbaum & Khesin 2012); concerning the geology of the Earth through its plate tectonics (e.g., Le Pichon 1968; Dietz & Holden 1970; Lilienberg & Shirinov 1977; Oreskes (Ed.) 2003; van Thienen *et al.* 2004; Keary *et al.* 2009).

The geology and evolution of the Caucasus, particularly the Greater Caucasus range has been studied for at least 150 years (Saintot *et al.* 2006).

The Caucasus (the Caucasian-Arabian belt) is a central part of the huge Late Cenozoic, Alpine-Hymalaian, orogenic belt and its complex structure and landscape of high mountain ranges, intermontane depression and volcanic highlands is a result of the continuous collision (convergence) of continental, that is the Eurasian and the African-Arabian, plates (Adamia *et al.* 2011a; Sharkov *et al.* 2015).

The Greater Caucasus – the backbone of the Caucasus mountain systems, as already mentioned, is Europe's largest and highest mountain belt (Bochud 2011). Saintot *et al.* (2006 and references cited in) comment: 'The Greater Caucasus is located in the Black Sea-Caspian Sea region, which is regarded as a mosaic of terraines, of Gondwanan, Tethyan and Eurasian affinity that are sometimes controversial in origin'.

The Caucasus region in the Late Proterozoic, Paleozoic, Mesozoic and the Early Cenozoic eras located in the now-transgrassed Tethys Ocean (Prototethys, Palaeotethys, Tethys, Paratethys) and its Eurasian and Gondwanian (Africa-Arabian) margins. Within this ocean-continent convergence zone, there existed a system of island arcas (Adamia *et al.* 2011b). 'The structure and geological history of the Caucasus, - as Adamia *et al.* (2011b) indicate, - are largely determined by its position between the still-converging Eurasian and Africa-Arabian lithosphere plates, within a wide zone of continental collision'.

The convergence of the Arabian Plate and Eurasia created the Lesser Caucasus, which is being pushed to the north by the Arabian Plate, stamped against the continental crust and this initiated the development of the Greater Caucasus (Gamkrelidze 1986, 1997; Volodicheva 2003; Philip *et al.* 2009). The Greater Caucasus uplifted simultaneously with the circum-Mediterranean alpine mountain belt: the Alps, the Balkans, the Carpatians, and the Atlas with the Alpine folding from the Oligocene (Vincent *et al.* 2007) till the Pliocene (Adveev & Niemi 2011).

In the course of a few million years the Caucasus thus experienced a dramatic transformation from a marine basin and lowlands to mountain ranges exceeding 4,000 m in altitude.

In the Mesozoic and Early Cenozoic, the Greater Caucasus and South Caucasus represented the North tethyan realm - the southern active margin of the Eurasian lithospheric plate (Adamia *et al.* 2011b).

The Caucasian mountain system belongs to the middle sector of the Apline-Himalayan mountainous belt that was created 28.5-23.8 million years ago by continental collision between the above mentioned plates (Gamkrelidze 1986, 1997; Golonka 2004). The northward motion of Arabia during the Oligocene and Miocene Epochs, and, therefore, collision between the Afro-Arabian and stable Eurasian lithospheric plates caused neotectonic faulting in the Caucasus (Tutberidze 2012). The folded structure of the Alpine system, coinciding with the Alpine orogeny (Alpide orogeny) started to form in the above mentioned Alpine belt. Among others these mountain systems include the Caucasus Mountains (Moores & Fairbridge (Eds.) 1998).

So, the orogenic processes, due to plates colliding, led to the present mountain chains starting in the Early Tertiary, accelerated during the Plio-Pleistocene, and they are still active nowadays (Adamia *et al.* 2011b). Today the rate of convergence is estimated at approximately up to 30 mm per year (Eppelbaum & Khesin 2012).

The largest part of the Caucasus is overlain by Jurassic and Cretaceous rocks (Figure 12). The section west of Mt Elbrus is predominantly limestone. Sediments of the Jurassic Period (mostly slates) are from the main ridge of the East Caucasus, Inner Dagestan and the Skalisty Khrebet (Ruban 2006). Cretaceous rocks are represented by marl, sandstone, claystone and limestone. Limestone is found in north Dagestan and in the east and west ends of the Greater Caucasus. Kartli and Kakheti (East Georgia) are composed of Tertiary rocks (Saintot *et al.* 2006; Mosar *et al.* 2010). Mt Elbrus and Mt Kazbek are volcanoes and stand aside of the main ridge.

Figure 12. Dinosaur track site from Sataplia site (Imereti province, W. Georgia), GCR. Such ichnological remains are exclusively rare in the Caucasus.

The Lesser Caucasus is formed of volcanic and sedimentary rocks. The northern part contains stratified limestone of the Cretaceous Period, covered by Tertiary sediments, slates, andesites, tufas, sandstones and marls. The south-eastern part of the Lesser Caucasus consists of Jurassic and Cretaceous rocks such as limestone and sandstone. Volcanic plateaus, ridges and massifs reach to the north-east Armenian highlands (Gamkrelidze 1949; Adamia *et al.* 1992, 2010; Gamkrelidze 1997; Sosson 2010).

At present, sedimentary rocks are overlined by lava flows whereas valleys are filled with alluvial-deluvial sediments. The largest volcanic highlands are Samarski, Ketchutski, Armyanski, Goktschinski, Bezobdalski and Mischanski. Individual mountains can reach high altitudes such as Mt Abul (3,300 m), Mt Akhdag (3,570 m) and Mt Aragats (4,090m) (Milanovsky & Khain 1963).

The tectonics and geology of the Caucasus have been extensively studied and many papers on the subject have been published in English (Khain 1975; Khain & Koronovski 1977; Skhirtladze *et al.* 1977; Adamia *et al.* 1981, 1991, 1992, 2010, 2011a; Gamkrelidze *et al.* 1977; Gamkrelidze 1986, 1977; Zonenshine *et al.* 1990; Volodicheva 2003; Golonka 2004; Saintot *et al.* 2006; Philip *et al.* 2009; Mosar *et al.* 2010) and in Russian (Gamkrelidze 1949, 1960; Gvozdetski 1958, 1960; Paffengolts 1959; Milanovski 1968, 1977; Milanovsky & Khain 1963).

1.3. The Modern Climate

Almost the entirc territory of the Caucasus is located within the temperate climate zone of Eurasia. According to another definition the Caucasus Isthmus is located on the border of the temperate and subtropical climatic zones (Gvozdetsky 1954; Chubukov 1966). Its large orographic heterogeneity, however, provides the region with a unique combination of a very diverse climate from the humid, almost subtropical areas of the West Caucasus and the arid, desert and semi-desert climate of the East Caucasus, to the alpine and nival zone with glaciers and eternal snow.

The Greater Caucasus mountain system with its lofty peaks sharply delimits the boundary between two climatic zones by protection from the excesses of the northern continental climate, i.e., by impeding the flow of cold air masses from north to south into Transcaucasia and warm air masses from south to north into Ciscaucasia. The mountain barrier formed by the Greater Caucasus is particularly appreciable in winter when Ciscaucasia is filled with cold air masses arriving from the north and northeast, while Transcaucasia is protected from their intrusion. The average temperatures for January are -2^0 to -5^0C in Ciscaucasia; 4.5^0C to 6^0C in Western Transcaucaisa (Colchis Lowland), and 1^0C to 3.3^0C in eastern Transcaucasia (Kura-Aras and Lenkoran Lowlands). In summer the temperature differences between the northern and southern parts of the Caucasus diminish. The disparity is more noticable between the temperatures of the western (with a more maritime climate) and the eastern (continental) parts of the Caucasus. The average July temperatures are 23^0C-24^0C in the west and 25^0C-29^0C in the east (Gerasimov (Ed.) 1966; Volodicheva 2003).

The Caucasus is influenced by wet oceanic air masses from the Atlantic Ocean and the Mediterranean Basin from the west and by dry continental air masses from Asia Interior, Central Asia and Siberia. The Black Sea increases the moisture of the Western Atlantic winds. In Transcaucasia the most frequent air movement is from the west bringing moisture from the

Atlantic Ocean and the Mediterranean Sea. On the other hand, the Caspian Sea is not large enough to moisten the dry air from Central Asia sufficiently, so it brings moisture only to the easternmost ridges, i.e., to the Talysh in Azerbaijan and the eastern slopes of the Greater Caucasus. A high pressure above the North Caucasus as an effect of the Siberian anticyclone causes the North Caucasus winter to be mostly frosty and dry. In spring the influence of the Siberian anticyclone slowly decreases in favour of the Azores cyclones that bring wet weather. In summer, continental hot dry air from the steppe zone of Asia moves westwards and facilitates the occurrence of steppe vegetation in the North-East Caucasus (Gerasimov (Ed.) 1966; Volodicheva 2003).

The climate of the high mountains results from the interaction of the free atmosphere and the topography. Altitude controls isolation and temperature, whereas the overall dimension and the orientation of the ridges in relation to the western winds are important for the distribution of precipitation and the accumulation of snow and ice. Relative elevation, terrain shape and slope angles result in strikingly different regional and local climates (Volodicheva 2003).

The barrier effect of the Caucasus is reinforced by two mountain ranges: the Stavropol elevation in the North Caucasus and Likhi (Surami) Range in Transcaucasia. They decrease the influence of western winds and therefore contribute to the creation of a dry climate in Dagestan, desert conditions along the Caspian Sea coast and a similar dry climate in the eastern part of the South Caucasus (annual average precipitation in Yerevan 304 mm, in Baku 206 mm). As a result, desert and semi-desert vegetation prevails in the lowlands and savanna-like arid open woodland in the foothills. The orographic conditions are very diverse and cause many regional differences in climate and irregularities in vegetation zonation (Gvozdetsky 1963).

Both temperature and precipitation vary strongly across the Caucasus, mainly as a function of relief. The temperature-altitude gradient is different in various parts of the Caucasus. In the western Caucasus, under the influence of the Black Sea, the temperature gradient is 0.50/100m. Annual precipitation varies from less than 2,000 mm to over 4,500 mm. Precipitation is influenced by altitude in each locality, but not in a uniform manner across the entire Caucasus. For instance, along the main ridge maximum precipitation occurs around 2,500m, whereas in the Adjara-Imeretian (Meskhetian) part of the Lesser Caucasus the maximum precipitation is at a lower elevation (Gvozdetsky 1963).

Most precipitation falls in the West Caucasus (Batumi 2,465mm, Sochi 1,410mm, Sokhumi 1,370mm). This high amount results in the prevalence of lush mesophytic forest in the lowlands and at lower altitudes, particularly in south-western Colchis. Two South Caucasus regions, Colchis and Talysh, have to be highlighted because they are characterized by an almost humid subtropical climate. This climate must have existed over a long time, because both regions are refugia of flora and fauna elements of the Tertiary. Their special character within the moderate climate zone of the Northern Hemisphere can be explained by the warming effect of the Black and Caspian Seas, especially in winter, and the barrier role of the mountain ridges. The combination of the Caspian Sea and the arid Central (Middle) Asian Kopet-Dagh and Alborz (Elburz) Mountains in Iran is essential for the existence of the Talysh (Hyrcania in general) refugium, whereas the Black Sea and the Greater Caucasus play a similar role for the Colchis refugium. The Greater Caucasus prevents the penetration of cold air masses from the north Boreal (Euro-Siberian) centre into Colchis, whereas the above-mentioned Asian mountains perform the barrier role against penetration of hot air masses in the Talysh refugium. In this way, natural greenhouse conditions maintained two famous shelters for ancient organisms in the South Caucasus. The exceptional climatic position is particularly clear in the

Colchis refugium. The average temperature in January, the coldest month of the year, varies from -20 C to +5^0 to 7^{0}C, average annual precipitation equals 2,000 mm, with a maximum of 4,500 mm at Mt Mtirala (Adjara). Both values constitute record data for the territory of the former USSR (Alpaidze & Shengelia 1989).

The highest degree of continentality is found on the landlocked Armenian-Javakethian Volcanic Plateau. Under a clear sky, a local high pressure cell develops over Armenia in winter and radiation cooling brings about temperatures that are very low for this latitude. The plains and foothills of the eastern Caucasus foreland are characterized by a dry climate and frequent droughts with dry winds and dust storms. Dry summer climates are typical of the Kura-Aras Depression (Gerasimov (Ed.) 1966).

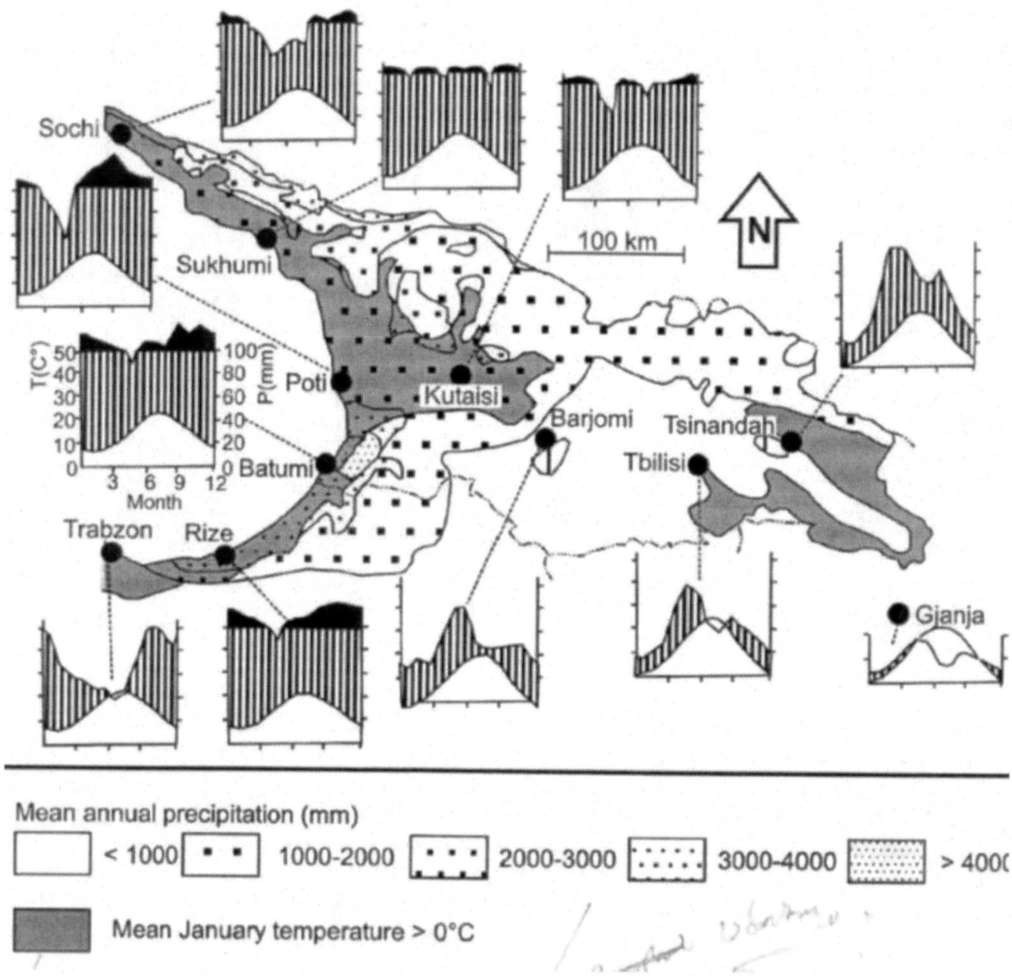

Figure 13. Climate characteristics of the Caucasus (Walter & Lieght 1960-1967).

Thus, the western part of the North Caucasus is characterized by a moderate continental (steppe) climate; the eastern part of the North Caucasus by a drier (desert - semi-desert) climate; a small part of the Black Sea coastal zone (north and northeasternward of Sochi) in the western part of Transcaucasia by a sub-mediterranean type climate; the area south of Sochi, including Batumi, and in general the Colchis Lowland by an (moist) almost subtropical climate; the Kura-

Aras Lowland in the eastern South Caucasus by a dry steppe, semi-desert and desert climate; the Lenkoran Lowland again by a moist almost subtropical climate with mild winters and an abundance of solar heat in summer; the middle part of the Aras River Valley by a dry subtropical climate; and finally the Armenian-Javakhetian Plateau by a dry climate with mainly low moisture and low temperatures in winter (Gerasimov (Ed.) 1966).

1.3.1. Current Glacial Fluctuations of the Greater Caucasus

Glaciers are located mainly in the Greater Caucasus where there are 1,521 of them (Penov et al. 2008) with a combined area in excess of 1,780 km^2. Most glaciers are located in the central sector, which is the highest, in the Glavny and Bokovoy Ridges (Kotlyakov 1966).

The north-facing slopes of the Greater Caucasus develop more glaciers than the southern slopes. The high altitudes and humid climate predetermine the wide-spread development of contemporary glaciation (Kotlyakov 1966).

Extensive plateau, intermontane depressions, and valleys, associated with the ancient drainage network, occur between 2,700 and 3,900m providing ideal conditions for the accumulation of snow and ice and the development of large glaciers of a branching mountain valley type. The north-facing slopes of the Greater Caucasus develop more glaciers than the southern slopes (Kotlyakov 1966).

Out of 215 mountain valley glaciers, known in the Caucasus, 130 occur in this region. Seven glaciers, whose terminations descend to 1,800-2,000m, exceed 20-30 km in length and 20-30 km^2 in area, accounting for 40 percent of the glaciated area of this region. However, the largest glaciers develop on the volcanic cones of the Elbrus (123km^2) and Kazbek (71km^2). Cirque and hanging glaciers are numerous in the central Greater Caucasus. There are also a few surging glaciers. The most dramatic surges were registered at the Kolka glacier, located on the northern slope of the Kazbek massif (Rototaev *et al.* 1983 cited in Volodicheva 2003).

Current knowledge of the dynamics, structure, thermal regime, and mass balance of glaciers is mainly limited to a number that have been selected as representative ones, and have been monitored in the course of national and international research programs. These include the Maruh glacier in the western Greater Caucasus, the Bashkara, Shkhelda, Dzhankuat, Bezengi, and the glaciers of the Elbrus and Kazbek in the central Greater Caucasus. With regard to the thickness of ice, radiosounding has shown that the thickness of typical mountain-valley glaciers of average size (e.g., the Dzhankuat and Maruh) is about 100m while the depth of the Bezengi, the largest in the Caucasus reaches 350m (Tushinsky 1968; Macheret & Luchininov 1973 cited in Volodicheva 2003) (Figure 14).

The later stage of glacial advance resulted from the combination of colder and wetter weather. Evidence of a warm interval (e.g. 'Arkhyz break in glaciation') preceding the Little Ice Age in this region is based on geochemical, pedological, palynological, tree-ring-based summer temperature reconstruction and archaeological data (Kotlyakov 1966).

Glaciers reached their maximum extent in the 1680s, 1750s, and 1850s. Currently, glaciers are in a state of retreat which began at the end of the 19th century. For example, the Dzhankuat glacier has been estimated that between the 1880s and the 1970s, the glaciated area has declined by 600km^2 (or 29 percent) and 25 percent of the ice has been lost. The number of glaciers has increased by 31 percent as a result of disintegration of large glaciers. Deglaciation is more intense in those regions where precipitation is less abundant and glaciers have been retreating faster on the northern macroslope. The intensity of deglaciation also increases from west to east: while in the western Greater Caucasus the glaciated area had declined by 21 percent. In the central and eastern sectors larger reductions of 30 percent and 40 percent have occurred.

The contemporary glacial retreat is not unique. Extensive degradation of glaciers occurred in the Greater Caucasus between the 3rd and the 13th centuries. The retreat of glaciers was so strong that this period is often referred to as 'the Arkhyz termination' (although it can be argued that 'termination' is not a valid term as glaciers declined but did not disappear altogether). In many regions the tree line was positioned higher than now by 200-300 m and many mountainous passes were free of snow and ice (Kotlyakov 1966; Tushinsky (Ed.) 1968; Volodicheva 2003 and references therein; Gobejishvili & Kotlyakov 2006; *www.rusnature.info/reg/15-4.htm*).

Figure 14. Weichselien glacier lowering from Bezengi Wall (Balkaria, S. Russia). It is one of the largest snowdrifts of the region (length 17 km, depth – 350m) (1976).

In the Northern Caucasus a wide scale glacier retreat occurred in the late 1840s, with four to five minor readvances in the 1860-1880s and three readvances or steady states in the 20th century (1910s, 1920s and 1970s-1980s) (Solomonia *et al.* 2016).

Khromova *et al.* (2014) found that glaciers lost 4.9% of the area between 2001 and 2010. Glacier wastage was higher in the southern slopes at 5.6% vs 4.3% in the northern slopes glaciers. To compare to the European Alps, glacier reduction in the Caucasus Mountains appears to be slower. For example, Fischer *et al.* (2014) reported 33% and 11% reduction for the eastern Swiss Alps for the 1973-2003 and 2003-2009 periods, respectively (Tielidze 2016).

In the Lesser Caucasus (Zangezur Range, Aragats Mts), due to a more continental climate, glaciers are only found on high mountain peaks (Ivankov 1959b cited in Kotlyakov 1966).

During the Late Pleistocene (115-10 ka), some of the largest glaciers in the Central Greater Caucasus descended to 600-750 m a.s.l. It means that the foothills and lowlands of any regions of the Caucasus were ice free. Glaciers on the Lesser Caucasus and the Armenian Highland were insignificant and the lower altitudes reached by glaciers in the Lesser Caucasus were about 2,500-2,700 m a.s.l. In Alborz, the snowline was about 600-1,100m lower than the present level and no traces of the Pleistocene glaciers have been found in the Talysh Mountains (Shcherbakova 1973; Milanovsky 2008; Gobejishvili & Kotlyakov 2006).

1.4. Hydrology and Main Water Bodies

1.4.1. The Black Sea

The hydrology of the Caucasus and especially its two seas, the Black Sea and the Caspian Sea, historically have a decisive importance in formation of the region's biota, first of all its two outstanding refugia – Colchis and Talysh.

The modern name of the Black Sea goes back only to the 13th century but the appellation is actually much older. It is generally accepted that the primary Greek name *Pontos Axeinos* (*Axenos*), i.e., etymology to á – xeinos – 'inhospitable'. This ominous name was later (after AD) changed to the euphemism *euxeinos* (*Euxinus*), i.e., Eú-xeinos – 'hospitable' (*Encyclopedia Iranica: www.iranicaonline.org/articles/black-sea*).

The Black Sea, like the Caspian Sea, Aral Sea, and Lake Urmia, is a remnant of the ancient Paratethys Sea.

The Black Sea biogeographical region consists of two coastal bands encompassing the southern half of the Black Sea. The longest east-west extent is about 1,175 km. The western part stretches from the delta of the Danube extending east towards the Bosphorus outlet. This stretch is 530 km long and its width varies between 20 and 60 km. The other part – 1,400 km long and between 10 and 160 km wide – streches from east of the Bosphorus over the various mountain ranges along the southern coast of the Black Sea and as far as the Caucasus mountain in the East (Liamine (Ed.) 1961) (Figure 15).

The Black Sea is the most isolated sea in the World. It is connected to the World Oceans via the Mediterranean Sea through the Bosphorus, Dardanelle and Gibraltar straits and with the Sea of Azov in the northeast through the Kerch Strait.

The maximum depth of the Black Sea is 2,212m., the surface area is 432,000 km^2 with a total volume of 547,000 km^2 (State of the Environmental of the Black Sea 2002). The replenishment of the bottom waters of the Black Sea with new seawater from the Mediterranean takes hundreds of years. This very slow rate of replenishment and the large input of freshwater have led to a stratification of the Black Sea. The thin upper layer of marine water (up to 150 m) supports a unique ecosystem. The deeper and denser water layers are saturated with hydrogen sulfide (H_2S), that has been accumulated from decaying organic matter. The slow replenishment and bad mixing of waters does not provide enough oxygen for the process of decomposition and the bacteria in the lower layers use it up entirely. Consequently, the Black Sea is virtually dead below a depth of about 180m and this boundary is being pushed upwards. Moreover the metabolism of some bacteria generates hydrogen sulfide, a soluble, poisonous gas associated with the small of rotten eggs. Hydrogen sulfide is present in the entire layer of seawater in the Black Sea (Black Sea NGO Network 2004).

Figure 15. The Black Sea. Panoramic view from Cape Bichvinta, Abkhazeti province, NW Georgia. In foreground: Pinus brutia var. *pityusa* (1975).

The catchment drainage area of the Black Sea is 2,000,000km^2 with the total river inflow 340.6 km^3. The Rioni, Kodori, Inguri, Chorokhi (Gorukh), Kyzyl-Irmak, Sakarya, Southern Bug, Dnister, the Danube, Dnieper and Don via the Sea of Azov are the main rivers that flow into the Black Sea.

The Geography and macro-circulation processes existing in the Mediterranean Basin influence the climate of the Black Sea Basin which, in the majority of the sea, is similar to the Mediterranean climate (warm humid winters and hot dry summers). The south-eastern part, surrounded by the mountains is characterized by a humid, almost subtropical climate – abundance of precipitation, warm winter, hot summer. Total amount of precipitation from the Bosphorus to Varne is about 500-700mm per year, in the north near Odessa – 300-400mm in the southern coast of Crimea – 586mm, the amount of annual precipitation increases eastward – 1,600mm between Novorossiysk and Sokhumi to 2,465mm in Batumi (State of Environment of the Black Sea 2002).

The Black Sea is now the largest natural anoxic water basin in the world. This means that 87% of its volume is practically devoid of marine life, except for some forms of bacteria. However, the sea is comparatively rich in living resources. Also, the Black Sea shelf and river deltas are important spawning grounds for sturgeon and other fish species, and the coastal wetlands are migration and breeding grounds for numerous rare and endangered European birds (Black Sea NGO Network 2004).

The thin upper layer of marine water supports the unique biological life in the Black Sea ecosystem. The deeper and more dense area is (450-2,250 m) inhabited by more specific organisms, basically, as was pointed out, at the level of bacteria, protozoa and some multicellular invertebrates (Ciliate, Foraminifera, Nematoda, Kinorhyncha, Ostracoda and others). There are six species of seagrasses in the Black Sea (State of the Environment of the Black Sea 2002) (Hsü *et al.* 1978; Hsü & Bernoulli 1978).

The Black Sea, along with the Caspian Sea and Aral Sea, in a broad sense is a remnant of the Paratethys (Hsü & Giovanoli 1979; Rögl 1999).

Due to a large catchment area compared to surface area the Black Sea is very vulnerable to pressure from land based human activity and its health is equally dependent on the coastal and non-coastal states of its basin.

Eutrophication, pollution and irresponsible fishing resulted in an overall decline of biological resources, the diversity of special and the recreational values of the Black Sea.

The Bucharest 'Convention on the Protection of the Black Sea Against Pollution' was signed in 1992 and ratified by Bulgaria, Romania, the Russian Federation, Ukraine, Georgia and Turkey in 1994. It was designed to initiate a regional approach for nature conservation of the Black Sea. The Black Sea Environmental Programme was officially established in 1933 by UNDP/GEP (Bucharest Convention). The programme is funded by GEP with additional cost sharing contributions from GEC's PHARE and TACIS programmes as well as bilateral contributions from Canada, the Netherlands, Switzerland and France.

The strategic Action Plan for the Rehabilitation and Protection of the Black Sea was adopted in Istambul, Turkey (1996).

1.4.2. The Caspian Sea

The Caspian Sea is the world's largest lake of inner drainage, with no outlet to the ocean (Figure 16). The word Caspian is derived from the name Caspi, an ancient people who lived to the southwest of the sea in Transcaucasia and has now disappeared. In classical antiquity among Greeks and Persians it was called the Hyrcanian Ocean (*https://en.wikipedia.org/wiki/Caspian_Sea*). The sea extends over 1,200km from the north to the south. The sea has a surface area of 371,000 km^2 and a volume of 78,200 km^3, maximum depth 1.015m. The Caspian drainage basin covers more than 3 million sq km, and included the Volga, Europe's largest river, which contributes more than 80% of the total river inflow, as well as several smaller rivers. The brackish water of the Caspian Sea contains three times less salt than that of the World Oceans. Presently the sea-level stands at 27m below the ocean's level, but the peculiarity of the Caspian Sea consists of repeatedly occuring oscillations of its level (Nikolaeva 1966, in: Gerasimov (Ed.) 1966).

The Caspian Sea, is a landlocked basin, which became landlocked about 5.5 mya due to tectonic uplift and a fall in sealevel. It is more sensitive to environmental changes compared to the world's oceans. As the Caspian is an inland water body, man-caused impact on the catchment area (about 3.5 million km^2) is accumulated here.

Figure 16. The Caspian Sea (schematic map).

The north Caspian is the shallowest portion of the sea, with an average depth of 4 to 6 meters, the Middle part has an average depth of 190 m and the South Caspian Sea contains a depression reaching a maximum depth of 1.025 meters (Voropaev (Ed.-in-chief) 1986).

The biodiversity of the Caspian Sea is 2.5 times poorer than that of the Black Sea (Zenkevich 1963). The main reason of this is probably its variable salinity. According to Mordukhai-Boltovskoy (1979), for the freshwater fauna and flora the Caspian water salinity is too high, and for the current marine species it is low. So, the contemporary Caspian Sea has an ideal environment for brackish water species originating both from marine and from freshwater bodies. Hence, due to their very good osmoregulatory abilities, fishes and crustaceans have the greatest diversity in the Caspian (63% of the total number of species) (Zenkevich 1963). Therefore, the modern biodiversity of the Caspian Sea reflects a complicated history of paleo-Caspian transgressions and regressions, desalinization and salinization (Abramova, 1972; Aladin & Plotnikov 2004).

Environmental problems of the Caspian Sea are multiple and characterized by heterogenity of their origin. From anthropogenic activity should be marked out the artificial regulation of Caspian rivers' discharges. Man-built dams have reduced water input into the Caspian, huge water reservoirs (e.g., the Volga River) prevent a high amount of river fresh waters flowing into the sea. Regulation of rivers that flow into the Caspian is one of the most significant human impacts on the biodiversity of the water body (man-made canals that were built to link the Volga and then the Caspian with other rivers and with other seas via the rivers). Accidents on oil tankers, a high amount of pollutants which are carried by Caspian rivers etc., make significant damage to the biological diversity of the aquatic community. Overfishing and illegal fishing with other factors mentioned earlier resulted in the reduction of populations or the complete loss of some species of fish and other groups of organisms. About 90% of world sturgeon reserves occur in the Caspian Sea therefore it can be considered a global, genetic fund for these species. Nowadays the catch of sturgeon in the Caspian Sea has reduced from 25,000 tonnes per year to 1, 000 tonnes.

Five countries (Azerbaijan, Iran, Kazakhstan, Russia and Turkmenistan) share the natural heritage of the Caspian Sea. Today the Caspian faces significant environmental problems, many of a transboundary nature, that arise from both anthropogenic and natural causes. The Caspian Environmental programme (CEP) is a regional umbrella programme established by the Caspian littoral states and aided by the international agencies. Due to the necessity for regional cooperation, expressed through a number of regional agreements including the Almaty Declaration of Environmental Cooperation (1994), the CEP is involved in a joint mission with the World Bank, UNEP, UNDP (1995, 1998), where UNDP plays a central role in inplementation of the CEP.

1.4.3. A Brief Review of Freshwater Bodies of the Caucasus (Rivers and Lakes)

An exceptionally large diversity of landscapes is typical of the Caucasus because of the complexity of its orography and its climatic contrasts, as well as the individual features of the history of the formation of the different parts of the Caucasus and the influence of the neighbouring territories.

The rivers of the Caucasus belong to the basins of the Caspian Sea (The Volga River, the Kura and Aras, Sulak, Terek, Kuma and others), the Black Sea (Rioni, Enguri and their tributaries), and the Sea of Azov (the Kuban and its tributaries) (Ioganson *et al.* 1966 in Gerasimov (Ed.) 1966) (Figure 17).

Among lakes should be noted numerous glacial lakes dammed by glaciers and marine ridges (Afkhazava 1969; Tatashidze 2000; Eppelbaum & Khesin 2012).

At present there are nearly 1,900 lakes with a total area of 95.8 km^2 in the Caucasus. As it is usually known the chemical composition of the water of lakes is controlled by geology, precipitation, vegetation, and drainage type (Rosssolimo et al. 1966 cited in Gerasimov (Ed.) 1966; Efremov 1988). Lakes appear in all altitudinal zones of the region. Some of the lowlands are represented by a number of coastal lagoons and limans (e.g., the brackish lake of Paleostomi, remnant of the Black Sea), mountain lakes (e.g., lake Sevan, the largest lake in the Caucasus (Armenia) and one of the largest (total surface is about 5.0 km^2 freshwater high-altitude (1,900 m a.s.l.); lake Ritsa (Abkhazeti (Georgia), located in the Greater Caucasus at a height of 950 m a.s.l. and known as 'the Pearl of the Western Caucasus'. Most of the high mountain lakes have a glacial moraine or karst genesis (Figure 18). Mainly they are small water-bodies. Today, as glaciers retreat, hundreds of new periglacial lakes have appeared in the Caucasus. Among karst lakes should be mentioned the blue lake of Arkhyz, lake Badukskoe (Western Caucasus, Karachay-Cherkessy Republic in southern Russia). The largest net of volcanic lakes occurs in the volcanic plateau of Samtskhe-Javakheti region (1800-2000 m a.s.l.) (the Lesser Caucasus) of southern Georgia.

Figure 17. River Rioni (Guria province, W. Georgia), named Phasis by the Ancient Greeks. Rioni is the major water catchment area for the SW Caucasus (2006).

Figure 18. The Green Lake in the Arsiani Range, WLC, Adjara, W. Georgia, 2,058 m a.s.l. In the background: *Picea orientalis* mixed with *Fagus orientalis* (1981).

According to ECPC (2012), freshwater habitats cover 8.5% of the Caucasus Ecoregion and are crucial for water conservation, spawning of fish and for breeding and migratory birds.

1.5. Pedology

The founder of the Russian school of genetic soil science, Dokuchaev, appropriately called the Caucasus a natural soil museum (Dokuchaev 1898, 1899, 1900 in Gerasimov 1949). As a result of complex bioclimatic, lithological and geomorphological conditions more than 50 soil types have been identified on its territory (Fridland 1966). The diverse natural conditions of soil formation are furthermore modified by human activity, e.g., by transforming natural into arable lands. In some regions of the Caucasus the cultivation of soil with the frequent use of irrigation has a history of thousands of years, whereas in other regions soil cultivation has only been conducted in the last few centuries. Closely connected with the climatic regime, the soil has a considerable importance for the development of vegetation.

The study of soils of the Caucasus has been instrumental in the formulation of a number of main principles of modern soil science, including the vertical zonality of soils, the theory of provinciality (Zakharov, 1914, 1934, 1938 in Sabashvili 1948), and the distinction of some new types of soils (Gerasimov 1949; Rosanov 1954). Various fundamental monographs were dedicated to the soils of the Caucasus, including in Georgia those of Sabashvili (1936, 1948, 1965), Talakhadze (1964), Urushadze (1987), Urushadze & Blum (2013), in Azerbaijan - Aliyev *et al.*(1953) and Salaev (1966), in Armenia - Galstian (1930), and in Ciscaucasia - Gavriljuk (1947); Zonn (1950) and Fridland (1950).

Figure 19. Red soils, Colchis (Guria province, W. Georgia) (1969).

The soils of the Caucasus are extremely diverse and patchy, as local processes overlay the general vertical zonality (Gegechkori & Joosten (Eds.) 2009).

Different types of chernozems occupy a large territory of lowlands, foothills and plateaux of the North Caucasus, the South Caucasus Upland, the plateau of the Shekino-Adjinouri low-mountains and the Shiraki-Garedji steppe in the South Caucasus.

The eastern slopes of the Stavropol Elevation are covered by dark-chestnut soils and further eastward by light-chestnut soils. The same type of soils occurs in the foothills of the Greater and Lesser Caucasus surrounding the Kura-Aras Lowland. The mountain-chestnut soils in the Armenian Upland that co-occur with mountain-chernozem soils are strongly developed. The coastal zone of the Caspian Sea and the Kura-Aras Lowland are covered by grey semi-desert soils, frequently with saline soils and true solonchaks, alternating with sandy grounds. Grey soils have developed in the lower parts of the Aras Lowland.

Forest soils of moderate climate are found over the extensive north slopes of Ciscaucasia and in the forest zone of the South Caucasus, where mixed deciduous forests prevail under colder climatic conditions. Depending on the dominant species in the forest stand, these forests have grey forest or dark-grey forest soils, whereas at the same altitudes long-fallowed mountain podzol forest soils have developed.

Various types of cinnamonic and mountain-cinnamonic soils occur in areas with a Mediterranean climate. In the Caucasus such soils are present in the south-western Caucasus and in the mid-Aras Depression, small areas are also found in the north-western Caucasus and in the warmest places of south-eastern Ciscaucasia. The limestone ranges of the mid- and high altitude areas have humid-carbonate soils. In the high mountain areas of the Greater and Lesser Caucasus mountain-meadow soils occur. The most conspicuous soil types are present in the most humid regions of the South Caucasus, i.e., red soils and yellow soils in Colchis (Figure 19) and yellow soils in Talysh (usually characterized by a red or yellow color, clayed and a well-developed profile). Meadow, wetland and floodplain soils are confined to water saturated areas and mainly consist of organic soils with undifferentiated profile, strong decomposition and sometimes signs of glaciation (Fridland 1966, in: Gerasimov (Ed.) 1949).

1.6. Biodiversity of the Caucasus Ecoregion Hotspot

'To prevent language extinction, we must redouble our efforts to prevent biodiversity hotspots and wilderness areas from being converted for short-term gain' (Mittermeier 2012).

The Caucasus as a whole is isolated from other mountains by seas and plains. This high degree of isolation - transit position of the Caucasus between two climatic zones, its extraordinary geographic range location that connects Asia to Europe and at the same time the region's extraordinary position among far more large biogeographical regions (Irano-Turanian, Euro-Siberian and Mediterranean) from east to west and from north and south and *vice versa* - creates favorable conditions for the successful evolution of a wide spectrum of flora and fauna with an extremely high percentage of endemism. No temperate climate zone has more biodiversity per square kilometer than the Caucasus (Caucasus Nature Fund 2013). The richness of biodiversity of the region increases within the boundary of the Caucasus Ecoregion.

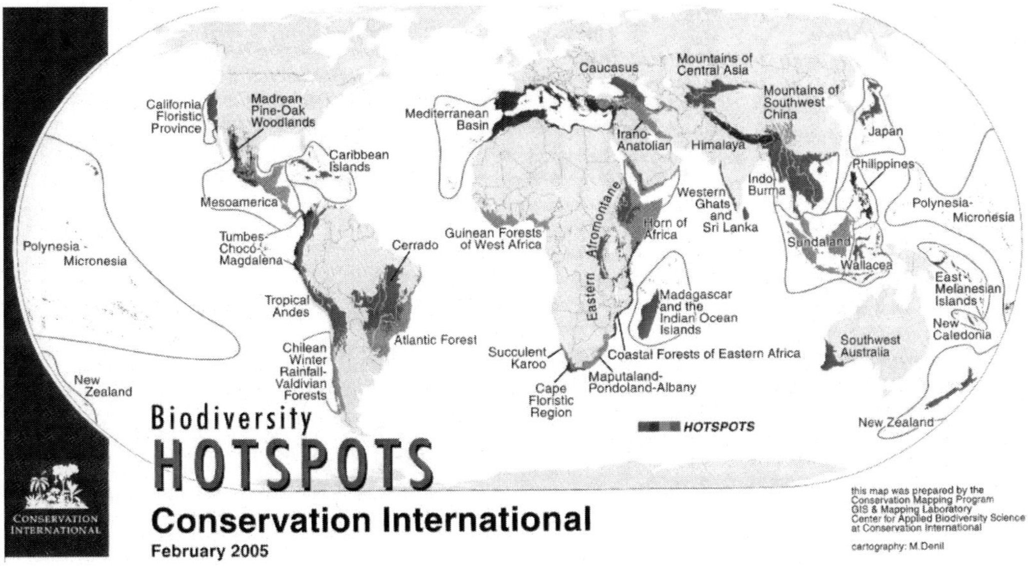

Figure 20. Biodiversity Hotspots of the Earth (Conservation International 2005).

The Caucasus is one of the most important biodiversity hotspots in the Northern Hemisphere (Myers *et al.* 2000). This fact, that the Caucasus biota raises special interest of scholars, here it is worth mentioning the following opinion of Myers *et al.* (2000): 'The hotspots' boundaries have been determined by 'biological communities'. Each of the areas features a separate biota or community of species that fits together as a biogeographic unit. This is apparent in the case of islands or island groups such as New Caledonia, New Zealand, the Caribbean, Polynesia/Micronesia, Madagascar and the Philippines. Much the same applies to 'ecological islands' in clearly defined continental units such as the Cape Floristic province, the eastern Arc and Coastal Forests of Tanzania/Kenya (hereafter abbreviated to 'Eastern Arc'), southwestern Australia and Caucasus'.

The Caucasus Ecoregion covers 580,000 km^2 in six countries: Armenia, Azerbaijan, Georgia, the North Caucasus part of Russia, northeastern Turkey, and part of northwestern Iran (Zazanashvili & Mallon (Eds.) 2009). As one of the biologically richest areas on the Earth, the Caucasus is included in WWF's list of Global 200 Ecoregions, ranked among the planet's 35 most diverse and endangered hotspots by Conservation International (Meyers *et al.* 2000), and one of the world's 15 'Biodiversity Hotspots' of the Critical Ecosystems Partnership Fund (CEPF) (Krever *et al.* (Eds.) 2001; Williams *et al.* (Eds.) 2006) (Figure 20). For a non-tropical region, the Caucasus has a high species diversity and endemism (Tarkhnishvili & Kikodze (Eds.) 1996; Zazanashvili & Molles (Eds.) 2009). Only 10% of its primary vegetation cover remains from the 500,000 km^2 it once occupied (Myers *et al.* 2000). No temperate climate zone has more biodiversity per square kilometer than the Caucasus. Despite some threats, about a quarter of the territory remains in environmentally good condition, and as much as 12 percent of the original vegetation, including forests, is considered (almost) pristine, compared to less than one percent in Europe (CNE 2012). Review works on the nature of the Caucasus include 'The Caucasus' (Gerasimov (Ed.) 1966; Schmidt 2005a, b; Sokolov & Syroechovski (Eds.) (1990); Khain & Koronovski (1997); Volodicheva (2003); and the South Caucasus Republics: Biodiversity analysis Updates for Armenia (2009), for Azebaijan (2010) and for Georgia (2009). Potential analysis for further nature conservation in Azerbaijan (2009), especially protected nature areas of Armenia (Khanjyan 2004), National Biodiversity Strategy, action plan-Georgia (NACRES 2005) and the the Hyrcan forest-restoration of forest landscape in the Talysh Region, Azerbaijan (2010) are in progress. Problems of biodiversity and environmental protection are outlined for the South Caucasus adjacent regions: 'The Western Caucasus, Russian Federation' (UNEP, WCMC 2010), series of letters - 'West Caucasus property in danger – in local newspapers (e.g., '*Novaja Gazeta Kubani* (New Newspapers of Kubani' etc.) (Rudomakha 2005), 'Assessing the effectivness of protected area management in the Turkish Caucasus' (Kurdoglu *et al.* 2011), 'Camili Biosphere Reserve, Turkey. GEMCONBIO Case Study Report' (Adem *et al.* 2007); for the Caucasus Ecoregion by Zazanashvili *et al.* (Eds.) 2012.

CEPF started implementation of its regional program on biodiversity conservation in the Caucasus in 2004. Within the CEPF project 'Amazing Caucasus Hotspots', as it is already mentioned, the book-album 'Caucasus - Treasure of Nature' was published (Gegechkori 2007) to contribute to stronger eco-politics and to enhance the local ecological capacity of the Caucasus Ecoregion. It identified the ecosystem value of some of the most precious reserves and National Parks of the eco-region, including the Vashlovani open woodland (Georgia), the Khosrov open woodland (Armenia), and the Talysh (Azerbaijan) and Colchis (Georgia) hardwoods with their Arcto-Tertiary floras. Much of the Caucasus Ecoregion is mountainous, but there are also extensive lowlands and coastal areas. The region stretches from the lush broad-leaved forests with evergreen understory in the moist, almost subtropical climate of the southern Black Sea to the clearly expressed nival zone of the Greater Caucasus and to the deserts and semi-deserts of the lowland of the Eastern Caucasus. The wide range of altitudes produce a climate diversity that, combined with a complicated relief, gives rise to a rich variety of ecosystems that — next to the above-mentioned — includes lowland, mountain, and high mountain forests, wetlands, riparian or flood forests, dry steppes, dry mountain shrublands, and coastal and marine ecosystems.

Among these ecosystems, the forests are the most important biome for diversity conservation, covering nearly one-fifth of the region. Mountain forests make up the majority of the forest biome and play a critical role in regulating water flow and preventing soil erosion.

The relict forests of Colchis and Hyrcan and some animals inhabiting these forests from pagan time are particularly valuable, as they harbour Arcto-Tertiary species that were widespread in Europe in Miocene-Pliocene times but have survived only here (Gegechkori 1984a, b, 2000, 2008, 2011a, 2012a; Nakhutsrishvili *et al.* 2011, 2015).

As is widely known recently, there had been close links between the ancient Hellenic World and the bordering Black Sea region, namely Colchis, land of the Golden Fleece. Priceless evidence of the Colchis culture is presented in the ancient Temple City of Vani located 40 km west of Kutaisi, West Georgia (Lordkipanidze 2000). A rich cultural history was discovered in Colchis Lowland as well as adjoining large mire complexes (Vickers & Kakhidze 2004; Papuashvili & Papuashvili 2005) or buried under the peat (Djanelidze & Tatashidze; 2010). The remains of human cultures in Colchis mires are stressed by Krebs and Joosten (2006). Recently most Colchis mires are protected in national nature reserves which is a real guarantee to conserve the mentioned unique ecosystem of Georgia.

The Caucasus Ecoregion is a biological crossroads, where plant and animal species from Eurosiberia, Central (Middle) Asia, the Middle East and North Africa mix with endemic species that are found nowhere else on the Earth.

Today the Caucasus Ecoregion (or its different countries) is included within various ecological networks, for example, ecoregional transboundary, a map of priority conservation areas (PCA) and corridors in the Caucasus Ecoregion have been prepared by WWF, as a guideline for future action towards an ecological network in the region; the same Ecoregion is included in the Emerald network, initiated under the Berne Convention on the Conservation of European Wildlife and Natural Habitats (equivalent to Nature 2000, to non-EU countries in Europe and northern Africa). Pilot projects have been implemented in the 12 new EU Member States before they joined the EU (including Turkey, the Russian Federation, and three countries of South Caucasus).

The Caucasus Ecoregion hotspot is home for over 7,000 vascular plant species, of which 2091 species are restricted to the region, i.e., the highest level of endemism in the Temperate zone of the Northern Hemisphere. 1,752 of them were evaluated for the Red List of the Caucasus (Solomon *et al.* 2014). One-third of the endemic plants in the Caucasus Ecoregion are thought to have originated in the Greater Caucasus Range. 17 endemic plant genera occur in the Caucasus, nine of which are associated with high mountain communities, 80% of the plants growing on the Colchis limestone scree are found nowhere else in the world. Similar values hold for the Caucasus proper, i.e., without the adjacent areas of Turkey and Iran.

Notable relict and endemic species include Persian ironwood *(Parrotia persica)* in Azerbaijan's Talysh and the Talysh-northern slopes of the Alborz (Elburz) in North Iran and the various rhododendron species (*Rhododendron caucasicum, Rh. ungernii, Rh.smirnowii)* in Colchis Triangle (Georgia, Turkey).

Many ancient species are still dominant and co-dominant in the plant communities of the hotspot, e.g., *Castanea sativa* and *Buxus colchica.* Of others, the populations are very small and fragmented, e.g., of Pontian oak (*Quercus pontica),* Medwedew's birch (*Betula medwedewii*) (=*B. medwediewii*) (DD) and Megrelian birch (*B. megrelica*) (EN).

The presence of tropical-subtropical tree species like the Persian silktree (*Albizzia julibrissin)* and Caspian locust (*Gleditsia caspica)* illustrate the outstanding character of the vegetation cover of the Ecoregion.

The highest species is calculated for Azerbaijan (4,500), Georgia (4,100) and northeastern Turkey (4,000), with the highest number of Caucasian endemics found in Georgia (1,320) and the Russian Caucasus (1,300) (Solomon *et al.* 2014). The most abundant families in numbers of species include Apiaceae, Asteraceae, Brassicaceae, Campanulaceae, Caryophyllaceae, Fabaceae, Poaceae and Rosaceae with more than 100 species represented in the Caucasus (Solomon *et al.* 2014).

The region also has a remarkable concentration of economically important plants, particularly wild crop relatives such as wheat, rye and barley, and nuts and fruits like wallnuts, apricots and apples (Ecoregion Conservation Plan for the Caucasus, Zazanashvili *et al.* (Eds.) 2012).

The Ramsar Convention (Ramsar, Iran 1971) on Wetlands characterizes the Colchis wetlands in Georgia as being of international importance for biodiversity. Many of the bird species observed in the Ecoregion nest, winter or migrate through here. In addition, there are several Important Bird Areas (IBAs). About 60% of the Colchis wetlands have been drained for agricultural production (especially tea and citrus plantations) and for rural development projects. The greatest threat faced by the Colchis wetlands is from the transportation of oil products and the operations of the oil terminals on Georgia's Black Sea coast. Similarly, along Turkey's eastern Black Sea coast, the 'Black Sea Motorway' has destroyed patches of coastal vegetation (WWF, 2011). By the same importance are characterized the wetlands and their wildlife of Iran's and Turkey's parts of Ecoregion.

Phytogeographically the position of the Caucasus was a subject of long discussions. Some authors (Boissier 1867; Il'insky 1937; Rikli 1943-1948; Wulff 1944; Grossheim 1948, etc.) referred the Caucasus to the 'Mediterranean region'. Other authors (Meusel *et al.* 1965) include it into the Macaronesian-Mediterranean Floral Region; representatives of another group of scientists did not agree with this viewpoint, referring the mentioned region to botanical-geographical areas: European broadleaf and Afro-Asian desert (Lavrenko 1960; Isachenko & Lavrenko 1980). In his chorionomic system, Takhtajan (1978, 1986, 1988) referred to the Mediterranean area as only the foothills part of northwestern Caucasus between Anapa-Tuapse, but the rest were parts of the Caucasus to the Euxinian province of the Boreal (Euro-Siberian) subrealm, and Armenian-Iranian, Hyrcanian and Turanian provinces – to Irano-Turanian region of the Tethyan (Ancient Mediterranean) subkingdom within the Holarctic kingdom.

There are about 153 mammal species in the Caucasus hotspot, of which one-fifth are endemic and several are threatened. There are both newly evolved species as well as ancient relict species. Ancient and primitive small mammals include the unusual long-clawed mole-vole (*Prometheomys schaposchnikowi*) that is the sole representative of a monotypic genus which is endemic to the region, and species of the genera *Sicista, Chionimys, Meriones* and *Apodemus.* Among small mammals threatened are also several bat species (for example, the Mediterranean horseshoe bat (*Rhinolophus euryale,* VU), Mehely's horseshoe bat (*Rh. mehelyi,* VU) and others.

Large threatened species are confined to a small part of their former ranges: the Persian (Caucasian) leopard, *Panthera pardus ciscaucasica* (= *P. p. saxicolor*) and the striped hyena (*Hyaena hyaena*). Both are on the verge of extinction in the Ecoregion. Other species are represented by the Caucasus High-Mountian goats (turs - *Capra caucasica,* EN and *C. cylindricornis,* EN) and the Caspian monk seal (*Phoca caspica,* VU). The goitered gazelle (*Gazella subgutturosa*, CR), the Bezoar goat (*Capra aegagrus*, VU), mouflon (*Ovis ammon gmelini*, VU), and Caucasian chamois (*Rupicarpa rupicarpa caucasica*), the European bison has been (*Bison bonasus*) reintroduced to the North Caucasus.

Mature intact forests provide an important habitat for large mammals, including red deer (*Cervus elophus maral*), roe deer (*Capreolus capreolus*), and one of the largest populations of brown bear (*Ursus arctos*) in Europe.

More than 400 bird species reside in the hotspot, four of which are endemic: the Caucasian snowcock (*Tetraogallus caucasicus*), the Caspian snowcock (*Tetraogallus caspius*), Caucasian black grouse (*Tetrao* (*Lyrurus*) *mlokosiewiczi*) and the Armenian gull (*Larus armenicus*). Several water bird species are globally threatened, such as the marbled duck (*Marmaronetta angustirostris,* VU), the lesser white-fronted goose (*Anser erythropus,* VU) and the white headed duck (*Oxyura leucocephala*, EN). Significant numbers of raptors breed in the hotspot, whereas millions of migratory birds pass through along two major migration routes on the east coast of the Black Sea and the west coast of the Caspian Sea.

Of the 87 species of reptiles, about 28 are endemic. Species of rock lizards (*Darevskia* spp.) are particularly interesting. Of the 32 species known to exist in the world, the majority of them occur in the Caucasus with a higher ratio of endemic species. It must be stressed that some species of this genus (e.g., *D. dahli, D. armeniaca*) are parthenogenetic. The genus *Lacerta* has many endemic species as well, whereas the vipers are represented by the endemic Caucasian vipers (*Vipera* (*Pelias*) *kaznakovi*, *Vipera* (*P.*) *orlovi*, etc.).

The number of species of amphibians (14 species) is relatively low; four of them are endemic. The Caucasian salamander (*Mertensiella caucasica*) is endemic to the West Caucasus and penetrates into the Turkish Caucasus, whereas the northern Iran salamanders *Iranodon* (*Batrachuperus*) *gorganensis* (CR), *I.* (*B.*) *persicus*) (NT) are residents of *the* northern provinces of Iran. The Caucasian mud-diver (*Pelodites caucasicus*) and Caucasian toads (*Bufo verrucosissimus* (NT) and *B. eichwaldi* (VU) should be mentioned also.

In the region 130 fish species inhabiting the lakes, rivers and seas can be found in the Caucasus of which over a third are found nowhere else. Three species of scaleless and jawless fishes are endemic, including *Caspiomyzon wagneri, Eudontomyzon mariae* and *Lampetra lanceolata*. Lampreys belong to the most ancient of fish taxa, dating back to the Permian Period (280-225 million years ago).

The Caspian Sea, the Sea of Azov and the Black Sea host, as mentioned, the world's largest diversity of another ancient group of fish, the sturgeons, including the largest freshwater fish the beluga sturgeon (*Huso huso*). The populations of all sturgeon species have been reduced through overharvesting, primarily for caviar, water pollution and river damming, which restricts anadromous migration. The Atlantic sturgeon (*Acipenser sturio*), which spawns only in rivers in the Colchis Lowland in Georgia is Critically Endangered (Tarkhnishvili & Kikodze 1996; Williams *et al.* (Eds.) 2006; Gegechkori 2007).

Possibly a no lower percentage of endemism occurs in terrestrial invertebrates. For example, 75% of molluscs (Mumladze *et al.* 2008) and 25% of psyllids (Gegechkori 1984a) are endemic for the Caucasus.

Unfortunately, the natural ecosystems of the Caucasus Ecoregion and their special fauna and flora are rapidly disappearing due to anthropogenic factors. Most important are deforestation caused by unsustainable logging, fuelwood harvesting and legal or illegal timber trade and overgrazing of grasslands and highlands meadows by livestock.

Zoogeographical division of the Caucasus is referred to in numerous works of zoologists occupied by investigation of vertebrate and invertebrate group of animals (e.g., Vereshchagin 1959; Tuniyev 1990, 1995; Kryzhanovsky 1965, 2002; Gegechkori 1984a). Wallace (1876) divided the Isthmus along the crest of the Greater Caucasus, referring Ciscaucasia to the European subregion and Transcaucasia to the Mediterranean.

The most zoogeographical classifications of the Caucasus lie within the Palaearctic (e.g., Wallace 1876; Darlington 1957; Cox 2001), or the Holarctic region. On the contrary Smith, (1983) based on mammal families occurrence, included the whole circum-Mediterranean area and south-west Asia overlapping the Caucasus and the Middle East, into the Afro-Tethyan region. According to Kreft & Jetz (2010), the mammal species-level boundary between Africa *sensu lato* and the Palaearctic goes along the Greater Caucasus Range, while on the family and genera levels the Palaearctic includes the Caucasus and Asia Minor (compiled from Naseka 2010).

There is another scheme of division of Caucasian fauna. Within these works the Caucasus referred either to the Mediterranean (Severtsov 1877; Menzbir 1934; Vereshchagin 1959, Tuniyev 1990, 1995), or to the province of desiduous forests of Europe (Kuznetsov 1949; Bobrinsky 1951), or to Boreal and Sethian (Saharo-Gobian) subrealms (Gegechkori 1984 a, b). The most outstanding work concerning zoogeographical division of the Caucasus Isthmus belongs to the famous researcher of the Caucasian fauna, Vereshchagin (1958a, b). From his viewpoint, taking into consideration the palaeogeographical peculiarities of the mentioned region, should be meant the southern vicinities of Europe (including the south of the Ukraine and the Crimea), North Africa, Asia Minor, Caucasian Isthmus (excluding Premanych), Talysh and Elburs Ridges. Vereshchagin's opinion of the Mediterranean biogeographical unit coincides with the boundaries of the floristic Mediterranean of Rikli (1946).

The most complex description of a rich and heterogenous vertebrate fauna of the Caucasus, its origin and zoogeographical division was made by Satunin (1910, 1912). With respect to invertebrate fauna the subject of the same multidisciplinary description became psyllids (Insecta: Hemiptera, Psylloidea) (Gegechkori 1984a, b).

Throughout the region the species are threatened by poaching (some big game ungulates), illegal wildlife trade (butterflies, salamanders, viper snakes, etc.), and overfishing (especially the sturgeons). The rarest mammal in the hotspot, the leopard (*Panthera pardus),* was once widespread and celebrated in many local poems, fairytales and songs, but only a few individuals are surviving in the Caucasus today. Its main threats are poaching and the overhunting of its natural prey, the wild ungulates.

Realization of the positive restoration of the mentioned problems requires multilateral support and collaborations among the many governmental and nongovermental organizations and authorities responsible for nature protection and biodiversity conservation. These include the recent legitimized environmental agreements and in particular the strategy for Biodiversity Conservation 2011-2020 and its associated 'Aichi Targets' which were adopted by the Parties to the Convention on Biological Diversity in Nagoya, Japan in 2010 (Ecoregion Conservation Plan for the Caucasus; Zazanashvili *et al.* (Eds.) 2012).

Fortunately, conservation experts from all over the region have come together to create a comprehensive long-term strategy to protect, manage and restore the natural resources of the Ecoregion. In the Caucasus, a number of priority conservation areas and priority conservation corridors have been identified, concentrated in four priority biomes (forests, freshwater habitats, marine and coastal ecosystems, and high mountain habitats) where the bulk of biodiversity is found and where the threats are largest (Williams *et al.* (Eds.) 2006).

Currently there are 224 protected areas in the Caucasus Ecoregion, including two biosphere reserves (Kavkazsky and Camili Biosphere Reserves), 47 strict nature reserves (Zapovedniki, covering 2.23% of the Caucasus area), and 32 national parks.

The Caucasus Nature Fund is currently supporting nine of the region's 25 main parks, safeguarding over 400,000 hectares (CNE 2012).

Other protected areas include hundreds of multipurpose, zoological, botanical, geological and hydrological reserves (zakazniki), and natural monuments, covering about 10% of the Ecoregion's area (Zazanashvili & Mallon 2009). Equivalent to the zakazniki are nature parks and protected landscapes. Next to the reserves, 56 priority conservation areas and 60 corridors have been identified (An Ecoregional Conservation Plan for the Caucasus; Williams *et al.* (Eds.) 2006).

The strict nature reserves (zapovedniki) correspond to category I of the protected area classification system of IUCN.

In Georgia the first strict nature reserve in the Caucasus Ecoregion was founded. This is Lagodekhi Strict Nature Reserve established in 1912. Therefore, in 2012 100 years of this outstanding monument (living, unliving) of the Caucasus nature was celebrated. According to data collected by Chape *et al.* (2008), it means that Lagodekhi gorge on the southeastern slopes of the Greater Caucasus Range was heralded as the first protected area just when the first national parks were established in Europe and earlier than protected areas were created on the other continents excluded North America.

An essential role in the establishment of this reserve was played by the Polish explorer, botanist and zoologist Ludwig Mlokosievich who convinced the Russian Royal Geographical Society of the necessity to create a sanctuary in the gorge with one of the richest, beautiful and typical examples of mesic Caucasian forest.

The Caucasus has from time immemorial been at the crossing-place of different cultures and religions, for example, Christianity (Orthodox Christian; Armenian Apostolic, Roman Catholicism, etc.), Islam and Judaism, and political systems ('geopolitical seismic zone') (de Waal 2010). During two millenia, the human populations were admixed and had experienced gene flow from both Europe and West Asia (Nasidze 1995).

The Caucasus is one of the most complicated regions in the world with many different peoples ('Refugium of peoples') and ethnolinguistic variations. According to this complexity, especially in later terms, the Arabs in the medieval ages called the Caucasus *jabal al-sun*, or the Mountain of Tongues (Coene 2010). In Dagestan alone in the North-East of the Caucasus mountain range with 50,300 sq. km. and counting about two million inhabitants is one of the most ethnically diverse regions in the world, counting 30 ethnic groups and 80 - odd nationalities. Linguists distinguish 29 different languages in Dagestan, where the Russian and Avar languages serve as a *lingua franca* between different Dagestani peoples. 'Multi-lingualism is common throughout the Caucasus but can take formidable proportions in Dagestan. . .' (Hewitt 1994). The highly complex landscape of the Caucasus determines the relative isolation of ethno-cultural groups and their mozaic-like settlement patterns (O'Loughlin *et al.* 2007 and references therein).

If 'all roads lead to Mecca', today the Caucasus region is a kind of 'Mecca' of energy resources (it is generated by oil and gas pipelines to Europe from Azerbaijan, Iran and Central (Middle) Asia). After construction of the Baku-Tbilisi-Ceyhan and Baku-Tbilisi-Erzerum pipelines, the 'Nabucco' project was conceived.

Due also to the mentioned climatic and orographic features, the Southern Caucasus has been identified as one of the centers of origin of many domestic plant species, such as grapes, soft wheat, and a center of genetic differentation. In this region agriculture and breeding were first developed about 10,000 to 5,000 years ago, in the Neolithic period (Batello *et al.* 2010).

The striking complexity of not only the ethnolinguistic variations, such rich historic and architectural monuments and other cultural heritage but also, as previously mentioned, the extremely diverse landscapes and biodiversity has attracted many researchers.

Table 2. Biodiversity and endemism of the Caucasus Ecoregion (from biodiversity hotspots 2012)

Taxonomic Group	Species (N)	Endemic species (N)	Endemism (%)
Plants	7000	1600	22,85%
Mammals	153	18	11,76%
Birds	378	4	1.05%
Reptiles	87	28	32,18%
Amphibians	14	4	28,57%
Freshwater Fish	130	12	9,23%

1.7. Main Types of Vegetation Zonation of the Mountains of the Caucasus

Almost all of the altitudinal belts common to the western part of Eurasia are represented in the Caucasus Mountains (Nakhutsrishvili 1998, 2013; Zazanashvili *et al.* (Eds.) 2000). Four main types are distinguished in the Caucasus mountain zonation.

1.7.1. West Caucasian (Eastern Euxinian or Colchis) Type

This type is characteristic of the western sections of the Greater Caucasus Range and of the Lesser Caucasus mountain chain, mainly where one of the refugia of hygro-thermophilous representatives of the Tertiary flora existed during the Pleistocene's Ice-Ages. This type was formed under humid conditions. The main characteristic of this type is a wide distribution of Colchic relicts along the whole zonal altitudinal profile, almost from sea level up to 2,300m.

Vegetation of the Black Sea coastal dunes, wetlands and riparian forests (almost from sea level up to 400-500 m).

Humid thermophilous Colchic broad-leaved forest belt (zone). This zone occurs from 500-600 up to 1,000-1,200 m. a.s.l.

Mixed broad-leaved forest belt. This belt occurs up to 500 (600) m and includes *Castanea sativa, Carpinus betulus, Fagus orientalis, Quercus hartwissiana* and *Zelkova carpinifolia*, with a Colchic understory. In relatively dry habitats thermophilous hornbeam-oak forests occur with *Quercus petrae iberica* (= *Q. iberica*), *Carpinus betulus* and *C. orientalis*, in the South Colchis (from 200m up pine-oak forests with *Quercus dschorochensis* (some botanists treat as subspecies of *Q. petraea dschorochensis*) and *Pinus kochiana* occur as well as Colchic thickets.

Chestnut forest belt. This belt is characterized by forests with *Castanea sativa* and *Fagus orientalis*, occurring from 500-1,000 (1,200m) with a Colchic understory, further by thermophilous oak forests and Colchic thickets.

Oriental beech forest belt. A zone between 1,000 (800) – 1,400 (1,800) m with *Fagus orientalis* forest often with a Colchic understory, and dark coniferous/beech forests partly with a Colchic understory.

Dark coniferous forest belt. This zone, from 1,400 (1,000) – 1,800 (2,100) m, includes forests with *Abies nordmanniana, Picea orientalis* and *Fagus orientalis*, partly with Colchic understory.

Subalpine elfin wood and meadow belt. This zone occurs from 1,800 (1,600) – 2,400 (2,700) m.

Lower subalpine belt. This belt occurs from 1,800 (1,600) – 2,100 (2,200) m and includes *Fagus orientalis*, *Quercus pontica* and *Betula medwedewii* elfin woods, often with a Colchic understory; tall herbaceous vegetation, Colchic thickets, meadows.

Upper subalpine belt. This belt, from 2,100 - 2,400 (2,700) m, includes birch/ash-birch elfin woods with *Betula pubescens litwinowii* and *Sorbus caucasigena*; *Rhododendron caucasicum* thickets; meadows.

Alpine grassland and thicket belt. This zone appears between 2,400 and 2,900 (3,000) m.

Lower alpine belt. This belt occurs from 2,400 to 2,750m and includes alpine turfs with grasslands and *Rhododendron caucasicum* thickets.

Upper alpine belt. This belt occurs from 2,750 to 2,900 (3,000) m and with grasslands and mats; rock and scree vegetation.

Subnival belt. This zone occurs from 2,900 - 3,700 (4,000) m and includes open plant communities (*Cerastium polymorphum* and *Minuartia trautvetteriana*); fragments of mats and grasslands (up to 3,000m).

Nival cryptogam belt. This zone exceeds up to 3,700m.

1.7.2. East Caucasian (Irano-Turanian) Type

This type is characteristic of the eastern section of the Greater Caucasus Range and of the Lesser Caucasus mountain chain. The climate has continental features over most of the area: the mean annual precipitation varies from 600 - 1,000mm. In addition, the northern slopes of the Eastern Greater Caucasus and the Lesser Caucasus are drier than the southern slope of the Eastern Great Caucasus, which is reflected in the zonation sub-type level. Furthermore, in comparison with the humid Colchis, corresponding zones are located 100 - 200 m higher here.

Riverside and foothill forest belt. This belt, at < 500 - 600 (1,000) m, includes oak/poplar forests with *Quercus robur pedunculiflora* and *Populus hybrida*, thermophilous hemixeric hornbeam-oak forests on the slopes (*Quercus petrae iberica, Carpinus betulus*); arid woodlands (*Pistacia atlantica* (=*P. mutica*), *Juniperus polycarpos*); shibliak (*Paliurus spina-christi, Rhamnus pallasii*).

Lower mountain belt. This belt, from 500 - 1,000 (1,200) m, includes oak/hornbeam forests (*Quercus petrae iberica, Carpinus betulus*); beech and hornbeam-beech forests (*Fagus orientalis, Carpinus betulus*).

Mesic beech forest belt. This zone occurs from 1,000 - 1,800(2,000) m.

Middle mountain belt. At 1,000 - 1,500 m this belt includes *Fagus orientalis, Pinus kochiana* forests.

Upper mountain belt. This belt, at 1,500 - 1,900 (2,000) m, includes *Fagus orientalis* forests; *Quercus macranthera* forests; *Pinus kochiana* forests.

Subalpine elfin wood and meadow belt. This zone occurs between 1,900 (2,000) and 2,500 (2,700) m.

Lower subalpine belt. This belt, at 1,900 - 2,200 m, includes *Quercus macranthera, Pinus kochiana* and *Acer trautvetteri* woodlands; elfin woods (*Betula pubescens litwinowii, B. raddeana*) (NT); tall herbaceous vegetation (*Heracleum sosnowskyi, Aconitum orientale*); low open juniper communities (*Juniperus hemisphaerica* (=*J. depressa*) on the rocks and scree; *Rhododendron caucasicum* thickets; meadows, meadow steppes.

Upper subalpine belt. This belt, at 2,200 - 2,500 (2,600) m, is characterized by birch/ash-birch elfin woods (*Betula pubescens litwinowii, B. raddeana, Sorbus caucasigena*); *Rhododendron caucasicum* thickets; meadows.

Alpine grassland and thicket belt. This zone occurs between 2,500 - 3,000 (3,200) m.

Lower alpine belt. This belt, found between 2,500 and 2,800m, includes grasslands and *Rhododendron caucasicum* thickets.

Upper alpine belt. This belt, from 2,800 - 3,000 (3,200) m, includes grasslands, with *Kobresia schoenoides*) and mats (*Sibbaldia parviflora, Carum caucasicum*).

Subnival open belt. In this zone, at 3,000 - 4,000 m, are open plant communities with fragments of mats and grasslands (up to 3,300m).

Nival cryptogram zone. This zone is found at > 4,000m.

1.7.3. South Caucasian (Asia Interior or Armeno-Iranian) Type

This type is characteristic of the uplands and mountains of the southern Caucasus mainly composed of volcanic sediments. Here representatives of the Caucasus relict flora do not occur: Anatolian-Iranian components predominate in the plant communities' (floristic composition); the typical forests are not characteristic of the zonal altitudinal profile which is formed in xerothermic, continental conditions: the mean annual precipitation varies within the 250-500 mm limit and increases in high-mountain regions. In comparison with humid regions in the Caucasus the corresponding zone limits are situated 300-400 m higher. The typical zonal altitudinal profile is as follows.

Semi-desert and desert zone. This zone, at <800 m, includes dwarf semi-shrubs (*Artemisia fragrans, Salsola* spp. with ephemerals and ephemeroides; deserts with *Halocnemum strobilaceum* and *Suaeda microphylla* on saline soils; thorn-cushions (*Astragalus microcephalus*).

Xeric grass and semi-shrub belt. This zone, at 800 (1,200) – 1,600 m, includes hamada, frygana, shibliak, thorn-cushion (tragacanthic), andmountain-xerophyte vegetation communities.

Hemi-xeric woodland belt. This zone, at 1,600-2,300 (2,400) m, includes *Quercus macranthera* woodlands: mid-story woodlands (*Pyrus* spp., *Acer hyrcanum, Juniperus polycarpos*); hemi-xeric shrublands (*Cotoneaster* spp., *Sorbus graeca*); steppes, thorn-cushion steppes and meadow steppes.

Subalpine belt. This zone, between 2,300 and 2,800 (2,900) m, usually includes *Quercus macranthera* woodlands, steppes, meadows, meadow steppes and thorn-cushions.

Alpine belt. This zone, at 2,800-3,400 (3,600) m, include alpine grasslands, mats, rock and scree vegetation.

Subnival open belt. Between 3,400 and 4,300 (4,400) m open plant communities.

Nival cryptogram belt. This zone is found at >4,200m.

1.7.4. Southeast Caucasian (Hyrcanic) Type

This type is characteristic of the extreme southeastern part of the Caucasus, southeast Azerbaijan and the northwest Iranian mountains along the Caspian Sea coast. Here the other refugium of the South Caucasus from the Tertiary flora, the Hyrcanic refugium, occurs.

Humid thermophilous (Hyrcanic broad-leaved forest belt). This zone occurs at altitudes < 1,000 (1,200) m.

Mixed broad-leaved forest belt. This belt is found up to 600m and includes several oak forests with *Quercus castaneifolia, Parrotia persica, Zelkova carpinifolia, Albizia julibrissin, Diospiros lotus* and other trees and also with shrubs/semi-shrubs.

Oak forest belt. This belt occurs from 600 to 1,000 (1,200) m and includes *Quercus castaneifolia* forest, beech and beech/hornbeam forests (*Fagus orientalis, Carpinus betulus*) and *Quercus petrae iberica, Carpinus betulus* forests.

Oriental beech (*Fagus orientalis*) forests belt 1,000-1,600 (1,800) m.

Steppe and xeric dwarf semi-shrub zone. From 1,600 to 2,300 (2,500) m.

Compiled from Grossheim (1948); Ketzchoveli (1960); Gulisashvili *et al.* (1975); Gagnidze (1970); Nakhutsrishvili (1998); Zazanashvili *et al.* (Eds.) (2000) with some modifications.

CONCLUSION

- The Caucasus (between latitudes 38^0 25' and 47^0 15'N and longitudes 36^0 30' and 50^0 20'E) is a region in the Northern Hemisphere. This region is physically situated between the Black and Caspian Seas, therefore it is often referenced in literature as the Caucasus Isthmus. It covers an area of approximately 441,000km^2.
- The Caucasus being a natural bridge between Europe and Asia has an extraordinary crossroads, geopolitical location, at the same time it has an outstanding cultural and remarkable biodiversity intersection position, correspondingly.
- The Caucasus as a mostly mountainous terrain, is isolated from other mountain systems by surrounding seas and plains (with semi-arid and arid landscapes). This high degree of isolation pattern ('islands in the sky') for the region causes insularity of native floras and faunas (i.e., coincides with 'non-equilibrium insular biogeography' theory).
- Biota of the same chorological features of the Caucasus is inheredited from time immemorial. Historically its insular and remote location has existed as a Caucasus Island from the Early Cretaceous (approximately 130 mya). During Cretaceous (136-65 mya) it was represented in the Tethis palaeocean between two palaeomegacontinents, Laurasia (modern North Hemisphere) and Gondwana (nowadays South Hemisphere). This isolation position apart from mainlands has been largely attracted to high level autochtonous processes of its living organisms.
- The Caucasus with its highly heterogenous fauna and significantly flora besides the vegetation of recent geological time, encompasses relictual elements of the former Laurasian ('flora of Ginko'), Gondwanian ('flora of Welwitschia'), and Tethian (Saharo-Gobian) roots.
- From the vicariant biogeographic point of view today in the Caucasus are present derivative forms of past geological epochs - relict and relict-endemic genera and species of quite different origin. For example there are remnants of moist tropical-subtropical, Mediterranean subtropical, initial temperate broadleaf deciduous forests (TBDF) derivatives, Pleistocene glaciations and hyperxerophytic desert biomes. Being restricted within refugial areas (frequently patches), these striking representatives of natural heritage are the key species to understand the past biotas.
- Endemism and range disjunction patterns of living organisms in the Caucasus are largely caused by crucial abiotic events. For example, by Miocene's Alpine-Himalayan orogenesis; then, in the Early Pliocene, the Messinian Salinity Crisis of the

Mediterranean Sea, and, finally, four major glaciations arose in the Pleistocene, which among other regions caused profound changes within the thermophilic biochore of the Caucasus.

- Therefore, it is without doubt that the modern-day distribution and abundance or scarcity of biotas throughout the different regions of the Northern Hemisphere reflects 'on taxa distribution, migration, speciation, evolution and extinction events' (Milne 2004), which enveloped the Caucasus – crossroad region - as well (Gegechkori 2016).
- The Caucasus with its highly topographic and climate diversity, and habitat heterogeneity is famous for having one of the world's richest biodiversities in temperate climate Eurasia, making it one of the 34 'Mostly Diverse and Endangered Biodiversity Hotspots' in the world, among four 'Ecological Islands' (South Africa, Tanzania, Kenya, SW Australia, and the Caucasus). The region occurs among the lush broad-leaved forests with mainly evergreen undergrowth along the Black Sea coastal area to the well expressed nival zone of the Greater Caucasus and to the deserts and semi-deserts of the eastern and extremely southern Caucasus.
- The flora of the Caucasus is diverse with about 6,300 species of vascular plants 1,600 of which are endemic (25.3%). It is the result of in situ transformation of some ecologically different flora of the mainly Neogene age.
- Biota of the Caucasus, with its high percentage of endemism and relict endemism, significantly wildlife of two famous climatic refugia - Colchic and part of Hyrcania (Talysh), has always aroused the special interest of scholars. Study of biodiversity of the Caucasus by botanists and zoologists, started in the very beginning of the 18th century.
- Such kinds of research activities were based solely either on plant or animal investigations. On the other hand, long-term comprehensive (floristic, faunistic, biogeographical) studies of the wildlife of the Caucasus in all biomes and altitudinal zones and in all vegetation season were carried out by G. Radde (the late 19th and early 20th centuries) and A. Gegechkori (the second half 20th and early 21th centuries).
- In the framework of the Caucasus biota the main target of research activity of Dr. Gegechkori was psyllids (Insecta: Hemiptera: Psylloidea). Psyllids are typically narrowly oligophagous insects, feeding almost exclusively on flowering plants (angiosperms). Low mobility of psyllids and, therefore, poor powers of dispersal allows them to be used successfully as a highly indicative (bio-indicator) group for environmental research.
- Various international and regional institutions and research centers (GEF, WB, USAID, CNF, IUCN and WWF) are involved in local biodiversity conservation activities.
- According to archaeological data, the territory of South Caucasus was settled by modern humans during the Holocene more than 3,000-5,000 years ago. Consequently, there are many human-influenced ecosystems. Transcaucasia, alongside the Fertile Crescent, is considered to be one of the main centers of the origin of cultivated plants. Nevertheless, in the upper reaches of native river gorges, sporadically there still remained almost pristine habitats.
- The Caucasus is one of the most complicated regions in the world with many different peoples and ethnolinguistic variations.

- Far earlier, about 1.8 mya ago, it seems that some abiotic and biotic factors were triggered, perhaps the most important events in human prehistory. In the early Pleistocene *Homo ergaster* (*H.* e. *georgicus*) left its African homeland and appeared at the Dmanisi town site (East Georgia, the Caucasus). It means that the first cosmopolitian (out-African) ancestor of our species (*H. sapiens*) had settled in temperate Eurasia about two millions years ago (see Box. 10, p. 272-275).

Chapter 2

BIOMES OF THE CAUCASUS

2.1. INTRODUCTION TO THE BIOMES OF THE CAUCASUS

The biome concept was developed as a way to classify the terrestrial communities of the Earth in major ecological units that correlate with regional climate types (Clements & Shelfold 1939; Voronov *et al.* 1985; Woodward 2003; Gegechkori 2008). A key element in the biome concept is the idea that similar climates produce similar morphological adaptations in plants and animals (plant and animal physiognomy) and similar structure in vegetation, regardless of the actual species present. It is this emphasis on the characteristic appearance of life instead of its taxonomic relationships that allows the delineation of intercontinental regions so similar in climate and corresponding living communities: the biomes (Woodward 2003).

The biomes of the Caucasus developed as a manifestation of the long history of the Caucasus Isthmus, its climate and relief (orography) and the formation of its native wildlife of various historical-genetical origins (Gegechkori 1984a, 2008).

The concept of biomes is closely connected with the concept of geographical zonality as documented by the investigation of Humboldt (1805); Dokuchaev (1948) and Berg (1947, 1952), and applied in a wide range of the Earth sciences including geobotany, soil science, biogeography, and geochemistry.

Geographical zonation is an embodiment of the climatic zonation of the Earth, in combination with the regional particularities of orography and the shape of the continents (Grigorjev 1960, 1964; Isachenko 1971, 1973; Budyko 1980; Isachenko & Lavrenko 1980).

In the mountains, analogue and derivative forms of zonations are present as altitudinal zones (belts), which illustrate the universal character of the concept of zonality (Ivanov 1948; Stanjukovitch 1955; Grebenshikov 1957; Lavrenko 1964; Prokaev 1967; Grigorjev 1960, 1964; Stanjukovitch & Stanjukovitch 1972). The well-defined character and complexity of altitudinal zonation, i.e., landscape-climatic zones (Ivanov 1948) are a function of the latitudinal position of the mountain ranges (Prokaev 1967).

Zonation is largely determined by the distribution of vegetation, as plants constitute the fundaments of all biotic communities and also indicate the conditions in which the fauna subsists. This does not, of course, exclude the fact that some animal groups are very good indicators of overall natural conditions. In general, zoogeography laggs behind phytogeography. The reason is clear: over 70% of all known species are animals, and most of these are insects. Until the mid-nineteenth century zoologists tried to study – next to taxonomy

–the general biogeographical patterns for better understanding of animal diversity and distribution. In 1866 Ernst Haeckel not only proposed the concept of ecology, but also heralded the term chorology, the background of biogeography, as a new discipline. Philip Lutley Sclater (1858) and Alfred Russel Wallace (1876) made a major contribution to biogeography. In his article 'On the general distribution of the members of the class Aves' Sclater discerned six global biogeographic regions that were further used as a basis for the zoo-, phyto-, and biogeographical division of the Earth. Darlington (1957), an eminent coleoptorologist and zoogeographer like Wallace, did not limit his work to one class but successfully included all five classes of vertebrate animals. The general characteristics of geographical zonality and distribution with respect to insects were considered by the Russian entomologists, Arnol'di (1957, 1968) and Bei-Bienko (1966); a proposal on the classification and nomenclature of ranges (*ареалы* in Russian) and historical and ecological approaches to biogeographic zonation of the Palaearctic – by Emeljanov (1974).

Walter (1968) distinguishes zonal and azonal vegetation types. He recognized 9 climatic biomes ('zonobiome'): equatorial, tropical, subtropical, Mediterranean, warm-temperate (oceanic), nemoral (moderate temperate) continental (dry temperate), boreal (cold temperate), and polar climate zones. In each zone the vegetation would develop to its typical climax condition (*www.vegetation-ecology.info/walter_physical_vegetation_classification_...*).

In his classification of the zonal-chorological specificity of terrestrial animals, Chernov (1975) distinguished three types by their distribution: zonal, zonal-intrazonal and intrazonal.

The zonal (or euzonal) distribution refers to species distribution which closely coincides with dominant abiotic (relief, soils, climate) conditions. The relevant species are called inhabitants of tundra, taiga, steppes, deserts, high mountains. Species or communities, distribution of which correlates less strongly with the dominant abiotic conditions of the zones, are called intrazonal (or azonal). Usually they are absent in the dominant communities of the zone and occur in special, local habitats, such as riverside forests, wetlands, coastal sands, rocky-scree slopes and anthropogenic landscapes. Some species are more or less equally represented in both zonal and intrazonal types of habitat. In one zone they occur within the zonal communities, whereas in the adjacent zone they are absent from the dominant formations but do occur in intrazonal habitats. These species have a so-called zonal-intrazonal (euzonal-azonal) distribution. For psyllids of the Caucasus this system has been elaborated by Gegechkori (1984a).

The communities of the various biomes of the Caucasus can be divided into zonal (deserts and semi-deserts, steppes, wooded grasslands, deciduous and coniferous forests, the high mountain biome) and azonal (ruderal communities, inland rock outcrop and scree habitats, mires, wetlands, riverside forests). The riverside forests with characteristic hygro- and hydrophilous communities (with species of *Alnus, Salix, Carex, Juncus* with their specific faunas) can occur in any of the above-mentioned biomes, e.g., *Alnus* with its psylloid consumer *Baeopelma forsteri, Salix* with *Cacopsylla memor, Juncus* with *Livia juncorum*, and *Urtica dioica* (in ruderal communities) with *Bactericera urticae* (Gegechkori 2013).

The biomes of the Caucasus, as emphasized earlier, through the psyllids have been well studied (Gegechkori 1984a; Gegechkori 1997b). The biomes of the Caucasus, except psyllid fauna, include much information concerning botanical, zoological, palaeobiological and biogeographical materials which are monographed for first time. The primary aim of the present monograph is a comprehensive analysis of each of the biomes of the Caucasus Isthmus, but in the very near future - the main biomes, refugial centers and their specificity of other countries and regions of the Caucasus Ecoregion (Euxine-Hyrcanian Province) willl be analysed.

2.2. Deserts and Semi-Deserts

2.2.1. Drylands of the Caucasus

2.2.1.1. General Characteristics

According to Dr. Farouk El-Raz (1988), the word 'desert' originated from an ancient Egyptian hieroglyph *tesert*. It means a place that was left behind. From *tesert* derived Latin *deserere*, i.e., abandoned. From this latter word came *desertum* (a waste – deserted place), and *desertus* (meaning abandoned). Dry (non-polar and cold high mountains) deserts cover almost 23% of the terrestrial surface of the planet (Stamp 1964 cited in Zaletaev 1976).

Geographically the biome of the deserts and semi-deserts is found between the vicinity of the Tropics of Cancer and Capricon. Therefore, deserts encompass arid subtropical area at 25-35^0 latitude in both hemispheres. So, the major world of deserts is found in these two discontinous belts which is 'ringing the globe' (Stamp 1964 cited in Zaletaev 1976).

There are some global reguliarities of deserts' formation. Firstly, in low latitudes the sun's effect is stronger at the equator rather than towards the north and south poles. In these areas constant high-pressure systems drive away the rain clouds (air flows from high-pressure zones to the low-pressure zones). Therefore, these low latitude areas on both sides of the equator receives far more lowered precipitations. These 'extraordinary' features of distribution of rainfall became the origin of deserts on the planet's surface. This global meteorological event is known as the Coriolis Effect, and it is produced by the Earth's rotation on its axis in space (Monroe & Wicander 1998; Breckle *et al.* 2002; Brown & Lomolino 2008).

Secondly, when the continents are large their centers are represented somewhat far away from the world oceans. Hence, the winds lose moisture long before reaching the deep interiors of large continents (e.g., the deserts of Mongolia and China). A rainshadow effect in some coastal mountains is another reason for desert origins (e.g., the western USA, some Australian deserts).

There are other (mostly regional) regularities of arid climate formation. As an illustration there should be mentioned some cold ocean currents. For instance, the Humboldt (Peru) current of cold Antarctic water. It has a profound cooling influence on the climate of the western coast of South America, along the Pacific Ocean shoreline. It is largely responsible for the severe drought conditions of coastal areas of Chile, Peru and Ecuador including the Atacama Desert in northern Chile. This latter is known as hyperarid (the average rainfall is about 15 mm and less) and is, therefore, the driest non-polar place in the world (Ezcurra *et al.* 2006; Monroe & Wicander 1998; Gegechkori 2008 (115-170pp).

Hyperarid deserts, as stressed, receive less than 25mm of precipitation per year. Usually an arid desert may receive nearly 100-200mm of rainfall annually. There are tropical-subtropical (warm-winter) and temperate (cool or cold-winter) deserts. Anyway, in deserts, water loss due to evaporation and transpiration by plants exceeds precipitation during most years. This deficit in available water balance determines the degree of aridity throughout the year (Molles 1999).

According to P. Meigs (Meigs 1953), hyperarid deserts occupy 4% of the terrestrial surface, arid – 15% and semiarid – 14.6%. Thus, drylands of the Earth, if we take into account semi-deserts as well, are estimated to be about 33.6% of the land surface of the Earth. Russian geographers consider that hot deserts and semi-deserts occupy about 22% (31.1 mya sq. km.) (e.g., Petrov 1973).

2.2.1.2. Cold-Winter Deserts and Semi-Deserts of the Caucasus as a Part of the Irano-Turanian Floristic Region

Non-tropical deserts and semi-deserts (Walter 1975) in the biome of the Old World are widespread in the continental arid belt of the Northern Hemisphere (Babaev et al. 1986). There is a subdivision of particular areas of land of various ranks, although 'Biogeography, and zoogeography as its parts, use a specific body of concepts and terms. 'However, the usage of many, if not most, terms varies between countries and even between different authors within a country' (Krivokhatsky & Emeljanov 2000). The same viewpoint is expressed by Manafzadeh *et al.* (2016). Taxa of desert and semi-desert biomes of the Caucasus currently is restricted in the Palaearctic and the Ethiopian desert areas.

The Palaearctic-Ethiopian desert belt can be subdivided into six large regions (Walter 1975). One of them, the Irano-Turanian Deserts, includes the plateaus of northern Iran and the southern part of the Aralo-Caspian lowlands. According to Lavrenko (1960, 1965) the deserts and semi-deserts of the Caucasus (without the East Ciscaucasia) as Turanian Province is included in the Irano-Turanian Subregion (Subdistrict) of the Saharo-Gobian biogeographic Regnum of the Palaearctic Realm.

It is worth mentioning that Grossheim (1936) in his handbook 'Analysis of the flora of the Caucasus' referred the drylands of the Kura-Aras Lowland to the Turanian Province together with the East Ciscaucasia and Absheron Peninsula. Emeljanov (1974) in his Palaearctic biogeographic divisions considered the drylands of the Caucasus in the Sethyan Desert Region and among two subregions – Saharo-Arabian and Irano-Turanian showed the Caucasus to be the latter one. On the other hand, the deserts of central and eastern Kazakhstan and the Caucasus with Caspian Lowland, belong to the temperate deserts and semi-deserts of the North Hemisphere characterized by cold winters (Lioubimtseva 2009).

Worthy of note, the general hierarchical, biogeographical scheme of the Palaearctic Kingdom (Realm) was proposed by Emeljanov (1974) and Krivokhatsky & Emeljanov (2000) and in a comprehensive (biogeographical) review by Manahzadeh *et al.* (2016). The latter authors pointed out the following: 'Covering approximately 30% of the surface of Eurasia (16.000.000 km^2), the Irano-Turanian (IT) region is one of the largest floristic regions of the world. The IT region represents one of the hotspots of evolutionary and biological diversity in the Old World and serves as a source of xerophytic taxa for neighbouring regions. Moreover, it is the cradle of the numerous species domesticated in the Fertile Crescent'.

Grossheim (1948) defines desert vegetation as 'a phytocoenosis that grows in a dry environment and that does not form clear communities. The vegetation is fragmentarily distributed and plants are present as solitary individuals (specimens) that do not impact on each other either with their overground or with their underground organs'.

According to some botanists, including Grossheim (1948), Gulisashvili (1964), Walter (1975), Gulisashvili *et al.* (1975), semi-deserts are the natural vegetation in the lowlands and foothills of the eastern part of the Caucasus where desert vegetation is fragmentarily distributed.

On the contrary, Berg (1952); Yaroshenko (1956); Lavrenko 1962; Sakhokia 1958; Barseghian 1966 cited in Lachashvili *et al.* 2007) suppose that the natural vegetation cover of the Kura-Aras Lowland belongs to desert communities. Succow (1995) considers the plains of Kura-Aras Lowland hosting salt-tolerant plants (halophytes) to 'halophitic deserts', which resemble the same type of deserts represented in Middle (Central) Asia.

The presence of characteristic desert psyllid taxa and their host plants in the arid regions of the Caucasus confirms that the Kura-Aras Depression and the middle reaches of the Aras River (in Azerbaijan's Nakhichevan and adjacent South Armenia) belong to the Irano-Turanian type of deserts (Gegechkori 1984a, b, 1985, 1999, 2004b, 2008; Gueguechkori, 1998).

In the eastern part of the North Caucasus (Ciscaucasia), deserts are present in the Terek-Kuma Lowland and the northern part of the Terek River delta. To the east the desert biome continues within a narrow area of the Caspian Sea (south of Makhachkala city up to the river Samur). In the South Caucasus (Transcaucasia) the vast deserts and semi-deserts of the Kura-Aras Depression extend westward and change into semi-deserts near Tbilisi (East Georgia).

Although a typical desert vegetation community does not occur in East Georgia (although Lachashvili *et al.* (2004, 2007, 2010) have a contrary viewpoint), it should be pointed out that some solonchak soils can be found locally in the eastern part of the country (Grossheim 1948). Next to the floristic elements (Ketskhoveli 1960), the impoverishment of hyperxerophilous taxa of psyllids (Gegechkori 1984a, b; Gegechkori & Loginova 1990) illustrates the gradual transition of Turan - eastern South Caucasus deserts and semi-deserts into semi-deserts in East Georgia, i.e., typological impoverishment of desert vegetations in South Caucasus in general and particularly in East Georgia.

Taking into account the above mentioned controversial botanical viewpoints which exist in both research literature and legislative terminology ('semi-deserts' and 'deserts'), the best variant appears to be use of both terms in a complex way – 'deserts and semi-deserts' (Gegechkori 2008).

In the two driest regions of the southern part of Transcaucasia, South Armenia and Nakhichevan (Azerbaijan), desert and semi-desert biomes reach an altitude of 1,000-1,200 m a.s.l. In the less arid regions of the Kura-Aras lowland and in the North Caucasus (except inner-mountain Dagestan) they only occur at 400 to 500m.

The Eastern Caucasus deserts and semi-deserts (Turanian type) were formed mainly after the transgression of the Caspian Sea and therefore, these deserts adjacent to the seaside originated only recently (some thousand years ago). The deserts of the southern part of the South Caucasus (Aras River Valley) are more related to the deserts of Asia Interior (Near East in general) and more closely resemble tropical-subtropical warm winter deserts (Gegechkori 1985).

The annual effective temperature sum in the Kura-Aras Lowland is 4000^0-4600^0, the absolute maximum temperature 40^0-43^0, the absolute minimum -160 - 260. The annual average precipitation in different parts of the region varies from 200 to 400 mm. The climatic characteristics of the North Caucasus are almost the same (Gerasimov (Ed.) 1966; Gulisashvili et al. 1975; Gadjiev & Ragimov, 1977; Isachenko & Shlyapnikov 1989).

The climate of the middle reaches of the Aras River differs considerably from the above-mentioned regions. The annual effective sum reaches 5000^0, the summer is very hot with a maximum temperature of 43^0, whereas the absolute minimum can drop to below – 30^0. Annual precipitation varies from 200 to 300 mm (Gerasimov (Ed.) 1966; Gulisashvili *et al.* 1975; Isachenko & Shlyapnikov 1989).

The Caucasus deserts and semi-deserts hold characteristic zonal soils, including grey soils, solonchaks near the Caspian Sea coast, and plaster (gypsum)-containing soils in the Aras River Gorge. Various types of desert can be identified on the basis of the predominant substrate: sand desert with dunes, solonchak, rock and stone desert (hammada—especially in the Aras River Valley), clay desert, clay-takyr and clay-badland soils.

2.2.1.2.1. Solonchak Deserts and Semi-Deserts of the Lowland (Flat) Area of the Eastern Caucasus

The word is Russian and Ukrainian – Солончак for 'salt marsh'; from Russian соль, 'salt' (*tttps://en.wikipedia.org/wiki/Solonchak*).

Solonchak soils are formed under the influence of saline rocks or highly mineralized groundwater. These soils as edaphotypes are divided into automorphic and hydromorphic. The former group refers to the outcrops of ancient saline rocks, mainly clay. In the surface horizon the content of soluble salts is not less than 1%. These soils are found along the slopes of hills. Hydromorphic solonchack soils pertain to the shallow ground water table and refer to depressed, flat and least drained surfaces. As well as the aforementioned hydromorphic solonchaks, there are marsh solonchaks and seacoastal (marine) solonchaks, originating from marine sediments and bearing the highest salt content of all soil types of certain desert area (Ospanov & Stamkulova (Eds.) 2012).

2.2.1.2.1.1. General Characteristics

Arising from to Grigore *et al.* (2014) and references therein, between 7% (about 930 million ha) and 10% (nearly 954 million ha) of terrestrial areas are salt affected.

There are some definitions of halophytes. For example, 'Ecologically speaking, we think that halophytes must be considered as all species of plants in saline habitats' (Grigore 2008a, b.; Grigore & Toma 2010a cited in Grigore *et al.* (2014). High levels of sodium (NaCl), a mineral nutrient is optimal to halophytes (Marschner 1995 cited in Grigore *et al.* 2014).

World resources of halophytic vegetations are characterized by large generic species and ecotype population diversity. The world halophyte gene pool numbers 2,000-2,500 species (Aronson 1989). Within the neighbouring eastern Caucasus, Middle Asia's 700 halophytic species belong to 34 families and 214 genera, of which about 30% are endemics (Akjigitova 1982). Halophytic species are the rich genetic resource of fodder, food, oil herbs and bioreclamation plants, containing plant species able to develop such ecological niches as saline (or alkaline) and solonetzic soils, coastal saline sands and dry takyr-like lands, where traditional crops do not grow. Therefore, they (Amaranthaceae, in particular) are successfully used in ecological restoration of saline lands in Middle Asia (Mamedov et al. 2009) and in the southern Caucasus (Babayev & Gerayzade 2010).

On the fairly flat Caspian Lowland (Middle Asia and Kura-Aras Lowland), with its numerous shallow temporary lakes and shallow groundwater lenses, the soil and the vegetation pattern typically follow the micro-relief. This situation determines considerable, mosaic-like variations of desert and semi-desert vegetation in various litho-edaphic conditions. Therefore, steppe, semi-desert and desert communities develop here side-by-side following the more or less similar arid climatic conditions (Grossheim 1948; Gegechkori 1985; Lioubimtseva & Henebry 2009).

Solonchak deserts and semi-deserts with halophitic vegetation occupy a vast area of the Kura-Aras Lowland near the Caspian Sea coastal area and in other plains (especially Saljan steppe) of Azerbaijan (Heideman & Doroshka 1931; Prilipko 1945; Grossheim 1948; Lavrenko 1962; Schmidt 2007; Atamov 2008). They are more diverse than the wormwood (Artemisieto) communities (Agaguliev (Ed.) 2000) because of their ability to grow in a wide range of salinity that leads to polydominant vegetation. On hypersaline soils in the Mugan steppe, Shirvan steppe, and Mil steppe in Azerbaijan the low shrub *Halocnemum strobilaceum* produces hills – hummocks - on the ground. In potash-containing soils of the eastern North and South Caucasus the tiny shrub *Kalidium caspicum* also makes hummocks around itself. The rather tall (60-250cm) halophyte *Halostachys belangeriana (= H. caspica)* which is important in the solonchak desert formation in Azerbaijan does not produce hills. When compared to the above mentioned desert formations, up to 5 or 6 ephemeral and sub-ephemeral species are observed among the Halostachydetum. In the central part of the eastern South Caucasus the dominant position is sometimes occupied by *Suaeda microphylla.* In the lowlands of the South Caucasus *Salsola ericoides* and *S. nodulosa* are widely distributed. When soil salinity is low, *Salsola dendroides* is the most common, growing independently or with the *Artemisia fragrans* and contains dozens of ephemeral and subephemeral plants species. There are 22 species of *Salsola* in Azerbaijan which populate certain habitats. Communities dominated by *Artemisia* often have *Alhagi pseudalhagi* and *Glycyrrhiza glabra* as co-dominants (Grossheim 1948; Aliev 1954; Agadjanov 1962; Prilipko & Agadjanov 1972; Abutalibov 1976; Agaguliev (Ed.) 2000; Abdieva 2005; Ibragimov, 2005; Atamov 2008; Bakhsiev 2009) (Figure 21).

Figure 21. A stand of sweetwood (*Glycyrrhiza glabra*). The Iori Upland, E. Georgia (2016).

Frequently the saline soil deserts are populated by *Nitraria schoberi*. In coastal areas this shrub sometimes colonizes sand dunes and contributes significantly to stabilizing the soils that are susceptible to degradation. Investigation concerning saltwort vegetation of the Ararat Plain (southern part of the South Caucasus) has been carried out by Barseghian (1966).

Halophytes as salt tolerant plants play a great role in solonchak desert vegetation in Ciscaucasia, namely in the Northern Caspian Region, including the Caspian Lowland and the

Mangyshlak – along the Kuma-Manych Depression (Yuritsina 2000). Halophytes form communities that are confined to salted soils and to salted variants of zonal soils. Resemblance to the South Caucasus, the most numerous is the group of dwarf semi-shrub halophytes of the family Amaranthaceae mostly of the same species composition (Safronova 2008).

80% of the Amaranthaceae (predominantly members of the genera *Aellenia, Anabasis, Haloxylon, Salsola* and *Sueda*) and all species of the genus *Calligonum* show occurrence of C_4 photosynthesis (Winter 1981).

The region of the eastern part of the Caucasus (Kura-Aras Lowland and Northern Caspian Region) is typical cold-winter deserts and semi-desert regions. It is populated by halophytes and phytogeographically belongs, as mentioned, to the Turanian Province of the Irano-Turanian Subregion (Subdistrict).

The central and eastern areas of the Kura Depression – part of the Aral-Caspian Depression - consist of alluvial plains and the low delta of the Kura River along the coast. Much of this low-lying area is threatened by rising levels of the Caspian Sea (Chapt. 1.4.2.). It is deliminated by the Greater Caucasus from the north, Lesser Caucasus - from the west, and the Talysh Mountains - from the south. In the central part of the Kura Depression, the Kura and Aras Rivers divide the lowland into plains, including the Karabakh, Mil, Mugan and Saljan and southeastern – Shirvan (Gvozdetsky 1958). Frequently in scientific literature (e.g., Geideman & Doroshka 1931), all five plains are called 'steppes' (vs the above labelled semi-deserts or deserts).

Even today, large parts of the eastern Kura-Aras Lowland are still subject to Caspian Sea level transgression (Chapt. 1.4.2.). At present it is an accumulative plain falling dry with sinking sea levels and being filled with sediments of the rivers Kura and Aras (Walter 1975 Gulisashvili *et al.* 1975; Walter & Breckle 1991).

The Kura-Aras Lowland belongs to a subtropical dry climate (Figurovski 1926-1927).

According to Russian nomenclature of soils for this region of Azerbaijan they are characteristic 'brown-meadow serozems', 'serozems', 'solonchaks and solonetz' types of soils (State soil map legend of Azerbaijan 2003). Rosanov (1954) suggests that deserts - steppes constitute a group of grey-cinnamon soils. In the Kura-Aras Lowland a considerable number have solonchak (saltwort) deserts (Bakhshiev 2009). In depressions scattered throughout the wider area, the pedogenetically undeveloped soil like the formation 'takyr' is represented (Burmester 2005; Schmidt, 2007 and references therein).

In the Caspian Sea coast area sand dunes occur. As a result of the wind's influence they make a chain of hills which form one of the main natural landscapes of the Lowland.

Except for the rivers, artificial canals cross most of the Shirvan, Mil and Mugan plains. Apart from large lakes like Sarisu (or Sari Su, 'yellow lake' in Azerbaijanian) and Ag-Gel ('white lake' in Azerbaijanian) temporary salt lakes occur on heavy loam ground, but all of them dry out in summer.

2.2.1.2.1.2. Floristic Composition and Classification of the Vegetation of Salt Affected Hydromorphic Lands in Relation to Soil Salinity

Among the main categories of desert and semi-desert ecosystems of the Caucasus (Chapt. 2.2.1.) saline (solonchak) desert occupies a distinguished place. The Kura-Aras Lowland's halophitic vegetation is the nucleus of saline soil tolerant communities in the eastern Caucasus.

The native halophilic shrubland consist of polydominant shrubs and semi-shrubs and is mainly restricted to depressions where the groundwater is close to the surface and in zones around semi-salt and temporary salt lakes where salt-rich soils are known as a solonchak habitat. The area of the region mentioned is a large mosaic complex of the patches represented below diverse halophytic habitats, temporary and permanent wetlands and the similar semi-salt and salt lakes. As already pointed out, chenopods are the predominant group being some of the more characteristic vegetations. Based on soil characteristics, water table level, mineral composition of soils above biomass, etc., various ecological groups of halophytes are differentiated. For example, Schmidt (2007), Bakhshiev (2009) for Transcacuasia (Azerbaijan), Safronova (2008) for Ciscauasia, Yuritsina (2000) for the northern edge of the Northern Caspian region, Grossheim (1948), Gulisashvili *et al.* (1975) for the Caucasus, Lysenko *et al.* (2011) for Ukraine, southern Russia and Kazakhstan, Kapustina (2001); Toderich & Tsukatani (2007); Mamedov *et al.* (Eds.) 2009) for Middle Asia, Golub (1995) for the territory of the former USSR; Akhani & Ghorbanli (1993); Akhani (2006) for Iran, and Kefu *et al.* (2002) for China and others.

2.2.1.2.1.2.1. The Main Ecological Types of Vegetation (Semi-Shrubs and Shrubs) Adapted to Saline Soils

In the deserts and semi-deserts of central and eastern Azerbaijan's plain, species composition of vegetation cover and its external land feature, except the mechanical texture, are strongly correlated with the chemical composition of soils. Soil salinity is a major abiotic factor in wild vegetation worldwide, demanding very specific adaptation ability. At the same time, salt tolerance might have much wider implications because a salt-tolerant (halotolerance) plant often also tolerates other stresses including chilling, freezing, heat and drought (Kapustina 2001).

On the base of the results of investigations carried out by various authors (Bikov 1981; Kapustina 2001; Schmidt 2007; Mamedov *et al.* (Eds.) 2009) the following edaphotypes can be distinguished: (1) euhalophytes, which have a wide tolerance range and are very well adapted to the average and high saline soils. Plants are succulent (e.g., *Halocnemum strobilaceum*); (2) hemihalophytes (mesohalophytes), vegetation with less salt in the soil profile (e.g., *Salsola dendroides*); (3) haloglycophytes are shrubs with weak tolerance to very saline soils. Sometimes communities of these species are found in almost saline-free soils (e.g., *Artemisia fragrans*).

Toderich and Tsukatani (2007) as well as noting, the chemical compositions of soils, pointed out the water table variation and proposed a classification the halophytic vegetation as follows: (1) hyperhalophytes; (2) euhalophytes; (3) haloxerophytes; (4) halogemimezophytes; (5) halo-gemipetrophytes; (6) hydro-halophytes, and (7) metallo-halophytes.

In the present work concerning the relation of plant communities to soil and hydrological properties we are following the classification of the above mentioned authors to take into account as well the geobotanical approach of the vegetation cover of deserts and semi-deserts (e.g., Grossheim 1948; Gulisashvili *et al.* 1975; Aliev & Gajiev 1986; Schmidt 2007) with some modifications (Gegechkori 1984a).

So, desert semi-shrubs and shrubs confined to salted and semi-salted soils can be divided into four groups according to their biological structures and ecological peculiarities: hyperhalophytes (-xerophytes), euhalophytes (-xerophytes), mesohalophytes (-xerophytes) or haloglycophytes, and hydrohalophytes (many plant ecologists consider halophytes a particular form of xerophytes) (Grigore *et al.* 2014).

Hyperhalophytes true, obligate halophytes, communities in hydromorphic solonchak soils). They have two types of adaptive strategy: (1) have succulent assimilative organs and are microphyllous or aphyllous (Figure 22); (2) have a thick and complicated structure of outer epidermis walls and other adaptive features and are sclerophyllous. Both strategies are designed to prevent loss of water through evaporation. The most obvious examples of the above mentioned adaptations is characteristic of the genus *Haloxylon*, which is not found in the arid regions of the Caucasus. Some genera with transformed assimilative organs are represented by *Salsola* and *Anabasis*. All three genera belong to the Amaranthaceae family.

Euhalophytes, communities on high saline dry soils. They solve the problem of evaporation decrease in the drought period mainly by dropping 70% of their leaves (Butnik 1991). Mention should be made of *Astragalus turkestanicus*.

Mesohalophytes (facultative halophytes or haloglicophytes). They drop 80 to 100% of their leaves in the hottest period of summer. Some elements of xeric structure include thin leaves, covered with hairs and they use essential oils for protecting the plant from overheating due to very high temperatures. Semi-desert *Artemisia* shrubs (semi-shrubs) of the subgenus *Seriphidium* belong to this group.

Figure 22. Succulent shoots of *Anabasis aphylla* (Amaranthaceae), they lack thorns but their juice consists of poisons, which discourage consumers (1972).

Hydrohalophytes. They occur mainly in wet to standing water, frequently adapted to brackish water (e.g., species of *Arundo*).

2.2.1.2.1.2.1.1. A Contribution to the Biota and Soils of Solonchak Deserts of the Eastern Caucasus (Chiefly the Kura-Aras Lowland)

2.2.1.2.1.2.1.1.1. Hyperhalophytes (True (Obligate) Halophytes); Communities in Hydromorphic Solonchak Soils

2.2.1.2.1.2.1.1.1.1. Introduction

Plants that can survive in a high concentration of excessively saline soils are called *halophytes* (Parida & Das 2005). *Saltwort* is the name for some wet-desert's vegetation plants which never dry out. On the contrary, they always remain fresh and juicy, therefore this type of desert differs strongly from other types of drylands. In the Kura-Aras Lowland saltwart deserts show a considerable distribution is in the central part of Shirvan Plain (Bakhsiev 2009).

The most characteristic components of salt-tolerating plant species should be mentioned: *Salsola crassa, S. soda, S. glauca, S. ericoides, Limonium caspium, L. meyeri, Petrosimonia brachiata, Kalidium caspicum, Halostachys belengeriana, Halocnemum strobilaceum*, etc.

The most typical saltwart communities of the Kura-Aras Lowland are given below.

This group of halophytes consists of plants of excessively saline (saline-sodic) soils. They develop in conditions of raised mineralization of soil solution, at the same time keeping their ability to renew and develop cenoses on strongly saline soils with a dry residue of 2.3-3.0 (3.5%) (Mamedov *et al.* (Eds.) 2009).

Irano-Turanian type of halophylic shrubland is mainly restricted to depressions, where the groundwater is close to the surface and in zones around salt- and semi-salt lakes.

The soil accumulations often appear white with salt enrichments due to its extreme salinisation (Kunze *et al.* 1994 cited by Schmidt 2007).

In Middle Asia's analogue deserts the water table varies from 0.5 to 1.5 m depth (Toderich & Tsukatani 2007). The most abundant and characteristic plants, as in most salty places, are members of the goosefoot family (Amaranthaceae).

2.2.1.2.1.2.1.1.1.1.1. Halocnematum - The Most Developed Communities of the Strongly Salinated Deserts

Its obviously dominant species – *Halocnemum strobilaceum* (sarsazan) has a wide distribution range in the Irano-Turanian, Mediterranean and Saharo-Sindian regions (Graffari *et al.* 2000) (Figure 23). It is represented as one of the more typical and widespread of the succulent-halophytic desert communities and occurs on moister, salt rich areas of saline flats and small salt hills. Its habitat is in high ground - water level represented in both, inland salines and littoral marshes. But in peripheries of most of the inland salines, like certain Turanian deserts (Akjigitova 1982) and Iranian deserts (Akhani & Ghorbanli 1993; Akhani, 2006), it forms a pioneer halophytic community, or the second one after the *Salicornia* species. It can be hummocky and is species-poor with other halophytes such as *Limonium meyeri, Salsola paulsenti, Suaeda confusa,* and *Salicornia europaea*. Especially large hummocks are observed in the saline places of Lokbatar Mugan and East Shirvan (Gulisashvili *et al.* 1975).

The aeolian dust is a major factor in the genesis of hummocks. Through aeolian deflation (removal of loose materials) the fine material is scattered throughout the Kura-Aras depressions. Shrubs within flat plains are an initial barrier, catching fine material – sand and salt of aeolian origin. 'Material catcher' leads their accumulation on shrubs, the initial stage for the creation of the hummocks (Schmidt 2007; Lancaster 2009).

Very poor in species Halocnematum communities mainly occur as a monodominant halophytic species. Within the same habitat of Iran, to quote Akhani and Ghorbanli (1993), 'the associated of *Halocnemum strobilaceum* species occur on margins or in transition zones towards the neighbouring types of the saltland vegetation'.

H. strobilaceum – the pioneer plant – appears after sand has already been accumulated (Walter 1975). However, this shrub occurs on bare clay too and with its deep root system does not depend on sand layers (Schmidt 2007).

H. strobilaceum is a 20-40 cm tall plant which is able, as mentioned, to produce hills on the ground and its community frequently develop as a large hill solonchak. The largest hills occur in the south part of Mugan steppe (Grossheim 1948).

Figure 23. The Shirvan NP site with *Halocnemum strobilaceum* shrubs growing in scattered clumps. In this way they are able to avoid competition and use all available water.

H. strobilaceum develops its main rhizosphere within the uppermost centimeters of soils to avoid the high salinity of deep horizons (Schmidt 2007).

Kalidium caspicum – the other common species in saltwort communities. This species has a wide distribution from the Caspian area to most parts of Central Asia (Nikol'skaya 1972). Communities of small-shrub Kalidietum are scattered frequently as patches in both eastern Ciscaucasia and eastern Transcaucasia and occur on a strongly salinited plains (Grossheim 1948). The westernmost border of distribution of this shrub from the Kura-Aras Lowland occurs on the eastern edge of Georgia (Lachashvili & Khachidze 2010).

Floristically, however, this formation is relatively poor and includes the salt-tolerant species, *Salsola crassa, Nitraria schoberi, Suaeda microphylla, Atriplex cana.* In foothill areas like Gobustan (Eastern Azerbaijan), typical xerophytic grass species are *Hordeum murinum, Bromus japonicum* and *Torularia contortuplicata.*

The community with *K. caspicum* dominancy sometimes is to be found on less saline soils than those with *Halocnemum strobilaceum* and *Halostachys belangeriana* dominating communities. Kalidietum is developed mainly on the Near Caspian plain and in the Kura-Aras Lowland (Prilipko 1949; Abutalibov 1976; Agaguliev 2000).

K. caspicum is a dwarf semibrush of 30-80 cm in height. On some plots under the shrubs, small bush - area hillocks can be noticed. When branches are covered with sand, they can be implanted and form additional roots (Mamedov *et al.* 2009). *K. caspicum* belongs to salt-accumulating halophytes able to accumulate in its tissues up to 40% mineral substances (Akjigitova 1982).

Halostachydetum – the next hyperhalophytic communities. *Halostachys belangeriana* is distributed in the Caspian Sea coastal area, Middle Asia, East Turkey, Iran, Afghanistan and Central Asia (Mongolia, Jungaria, Kashgaria) to China (Hedge *et al.* 1997). In the South Caucasus it extends in the Kura-Aras Lowland, and its westernmost limit of distribution is observed in extreme East Georgia (Lachashvili & Khachidze 2010). It is a typical leaf-succulent shrub. Its ecological range is restricted, it populates ground with a well supplied groundwater table, although avoids a flooded area (Akjigitova 1973). *H. belangeriana* is a dark brownish-green, large shrub (60-250cm tall, and 20-170cm canopy) which grows on permanently wet soils, where it reaches the groundwater table as *H. strobilaceum* does. *H. belangeriana* is developed due to variability of relief-edaphic conditions within the distribution range, but another important factor of its relationship to the desert's habitats is contact with the riparian vegetation formations, on the one hand and different communities of hemixerophylous and xerophylous vegetation, on the other hand (Lachashvili & Khachidze 2010). It is most common at secondary solonchak among irrigated and abandoned fields (Mamedov *et al.* 2009).

As compared to the above-mentioned desert formations, the communities with participating *H. belangeriana* are represented sporadically throughout its wide range. Lower level species consisting of up to 5 or 6 ephemeral and annual plants reaching about 20cm in height are observed among the Halostachydetum (Grossheim 1948; Prilipko 1980).

It is able to accumulate in its tissues about 30% salts, when the alkaline process stops and salt content in soil decreases, the plant gradually disappears and it is substituted by other species (Akjigatova 1982). This plant is used as a source to produce soda and potash (Bakhshiev 2009).

All these chenopods, as mentioned above, have small leaves, hard or sometimes fleshly, as the presence of strong brine makes absorption of water difficult and they must conserve moisture. Saltwart deserts have considerable distribution in the central part of the Shirvan plain. This type of desert differs strongly from other types of deserts and semi-deserts in that semi-shrubs here never dry up. The local deserts are dotted with these small bushes. The color of the 'carpet' consists of semi-shrubs that are very attractive and vary greatly throughout the year. It changes from bright green or blue green (in spring) to yellowish and bright yellow (in summer), pink or purplish-brown (in autumn as they age and then blood-red), and at last, to violet (in winter) (Bakhshiev 2009). Between are stretches of greyish soil, or salt patches blindingly white in the sun.

- ***The Suaedetum desert***. Communities consisting of *Suaeda dendroides* sometimes occur as almost pure stands on saline hillsides and foothills but more frequently make mixed associations with *H. strobilaceum* (Abutalibov 1976). *Suaeda microphylla* occurs as an edificatory mostly in the Shirvan plain and partly in foothills, subshrub to 75cm tall. Stems erect much branched, branches spreading, rigid. Except saline - alkaline habitats, occur on dunes and lake shores.
- ***Nitrarietum community***. Representative of this community – nitrebush (*Nitraria schoberi*) (fam. Nitrariaceae) is distributed in the European part of the former USSR, Ukraine (Krym), southeastern Europe (Romania), the Caucasus, Middle Asia, Western Siberia, Iran, Afghanistan and Western China (Komarov (Ed.) 1934-1978).

This shrubland type in the South Caucasus has a wide range and is most typical in Azerbaijan, where it may acquire better growth and dominance except clay, solonetzic soils. Nitrarietum is found in an equally wide range of habitats, sometimes recognizably different from the type mentioned above such as sandy grounds (especially the Caspian Sea coastal area) (Figure 24).

Figure 24. Central Azerbaijan, the Mugani deserts and semi-deserts, Nitrarietum community (1974).

The former community occurs on plains, slopes, hills and in badlands, dry ravine-gullies on very saline grey soils. On these type of soils, salts are "rise-over, ground-up" and have dough rise features. Various authors have hypothesized that Nitrarietum shrubland type was a relict remnant once (Pliocene), more common in the South Caucasus (Lachashvili *et al.* 2007 and references therein). *N. schoberi* is recommended as a valuable, economy shrub, the fruit of which is rich in ascorbic acid (Mamedov *et al.* 2009).

Stands of *N. schoberi* also occur in transitional areas between main solonchack's halophytic communities of *Halocnemum* and, on the other hand, *Salicornia*. In central Azerbaijan and in the coastal areas of the Caspian Sea (Transcaucasia, Ciscaucasia) this shrub is represented as one of edificators – dominant of desert phytocoenoses. It sometimes grows to a height of 1.5 meters.

According to Akjigitova (1982), in plant's leaves up to 57% salts are accumulated with a predominance of chlorides (about 14%). The root system is of a universal type, the main root penetration to a depth of up to 3m. The shrub tolerates well a covering of sand. On ground adjacent to soil and covered with sand, stems with additional roots are formed, which are located in the superior soil horizons. Therefore, it can also be used for reclamation of sandy deserts as a good sand-fixer (Mamedov *et al.* (Eds.) 2009).

The shrub is not very palatable to grazing animals and is infrequently nibbled by sheep and goats (Sokhadze & Sokhadze 1991).

Munkhbaatar (2012) has hypothesized that in the sand dunes of the Mongolian Gobi there is an interesting mutualism between the Gobi racerunner lizard (*Eremias prszewalskii*) and the Siberian nitrebush (*N. sibirica*). The lizard, as well as insects and spiders eats the ripe nitrebush fruit, but it cannot digest the hard shelled seeds of the shrub. Later the seeds come out with its feces. As a result of such extraordinary mutualism of the animal's carpophagous diet and function of endozoochore these have a positive result on distribution and ability. It has been observed that plant growth of *N.sibirica* bushes in sandy habitats of the Gobi Desert grow thickly if the racerunner inhabits it and, on the contrary, it grows thinly without the lizard.

Until modern industrial processes were started the nitrebush containing alkall salts became important for some local people for soap production (soap boiling) (Bobrov 1946).

Halophytic tamarisk (Tamarix spp.) shrubs - Among the saltland vegetation of the Kura-Aras Lowland, a most important vegetation is a species of the genus *Tamarix* (or salt cedar), Similar to the previously mentioned hyperhalophyte semi-shrubs, it can exist in such an extreme habitat thanks to the ability of the roots of this species to reach down to the groundwater containing less saline.

Species of the genus *Tamarix* inhabit a variety of saline (haline) habitats, especially high-saline river beds, bank of irrigation cannals, saline and sandy soils, estuaries and vast areas of inland salines with a relatively high water table, dry rivers and plain deserts. Sometimes it forms a phytocoenosis (Akhani & Ghorbanli 1993; Lachashvili *et al.* 2007).

Species of *Tamarix* (*T. hohenackeri*, *T. ramosissima*) – ecologically well adapted to xeric habitats – have a number of features including reduction of leaf size, photosynthetic activity in cladophylls, a deep, extensive root system that extends to the water table. They are also capable of extraction of water from unsaturated soil layers (a facultative phreatophyte). Initially, the primary root grows steadily downward with little branching until it reaches the water table, which can be at the depth of 3m or more. Once the water table is reached, secondary root branching became profuse (Gambarova 2011).

Toderich & Tsukatani (2007) include species of *Tamarix* into a group of euhalophytes, where representatives of this genus coexist with some other species, *Suaeda, Kochia* and so on. More frequently species of tamarisk occur in tugai forests. They stretch as narrow, interrupted lines along the rivers with other riparian habitat's trees (willows, elms, poplars, etc.). On increasingly dry sites, the tamarisk thickets form a transition to the desert-like semi-shrub and shrub vegetation (Prilipko 1970) (Figure 25).

Figure 25. Tamarisk shrub (*Tamarix ramosissima*) in the Iori Upland frequently reaches giant sizes, 10 - 12m in height (2015).

It should be stressed that the genus *Tamarix* consists of about 54 species in the Old World (Baum 1978), several of which were cultivated into North America over 100 years ago as ornamental, but sometimes planted as a wind break, and for stabilizing sand and sandy soils. Over a million acres are today infested with saltcedar monocultures along streams and riverbeds. The salt cedar's long taproots suck up salty ground water and drop salt-crusted leaves on the soil surface. This makes it almost impossible for native plants to take root. The loss of native plants also decreases the insect and bird biodiversity (Brotherson & Field 1987).

As a biological pest control agent against tamarisk beetles of the genus *Diorhabda* can be used. Larva and adult stages of the beetles feed primarily on saltcedar leaves and cause defoliation of its host plant and death of the tree within several years (DeLoach & Carruthers 2004; Delin *et al.* 2009).

Takyr (takyr-like) desert and its vegetation. Takyr – the term is originally from Kazakh or another Turkic language and means 'smooth', 'flat' or 'barren land' (Figure 26). Takyr soils also belong to zonal ones. They are extremely low-humus, carbonate only from the surface, the gypsum content in them is small, rarely exceeds 1% (Mamedov et al. 2009). It is a specific soil formation in Central (Middle) Asian deserts, but takyr-like formations (claypans) are also found in deserts of some other countries with temperate deserts including the eastern Caucasus.

Takyr-like soils are mainly developed on shallow basins and depressions with heavy clay soil and ancient delta plains of the large rivers. They are varying in size ranging from 0.5 to tens of km^2. For example, in Turkmenistan former shallow and today dry lake beds reach huge sizes.

Takyr-like soils are usually characterized with a shallow water table and highly mineralized underground water. Takyr has a bare smooth surface, broken up by a network of splits into numerous polygonal aggregates. The thickness of the soil profile is 15-12cm (The Great Soviet Encyclopedia, vol. 25, 1978).

Figure 26. Clayey takyr-like soils with a network of polygonal cracks. Shirvan, Azerbaijan (1974).

Walter (1975) considers that takyrs can be represented within all transition stages from barren ground over solonchak halophytes to ephemeral desert often leaching, but Russian and Azerbaijanian scientists always extend the interpretation of takyrs into the solonchak soil type (e.g., solonchak, takyr-like soils) with different interdigitations (Schmidt 2005b and references therein).

The takyrs after seasonal rains are flooded sometimes several times during the year, but the water usually evaporates quickly and a dried crust with fissures forms on their surface with a characteristic texture. Salinization of Takyrs is the dominant process in summer months. Soils of takyrs are medium or strongly saline (Billwitz 1997).

Usually takyrs have no higher vegetation, but according to Mil'kov (1977), in Middle Asia due to the density of takyr/takyr-like soils different halophyte vegetations dominate: (1) on the dense solonchak-solonets takyr soils lack the vascular vegetation and are covered by communities of blue-green algae and lichens with sparcely scattered annual saltworts (species of *Salicornia, Petrosimonia, Salsola*); (2) on less dense saline takyr and takyr-like soils, which are far more diverse with regard to their mechanical composition include widely scattered patches of perennial salt tolerant dwarf shrubs (*Haloxylon, Tamarix*) and semi-shrubs of saltwarts (*Salsola* spp., *Artemisia* spp.).

Except for the genus *Haloxylon* (saksaul) - dominant shrubs or small trees in Middle Asian sand deserts, and semi-shrubs of wormwoods, takyr-like soils of the eastern Caucasus contain taxonomically and physiognomically the same hypersaline-loving and drought-resistant halophytic vegetation found in the Turan Hollow.

The Kura-Aras Lowland takyrs, except for the high number of algae, are hosted by *Salsola* spp., *Halocnemum strobilaceum* and *Salicornia europaea* (Mikhailovskii 1972; Succow 1995).

Among vertebrate animals edaphically closely connected with the takyr deserts of Middle Asia (Turan) and southeastern Caucasus (Kura-Aras Lowland) it should be emphasized, the agamid lizard – toad-headed agama (*Phrynocephalus helioscopus* (=*Ph. persicus*).

Annual hygro-hyperhalophitic communities. High salty and wet soils of the margins of semi-salt lakes, temporary salt lakes, salt marshes and saline littoral marshes along the Caspian Sea are a home for high salt tolerant annuals and ephemerals. In summer significant number of such plants make their appearance.Worthy of mention here are *Petrosimonia brachiata, Salsola crassa,* more sporadically *Salicornia herbacea* (= *S. europaea*), *Salsola soda* and *Suaeda salsa* (Gulisashvili *et al.* 1975; Schmidt 2007; Bakhshiev 2009).

It should be stressed that the saline vegetation is developed in more numerous options than the wormwood one (Grossheim 1932a, 1949).

2.2.1.2.1.2.1.1.2. Euhalophytes, Shrub and Semi-Shrub Communities in Automorphic, High Saline, Solonchak Soils

Euhalophytes are characterized by a wide range of tolerance to mineralization of soil solutions, within which they can successfully develop and preserve prevailing role in phytocoenoses. These plants are also well adapted to strongly saline soils, however, they develop better and prevail in phytocoenoses with lower salinity of the substrate, dry residue – 1.8-2.0% (Mamedov *et al.* 2009).

Figure 27. *Salsola ericoides* in Gareji Hollow, E. Georgia (2018).

***Salsoleta ericoides* formation.** *Salsola ericoides* occupies significant areas in the Eastern Caucasus. In the South Caucasus it is distributed mainly in the Kura-Aras Lowland, Gobustan and Djeiranchel. The common distribution is in: the Caucasus, eastern Anatolia and Iran (Akopian 2011). It is a dwarf leaf-succulent semi-shrub. It normally develops and dominates in the cover on saline grey soils, rarely – saline clayey hills and badlands. In the North Caucasus (Inner Dagestan) it grows on gypsum-bearing soils. In plain deserts it is represented as one of edificatory-dominant species, rarely occuring as a mixed stand in Pistacietum and phryganoid vegetation. It is a good fodder for animals, in the autumn-winter season, in particular (Akhmedova 2004, 2005; Asadova 2008; Lachashvili & Khachidze 2010) (Figure 27).

***Salsoleta nodulosae* formation**. *Salsola nodulosa* is widely spread in arid regions of the South Caucasus, especially in Kura-Aras Lowland, in Gobustan and Djeiranchel; to the west it occurs in the Kura river gorge (East Georgia, until Kaspi town). It is a dwarf semi-shrub, which populates plains with gypsum containing saline grey soils and clayey bedlands. *S. nodulosa* is one of the edificatory dominants in deserts and phryganoid communities (Asadova 2008; Lachashvili & Khachidze 2010).

The normal distribution: the Caucasus, Anatolia, south-western part of the Caspian Depression (the northern part of the Caspian Sea) and the north-western part of Iran (Akopian 2011).

***Anabasietum aphylla* formation**. The range of *Anabasis aphylla* consists of Kazakhstan, Middle (Central) Asia and the European part of the Russia (Lachashvili & Khachidze 2010). It is represented as one of the most common communities in Turanian and the eastern Caucasian plains with very saline grey soils, frequently it is a halophyte shrub component of a Salsoletum dendroides community. It is fleshy with a succulent stem. The water stored in the shoot is apparently a valuable water supply during the dry part of the year. *A. aphylla* is in its best vegetative form in the late summer and early autumn. Taking into consideration the species' specific, *A. aphylla* can be considered as a valuable indicator of the ground waters' level change and desertification (Mil'kov 1977).

The extreme westward distribution of *A. aphylla* occurs on the boundary of Azerbaijan and Georgia (Lachashvili & Khachidze 2010).

***Atriplexeta canae* formation.** *Atriplex cana* is subshrub, 20-25cm tall, sometimes somewhat cushion-shaped. Stems are much branched, woody, bark gray-brown. It inhabits clayey deserts and semi-deserts, where it is represented as a transitional form solonchacks to solonetzs, on crust solonetzs, arid slopes (argilloideus badlands); it occurs in abundance and is one of the typical landscape - forming deserts and semi-deserts plant (Il'jin 1936; Grossheim 1949) (Figure 28).

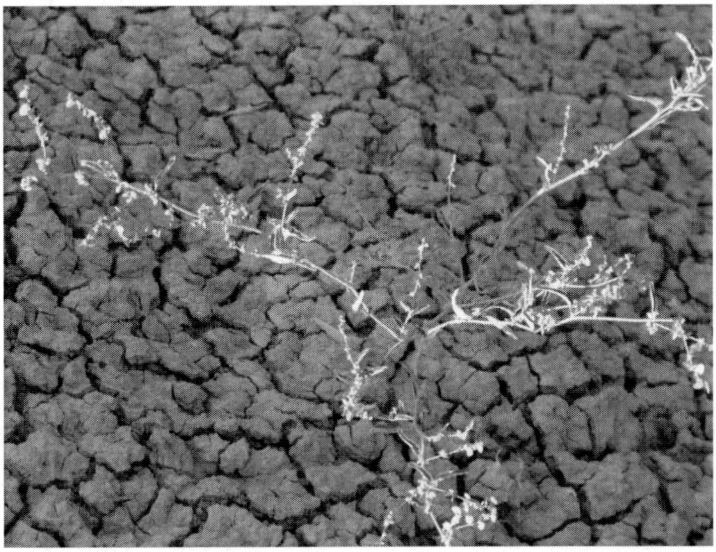

Figure 28. The sole *Atriplex* cana that grows in the badland area of Adigeni environs (S. Georgia).

In deserts it is a perfect fuel and burns well even fresh; as a forage – eaten mainly by camels, satisfactorily by sheep and horses (Goloskokov 1969; Ryabushkina *et al.* 2008).

In the South Caucasus from Azerbaijan it penetrates in eastern Georgia (Kakheti, Shida Kartli, where in Vashlovani Hollow it creates microcoenoses (Lachashvili & Khachidze 2010). So, euhalophytes, mainly form the communities of the halophyton of Irano-Turanian type, have a wide tolerance range and are very well adapted to the average and high saline soils, containing magnesium-calcium sulphate (Kapustina 2001).

2.2.1.2.1.2.1.1.3. Mesohalophytes (Facultative Halophytes) or Haloglycophytes

The third group–haloglycophytes, consist of plants with low salt-resistance, which develop an adaptive reaction to low saline soils. They normally develop and renew with dry residue of 0.3-0.8%. Representatives of this group can often be met on non-saline soils (Mamedov *et al.* 2009).

2.2.1.2.1.2.1.1.3.1. Semi-Shrub Communities on Lower Saline Soils. Clay- (Semi)-Deserts

Both in the North and South Caucasus, the deserts and semi-deserts are generally characterized by wormwood semi-shrub (*Artemisia*) growing on slightly saline grey-brown, sometimes rocky and stony soils (Grossheim 1948; Gulisashvili *et al.* 1975). Floristically the wormwood formations (Artemisetum) are the most widely spread type of semi-desert and steppe vegetation. All kinds of Artemisetum communities of the South Caucasus - with 532 species - are quite rich (Aliev 1971). In the North Caucasus wormwood communities are composed of *Artemisia fragrans* (subgenus *Seriphidium*), *A. taurica* and *A. arenaria,* in the South Caucasus - by the fragrant wormwood - *A. fragrans* (Grossheim 1948; Gulisashvili *et al.* 1975; Valles *et al.* 2003). Species of the mentioned subgenus make the dominant features in steppes and deserts and semi-deserts ranging from Asia Minor and Middle Asia up to the Hindu Kush (Peer et al. 2007). In various regions of the Kura-Aras depression wormwood deserts and semi-deserts occur from 28m b.s.l. to 400-500 m a.s.l. (Grossheim 1948; Prilipko 1970; Aliev 1971; Gribova *et al.* 1980; Schmidt 2007; Peper *et al.* 2010). In the Aras River Valley, arid landscapes with the same wormwood communities occupy altitude from 700 to 1,000 m (Melikov & Hatemov 2002). The same communities also constitute an important part of the landscape westernward of the Kura-Aras Depression, namely in East Georgia, for example, on the Eldari Lowland, Iori plateau and Kvemo (Lower) Kartli (Ketskhoveli 1960; Lachashvili 1989; Lachashvili & Khachidze 2005; Lachashvili *et al.* 2007). In the lower zones (-28–300 m a.s.l.) the wormwood communities frequently change into wormwood-saltwort and saltwort deserts (Artemisieto – *Salsola dendroides*) or perennial cereals (Gulisashvili *et al.* 1975).

Unlike the warmwood formations, the wormwood-saltwort communities are restricted to clay-brown soils and chernozems and floristically are very poor (Sokhadze & Sokhadze 1991; Lachashvili *et al.* 2007).

All kinds of Artemisetum include 30-35 or even 50-55 species of ephemeral and sub-ephemeral plants. In spring they are dominated by ephemers and annuals like *Adonis aestivalis, Koelpinia linearis, Poa bulbosa* and *Papaver minus* (Takhtajan 1941; Sakhokia 1958; Sokhadze 1977). In East Georgia some semi-deserts are dominated by *Limonium meyeri* (Lachashvili *et al.* 2007).

A phytocoenosis dominated by *Salsola dendroides, S. ericoides* and *S.nodulosa* semi-shrubs occurs on plain desert localities with slightly and moderately saline clay grey soils. On less saline soil, *Artemisia fragrans* occurs again. In the uplands and on low mountain slopes the saltworts and wormbushes face serious competition from dry steppe elements (*Stipa* spp.) and shibliak vegetation (*Paliurus spina-christi, Rhamnus pallasii, Berberis iberica,* etc.) (Ketskhoveli 1960). The adaptation peculiarities of xerophytes of the Ciscaucasia were researched by Snisarenko (2008).

***Artemisietum fragrans* formation.** The wormwood community is the most frequently occurring type of desert and semi-desert vegetation of the Kura-Aras Lowland and the South Caucasus in general (Grossheim 1932a, 1948; Grossheim & Sakhokia 1929; Walter 1975; Prilipko 1970, 1980b, etc.). The deserts and semi-deserts consisting of a dominance of wormwood belong to zonal types of deserts and semi-deserts in the Caspian Sea coastal regions, where they completely cover whole areas (Safronova 1980). Artemisietum is composed of the mentioned species (*Artemisia fragrans*) and is also very common in eastern Georgia, where it is represented as the main canopy covering building species (Kakulia 1952; Ketzkhoveli 1960; Khachidze 1984; Lachashvili & Khachidze 2010) (Figure 29). It also occurs occasionally in conjunction with less-saline loving species of the genus *Salsola* (*S. dendroides, S. nodulosa*), *Kochia prostrata, Kraschennikovia ceratoides* (=*Ceratoides paposa*), arid woodland and phryganoid vegetation (Lachashvili & Khachidze 2010 and references therein). The existence of ephemers and ephemeroids are characteristics: *Tulipa eichleri, T. biebersteinii, Iris iberica,*

Poa bulbosa, Bromus japonicum, etc. (on the example of Iagluja, Kvemo (Lower) Kartli, E. Georgia).

A. fragrans grows as on foothills, so in denudation plains and accumulation plains. It is a haloglycophyte plant (Barseghian 1966) and is distinguished by good adaptation to the soils of varying salinity, preferring slight to average saline grey soils, sometimes found on almost saline-free soils.

A. fragrans is known from northern Turan, the Caucasus, partly in Turkey and northern Iran (Lachashvili & Khachidze 2010).

Figure 29. Deserts and semi-deserts covered with wormwood (*Artemisia fragrans*), East Georgia (the Iori Upland). Under summer's long drought the soil cracks slightly into polygonal blocks ('thermal cracking') (2017).

Desert communities of the last two groups (mesohalophytes and Haloglycophytes), which contain camel prickle (*Alhagi pseudalhagi*) and weeds such as *Carthamus glaucus* and *Cichorium intybus*, are indicative of disturbance and are found where anthropogenic influence was great (Gambarova 2011).

***Salsoleta dendroides* formation.** The range of *Salsola dendroides* includes the Caucasus (Ciscaucasia, Transcaucasia), Eastern Europe, the area downstream of the Volga, Asia Minor (Anatolia) almost all the Near East and the eastern part of South Caucasus (Lachashvili & Khachidze 2010; Akopian 2011). The largest area of *S. dendroides* occurs in Kura-Aras Lowland, where, as a monodominant or with *Artemisia fragrans,* it forms solid cover (Grossheim 1930, 1948; Gulisashvili *et al.* 1975). In the North Caucasus (Inner Dagestan) *S. dendroides* formations are distributed on gypsum containing clayey badland (Tumadjanov 1966). It is a dwarf semi-shrub (subshrubs), 60-80 cm high (sometimes 150 cm). Due to edaphic conditions it belongs to the group of sulphate- sodium plants (Mamedov *et al.* 2009). It occurs on plains with moderately and very saline grey soils, rarely saline clayey hills and badlands. In plain deserts it occurs as one of the edificatory-dominants, rarely – Pistacietum and phryganoid vegetation (Lachashvili & Khachidze 2010) (Figure 30).

***Kochieta prostratae* formation.** Forage kochia (*Kochia prostrata*) is semi-shrub 30-150 cm in height that is native to areas from the Mediterranean Basin to Siberia. On dry plains, slopes and hills with slightly and moderately saline grey and open woodland's cinnamonic soils, on clayey and clay-sandy badlands. The plant is sustainable in weak salination; mainly in Artemisietum, sometimes forming independent phytocoenoses (Lachashvili & Khachidze 2010). Forage kochia is palatable and has been used by livestock during the whole vegetation period in Kazakhstan, Middle Asia and Caucasus during autumn and winter. It was introduced and has been seeded on over 200,000 ha throughout the Intermountain West (Idaho, USA) to provide fuel breaks and winter forage, and to compete with invasive plants (*Halogeton glomeratus, Bromus tectorum*) and it is one of few species that can be successfully established on severaly degraded rangelands (Gray 2011, Warldon 2011) (Figure 31).

Figure 30. The vegetation cover consists of saltwart (*Salsola dendroides*). In the background: gentle hills enclosed with grassland. Kvemo Kartli province, Jagluja site, E. Georgia (1978).

Figure 31. Forage kochia (*Kochia prostrata*) semi-shrubland, environs of Gareji Hollow, E. Georgia (2014).

***Poaeta bulbosae* formation.** *Poa bulbosa* is a perennial bunch grass (ephemeroid) from Poaceae. It grows on clay, loamy, dense, sandy, crushed-stone soils. It can grow in semi-deserts, deserts and mountains on subalpine soils. It is rather saline tolerant, but prefers weak soil salination. It is interesting as a plant preventing from erosion the soil cover as well; it is also a good fodder for livestock (Asadova 2008).

Clay (semi-) deserts as a rangeland in eastern Caucasus on the example of Gobustan National Park. Arid and semi-arid rangelands (pasturelands), which are not used for good agriculture practices (sowing of seeds, irrigation, use of fertilizer, etc.), are located in the eastern Ciscaucasia, eastern and southern Transcaucasia.

Drylands in eastern Transcaucasia cover a vast territory, are classified as rangelands and are largely inhabited by nomadic pastoralists who since the remote past are largely dependent on livestock (Pietzsch 2008; Peper *et al.* 2010).

Rangeland in eastern Transcaucasia is an arid climate area consisting of the following biomes and ecosystems: grasslands, shrublands, arid open woodlands, deserts and semi-deserts.

The rangelands of eastern Transcaucasia are represented mostly in plains and foothills of the Kura-Aras Depression (central-southern Azerbaijan, including Gobustan National Park) and partially in eastern Georgia.

One of the type of winter rangeland as a vast natural landscape in eastern Transcacuasia (360,000 ha) is recognized as a deserts and semi-deserts biome made primarily of wormwood (*Artemisia fragrans*) (Gulisashvili *et al.* 1975). This classic rangeland type of this region is frequently mixed with species of saltworts (*Salsola* spp.).

Rangelands in Azerbaijan occupy approximately one third of the land area (Peper 2010).

Among the first scholars, who studied winter pastures in the eastern Transcaucasia and especially Azerbaijan were Mayer (1831) and Radde (Radde 1874, 1899 cited in Peper 2010). The next series of publications, according to Peper (2010 and references therein) with respect to vegetation structure, ecology and geobotany of rangelands began in 1920: Grossheim (1929, 1932b), Grossheim and Sakhokia (1929), later – Isaev (1957), Prilipko (1949), Aliev (1954), Aliev *et al.* (1965), Karjagin (1950-1961), Ponomarenko (1986), Atamov (2002) and Neudert and Allahverdiyeva (2004).

In the 2000s some scientific articles worthy of attention were dedicated to rangelands of this country by Azerbaijanian and German scientists.

Among the rangeland sites in Azerbaijan, the scholars focused considerable attention on two largest winter pastures at the foothills of the southeast ending of the Greater Caucasus in Azerbaijan, on the west bank of the Caspian Sea. They are Gobustan (1,780km^2) in the east, and Jeiranchel (1,300km^2) in the west (Mamedaliev 1963; Peper *et al.* 2010; Manthey & Peper 2010).

The territory of Gobustan is cut up with numerous, sometimes rather deep ravines (in Azerbaijani – gobu). Thus it is suggested as the origin of the Gobustan geographical name. Jeiranchel translates from Turkish as ‘gazelle steppe’.

Average annual precipitation ranges between 460-600 mm, but can be as little as 150-200mm. The winters are humid and cool and the summers–dry and hot (National hydro-meteorological service 2004).

Gobustan ranges from the Caspian shore up to its highest point, mount Gijäkiat (1,047m a.s.l.). Numerous riverbeds and gullies carve through the landscape. They mostly belong to the drainage basin of the river Jeirankezchmez. During summer, the local rivers are completely dry and salt crusts cover their beds.

The dominant soil type is grey-brown, carbonate-rich clayey to salty (Memedaliev 1963). Azerbaijan and its Caspian coastline are home to nearly 400 mud volcanoes, more than half of the total throughout the world. The majority of mud volcanoes are found in Gobustan, which makes a special feature of this area of the country.

In recent time, several mud volcano clusters have been protected as National Monuments (Kholodov 2001).

In 1966 Gobustan was declared a national historical landmark (today Gobustan National Park) of Azerbaijan in an attempt to preserve especially such famous carvings (see below), mud volcanoes and gas-stones in the region (*en.wikipedia.org/wiki/Gobustan_National_Park*).

The deserts and semi-deserts communities in the Gobustan represent a most, ecologically, important habitat from a botanical point of view. Some vegetation within this area is now being classified as either rare or threatened and is recommended for inclusion in an updated National Red List and some species are listed as globally threatened. The climate change and various anthropogenic activities are causing large losses to natural habitats in this area.

The importance of this habitat type is one of the reasons that the Gobustan desert has been proposed as a State National Park so that some level of protection is offered to this desert (Gambarova *et al.* 2010; Gambarova 2011). 45% of all species recorded in steppes and semi-deserts in Azerbaijan (Grossheim 1936) were found in Gobustan, which occur in high abundance only at higher altitudes (Gambarova *et al.* 2010).

As mentioned, perennial component of plants includes shrubs and semi-shrubs such as wormwood (*Artemisia fragrans*) and some species of saltwort (*Salsola* spp.).

In Salsoletum nodulosae – *Salsola dendroides* associations of Gobustan are distinguished two species of saltwort: *S. nodulosa* and *S. dendroides* (Huseynova 2004). Significantly *S. nodulosa* (gengiz) creates semi-deserts or deserts on developed soils, especially in foothills areas like Gobustan independently or with the *A. fragrans*. In mixed, warmwood-gengizian associations (Artemisia lerchianosa - salsola nodulosotum) 45 plant species participate (Huseynova 2004, Bakshiev 2009). *S. nodulosa* is an important highly productive arid livestock fodder of winter pastureland. The plant has great historical importance as a source of sodium, which is extracted from the ashes of *S. nodulosa* and other saltworts. On clayey bare slopes gengiz plays a soil protected role (Bakhshiev 2009).

This association common for this area is characteristic of solonchak soil, although *S. dendroides* and *S. ericoides* prefer soils containing relatively high levels of natural salts. *Reaumuria alternifolia* (=*R. cistoides*) also belongs to the salt-resistant species. Usually this vegetation forms insular patches enclosed by salt desert. Volcanic soils are populated by *Petrosimonia brachiata* and *Lepidium perfoliatum*.

Soils with low salinity and typically clay are populated by Artemisieto-salsoletum dendroides communities (Grossheim 1948).

The most characteristic species, which occurs in nearly all plant communities is *A. fragrans* which is typical above 300m a.s.l. A pure variant of *Artemisia* is mostly spread, but very often, there are also transitory types, diffusely or complexly mixed with options of saline deserts. Wormwood is frequently dominant within certain communities where species of the genera *Salsola* and *Suaeda* are also very common. However, dry stony or clay (i.e., not a true saline soil) semi-deserts with wormwood, sometimes are represented by patches of wet, salty land, where wormwood is mixed with dwarf shrubs. Here, co-edificator plants of *A. fragrans* occur - *Salsola gemmascens* and *Plantago ovata*. A mixed variant of Artemisieto-salsolatum dendroides, Artemisieto-salsoletum ericoides, Artemisieto-salsoletum gemmascens is also spread in central Azerbaijan. In these communities the dominant position is occupied by the following ephemers: *Bromus rubens, Medicago minima*, and *Filago germanica.* At about 550-750m a.s.l. *A. fragrans* participates into a belt of steppe vegetation, where the herbaceous plant communities are richer – 30-35 or even 50-55 species than those of semi-deserts/deserts association (Grossheim 1948; Gulisashvili *et al.* 1975; Schmidt & Uppenbrink 2009).

In Gobustan, to the relatively rare plants, belongs camelthorn (*Alhagi pseudalhagi*) – one of the most dangerous noxious weeds, which develops on terraces of dry ravines, along irrigation canals and on saline grey soils. Another rare plant of this area is represented by the tallest (1.5-4m) shrubs of the National Park, tamarisks (*Tamarix ramosissima, T. tetragyna*), which grow in river valleys (tugai shrubland), foothill deserts on dry ravines, banks of irrigation canals (Schmidt & Uppenbrick 2009) (Figure 32).

The plant species of the Gobustan National Park are mostly annuals or geophytes. They develop during March to October. The flowering time lasts until the summer drought. Species of Amaranthaceae (*Chenopodium ficifolium, Atriplex tatarica, Salsola crassa,* etc.) germinate in spring, flower in autumn and die with the first frost. From ephemers and ephemeroids characteristic for desert vegetation should be mentioned species of *Poa, Bromus, Alyssum, Helianthemum*, etc. They should not be regarded as xerophytes, because their short duration is connected with the vernal and autumnal vegetation period. Therefore, within this habitat plants are mostly of mesic nature. Low shrubs (*Salsola* and *Suaeda*) are also very common in these communities (Grossheim 1948; Schmidt & Uppenbrink 2009).

The pastureland of Gobustan is home to some endemic and other rare plants of the Caucasus dryland. In different habitats, frequently with restrict ranges, the following herbs unique to the Caucasus region should be mentioned: *Iris acutiloba, I. camillae, Gypsophila stevenii, Orchis caspia, Ophrys sphegodes mammosa* (= *O. cauasica*), *Veronica amoena, Linaria schirvanica.* All of them are strongly threatened by grazing (Shmidt & Uppenbrink 2009; Peper *et al.* 2010).

Figure 32. Wormwood community of Gobustan (Azerbaijan) (1978).

Animals of Gobustan National Park with a considerable attention to psyllids (Insecta, Psylloidea) fauna. About 110 wild bird species, 28 species of mammals, 2 species of amphibians and 20 species of reptiles have been recorded in the Gobustan region. Many of them are included in the IUCN and the Azerbaijan Red Data Book lists. The following rare species can be mentioned: lesser kestrel (*Falco naumanni*, VU), Egyptian vulture (*Neophron percnopterus*, VU), red-breasted goose (*Branta ruficollis*, EN), and the saker falcon (*Falco cherrug*) (Patrikeev 1991). Threats for bird hunting (for chukars, geese and little bustards) and habitat destruction through intensive grazing are usual in Gobustan's rangeland.

Common large carnivores (red fox, golden jackal, wolf) occur in Gobustan. Goitered gazelles (*Gazella subgutturosa*) – the national symbol of Azerbaijan – during the summer migrate from Shirvan National Park, where they are protected into Gobustan. They would probably inhabit the whole region if not for the strong poaching pressure, killing by dogs and highly desirable co-existence with livestock. From amphibians and reptiles *Testudo graeca buxtoni* is the only species included in the IUCN list and the Azerbaijan Red Book (Schmidt & Uppenbrink 2009; Peper *et al.* 2010).

Among the invertebrate fauna of clay (wormwood) deserts and semi-deserts of eastern Transcaucasia psyllids (Psylloidea) were investigated in detail. 40 species were registered. This is as much as 18,9% of all species known in the Caucasus and 39,6% of all species found in various drylands of this region (Gegechkori 1969, 1976, 1987a, b).

Psyllids of white wormwood semi-deserts are highly host-specific, mostly monophagous or narrow oligophagous and live on pasture dominant (wormwood) and non-dominant plant species.

13 species are associated with several species of white wormwood of different drylands of the Caucasus (Gegechkori & Loginova 1990). The nucleus of this fauna belongs to the genus *Craspedolepta* Enderlein (family Aphalaridae sensu Burckhardt & Ouvrard 2012), 11 species are known from *Bactericera* (Puton) and *Trioza* Förster (family Triozidae Löw). In deserts and semi-deserts of the Caucasus 8 species of *Craspedolepta* occur, all of them trophically linked with *Artemisia* spp. (*C. armazhi** Gegechkori, *C. azerbaijanica** Gegechkori, *C. gulisashvilii** Gegechkori, *C. indeserta* Bajeva, *C. laevigata* Loginova, *C. pilosa* (Oshanin), *C. megrica** Gegechkori, *C. setosa* (W. Wagner). The other 3 species on wormwoods (*C. caucasica** Gegechkori, *C. georgica** Gegechkori, and *C. malachitica* (Dahlberg) populated mountain-xerophyte phytocommunities, steppe and other semi-arid habitats.

The genus *Craspedolepta* is trophically linked with white wormwoods and distributed in the Caucasus, with about 55% of the species being endemics of this region and described by the author as a new species for science (species marked with asterisks).

In Gobustan National Park 4 species of psyllids were registered: *C. azerbaijanica, C. armazhi, C. laevigata,* and *C. setosa*. The latter two species are most xeric and restricted in their ranges from - 28m b.s.l. to 300-400m a.s.l. of the Greater Caucasus foothills and adjacent area in Azerbaijan. *C. azerbaijanica* and *C. armazhi* are characterized by a larger altitudinal distribution within the rangeland of the National Park.

In deserts and semi-deserts phytocommunities of Middle Asia and Central Asia, among host-specific insect herbivores with respect of heavy infestation of host plants, psyllids occupy an advance role (Loginova 1969a). *C. laevigata* and *C. setosa* are distinguished by rather dense populations, but as plant-juice sucking insects they should not be considered as a severe pest in pastureland in Gobustan and in the South Caucasus in general.

However, other severe pest species of arthropods should be taken into consideration, because arid and semi-arid communities in the Caucasus are composed of the semi-brushes of white wormwoods: *Artemisia fragrans, A. taurica* and *A. arenaria*. Their ability to grow well on severe sites and therefore on these natural pastures are the most important forage reserve and provide the basis for Azerbaijan's animal husbandry and maximal effect on the forage nutritive value for livestock in rangelands of the Caucasus in general.

Another major value of white wormwoods is their use in medicine as a tonic, stomachic, febrifuge and anthelminthic. These herbs held a high reputation in traditional Caucasian people's medicine. Some species are also valued for their essential oil, which is sometimes used in cosmetics and perfumes (Flora of the USSR, Vol. 26, 1961).

Among the other representatives of the herb layer mentioned, no lesser important is a complex of psyllids, which live exclusively on the other representatives of the family Asteraceae – *Achillea*. Across the whole area of rangeland of the Greater Caucasus foothills were found three species of the same genus *Craspedolepta*: *C. pontica* Dobreanu & Manolache, *C. nervosa* (Förster), and *C. bulgarica* Klimaszevski.

*Heterotrioza kochiae** (Gegechkori) and *Eumetoecus kochiae* sometimes coexist on xerophyte semi-brush, prostrate kochia – *Kochia prostrata*. These two species are not distinguished by dense populations and are found sporadically. *Kochia* is another representative of the family Amaranthaceae. Unlike the wormwoods it is not such a valuable forage plant in seasonal pastures in the Caucasus, due to its less frequent occupation, except when it is occurring as the dominant species population from the foothills up to the lower mountains.

Xenaphalara signata is a very characteristic species on *Ceratocarpus arenarius* (Amaranthaceae), but at the same time is the least commonly met species among rangeland's psyllid fauna.

Among the large shrubs and trees ecologically adapted to a xeric habitat of the Gobustan National Park should be mentioned genus *Tamarix*, in particular two species. About 54 species of tamarisks (Baum 1978) are distinguished by their rather specialized fauna, which includes numerous monophagous and oligophagous species (Zocchi 1971).

According to our survey (Gegechkori 1984a; Gegechkori & Loginova 1990), 9 species of two genera *Colposcenia* Enderlein (8) and *Crastina* Loginova (1) (Aphalaridae) feed on *Tamarix* spp. in the Caucasus. Among them are 6 species represented in the rangelands of Azerbaijan, 4 of which were found in the Gobustan National Park: *Colposcenia aliena* (Löw), *C. vicina* Loginova, *C. jakowleffi* (Scott) and one of the the rarest species – *Crastina tamaricina* (Loginova) (Aphalaridae).

Conservation activity of rangeland ecologists in Gobustan National Park (socio-economic issues). After the collapse of the USSR and Azerbaijan's independence livestock numbers increased, because large herds of sheep are a sign of societal influence and a good investment strategy in Azerbaijan (Neudert & Allahverdiyeva 2009). However, investment rarely goes towards the modernization of farm infrastructure, more often it is exclusively directed towards the enlargement of herds. Hence the number of small ruminant livestock nearly doubled from approximately 4 to 8 million animals in recent times (Azerbaijan State Statistical Committee 1995; Peper *et al.* 2010). Therefore, dryland rangelands are currently subject to heavier grazing pressure than was the case about 20 years ago. Because livestock is the major user of primary production, long term grazing leads to reduction of biodiversity and erosion, particularly on the slopes giving changed ecosystems. For instance, on the example of Gobustan steppes within some sites they have been changed into semi-deserts.

Many sites of arid landscapes, like the Gobustan region are surely one of the natural treasures of Azerbaijan.

The protected areas in Azerbaijan have increased, during the last 20-25 years. More than 10 new reserves were created and another 5 reserves were expanded and in total more than 2.5% of this country is now under the protection of the government, state reserves. Despite this, the measures are not enough to protect very valuable rangelands so important from a biodiversity point of view.

Box. 2. Petrogliphs of the Gobustan Area (Azerbaijan)

The Gobustan (or 'the land of dry riverbed') Rock Art is an open air archeological site. It is located about 65 km southwest of Azerbaijan and lying between the southern slope of the Greater Caucasus and the Caspian sea. Altitudinally it extends at an average height of 600-700m a.s.l.

In 2007, the Natural Reserve or National park was officially inscribed as Goravan Rock Art Culture Landscape and became listed as a UNESCO World Heritage Site due to very rich archeological monuments and an estimated 6,000 petrogliphs (rock engravings). They were incised on the rock some 5,000 to 4,000 years (Farajova 2017 and literature cited in). They were carved into blocks of stones. Huge boulders and rock formations were taken from the cliffs of local mountainous area. Large stony blocks and caves (including man-made ones) provide shelter for Late Pleistocene and Holocene. Therefore, this cultural continuity embodies Eurasian human history time between prehistoric and mediaeval times.

Due to archaeologists investigations (the first archeological investigation began by I. Jafarzade), the well-protected petrogliphs were revealed. From that time to nowadays thousands of rock pictures have been discovered and recorded in Gobustan (Figures 33, 34).

Figure 33. Gobustan covered with big lumps of stones, the 'sea of rocks'. This area is surrounded by characteristics of Azerbaijan drylands, semi-deserts and deserts and dry steppes (1972).

Figure 34. Monument to prehistoric human, Gobustan National Park (1972).

Ancient settlements of Gobustan display art culture of pre-historic human beings. Late Paleolithic engravingss include images of animals and humans. Among wildlife should be mentioned large predators, such as lions and a lot of grazing herbivores. For example, aurochs (local bison), onagers, deer, gazelles, bison goats, wild boars, etc. They were depicted either one by one or in groups. some of them being nearly life-size.Among late sequences of the Gobustan rock art, namely in the Early Holocene, anthropomorphic and zoomorphic creatures had appeared. They include domesticated oxen, wild horses, goats and camel caravans (Farajova 2017).

Regarding human beings, they settled in the Gobustan area during remote millennia and centuries. The rock petrogliphs of Gobustan are very diverse if we take into consideration themes and content of rock engravings. Among them should be pointed out ritual dances, bullfights, battle scenes, oarsman, female figures near need boats, on some of them the symbol of stars, woman dancing, hunting (with sacrifice hunting scenes), fishing, magic rites, etc. 'Therefore, it should be supposed that Gobustan was the earliest center in which navigation emerged in the Caucasian region' (Farajova 2017). On the other hand, the carvings depicting local animal species clearly indicate that the climate was warmer and wetter (with precipitation heavy enough) and resembled that of a savanna.

Gobustan cultural heritage is similar to that of the Pyrenees, the Sahara desert (Africa) and others representing the early stages of the history of civilization of humankind. Therefore, it has played an invaluable role in the field of reconstruction of pathway of European and Asian civilization (compiled from Farajova 2017 and scholars cited in).

In recent times, the dry grasslands and deserts and semi-deserts of Azerbaijan under the present regime (overgrazing) have come into the focus of conservation for rangeland ecologists. To keep the rangeland in good conditions, comprehensive rangeland researches have included and given information on the following situation. For example pastoral landscapes vegetation composition, patterns of species richness and species turnover along grazing gradients, selective impact over herbaceous species under grazing pressure, co-evolution of vegetation and livestock grazing (historically dry-season vegetation has its own defence mechanisms, such as the protection of the living parts behind the thorns or in the seeds) and so on. Finally, practical recommendations are for rangelands management and conservation, sustainability of pasturlands and their value for future generation in Azerbaijan.

The results of perspectives and ways of rational use of desert plants of Gobustan and Jeiranchel and grazing impact studies have been carried out by Azerbaijan and German (particularly scientists from the University of Greifswald) botanists and vegetation ecologists. Their findings are presented in some articles and papers. For example, Akhmedova (2005), Asadova (2008, 2013), Huseynova (2008), Pietzsch (2008), Schmidt & Uppenbrink (2009), Gambarova *et al.* (2010), Peper *et al.* 2(010), Manthey & Peper (2010), Gambarova (2011) and others.

The present monograph, according to its structure, does not aim to present state-of-the-art reviews of the results of the above mentioned scientists.

2.2.1.2.1.2.1.1.4. Hydrohalophytes

A characteristic feature of species within this group is growth in constant humidity, which is usually represented in wet to standing water, varying from freshwater to brackish water marshes, ditches, artificial lakes and canals. The main vegetation includes: *Phragmites australis, Arundo* spp., *Typha* spp., etc. Nearby occur: *Hippophae rhamnoides, Elaeagnus angustifolia, Tamarix* spp. (Schmidt 2007) (Figure 35).

Figure 35. A small lake surrounded by canes (*Phragmites australis* and *Arundo donax*), the Iori Plateau, E. Georgia) (2017).

Box. 3. Sakhare-Tba (Sakhare Lake) Covered with Sparce Stands of Halotolerant Plant Species on the Salt Deposits Surface

Among other small salt lakes of East Georgia Sakhare-Tba should be mentioned. Recently this site was visited by me and my students for observation of salt tolerant herbaceous and semi-shrub communities and some groups of native invertebrate and vertebrate fauna which found shelter in this extreme habitat.

The extraordinary nature of the lake, with its various saline ecosystems and associated halophiles provide an ecological uniqueness.

Sakhare-Tba is a small (600 m long and 200m width), shallow saline wetland situated not far from the David Gareja monastery complex. This physiographic unit covers an altitude of

800m a.s.l. Depending on rainfall, annual precipitation ranges between 250-350 mm. The soil in the extreme eastern part of East Georgia, where the lake is found, naturally contains salt (high level of sodium and potasium). The main reason has been weathering parent rocks and rising in the level of saline ground water tables, bringing salt to the surface. Hence, due to drought seasons, in the lake it sharply increases in salinity. Sakhare-Tba lacks any sort of permanent drainage system, therefore, provides an example of present closed, saline lake, where salinization of ground water occurs. Heavily eroded strata of the lake's basin consist of predominantly Mio-Pliocene sediments (Connor & Kvavadze 2008 and references therein).

The distance through the terrace upward from the lake to the dirty road is about 300-400 m. According to increase or decrease of salinity of the ground dramatically affects the composition of the vegetation community from extreme halophites on the surface and lake shore sites to glycophytes (non salt tolerant) on the top of the terrace. For example, true halophiles are represented by *Salicornia herbacea*, which usually forms dense mats up to 10-15 cm thick around the lake. On the surface of the white mini-salt desert of the lake grow stem- and leave halosucculentes. Here should be pointed out are species of *Salsola* (*S. crassa*, *S. soda*). Salt-core free sites of the lake are presented by the ubiquitous *Phragmites australis*. The shoreline of the lake is populated by other salt-tolerant species: *Sueda dendroides*, *Petrosimonia brachiate* and *Limonium meyeri*. Upper layer sites of the terrace are dominated by *Glycyrriza glabra, Salsola nodulosa*, and *Artemisia fragrans*. On the next, upper layer of terrace, near to the ground road distinguished by sites with lowest salinity, are distributed the wild cereals, species of *Festuca* and *Stipa* (Ketzkhoveli, 1960) (Figure 36).

As we see the halophytic species are indicators of the degree of salinity at their micro-habitats. In other words, changes in the salinity gradients may lead to gradual variation on rather restricted areas not only specimens of plant species, but their communities as well.

Figure 36. Part of the hypersaline lake of Sakhare (a landlocked body of water) near the Gareji Hollow, Kakheti province, E. Georgia. Plants (*Salsola soda*) are adapted to highly salty soil (2012).

Riverside (tugai) forests of eastern Transcaucasia. The floodplains of eastern Caucasus rivers border riparian, so-called 'tugai' type forests, although they differ significantly from tugai forest stands of Middle Asia by soil conditions and so by composition of tree layer (Ketskhoveli 1960; Gulisashvili *et al.* 1975).

Tugai is a 'Turkii' word to describe the unique desert flood-plain forests occuring in Middle Asian arid environments. 'Tugai' is generally accepted as referring mainly to the forested areas. However it is, from an ecological point of view, a more complex feature incorporating the transition from open water to reed stands, dense gallery forest, shrub (mainly *Tamarix*) communities, and halophyte vegetation which from a mosaic landscape arises depending on the ground water level, distance from the main watercourses, and groundwater salt content (Zerbe *et al.* 2010 and references therein). This diversity of habitats is matched by a corresponding biodiversity including a number of species which are restricted to only the tugai environment (e.g., Bukharian deer (*Cervus elaphus bactrianus*) (Conservation of 'Tugai Forest' in the Amu Darya Delta of Karakalpakstan, an autonomic republic of Uzbekistan).

Similar to other countries or regions (see Sharitz & Mitsch 1993), alluvial forests of East Transcaucasia occupy a variety of situations in riverside systems. They are usually intermixed with lower areas of floodplain swamp and high areas of bottomland forest, or upland hardwood forest.

In Kakheti, on the left bank of the Alazani River, the special geographic location of the mountains, the south slopes of the Greater Caucasus Range, namely Kakhetian Ridge and Tsiv-Gombori Ridge, trap the moisture (800-1000mm) brought by the prevailing winds moving from the Caspian Sea (in winter there are almost no negative temperatures). Consequent climatic conditions, like the Colchis topographical triangle, encouraged survival of riparian forest enriched by the Arcto-Tertiary relict species of trees, shrubs and vines. These resemble those of Colchis Lowland but differ markedly from those of the rest of eastern Georgia, and still less Kura-Aras Lowland and Middle reaches of Aras River (Ketskhoveli 1960). Along the Alazani River the moist riparian forest is composed of large size trees: *Quercus robur pedunculiflora, Pterocaria fraxinicolia, Populus hybrida* (trees of this species grow to height of 40 m, and attain a diameter of up to 150cm). Also mentioned should be: *P. nigra*, *Alnus barbata* and *Morus alba*. In the second lyer: *Coryllus avellana, Diospyros lotus*, *Staphylea pinnata* are common. Trees are covered with vines: *Vitis vinifera sylvestris, Hedera helix, Smilax excelsa, Clematis vitalba*, etc. (Grossheim 1948; Ketskhoveli 1960; Gulisashvili *et al.* 1975).

Lowland forests of East Transcaucasia are distributed in ancient alluvial plains where pedunculate oak (*Q. r. pedunculiflora*) dominates. Until recently *Q. r. pedunculiflora* forests covered vast areas of the plains adjacent to the Kura (Mtkvari) and Alazani Rivers and their tributaries, but today these forests only survive in East Georgia in the river valleys, especially in that of the Alazani. Isolated stands of *Q. r. pedunculiflora* can, however, be found almost everywhere in the lowlands of East Transcaucasia up to an altitude of 1,050 m a.s.l. (Gulisashvili *et al.* 1975).

Q. r. pedunculiflora forests are characterized by a rich flora of trees, shrubs and lianas and hold 30 to 40 species of arboreal plants per hectare. Several types of oak forests have been distinguished in East Georgia. *Q. r. pedunculiflora* forests with *Ulmus minor*, oak-elm forests with an undergrowth of *Sambucus nigra*, oak forests with species-rich underwood (*Prunus spinosa, Mespilus germanica, Cydonia vulgaris*), and oak forests with grasses (*Poa nemoralis, P. pratensis, Dactylis glomerata*) in the herb layer (Gulisashvili *et al.* 1975).

Figure 37. A student caught her first lizard (*Darevskia* sp.) in an oak (*Quercus robur pedunculiflora*) trunk. Gardabani Reserve, E. Georgia.

The Kura-Aras river basin is an internationally significant river system. The basin covers almost of all Armenia and Azerbaijan, and a large part of Georgia. These countries together with Iran and Turkey rely heavily on the Kura-Aras river system as a principal source of water for industry, agriculture, energy and residential uses (Kura-Aras River Basin, TDA 2007). Therefore, riparian gallery forests, which are represented as a marginal zone of specific vegetations in arid and sub-arid climate countries play an important role in influencing climate and keeping the sustainability of healthy ecosystems in semi-deserts and deserts harsh environment (Figure 37).

Riverside forests of the South Caucasus consist mainly of *Populus hybrida*, *P. nigra*, *Salix excelsa* (= *S. australis*) and *U. minor*. On the first terrace, very close to the river bank, *Tamarix meyeri, T. hohenackeri* and *Hippophae rhamnoides* occur.

At the transition of lowland forests to arid open woodlands (or in higher places deserts and semi-deserts) dense thickets of various tamarisk species can also be found.

The middle reaches of the Aras River (extreme southern part of South Caucasus) are famed for the prominent regional character in the Caucasus as well as the environment, including the largest stands within this region of the Euphrates poplar (*Populus* (*T.*) *euphratica*), willows (*Salix* spp.), oleaster (*Elaeagnus angustifolia*) and tamarisk (*Tamarix* spp.). Under the cover of dominant trees are shrubs such as *Hippophae rhamnoides, Berberis* spp., *Rosa* spp., *Halimodendron halodendron*. The herbaceous vegetation is represented by desert, steppe and swamp species. Therefore this riverside forest has an interesting component of similiarity attracting a woody genera of wide distribution in Middle Asia (sensu stricto) and Central Asia (Mongolia, part of China). However, during the past 40 years, during my field excursions in the Caucasus, because of immoderate utilization of water resources and logging by native settlements die-offs of significant population of poplar trees has been caused. It should be noted that on mentioned species of poplar, as in Middle Asia, are found strict olygophagous psyllids - *Egeirotrioza gracilis* (Bajeva) and *E. verrucifica (*Loginova) associated with *P.* (*T.*) *euphratica* found in the Caucasus exclusively in the middle reaches of Aras (Gegechkori 1978).

In Armenia, specific sites have been identified that support ecosystems of global or regional significance which are rich in endemics, relict or rare species. Among others the relict tree stands and forest patches consist of the Oriental plane tree (*Platanus orientalis*) which is represented in the Caucasus exclusively near the Tsav river valley (Gulisashvili *et al.* 1975).

This site occurs in southern Armenia, towards Shikahogh State Reserve bordering Iran (covering some 10,000 ha of land). The Tsav River flows in the foothills of the north slopes of the Meghri range, approximately 15-20 km from Aras River. The Oriental plane tree natural grove as a thin (100-150m) riparian forest covers 150 ha of land. The trees have a mid-height of 30-32m, with a mid-diameter of 80cm, and an estimated age 250 years. It is often found as a pure stand and rarely growing together with other hardwoods (*Juglans regia, Celtis caucasica* and *Populus hybrida*).

On the banks of Tsav River ia a shrub stratum composed of *Sambucus nigra, Rubus* spp., etc. Among the vines occur *Periploca graeca* and *Vitis vinifera silvestris*. The herbal cover is less characteristic (species of *Circaea, Arctium, Geranium*, etc.).

The Oriental plane is one of the major forest-building species within the riparian forests of bordering Iran and, on the other hands, in Turkey (Gulisashvili et al. 1975 and references therein).

Gardabani Managed Reserve as a characteristic riparian forest. The area of Gardabani Managed Nature Reserve established in 1996, encompasses 3,484 hectares. It is located near the Azerbaijan border and found in the Gardabani and Marneuli districts (it is at a distance of 39 km from Tbilisi). It was created with the aim of maintenance of the unique riparian forest groves as well as protection of the representatives of fauna which inhabit it.

The Gardabani Managed Reserve is a typical floodplain forest (tugai) across the River Kura (Mtkvari). It include relict nature floodplain forest, where the major forest-building species are either long-stalls (floodplain) oak (*Quercus r. pedunculiflora*) or hybrid poplar (*Populus hybrida*), or both species. A magnificant, relict long-stalls oak usually grows in a narrow strip along river. Besides the leading species, associated components of the riparian forests are composed of about 30 species of arboreal plants (trees, shrubs) including many relict species (*Hedera helix, H. pastuchowii, Vitis vinifera sylvestris, Smilax excelsa, Ligustrum vulgare*, etc.).

Among the co-edificator summer-green tree species should be mentioned *Populus nigra* and *Ulmus minor*. The shrub layer consists of *Salix wilhemsiana, Crataegus curvisepala, C. pentagyna, Tamarix ramosissima, Cornus mas, Prunus spinosa, L. vulgare, Elaeagnus angustifolia*, etc.

Such relict riparian forests are very seldom found in the arid climate zones of Georgia, therefore the Gardabani Managed Reserve has a very important conservational value in the Kura River basin of the South Caucasus.

The Gardabani Managed Reserve is inhabited by 21 species of fishes, 4 species of amphibians, 4 species of reptiles, 135 species of birds, and 26 species of mammals. This fact indicates a high level of biodiversity. The Gardabani Managed Reserve is the only riparian forest in the Kura River basin where red deer still can be found. Overhunting is the main cause of the drastic decline of deer. At the beginning of the 20th century there are some short film chronicles about the Gardabani riparian forest. Here is shown officials of the Russian Empire who had been visiting the forest for hunting purpose. In this drama film, in front of the hunters are deposited tens of dead red deer and wild boars, four wolves and four jackals. The next considerable loss of this disjunct population of mentioned ruminanat has been suffered in the1990s, during the first years of the disintegration of the USSR (Figure 38).

Figure 38. Riparian forest of the River Kura (Mtkvari) with white poplar (*Populus alba*); lianas, significantly, ivy (*Hedera pastukhovii*) are common. Gardabani environs, Kvemo Kartli province, E. Georgia) (2015).

Among large animals which inhabited the area are wild boar and jackal but the jungle cat, red fox and marten should be mentioned. Here also nest some species of the raptors: white-tailed eagle (*Haliaeetus albicilla*), imperial eagle (*Aquila heliaca*), Egyptian vulture (*Neophron percnopterus*), greater spotted eagle (*Aquila clanga*), Levant sparrowhawk (*Accipiter brevipes*), and Saker falcon (*Falco cherrug*). So, in this Managed Nature Preserve as well as scientific researches it is possible to organize bird watching and in general, animal watching tours as well as botanical and ecological tours (Kikodze & Gokhelashvili (Eds.) 2007).

The adjacent territories of the floodplain forest are covered with semi-desert and steppe plants.

It should be mentioned here that riverside vegetation was cleared for cultivation and inhabitation. Therefore, the Turanian (Caspian) tiger and its prey were deprived of their fragmentated and degradated habitat in the 1930s. A further environmental insult occured when the riparian forest and other riverside ecosystems (i.e., reed stand) were cleared to eradicate malaria in the southern shores of the Caspian Sea in the 1950s and 1960s (Azerbaijan). More or less the same situation occured in the southern Caspian littoral and foothills of the Talysh Mountains in northern Iran, within the nucleus area (Masandaran Province) of the tiger's distribution, where these majestic cat had once been numerous. Soon the Turanian tiger became an alien in its own territory and was targeted and hunted down as a menace to human settlements and a threat to livestock. The treatment of prey species like mainly wild boars and red deer were also targeted out of the extreme western margin of the natural range, including, as it seems today's Gardabani Managed Reserve. Clear confirmation of this suggestion is the following example: the last known Caucasus Turanian tiger was killed in 1922 not far from the capital of Georgia, Tbilisi in the village of Lelobi after killing domestic livestock. From the 1950s the Turanian tiger – one of the keystone species of the Caucasus Ecoregion - became extinct (Gegechkori 2010a and references therein).

The plain of Kvemo Kartli, which occurs in Marneuli and the Gardabani districts is crossed by the exotic riparian forest of the Mtkvari River. The Tugai forest, represented in the semi-arid environment has an important water regulation and water and soil protection significance. The Tugai forest of the Mtkvari River with its mild and mitigating environment is a habitat for many rare animals in the classic sub-arid landscape. This fauna within transitional habitat has a composite character between deciduous forests and semi-deserts. In other words such a defuse zone of vegetation, as broadly known, is called an ecotone.

Ecological relationship of distribution patterns of riparian softwood forests and their consumers – species of psyllids (Hemiptera, Psylloidea). Riparian forests are important areas for global biodiversity (Sala *et al.* 2000). Owing to their highly dynamic character caused by the flow regime, riparian woodlands exhibit much special and temporal heterogeneity (Mosner *et al.* 2011). The natural vegetation associated with waterways, particularly in arid and semi-arid landscapes, is mostly represented by riparian woodlands (Williams *et al.* 2006). Riverside forests in arid regions are frequently the only mesic ecosystem breaks in an extensive arid area (Roberts et al. 1977). A quote from Roberts *et al.* (1977) 'The microclimate in riparian forests is similar to that of a greenhouse'.

The vegetation surrounding the stream helps to shade the rivers´ water, mitigating water temperature changes. They contribute to water quality maintenance by filtering it, stabilize substrate with their large root systems and river bank stabilization (Tockner & Stanford 2002; Ward *et al.* 2002). The riparian habitats (woodland, wetland, grassland) largely influence the fauna. Ecologists and biogeographers recognized them as a priority area, sometimes vital for conservation of terrestrial animals, especially mammals. They provide habitats as well as dispersal corridors, supporting wildlife to move fluently along river banks (Kuyéma 2003 and references therein). As a striking illustration of confirmation mentioned from this point of view, as previously mentioned, it is very rare examples of recent centuries (19th, 20th) the migration of the Caspian tiger from Talysh (Azerbaijan) in East Georgia via the riparian corridor of the Kura (Mtkvari) River (Djanashvili 1963; Gegechkori 2010a).

Undisturbed riparian ecosystems with lush vegetation and well-developed overstory, midstory, and understory of the rivers Amu Darya, Syr Darya, and Tarim are typically characterized as those having a considerabe high spatial heterogeneity of plant communities, trees, shrubs, dense reeds and grasses, especially, dense forests dominated by relict poplars (*Populus* (*Turanga*) *euphratica* and *P.* (*T.*) *pruinosa*). Tugai forests have a high value of habitats for very rich and rare wildlife. For instance, the recently vanished big cat (*Panthera tigris virgata*), Bucharian deer (*Cervus elaphus bactrianus*), Goitered gazelle (*Gazella subgutturosa*), desert monitor (*Varanus griseus*), etc. (Bannikov (Ed.) 1969; Magin 2005).

Riparian softwood forests together with riparian hardwood forests form the potentially natural vegetation for the most part of the floodplain of the Caucasus. A considerable part of the lowland and riparian hardwood forests consisting of *Quercus robur* (Ciscaucasia) and *Q. r. pedunculiflora* (Transcaucasia) have been destroyed by timber logging. In the western part of Transcaucasia, species dominating the riparian zone consist of mainly Salicaceae (willow and poplars). In the eastern part of the region riparian softwood is enriched with Tamaricaceae (tamarisks) (Gulisashvili 1964).

Arboreal vegetation of riparian forests are largely populated by psylloid fauna. Among them are psyllids specific and intimately tied to their hosts taxa on a genera level. For example, narrowly olygophagous species of *Colposcenia* are restricted to*Tamarix*, *Egeirotrioza* Boselli with *Populus* (*Turanga*) are represented exclusively in arid zone´s – tugai forests; species of *Camarotoscena* Haupt are also found on another species of *Populus*, *Crastina* on *Myricaria* and *Tamarix* (Loginova 1968, 1969a,b; Gegechkori 1984a; Gegechkori & Loginova 1990; Gegechkori *et al.* 2013; Baeva 1985).

Table 3. Psyllid fauna confined to riparian forests of the Caucasus

Family, genera & species	**host plants**
Aphalaridae	
Aphalara polygoni Förster	*Polygonum* spp., *Rumex acetosella*
Camarotoscena fulgidipennis Loginova	*Populus pyramidalis*
C. libera Loginova*	*Populus* sp.
C. unicolor Loginova	*Populus* spp.
Colposcenia aliena(Löw)	*Tamarix* spp.
C. conspurcata Loginova	*Tamarix* spp.
C. jakowleffi (Scott)	*Tamarix* spp.
C. kiritshenkoi Loginova	*Tamarix* spp.
C. orientalis (Klimaszewski)	*Tamarix* spp.
C. osmanica Vondraček	Tamarix spp.
C. vicina Loginova	*Tamarix* spp.
Crastina dimorpha Bajeva	*Tamarix* sp., *Myricaria bracteata*
C. myricariae Loginova	*Myricaria bracteata*
C. tamaricana Loginova	*Tamarix* spp.
Psyllidae	
Baeopelma foersteri (Flor)	*Alnus* spp.
Cacopsylla abdominalis Meyer-Dür	*Salix spp.*
C. ambigua Förster	*Salix* spp.
C. compar (Loginova)	*Salix* spp.
C. fraterna Gegechkori*	*Salix caprea, Salix* sp.
C. hippophaes Förster	*Hippophae rhamnoides*
C. loginovae Gegechkori*	*Salix elbursensis*
C. memor Loginova	*Salix* spp.
C. montanica Gegechkori*	*Salix* caprea
C. moscovita Andreanova	*Salix* spp.
C. nigrita (Zetterstedti)	*Salix* spp.
C. saliceti Förster	*Salix* spp.
C. sibirica Loginova	*Salix* spp.
C. subklapaleki Gegechkori*	*Salix* spp.
C. submoscovita Bejeva	*Salix rosmarinifolia*
C. zetterstedti (Thomson)	*Hippophae rhamnoides*
Psylla alni (Linneaus)	*Alnus* spp.
Triozidae	
Bactericera albiventris (Förster)	*Salix* spp.
B. curvatinervis (Förster)	*Salix* spp.
B. daghestanica Gegechkori*	*Salix* sp.
B. magnisetoza Loginova	*Elaeagnus angustifolia*
B. melanoparia Loginova	*Salix spp.*
B. salicivora Reuteri.	*Salix* spp.
B. salictaria Loginova	*Salix* spp.
B. striola Flor	*Salix* spp.
Egeirotrioza (*Astutia*) gegechkorii Loginova*	*Populus? alba*
E. gracilis (Bajeva)	*Populus* (*Turaga*) *pruinosa*
E. populi (Horvath)	*Populus? alba*
E. verrucifica (Loginova)	*Populus* (*Turaga*) *pruinosa*
Total number: 43/7*	

* Endemics of the region.

Several species of psyllids are shown to be generally slightly injurious to riparian forest shrubs and trees. In some cases, population densities may reach conspicuous levels, but damage to these plants is usually not of significant economic and ecological damage although some of them are noxious to hosts through leaf and other organs causing galls. For instance, species of *Egeirotrioza gracilis* and *E. verrucifica* which are responsible for the damage events to the desert's poplar (Turanga) trees (*Populus* (*T.*) *pruinosa*) (roll-leaf galls on the mature leaves and young shoots).

Impacts caused by galling psyllids on their host plants (tamarisks) are more seriously expressed in several shoot- and bud-gall forming species of *Colposcenia* and *Crastina* including *Colposcenia aliena, C. kiristshenkoi, C. vicina* and *Crastina tamaricina* (Gegechkori 1984a).

There is a suggestion that gall-makers and hosts are co-evolutioned from Cretaceous (Kovalev 1995; Kovalev & Zaitsev 1996). 'Gall-forming psyllids – to quote Kovalev and Zaitsev – and other insects attacking tamarisks develop on a limited number of host-plant species and have been in existence without morphological change for million of years'. The absence of the galling ability in *Crastina montana* and *C. myricaria* on *Myricaria* spp., unlike *C. tamaricina which* feeds on the both genera of Tamaricaceae, suggest that the two former species have been exploited by phytophages comparatively recently (Kovalev & Zaitzev 1996).

Distribution patterns in the Caucasus with respect to azonal habitat type (i.e., riversides), can be residential for one group of species, transit for another and non-receptive for species being settled in zonal habitats (Gegechkori 1984a).

Via azonal habitats, there occurs a mutual enrichment with mesophilous and xerophilous species in both parts of the Caucasus with arid and temperate humid climate. Notable migration usually involves both the latitudinal and altitudinal dispersions (Chernov 1975; Gegechkori 1984a).

The psyllid fauna of the riparian forests of the Caucasus was collected by some entomologists (Triapitzin and especially Loginova (1968), St-Petersburg, ZIN). In the framework of a comprehensive work project on the psyllids of the Caucasus, this fauna was investigated in detail (Gegechkori 1984a). Research includes, besides the general biomes, riparian areas of main rivers and their tributaries during which target material – 41 species (7 species are endemic to the region) belonging to 9 genera have been collected by me and recently by my collaborators (Table 3) (Gegechkori 1984a; Gegechkori et al. 2013).

Via the intrazonal habitats, some species of psyllids are distributed through all the altitudinal zones (belts) of the mountains of the Caucasus reaching with their host plants from lowland to high mountain areas. For instance, *Aphalara polygoni* occurs on species of *Polygonum* and *Rumex*, *Crastina myricaria* on *Myricaria*, species of the families Psyllidae and Triozidae (*P. alni*, *B. foersteri* - on *Alnus*, *C. zetterstedti* - on *Hippopheae*, *C. ambigua, C. compar, C. memor, C. moscovita, Bactericera albiventris, B. salicivora, B. salictaria, B. melanoparia, B. striola* link on *Salix)* (14 species). The ranges of other species are bound in either lowland and foothills or mountains or high mountains area. To the former group belong species occuring in arid landscapes of the Caucasus: *Camarotoscena fulgidipennis, C. unicolor, C. libera, Egeirotrioza (Astutia) populi* linked with species of *Populus*; *Colposcenia orientalis, C. jakowleffi, C. kiritshenkoi, C. conspurcata, C. vicina, Crastina dimorpha, C. myricaria* restricted with *Tamarix* (11 species).

The following species settle in the mountain and high mountain area: *Cacopsylla abdominalis, C. fraterna, C. loginovae, C. montanica, C. nigrita, C. saliceti, C. sibirica, C. submoscovita, C. subklapaleki, Bactericera daghestanica.* All of them are recorded on *Salix* spp. (10 speceis).

The prevailing mitigating gradient of riparian woodlands (Chernov 1975) is used by psyllids settled under the generally cooler conditions of the highlands. Usually they are inhabitants of riparian *Salix* communities of the Caucasus: *Cacopsylla moscovita, C. nigrita, C. compar, C. memor, C. saliceti, Bactericera salicivora* (6 species). These riparian-dependent species can be met frequently, usually they are distinguished by population density. Psyllids feed on willows and use the riparian strip as a transit habitat occurring from highland landscapes in numerous altitudinal belts including lowland area, or *vice versa* (Gegechkori 1984a).

In the fourth group are connected species of psyllids occupying two or three altitudinal belts (zones): *Colposcenia aliena, C. osmanica* (on *Tamarix*), *Bactericera curvatinervis* (on *Salix* spp.), *B. magnisetosa* (on *Elaeagnus*) (4 species). Through riparian woodlands they are able to disperse in more than one zone following their hosts.

The most restricted group of psyllids in their dispersion, even via azonal habitats, is represented in semi-deserts, deserts and other arid landscapes closely connected with their hosts. For example, rare species of psyllids on *Populus* (*Camarotoscena libera, C. fulgidipennis, Egeirotrioza* (*A.*) *gegechkorii, E.* (*A.*) *populi*), on desert's poplar *Populus* (*Turanga*) – *C. unicolor*, and on *Tamarix* - *Colposcenia orientalis, C. conspurcata.*

- Protected areas and some keystone animal species of solonchak desert
- Shirvan and Ag-Gel National Parks – an emblematic solonchak deserts and semi-deserts of the Kura-Aras Lowland
- Shirvan National Park (SNP) was legally implemented in 2003. It was established on the base of the Shirvan State Reserve (founded in 1969) and neighbouring area. It is the first national park in Azerbaijan's history.

The Shirvan National Park is part of the geographical region of South East Shirvan. The SNP is located on Azerbaijan's coast of the Caspian Sea, its surface area is 65,536 hectares (543,735 km^2) in total. The main elevation of the park is – 27m a.s.l. The SNP's area is characterized by a semi-arid to arid climate. Summers are hot with a possible peak of 41^{0}C whereas winters are rather mild, rarely reaching 0^{0}C although they may drop as low as -22^{0}C. The mean annual precipitation of 286 mm is limited almost completely to spring and autumn (Schmidt 2007 and references therein) (Figure 39).

Shirvan National Park is considered to be the last stronghold of the goitered gazelle (*Gazella subgutturosa*) populations in the South Caucasus. Therefore the SNP's function is focused on the protection and reproduction of this endangered species of gazelle. Unfrozn water collectors and Flamingo Lake in the north of SNP fully supply the goitered gazelles with water. The reserve also makes an important contribution to the protection of waterfowl birds, their resting, wintering and breeding sites. There are 35,000 hectares of water reservoirs (especially Flamingo Lake, the General Shirvan Water Collector and the Caspian Sea) support very diverse waterfowl and resident water bird species, including the black francolin (*Francolinus francolinus*), great bustard (*Otis tarda*), little bustard (*Tetrax tetrax*), mute-swan (*Cygnus olor*), greater flamingo (*Phoenicopterus roseus*), red-breasted goose (*Branta ruficollis*), species of herons (*Ardea* spp.), etc.

Figure 39. Approaches to Shirvan NP with a dense network of small lakes and their surrounding wetlands. This environment is an all year round rangeland (1974).

Despite its location in a desert and semi-desert zone the National Park is populated by a large number of mammals. As well as the goitred gazelle the following should be mentioned: wild boar (*Sus scrofa atilla*), beaver (*Castor fiber*), wild (jungle) cat (*Felis chaus*), wolf (*Canis lupulus cubanensis*), jackal (*Canis aureus*), fox (*Vulpes vulpes alpherakyi*), badger (*Meles meles*), hare (*Lepus europaeus*), etc.

In south-east Shirvan (SES), where the SNP is represented, the floristic composition is dominated by Irano-Turanian floristic elements. Today from SES 157 species of vascular vegetation (Agaguliev (Ed.) 2000) are known in total 131 higher vascular plants species were investigated for Shirvan National Park (Schmidt 2007). The vegetation communities of the park are categorized into halophytic, psammophytic and steppe groups (Prilipko 1949; Abutalibov 1976; Aliev &Gajiev 1986; Agaguliaev (Ed.) 2000; Schmidt 2007; Bakhsiev 2009).

Recent research of the halophyte desert vegetation in the Northern Caspian Region (Caspian Lowland) belongs to Freitag *et al.* (2001) and Safronova (2008).

The Reserve is focused on the protection from human impact of typical solonchak and other type of deserts, coastal dunes and wetlands (MENR 2003). Its functions also envision the implementation of environmental education, as well as creating conditions for tourism and recreation. The National Park was officially opened for visitors in 2006 (Schmidt 2007).

Ag-Gel National Park is situated in the south of the Kura-Aras depression. The National Park occupies the western part of the Mil steppe (central Azerbaijan) and stretches between the Greater and Lesser Caucasus Mountains (Gribova et al. 1980). It is composed of the shallow lake Ag-Gel, which lies below sea level, adjacent deserts and semi-deserts and also dry scrabland landscapes (Figure 40).

The Ag-Gel was for the first time protected as a Zakaznik (lower protection status) in 1964 of about 9,200ha. From 1978, a part of the protected area (4,400 ha) had the status of Strict Nature Reserve (Zapovednik), and in 2003 an enlarged area of about 18,000 ha was designated to become a National Park.

Nowadays, the Mil steppe (stretches over 5,500km^2) represents a uniform plain with a slight decline towards north-east. The lowlands of the Mil steppe are bordered by the rivers Kura and Aras to the north and east, and by the northern slopes of the Karabakh mountains to the south-west. The Mil steppe, resembles the Shirvan National Park belongs to the subtropical continental dry climate zone. The lake and other waterbodies do not freeze in winter.

In the present hydrological system (until the construction of the Mingechaur reservoir in 1953), the lake was fed and drained by a system of feeding canals in the south and drainage canals in the north. It was built in the 1950s-1960s. The minor and major Ag-Gel lakes are separated by a middle wall. Today, the lake area is about 44km^2, average water depth is 0.8m, and maximum depth 2.5m. The water is mesohaline. Ag-Gel is an eutrophic lake (Strauss 2005 and references therein).

Many globally threatened bird species use the Ag-Gel as a breeding and wintering habitat. It has a high Transcaucasian value as one of the few wetlands in the arid Kura-Aras Lowland. Together with the neighbouring lake, Sarysu, it is the most important site ('ornithological oasis') for many rare and threatened species of waterfowl and shorebirds in central Azerbaijan. Because of its importance for birds the Ag-Gel is listed as an Important Bird Area (Birdlife International 2003) and the Ramsar List of Wetland of International Importance (Ramsar Convention on Wetlands 2001), (Gauger & Pietzsch 2005; Strauss 2005). The National Park is a part of the former range of the endangered goitered gazelle (*Gazella subgutturosa*) (Strauss 2005).

The objectives of the National Park are the reintroduction of goitered gazelle, the conservation of the characteristic flora and fauna of the lake Ag-Gel and the surrounding semi-desert and desert landscapes.

Figure 40. Dr. Arnold Gegechkori photographing in the Ag-Gel NP (1972).

Some indicator and flagship species of solonchak deserts and semi-deserts fauna.

Marbled duck (Marmaronetta (Anas) angustirostris). There are three distinct populations of marbled duck. The first has fragmented distribution in the western Mediterranean and winters in north and sub-Saharan West Africa. The second breeds in the eastern mediterranean and winters in Egypt, and the third breeds in western and southern Asia, wintering in North Africa, Iraq, Iran, Pakistan, northwest India and sometimes the southern regions of the Caspian Sea. The duck can also be found in the Caucasus, most notably around the desert lakes of Azerbaijan (Sari–Su, Ag-Gel and others), lake Sevan (Armenia) and to a lesser extent in the Aras Gorge and river area: sporadically in Georgia as well.

The species prefers steppes, salty deserts and small lakes with brackish waters and white canes thickets.

It is believed that somewhere between 9,000 and 19,000 ducks remain. The marbled duck appears to have suffered a rapid population decline, according to numbers in its wintering range, largely a result of extensive habitat destruction. Another major threat is poaching. The species is listed as Vulnarable by IUCN (Zhordania 1999; Gegechkori 2007).

White-headed duck (Oxyura leucocephala). It has wide range in the Palaearctic. Within the Caucasus Ecoregion significant populations of white-headed duck breeds are primarily in Russia, Turkey, Iran, and Armenia. In Georgia (Colchis, Paleostomi Lake) a few wintering birds are found. The duck is found in steppe and desert landscapes and freshwater, but it much prefers eutrophic lakes that provide plenty of vegetation cover.

The species' range and population size have declined as a result of habitat degradation and hunting. The white-headed duck is classified as an Endangered Species in the IUCN Red List and at the European level by Birdlife International included in the Red Data Books of the USSR, Russia, Armenia, Georgia (el. version) (Zhordania 1999; Gegechkori 2007).

The Persian or goitered gazelle (Gazella subgutturosa subgutturosa) or djeiran is found in northern Azerbaijan, part of Iran, southwestern Pakistan, southwestern Turkey and the Gobi Desert, in steppe, semi-desert, desert habitats and open juniper woodlands. In the South Caucasus, historically, the animal was found in the Aras Valley in Armenia, and in the eastern part of the Greater Caucasus Range. Today, the goitered gazelle remains only in the Kura Lowlands. At the end of 19th, 50,000-60,000 gazelles inhabited the open arid landscapes but in 1961 the population of this ungulate fell to catastrophically low levels, only 130 animals were left near the mouth of the Kura River. Some conservation measures, first of all the creation of the Byandovan Sanctuary in 1961 and the Shirvan Strict Nature Reserve in 1969, helped to save the population from extinction in the South Caucasus. Today within the protected areas, gazelle numbers have grown to more than 5,000 individuals. People began to introduce the animals into parts of their former historical range in Vashlovani National Park in Georgia and Azerbaijan (Borodin *et al.* (Eds.) 1984; Schmidt 2007. An ecoregional Conservation Plan for the Caucasus 2006) (Figure 41).

Figure 41. The goiterred gazelle roamed in a large area of the eastern Caucasus drylands up to the beginning of the 20th century.

In summer herds of goitered gazelle live in small, no more than 10 individuals, whilst in winter, large herds of dozens to hundreds or more (in different regions of cold and warm deserts of Asia) of animals congregate together. Mating is polygamous, females give birth to 1-3, usually – 2 newborns (Borodin *et al.* (Eds.) 1984).

Except for those mentioned, three more subspecies are recognized (*https://en.wikipedia.org/wiki/Goitered.gazella*).

The goitered gazelle is listed in the IUCN Red List as Vulnerable and in the national red list of Azerbaijan (Schmidt & Pietzsch 2007; ECPC 2012).

Saiga (Saiga tatarica). Two subspecies of saiga are recognized: *Saiga tatarica tatarica* (to which the majority of the global population belongs) and *S. tatarica mongolica* (endemic to western Mongolia) (Milner-Gulland *et al.* 2001). During the Ice Age of the Pleistocene the saiga as a part of a mammoth fauna ranged in the cold steppes from the British Isles through Central Asia and the Bering Strait into Alaska and the Yukatan. At the beginning of the 18th century, it was still distributed from the shores of the Black Sea through the edge of the northern Caucasus into Mongolia. In the 1920s the saiga's population was almost completely exterminated, but thanks to a strikingly high level of adaptation to extreme environments, after active conservation measures (limit hunting, creation strict Nature Reserves), they were able to recover and by 1950, two million animals were found in the arid landscapes of the USSR (Borodin *et al.* (Eds.) 1984). Since the break-up of the former Soviet Union there has been a dramatic fall in the saiga's population, due to uncontrolled hunting for meat and demand for horns, which are believed to be aphrodisiacs in Chinese medicine (selective hunting of young males). The next significant threat is the destruction of key habitats into traditional migration routes. An estimated total number of 178,000 saigas survive today in steppes, deserts and semi-deserts regions of south-eastern Europe and Central Asia (mainly in Kazakhstan) from the Precaspian open landscapes to Mongolia and China (Milner-Gulland (2001) and references therein).

Figure 42. The face of a stuffed saiga, with its bulging nose, the strangest of all antelopes. Georgian National Museum (2002).

The most notable feature of this animal is the inflatable, humped nose, which looks similar to that of a tapir's proboscis. This fleshly proboscis has a wide range of mobility and a unique internal structure with numerous hairs and mucous-secreting glands. The large nose is functional throughout the year, filtering out airborne dust during the summer migrations and heating the air before it reaches the lungs during the icy winters. Their diet includes herbs, lichens, plants containing salt and poisonous materials (Wilson & Reeder (Eds.) 2005) (Figure 42).

The saiga is a migratory species with widely separated summer (northern) and winter (southern) ranges. The species lives in large herds, usually up to thousand individuals. It has a strikingly high rate of reproduction and recruitment. In years of favorable climate, the population can increase by up to 60% in a single year.

As a species, the saiga is classified as Vulnerable by IUCN (Sokolov 1974; Milner-Gulland *et al.* 2001; Gegechkori 2007; Mallon 2008).

2.2.1.2.2. Foothill Deserts and Semi-Deserts with Saline Ecosystems – Discrete Habitats (Patches) of East Georgia – Patterns of Extraordinary Biodiversity

2.2.1.2.2.1. Introduction

Deserts and semi-deserts of the East Caucasus (the Kura-Aras Lowland and Northern Caspian Region) and the vegetations occurring in this biome are represented as two main ecological types – (1) plain deserts and semi-deserts and (2) foothill deserts and semi-deserts. At present the plain desert and semi-desert vegetations are most widely distributed in the eastern South Caucasus, in the Kura-Aras Lowland. In this geographical area typological compositions of this type of vegetation are relatively diverse.

Foothill (sub montane) desert and semi-desert vegetations (maximal elevations reach 400 m a.s.l.) mainly cover part of uplands and hilly badlands. According to Lachashvili and Khachidze (2010), the latter type of phytocoenosis is represented in East Georgia (e.g. the Iori upland, Vashlovani depression and hilly badlands, the vicinity of the David Gareja monastery complex, the Shio-Mgvime monastery complex, the environs of Kaspi town, the two sites of Meskheti – South Georgia) and Azerbaijan (Sheki upland and Gobustan (Kobistan) foothills).

In the southern part of the South Caucasus, in the middle reaches of the Aras River, an exclave of the Middle East deserts, eroded clayey badlands with a high content of gypsum are frequently found.

Distribution of foothill deserts and semi-deserts is azonal and is associated with specific relief-edaphic conditions. For example, phytocoenoses of saltwort (*Salsola nodulosa*) prefer sites with a high gypsum content. On the contrary, Artemisietum lerchianae found its shelter in areas where gypsum content is poor or completely absent and in stony-gravelly substrates, where soils are shallow to bedrock or have a stony subsoil, include generally wormwood (*Artemisia fragrans*) phytocoenoses. In Iori upland the foothill deserts and semi-deserts are penetrated in the arid open woodlands composed of the species junipers (*Juniperus* spp.) and pistachio tree (*Pistacia atlantica* (= *P. mutica*) (Lachashvili & Khachidze 2010).

The land type between the villages Aspindza and Rustavi includes very impressive rows of 'albino' clayey hills with hyper content gypsum. This almost revegetated area is settled with single specimen of *Peganum harmala* growing on the eroded slopes.

In the environs of Adigeni the same deserts and semi-deserts on gentle to moderately steep slopes intrude into other type of plant communities, open woodland consists of *Pyrus georgica* and *P. salicifolia*, therefore, the desert and open woodland vegetations in some micro-habitats are in direct contact with each other (Gegechkori, present work).

In total, 278 species of vascular plants were recorded in the deserts and semi-deserts of Georgia from which 93 species were found in the foothill deserts with 19 species (5.4%) endemic for the Caucasus (*vs* 162 species found in plain deserts and semi-deserts) (Lachashvili & Khachidze 2010).

Badlands are common in the research area of East Georgia, considerably in Vashlovani National Park (Chapter 2.4.4.2.). Soils are poorly developed and reflect the geologic parent material exposed by erosion. Vegetation is very sparce including *Salsola nodulosa, A. fragrans* and *Gamanthus pilosus* formations with poorly settled or revegetation sites. Phryganoid phytocoenosis on argillaceous (clayey) and argillo-aranaceous (clayey-sandy) badlands consist of *Reaumuria alternifolia, Caragana grandiflora* and *Atraphaxis spinosa*. Sometimes foothills are occupied by tragacanth plant communities (*Astragalus* spp.). On bare mother rock of sandstone (e.g., the cave town-monastery of Uplistsikhe) different communities of plant cover appear. This is a tomilliar type of plant species (*Ziziphora serpillacea, Thymus* spp.) (Lachashvili *et al.* 2005). A spectacular red tulip (*Tulipa eichleri*) and an outstanding with its color iris (*Iris iberica*) display the beauty of some of the ephemeroid spring vegetation.

It is worth noting that few taxa are unique to either habitats of foothill desert and semi-desert 'islands': the Aspindza-Rustavi patch until recent time was famous with *Nitraria schoberi*, The Gareji Hollow patch, represented with unique, for east Georgia, *Halimodendron halodendron*, Adigeni patch - with sometimes a mature canopy of endemic *Pyrus georgica*, in some clay-gipseous deserts and mixed clay-gipseous deserts and steppes habitats.

The valleys of sub montane rivers house riparian forests with already mentioned dominant tree species such as poplars, willows, tamarisks, shrubs (species of *Populus*, *Salix*, *Tamarix*, *Hippophae*, *Berberis,* etc.).

2.2.1.2.2.2. Patch of Painted Clay-Gipseous Deserts of Gareji Hollow (East Georgia) – Patterns of Extraordinary Ecotone Habitats

In East Georgia the driest - almost, or true desert habitats (*sensu* both sides of scientists pointed out their view regarding arid and sub-arid landscapes of the Caucasus) are found in the Kakheti province (Kiziki district), namely on the Iori Upland, in the flat Eldari Lowland and in the Gareji Hollow, where average annual precipitation varies from 250 to 350 mm (Lachashvili *et al.* 2007; Lachashvili & Khachidze 2010). Both endogenic and exogenic factors (inclination of strata, chemical and physical weathering, erosion and denudation processes) determine the peculiar terrain forms (turrets, pillars, skerries) and the typical colors, caused by intense sunburn of rock layers. This type of arid landscape is known as a painted desert (***пестроцветы*** – in Russian). It is represented in its perfect form in the Mio-Pliocene sedimentary rock strata in deserts and semi-deserts of Gareji Hollow (Maruashvili 1964; Gegechkori 1984a, 2008).

The length of Gareji Hollow from north-west to south-east reaches 2.5-3.0 km, by about 0.5-1 km wide and depth 200 m. It is one of the driest places in Georgia (Figure 43).

The hollow is delineated by the Udabno (Georgian - უდაბნო, desert in English). It ranges from the west with the highest peak of Udabno Mt (898m a.s.l.) and Iailaji and Dodo Ranges from south-east and north-east. The hollow resembles a geological fault, and a 'scar' of that can be readily seen from Georgian Orthodox David Gareja Monastery complex (which is an outstanding Medieval cave monastery monumental ensemble with Georgian inscriptions dating back to the sixth century, and original frescoes dating from 8-13 centuries) in uninhabited rocky mountains of Gareji (Vekua *et al.* 2012) (Figure 44).

Figure 43. Gareji Hollow - southernmost shelter of Irano-Turanian living organisms. Some of the natural factors impacting erosion, give the clay-enriched foothill desert their peculiar characteristics. In the foreground a tamarisk shrub growing in a dominant grassland landscape provides a visual artwork (Sagarejo district, E. Georgia) (2006).

A noticeable feature of this natural corridor of foothill deserts and semi-deserts is several hills continuously extended between the ranges′ walls. The hills of the hollow are built by Miocene-Pliocene fluvial (maritime, shallow fresh waters) and terrigenous sediments. They are characterized by rhythmically bedded sequences of thick sediments. The transition layers of hills consist of thick deposits (about 300-400m) of grey, purple, reddish, blue and green mudstone, sandstone, gypsum beds, conglomerate and breccia, patches of dark black and frequently round stone embankments deposited by river system. Salinized depressions are represented by patches of solonchak. The reddish and purple colors are created from oxidized iron and greenish hues come from white copper which provide the pigments for the various colors of the site. The reddish saline clay soil is frequently accompanied with striped layers consisting of whitish-grey volcanic ashes. Here and in patches of the stone embankments it is possible to encounter quartz, jasper, opal, plaster of Paris and other minerals (mineraloids) (Alekseev 1930; Andrianov & Larin 1935; Markeevich 1954; Buleishvili 1964; Tsereteli 1941 and others cited in Vekua *et al.* 2012).

Spectacular scenery of the dominant reddish and grey colored bends across the landform is typical of most area in the hollow as seen from the dirty road across the monastery area. During millions of years strong wind carried away dry soil. Therefore, among the most striking desert landforms – aeolian feature – were created by the chemical weathering processes. At the same time the temperature – induced weathering of nude rock-layers, another type of erosion occurred. The erosion of local rock layers has resulted in the formation of the colorful badlands and painted hills topography of the hollow (Tsereteli 1941 in Vekua *et al.* 2012).

In recent times the human impact via over-farming and over-grazing plays a significant part in the continuing problem of desertification in the Kakheti (Kiziki) district including Gareji Hollow (Vekua *et al.* 2012).

Within multi-colored layers of the Gareji Desert in the late 30s and early 40s of the 20th century, paleontologists discovered fossil remains of wildlife of the Mio-Pliocene time.

Dryopithecus garedziensis should be mentioned here (fossil teeth of ancient primates), a representative of one of the relict populations of its genus (Burchak-Abramovich & Gabashvili 1945 in Gabunia *et al.* 2001).

Other mammalian fossil remains (lower jaw) include *Deinotherium giganteum*. This 'terrible' large prehistoric proboscidean, a relative of modern-day elephants, appeared in the Middle Miocene and survived until the Early Pleistocene. *D. giganteum* the type species of the genus was the only species known from the circum-Mediterranean region (*en.wikipedia.org. /wiki/Deinotherim*).

An assortment of other fossilized prehistoric mammals is found in the hollow and include *Hipparion garedzicum, Udabnocerus georgicus* and others (Tsererteli 1941 in Vekua *et al.* 2012; Bukhsianidze, 2015).

All the above mentioned and other (invertebrate animals) massive treasure–trove of ancient fossils indicates that the climate of the Miocene was humid and tropical-subtropical, but in mid-Pliocene the latter biome was largely replaced by savanna landscape (Vekua *et al.* 2012).

In 80s of the 20th century signs of ancient habitation sites of early humans were discovered in the Gareji Desert. According to remains of an iron furnace, it has been described as one of the most significant discoveries of Iron Age metalwork in East Georgia. About 500-600m away from the metalwork area there was a rich oxidized iron ore (FeCO), which existed close to the surface of the ground in Gareji Desert (Natsvaltskali area) (Mgeliashvili 1991 in Vekua *et al.* 2012).

According to relevant published works and our long-term multidisciplinary studies from the 60s of the 20th century until today, Gareji Hollow should apparently be classified as a micro-refugium that supports some life forms mostly of Irano-Turanian origin in almost extremely arid deserts. They survived during post climate oscillation found their extreme eastern ward (from Turan Lowland) shelter in this site. At the same time this refugial site protected high genetic diversity and endemism of living organisms frequently in the area near the edge of the geographical distribution. The desertified hollow with adjacent David Gareja Cave Monastery should be prioritized as a new segment of protected neighbouring Vashlovani National Park. Gareji Hollow is a natural large desert patch, which values for its geological and geographical peculiarities, providing valued habitats for wildlife. Geomorphologically it is a highly heterogenous landscape and includes scientifically important biological and non-biological components. Surrounded by rugged mountains and ranges its extraordinary foothill desert and slightly expressed riparian ecosystems sustain a unique web of life. If we take into consideration the fossil materials, which have been discovered at this site, the open bare landscape of the Gareji Hollow with many textures and colors in its relief should be considered as a natural museum under the open sky (Figure 44).

Historically, transgression event of the waters of the shallow Sarmat Sea, depending to geological times, corresponded to cold and semi-arid climatic regimes.

This event is largely responsible for creating the bare bottom of the former sea covered in the Pliocene with savanna and later, in Pleistocene, desertified arid landscapes including today's hollow patch (and micro patches within it) of the Gareji Hollow. Abiotic events of ice ages starting about 2 million years ago greatly increased the excavation of the Gareji Hollow. During these collisions of the climate and relief, part of past semi-arid landscapes plant community assemblages (grassland, semi-desert) occurred within the 'geological fault' of the Gareji Hollow. From this time, it responds to fluctuations (mostly regarding a desertification trend) in an isolated 'island' site transforming semi-arid life forms into semi-arid and desert climate-tolerant living organisms. Therefore, during millennia it became a transition zone between local desert and adjacent to the Hollow semi-arid habitats. Overall, the hollow today needs to be viewed as a transition zone - an ecotone (edge effects) between two major landscape units – desert and semi-desert landscapes on the one hand, and adjacent grasslands and open arid woodlands on the other hand.

Figure 44. David Gareja Cave Monastery ensemble (the 6th century – the diadema of Georgian desert Cave monasteries with painted deserts in the background (1967).

The Gareji Hollow as an ecotone is characterized by a higher severity of climate, fire frequency and absence of fertile soils. On the other hand, we can see a high heterogeneity of relief created by numerous micro patches. Within this natural patch mosaic, each patch pattern favors independent, drought-tolerant plant communities and patterns including its own web of life, in general. At the same time the Gareji Hollow is distinguished with very restricted but shade tolerant species in a riparian-stream habitat (temporarily or seasonally small rivers), which evidently enrich the local biota. Gareji Hollow with other desert foothills represented in South Georgia (Aspindza-Adigeni municipalities) with their geomorphological structure and characteristic living organisms should be referred to as discrete habitat patches, frequently with allopatric speciation events and pattern of extraordinary biodiversity characteristic to each patch.

The modern vegetation of Gareji Hollow is found in the foothill desert. The latter is characterized by badlands. They are composed of the Upper Tertiary eroded saline clays, frequently with a high content of gypsum. Painted deserts of the hills are characterized by bare saline soil where no plants are found or solitary individuals grow at a large distance from each other. Some exceptions occur in depression among hills with comparatively deep soil layer or in the intermittent fringes on the banks of the seldom and small brooks, – where marine and terrestrial ecosystems intersect, - populated mainly by tamarisk shrubs (Ketskhoveli 1960; Lachashvili & Khachidze 2010) (Figure 44).

Within the Gareji Hollow and its environs biodiversity of flora and ecosystems in general consisting of xerothermophilous (xerohalophytes) species has been determined by the diversity of physico-geographic conditions, geological history, and habitat diversity of this site. Desert phytocoenoses are represented mainly on the slopes of badlands and hills. Species typical at dry or disturbed sites are common with *Salsola nodulosa, Artemisia fragrans, Gamanthus pilosus*, *Atriplex cana* and more rarely *Reaumuria alternifolia*. Syntaxomically they are ranked as Salsoletum nodulosae-artemisiosum lerchianae, Salsoletum nodulosae-reaumuriosum alternifolianae, Salsoletum nodulosae-artemisioso-ephemerosum, Salsoletum nodulosae-tamaricosum ramosissimae, Atriplexeta canae, etc. (Lachashvili & Khachidze 2010).

Figure 45. Disjunct stand and unique for eastern Transcaucasia shrubs, *Halimodendron halodendron* (monotypic genus of Fabaceae) in Gareji Hollow (2012).

Figure 46. Spiny atraphaxis (Polygonaceae), one of the most drought-tolerant shrubs. The Gareji Hollow (2017).

The most significant feature of vegetation species of local desert flora is the presence of common (silver) salttree (*Halimodendron halodendron*) (Figure 45). Gareji Hollow is the rarest site in East Georgia, where the species of monotypic genus of legume (Fabaceae) occurs. It is closely related to the genus *Caragana*. The range of this deciduous thorny shrub (0,5-3m tall) encompasses Middle (Central) Asia with Mongolia, Pakistan, Afghanistan, Iran, Turkey and Transcaucasia (Middle reaches of the Aras River, i.e., southern Armenia and Azerbaijan - Nakhichevan), south of European part of Russia, and south of Ukraine. It can be found on other continents as an introduced species, where it is often represented as a noxious weed. This is a deciduous spiny shrub up to three meters tall. It inhabits salt steppes, dry sandstone slopes, saline flood plains; leaves silvery-green, flowers in clusters, pink pea-like blossoms, scented; hermaphrodite flowers in June-July, melliferous; the fruits – inflated pod - ripen by the late summer-early autumn; ornamental (*https://en.wikipedia.org/wiki/Halimodendron*).

Phriganoid vegetation of Gareji Hollow is represented by *R. alternifolia, Atraphaxis spinosa, Rhamnus pallasii, Caragana grandiflora* (Figure 46); tragacant plant phytocoenosis composed of huge specimens of *Astragalus caucasicus, A. microcephalus* and *Acantholimon fominii*. They are very characteristic on dry sandstone slopes. Less saline sites are occupied by steppe dominant grass *Bothriochloa ischaemum*. The shrub layer is characterized by local occurrence of *Paliurus spina-christi* and *Tamarix ramosissima*. Steppe grassland vegetation with dominancy of *Stipa tirsa* co-dominants being *Glycyrrhiza glabra, Medicago coerulea* (Lachashvili & Khachidze 2010) is continuously present throughout surrounding vast areas of open land of the hollow.

In the northern side of the Gareji Hollow species of riparian forest fragments – azonal sites – as frequently isolated localities nearby of a natural water deficit ('brook-bedsides') in local dry areas and habitats consist of species of *Populus, Salix, Tamarix* and *Elaeagnus*. In the same azonal sites spontaneously spreading endemic species of Kiziki area of Kakheti district Georgian elm (*Ulmus georgica*) makes the Gareji Desert unique. Here occurs *Salvia garedji*, another plant species with restricted range (Georgia, Azerbaijan) (pers. observation).

Figure 47. The Georgian almond (*Amygdalus georgica*), endemic to eastern Georgia, Gareji Hollow (2004).

In some patches of the Gareji Hollow and the surrounding area of the David Gareja Monastery complex of Gareja Mount, its rock outcrops and dry slopes with shallow soil are covered by arid open woodland of relict junipers (*Juniperus foetidissima, J. polycarpos*). Deciduous trees occur by *Celtis caucasica*, sometimes growing as huge specimens which extend on a monastery area and, therefore, were probably influenced by humans (native monks). Celtisetum with an admixture of arid landscape wild pears is usual. Mention should be made of *Pyrus salicifolia* and endemic species of Kiziki *P. sachokiana* and endemic of East Georgia *Amygdalus georgica*. Among understory shrub *Spiraea hypericifolia* is common with an admixture of *Berberis iberica, Ephedra procera, Punica granatum* and others (Lachashvili & Khachidze 2010, pers. observation) (Figure 47) .

Between hills the ground land communities are dominated by species of *Bothriochloa, Festuca*, etc. (Sokhadze & Sokhadze 1991).

2.2.1.2.2.3. Patch of Clay-Gipseous Soil Deserts of Aspindza-Rustavi Environs (Samtskhe (Meskheti)–Javakheti, (East (South) Georgia)

Unlike the middle reaches of Aras River or the eastern part of the north slopes of the Greater Caucasus Range (Dagestan), the original type of halophilous vegetation, which is spread from 400-500 to 1,000-1,500 m a.s.l., is quite rare in the montane zone of Georgia. Within this altitudinal zone, between 1,200-1,400m a.s.l., in the upper part of the Kura (Mtkvari) River is represented a halophytic relict stand, an insular patch of desert community, surrounded by more hospitable habitats (mainly shibliak shrubland). This is a typical insular desert plant community with a striking, for this region, species composition.

Figure 48. The chain of hilly patches of clayey gypseous soil desert of Aspindza-Rustavi settlement approaches (S. Georgia). General view (2012).

Halophytic (xerohalophytes and halo-gypsophytes) vegetation phytocoenoses are developed in southern Georgia's district of Samtskhe-Javakheti (Aspindza-Rustavi environs), on the south and south-eastern slopes of the Takna-Tash Mountain, on badland, on the gypsous basic rocks and on salty clay soil. Badlands occur on various slopes of nearly all exposed salty clay soils on both banks of the Kura River, but a chain of hills covered by massive white and whitish-pink gypsum rock outcrops is especially represented on the left river bank near v. Rustavi and v. Aspindza eastern part of Akhaltsikhe depression (Khintibidze 1990; Lachashvili & Khachidze 2010) also northward from these village, near the town of Adigeni, Kura (Mtkvari) River Gorge (Figure 48).

Gypsum ($C_aSO_4.2H_2O$) is a common mineral in semi-arid and arid environments. Gypsum rich soils are either of pedogenic or geogenic origin, i.e., these types of soils could be attributed to geological or pedogenic processes (Shahid & Abdelfattah 2009).

Except common habitats of halophytes with soil salinization to different extent and alkaline lands of different genesis including takyr-like surfaces, they predominate also on relief of Tertiary-Cretaceous periods – gypsum deserts (Mamedov et al. (Eds.) 2009).

Halo-gypsophytes of Meskheti with other halophytes are represented by 39 species, among which should be mentioned *Nitraria schoberi*, *Reaumuria alternifolia* (=*R. kuznetzovii*) (endemic of Meskheti and Kartli, East Georgia), *Ceratocarpus arenarius, Camphorosma monspeliaca, Krascheninnikovia ceratoides* (=*Ceratoides papposa*), *Gamanthus pilosus, Peganum harmala,* etc. (Figure 49).

Above 1,400 m, the halophyte plants on the white and white-pink hills of Aspindza-Rustavi environs are replaced by shibliak formation consisting of *Cotinus coggygria, Atraphaxis caucasica, Astragalus microcephalus, Rhamnus pallasii, Berberis iberica* and so on. A very rare species growing on the surface of dry ravine is *Rhamnus spathulifolia* which is considered to be of hybrid origin (Bobrov 1946, Kikodze 1967, Khintibidze 1990).

Figure 49. Slope of a clayey gypseous hill populated with *Reaumuria alternifolia* (=*R. kuznetzovii*) (Tamaricaceae) shrubland. Aspindza environs, S. Georgia.

This tiny patch of hyper - and euxerophytes (Reamurioso-salsoleto-comphorosmetum) is disjunct hundreds of kilometers from widely distributed Nitrarietum shrubland of Azerbaijan.

The accumulation of mentioned genera and several species of a wide range within the arid habitats of eastern Palearctic and phyletic links with the same taxa in Middle Asia and Central Asia are considered to support the pre-Pleistocene origin of their distributions, associated with the Meskheti desert and semi-desert habitats over gypsiferous soils in the area since the Late Tertiary (Khintibidze 1990).

Two other patches of halophyte complexes participant in the first stratum of *N. schoberi* can be found in East Georgia – Shida (Inner) Kartli (near Kaspi town), and Iori Plateau (r. Alazani gorge) (Kikodze 1967; Lachashvili & Khachidze 2010).

These three isolated, probable relict communities, separated from each other geographically, but once characterized by a non-disjunct range have great importance in the study of South Caucasus biota formation history.

The recent excursions between the towns of Aspindza and Adigeni, the village Rustavi and the surrounding hill-lands in 2009-2018 show that because of continuous overgrazing by livestock during each season, significant destruction of the unique plant cover represented on gypsum containing hills of Meskheti occured. This negative anthropogenic effect on vegetation has been caused either by a lack of ability of regeneration of characteristic semi-shrubs (e.g., *Reaumuria*) or disappearance from the vegetation cover. According to our observation the former population of *N. schoberi* in this area from the beginning was so extremely small – a typical example of population (genetic) bottleneck – that under the excessive numbers of animals per hectare of pasture of local rangeland and a 'vivid' interest of botanists (!) this shrub became extinct during 1960s to 1970s (Gegechkori 1975b). Today such attractive and charismatic, for South Georgia, white gypsous hills of Meskheti are abandoned. In the same desert patch, near the mentioned hills, on the saline ground, one usually can find vegetation cover, composed of pure *Peganetum* (*P. harmala*), or badlands covered by sparce population of *Zygophyllum fabago*, which livestock do not eat (Figure 50).

Figure 50. Clayey gypseous hill of Aspindza environs, S. Georgia, with dense herb layer consist of wild rue (*Peganum harmala*) (Nitrariaceae).

2.2.1.2.2.4. Patch of Clay-Gipseous Soil Deserts of Adigeni (East (South) Georgia) Foothills and Its biota

Prominent geological features of the Gareji Hollow are outstanding among arid and semi-arid landscapes of Georgia. But the hollow has a more restricted analogue in the southern part of this country. The low mountain chains of Adigeni Municipality are built of volcanic colored mudstones, sandstones, andesites and basaltic andesites. They are typical, picturesque hilly patches of clayey and clayey-sandy (frequently with a high content of gypsum) eroded slopes and badlands with a lack of real soil cover. They are represented near to the small town of Adigeni. Geographically they belong to the western part of the Akhaltsikhe Hollow (in Samtskhe (Meskheti-Javakheti), in the Kvabliani River Gorge (1,220-1,400m a.s.l.), left tributary of the Mtkvari River. These reddish and purple colored hills are built by Eocene gypseous clayey-sandy layers and Olygocene mudstone and clayey-sandy layers (Maruashvili 1964). Frequently this area is represented by mosaic patterns of habitats: arid open woodlands consisting mostly of tree-like xerophilous Georgian pear (*Pyrus georgica*), shibliak and steppe (pers. observation).

The clayey and clayey-sandy hills of Adigeni environs represent a geomorphological phenomenon which is essentially an isolated ecotone represented on both sides of riparian forests of the Kvabliani River. This is quite an exotic site even within rich arid/sub-arid environments of Samtskhe-Javakheti. The hills of Adigeni vicinity consist of steep slopes of badlands composing clay-rich soils ended by rounded peaks frequently broken by gorge-like canyons. This area of badlands and clay-rich soils is completely useless for agriculture. Nevertheless, these hills with the similar disjunct patch of deserts and semi-deserts near town of Aspindza (occurring 59 km northwestern from Adigeni) are of major interest to the biogeographic science viewpoint. Unlike almost de-vegetated slopes of gypseus hills of Aspindza – the village of Rustavi surroundings even the steepest part of windblown badlands of Adigeni is dotted by unusual dwarf and sparce populations of forage kochia (*Kochia prostrata*). Higher up on adjacent rounded peaks, forage kochia partially avoiding severe wind erosion, grows as larger shrubs bowed down with numerous cone like shells of land snail (? *Micropontica* sp.) (Figure 51).

Figure 51. Hilly patch of clayey and clayey-sandy desert (Adigeni environs, S. Georgia) provided with a cover of dwarf shrubs of the *Kochia prostrata* (2014).

The predominant vegetation of terrigenous sediments of the tops of rounded hills and gorge-like canyons is endemic to the Caucasus wild pear (*Pyrus georgica*) from the pear tree complex (sect. Argyromalon Fed. Ex Kutateladze) (Gagnidze 2005) of arid landscapes. In the Adigeni hills, the Georgian wild pear (a small tree, from three to seven meters tall) forms open woodland in a large area. On the contrary, depending on topography, open woodland communities, dominated by tall specimens of *P. georgica,* are found in gorges among the hills but primarily along the bottomlands of the canyon-like gorges which usually are replaced by clayey and clayey-sandy hills with scanty vegetation.

In semi-arid and arid ecosystems, for a variety of reasons, the density, height and fruit crop of wild pear are different. In the higher, abandoned part of the hills, pear stands are more dwarf-like. This is the most xeric shrub-like open woodland.

In the bottomlands of gorges recovery processes of wild pear stands occur. It should be stressed two main reasons of the mentioned circumstances. Firstly, there is an underground water table and, secondly, comparatively little disturbance by overgrazing of these micro-habitats and surrounded environments by livestock herds as in a traditional pastoralist region (there is no tradition of goat herd keeping in Samtskhe-Djavakheti).

The wild pear is rich with fruits which attract numerous consumers including mammals (badgers, foxes, wild boars reaching hills from adjacent riparian forests and other habitats) and birds (jay, blackbird, magpie and others). On the open areas of desertified hills the fruits remained hard (or less soft), small and characterized by less palatability, while in most of the bottomlands and neighboring areas in later summer and early autumn the larger fruits of wild pears ripen, similarly to domestic ones, which are distinguished with sweetness and flavors edible for human beings.

In this disturbed 'island' of hilly deserts and semi-deserts, *P. georgica* is accompanied by co-dominant shrubs: *Crataegus orientalis, Lonicera iberica, Cotoneaster integerrimus, Berberis iberica*, etc. Palatable deciduous leaves and fruits have all of the mentioned co-dominants but the largest fruit crop in the mentioned habitat has wild pear. Hence, this tree within arid community of open arid woodland and partially clayey-gypseous hills in general produces much more valuable trophic resources in the form of large fruit crops during a summer drought.

This complex of wild fruit-bearing arboreal plants, relying on comparatively useful habitats, should be considered as a climax stage of the general trend of vegetation development in Adigeni environs badland habitats with saline soil.

The halophytic herbaceous over ground layer of Adigeni environs hills includes species of solonchak communities: *Ceratocarpus arenarius, Camphorosma monspeliaca, Limonium meyeri, Peganum harmala, Zygophyllum fabago* and others (Figure 52).

It is noticeable that the local phytocenosis lacks a unique, for East Georgia, relict of desert and semi-desert communities, *Nitraria schoberi* represented until over the last decade in neighboring Aspindza – Rustavi clayey-gypseous deserts and semi-deserts.

Figure 52. *Zygophyllum fabago* (Zygophylaceae) growing on a bare slope of clayey-sandy hill. Adigeni environs, S. Georgia (2018).

2.2.1.2.2.5. Some Features of Solonchak Deserts of the Kura-Aras Lowland Shown Through the Psyllid (Hemiptera, Psylloidea) Fauna

Until our research there were no papers dealing with Kura-Aras Lowland's psyllid fauna.

On the basis of original data collected between1972 and 1975, a brief overview is presented of the present knowledge of the psyllid fauna of the Kura-Aras Lowland from the faunistic and biogeographical points of view. The author has investigated the Shirvan, Mugan, Salian and Mil deserts and semi-deserts of Azerbaijan. The psyllids of investigated regions include at least 62 species (Gegechkori 1979). The updated fauna of the Caucasus, as mentioned, includes about 212 species (Gegechkori 1969, 1979, 1987a). In their geographical distribution, psyllids are strictly linked to their host plants, but some exceptions may occur to this general rule. For example, the majority of psyllid species often shows a more restricted dispersion, compared to their host plants, appearing to be frequent mostly in areas where their hosts have a center of origination or feeding plants growing abundantly (Gegechkori 1984a, 1985).

The best confirmation of the above mentioned are deserts and semi-deserts fauna of lowland and foothills of the Kura-Aras Lowland.

Much attention in the Kura-Aras Lowland has been given to the psylloid fauna that exist in the area represented by hyper- and euhalophytes. Psyllids of these communities are characterized by low species diversity, but very high specificity. All of them by their feeding habits are linked with halophytic plants.

From the Aphalaridae family an e particularly characteristic species is *Rhombaphalara halocnemi* (Loginova), which feeds on *Halocnemum strobilaceum*. A poor population of this species is known from Shirvan deserts and semi-deserts; lower populations occur from the same habitat of Ciscaucasia. It should be mentioned that within the geographic range of *H. strobilaceum* its consumer - *Rh. halocnemi* continuously decreases westward and is completely lacking at the western border of its range-boundary of two countries – Azerbaijan and Georgia.

In many areas of Kura-Aras Lowland another especially characteristic species is *Caillardia anabasidis* (Loginova) trophically linked with *Anabasis aphylla*. This species sporadically and with decreased population occurs in middle reaches of the river Aras (Nakhichevan) bordering on Iran. The presence of a population of *C. anabasidis* should be sought after in the extreme western part of the range of its host plant, where the occurrence of *A. aphylla*, as it was already mentioned, was discovered in recent times (Lachashvili & Khachidze 2010).

In the solonchaks of South Caucasus and especially in Kura-Aras Lowland the most common species is *Rhodochlanis orientalis* (Loginova) which feeds on species of *Suaeda,* and *Rh. salicorniae* (Klimaszewski) – a wide olygophagous organism occurring on different species of *Salicornia, Petrosimonia* and *Suaeda*. On the contrary, an insignificant population is represented by *Rhombaphalara halostachydis* (Loginova) feeding on *Halostachys belangeriana*, which is restricted to the extreme eastern part of Shirvan deserts and semi-deserts.

The stands of *Tamarix* provide a particular habitat for many insects and other arthropods including psylloid fauna. It should be mentioned these include species of the genera *Colposcenia: C. aliena* (Löw), *C. jakowleffi* (Scott), *C. kiritschenkoi* (Loginova 1960). However, the most striking Turanian species feeding on tamarisk – *Crastina dimorpha* (Bajeva) – was not found in such continuously solonchak deserts of Ciscaucasia and Kura-Aras Lowland. It was restricted to a comparatively minute area of deserts and semi-deserts area of Armenia (Oktomberian district). This species causes severe galling of the stem tissue of host plant and shows visible damage, hence to recognize *C. dimorpha* in field conditions is easy. Therefore, *C. dimorpha* lives allopatrically in a very restricted area of the South Caucasus, and on the other hand, populates far more vast area in Middle Asia (Tadjikistan, Turkmenistan) at the same time widening its host plant range on *Myricaria bracteata* (= *M. alopecuroides*) (Gegechkori & Loginova 1990).

The case of extraordinary disjunct distribution of *C. dimorpha*, leads the author to speculation supporting the continuity of its presence in the northward area, including the Kura-Aras Lowland's solonchaks throughout the Quaternary. But several times transgressions of recent geological time might have had a greater impact on the biogeographic history of the Caspian Sea coastal and adjacent areas' organisms, unlike the deserts of southern Armenia, free of transgressions. *C. dimorpha* is third after *Rh. halostachydis*, a species with such a limited range in deserts and semi-deserts biomes of the South Caucasus.

Within the mentioned biome, to the most widespread species belong *Heterotrioza dichroa* (Scott) and *H. obliqua* (Thomson) (fam. Triozidae) occurring on the species of *Atriplex* and *Chenopodium*. However, *Heterotrioza kochiae** (Gegechkori) and *Eumetoecus kochiae* (Horváth) appear to feed on *Kochia prostrata*, whereas *Cyamophila glycyrrhizae* (Becker) (fam. Psyllidae) feeds on *Glycyrhriza glabra*. Unlike the Artemisietum habitat, psyllids on *Kochia* in solonchak deserts are met sporadically.

Prostrate kochia (*K. prostrata*) by its nutritive quality as underlined, is a valuable Amaranthaceae forage in semi-arid and arid regions of the former USSR including the eastern Caucasus (Balyan 1972). The same should be said concerning this introduced and potentially useful shrub in the semi-arid intermountain west, USA (Moore et al. 1982). If the damaging potential of *H. kochiae* and *E. kochiae* is not clear enough for solonchak deserts of Ciscaucasia and Transcaucasia, then both species and *Heterotrioza bathiashvilii** (Gegechkori) in the Vashlovani Reserve (East Georgia) are rather important pests in semi-deserts with the dominancy of *Artemisia fragrans* and arid open woodland, where prostrate kochia frequently occurs as far more concentrated stands.

From the 62 species of psyllids collected in Kura-Aras Lowland, the majority of species (more than 50) are represented within this region, occurring in habitats and differ from those of solonchak proper deserts, for instance, riparian forests and arid open woodlands (Turian-Chay State Reserve, Azerbaijan).

The richness of psyllid species on host plants of solonchak proper deserts increases from the eastern Caucasus to Turan (Middle Asia) and Central Asia (Kazakhstan and Mongolia). For example, the genus *Rhombaphalara* Log. in the eastern side of the Caspian Sea includes three species, instead of the two species of the Caucasus (3/2), genus *Caillardia* Bergevin has seven species in Central Asia (sensu lato) and one species in the Caucasus (7/1). Six more species in the former region feed on *Haloxylon*, which has not penetrated within the deserts of the Caucasus. Genus *Rhodochlanis*, correspondingly, 4/2/; the genus *Colposcenia*: 10/7; *Crastina*: 4/2 (taking into account species of psyllids found only in arid habitats); species of *Heterotrioza* feed on *Atriplex/Chenopodium*: 6/4; species of *Heterotrioza* and *Eumetoecus* occur on *Kochia prostrata* in semi-arid and arid habitats, a *vice versa*, decrease from Turan/Central Asia to the Caucasus: 2/4 (according to materials represented in Gegechkori & Loginova 1990).

The same data for Iran and the Caucasus, *Rhombaphalara*: 0/2, *Caillardia*: 5/1 (all five species common with Turanian fauna feed on *Haloxylon* spp.), *Rhodochlanis*: 1/2, *Colposcenia*: 7/7 (three species are endemic to Iran), *Crastina*: 1/2; *Hererotrioza* (occurs on

Atriplex/Chenopodium): 2/4, *Heterotrioza* and *Eumetoecus* (feeds on *Kochia*): 0/4 (data with respect to Iranian fauna, which seems to be incomplete, has been used from Burckhardt & Lauterer 1993).

The possible way of migration of desert fauna from the Irano-Turanian center in the geological past (Gegechkori 1984 a,b, 1985), show that expansion of some desert proper genera or their species into the Caucasus is closely connected with palaeogeographical events.

This expansion was restricted at the first stage from Iran concerning uplifting of the Zagros Mountains in the Late Miocene or Early Pliocene, and then from Turan with respect to climatic events. There is a higher level of precipitations in arid habitats of the eastern Caucasus *vs* low moisture limits of Central (Middle) Asia in Late Neogene. Therefore, some true desert taxa consequently drop out from the hyperxerophilous fauna of the Caucasus. For example, despite its presence in arid regions of the Caucasus following ancient deserts, a genus, such is *Nitraria* – host of *Brachystetha* Loginova with three species in Turan, this shrub in East Caucasus is lacking host-specific phytophagous group of insects – psyllids.

2.2.1.2.2.6. Patterns of Biogeographic Filter Effects – Evidence of Impoverishment of the Turanian Biota in Drylands of the Eastern Caucasus

Deserts and semi-deserts of the Caucasus constitute the extreme western (from Turanian Basin) and northern (from arid regions of Iran) extension of the natural range of plant and animal taxa of the Irano-Turanian origin

Depletion of characteristic floras and faunas of solonchak and sand deserts of the eastern Ciscaucasia and the core area of saline deserts in the eastern Caucasus should be explained by patterns of biogeographical filtering. In other words, the mechanism of environmental (habitat) filtering affects species immigration (ineffectual colonization) from the Turanian center of biodiversity. Identifying the causal factors that determine filtering barriers should be mentioned by some major causal explanation. According to plate tectonic theory, Afro-Arabian and Eurasian modern plates' collisions took place from mid-Miocene until today. As a result, by the Late Pliocene (4-3 mya ago) the Caucasus Isthmus has been formed (Gegechkori 1984a). Taxa of plants and animals of distinct ecological demands had occupied the entire Caucasus. Arid elements of the Turanian biota entered the Caucasus in waves of migration over the Volga River Delta located in the Caspian Depression, the northwestern shoreline of the Caspian Sea (Russia's Astrakhan Oblast') (Gegechkori 1984a). In this direction of movement to the adjacent isthmus area (this trend of migration continuing currently) was represent by a comparatively narrow landform (the narrow-bottleneck biogeographic site) of the north-eastern Ciscaucasia with a web of ecotonal transitions. The presence of the deserts' origin plants and animals shows that dispersal of the Turanian organisms across the mentioned biogeographic filter has occurred.

Such a kind of filter environment has been established mainly due to following two general reasons: (1) the solonchak and sand deserts occur on the westernmost range edge of the huge deserts of the Middle Asia generally provide important habitat filtering effects; (2) The Caspian Sea's frequent level fluctuations (transgression and regression) in the Holocene (Gegechkori 1984 a, b, 1993).

Salt-desert shrublands as a rangeland. As mentioned above, shrubs of salt-deserts are generally better suited to harsh conditions of semi-deserts and deserts than grasses or forbs because of their deep root system. Therefore, these shrubs have a high nutritive value in winter.

Consequently, salt deserts are an excellent winter range for winter cattle and especially sheep grazing. Today most of the salt-deserts of the Eastern Caucasus lowland foothills have been converted to winter farmland. This landscape is a good winter habitat for goitered gazelle (Schmidt 2007).

In some areas of the region the rangeland is heavily infested by *Peganum harmala* (fam. Nitrariaceae). This perennial herbaceous plant is considered as noxious and maybe the worst weed of rangelands of many countries with dry-grasslands and saline waste areas, and run-down pastures caused the deterioration of their forage value. It is toxic and unpalatable to grazing of wild animals, and is toxic to humans. Once established, the weed is persistent, dominant and difficult to control (Flora of the USSR, vol. 15 (1949); Abott *et al.* 2006; Dugarjav *et al.* 2010).

2.2.1.2.3. Deserts and Semi-Deserts of the 'Southern Type' – the Middle Reaches of the Aras River Valley – a Brief Review

The deserts and semi-deserts of the middle reaches of the Aras River Valley (in the borders of South Armenia, Nakhichevan AR, Azerbaijan) – the landlock region – is considered as a biome of deserts and semi-deserts of the 'southern type' (Grossheim, 1948). 'It seems, as pointed out by Grossheim (1948), such a type of arid landscape is the most northern outpost of the plant communities' characteristic to the countries of eastern Mediterranean and Asia Interior.

Accordingly, the deep river gorge of the Aras River is believed to be transitional, a crossroad, i.e., a bridge terrain between North Turanian Lowland (with Caspian, Aral, Balkash coastal and adjacent areas sites) and northern Iran and through this latter region – the Near East. The Aras River Valley with characteristic plant communities (see below) should be exemplified as the north-easternmost exclave of the Near East region.

These 'southern type' of deserts and semi-deserts are nutrient-poor represented on the pebble-gravel, pebble-sand and sandstone soil-grounds. Phytocoenosis with *Achillea tenuifolia* association are characterized by a clear predominance in the Yerevan Hollow, and *Stipagrostis (=Aristida) plumosa* – in the Djulfa Hollow. According to Grossheim (1948), incredibly rare species can still be found in this region. Among others, should be mentioned communities with *Stachys inflata, Thymus kotschyanus, Salvia dracocephaloides* (fam. Lamiaceae) (tomillares community). A sparce community of plant cover is positively correlated with ground bareness. Typical feature of native phytocoenosis is a rich annual (ephemers) vegetation but seldom participation species of Poaceae. On the slopes of foothills and low height mountain belts thermophilus communities – hammada and xerophytic cushion thorn phrygana (phrygana plants) are developed.

In conclusion it should be pointed out that the harsh environments of the Aras River Valley in the South Caucasus provide a stable refugium for many plant and animal species.

Detailed phonological investigations of psylloid (Hemiptera, Homoptera) fauna of the Aras River Valley - patches of sandy and solonchak deserts, very dry rocky habitats with hammada, frigana, tomilares, cushion-thorn communities, and tugai type vegetations of rivers, revealed harsh climate-tolerant complex of psylloids – 130 species (61% from about 210 species known today in the Caucasus).

These data with 36 species (27%) inhabiting exclusively drylands of this region, and 13 (10%) of the Caucasus new for science species, which were described by the author, obviously prevailed on the same data of any other regions (including the other desert regions of the Caucasus). Richness of xerophytic plants support the considerable generic and species richness of native Psylloidea. It should be mentioned psylloids associated with *Calligonum* in the minute area of the patch of sandy desert of Goravan Sands, Armenia (3 genera, 4 species – unique for the Caucasus. Local stony habitats constitute the eastern limit dispersal area of relict populations of the *Diaphorina kopetdaghi* Loginova (found only in this region) which feeds on *Zygophyllum* (*Halimiphyllum*) *atriplicoides*). This species lives allopatrically in a very diminished area in the Caucasus, and after a large distribution gap, presents in the Middle Asia (Turan Hollow) and Iran (Gegechkori & Loginova 1990). The fauna of the tugai environment includes species associated with *Populus euphratica* (*Egeirotrioza gracilis* (Bejeva), *E. verrucifica* Loginova, *Camarotoscena unicolor* Loginova), reaching the northern limit of their geographic range (from adjacent Iran). All of these and closely related species have now been demonstrated to be far more widespread in Middle Asia, but absent in other regions of the Caucasus. Autochthonous oreoxerophylic species of psyllids of the Aras River Gorge reveal ancient pathways of connections with those of the Anterior Asia. Example are *Cyamophila astragalicola* (Gegechkori) (host species – cushion species of *Astragalus*), described from southern Armenia and later found in Iran; *Aphalara grandicula* (Gegechkori) on *Rumex scutatus* is replaced in Iran by close (vicariant) species *A. loginovae (*Burckhardt & Lauterer). There are several other stenochorous and xerophilic species. For example, species of *Psyllopsis* (*P. machinosus* Loginova, *P. repens* Loginova, *P. securicola* Loginova) feed on the ash tree (*Fraxinus* spp.). These species are found on both sides of the Aras River (the extreme Southern part of the South Caucasus and Iran).

Biogeographical analysis of local separate (insular) psylloid fauna show its undoubtedly relict character. Here should be speculated that this insularity of local fauna from the ancestor clades plays a decisive role in the following abiotic factors. Firstly, collision between the northward- moving African, Arabian and Indian plates with the Eurasian plate, resulting origin of the Alpine-Himalayan orogenic belt, including he Zagros fold-thrust belt creation. This Mesozoic-Cenozoic to recent event caused delimitation of areas of the middle reaches of the Aras River Valley, which, as mentioned, has been turned into exclave.

2.2.1.2.3.1. The Psammophilous Plant Communities of the Eastern Caucasus

Sand deserts and semi-deserts are very common in the Caspian Sea coastal regions (Azerbaijan, North Caucasus). They further occur in Azerbaijan in the Shirvan, Mil and Mugan 'steppes'. The lower parts of mentioned sites are considered as an old lagoon or a shallow bay (Gulisashvili *et al.* 1975). In the Aras River Valley (South Armenia) occur the Goravan relict sands (Prilipko 1945). The composition and structure of the vegetation depends on the nature of the substrate that may range from moving dunes to desertified steppes with firm soils.

Sarikhum Barchan (part of the Sulak Lowland) area is a sandy dune in north-western Caucasus. Along with the large sand site, the Kumtorkala sands, it is the largest dune in Russia and Europe (Zonn *et al.* 2010). Xeropsammophytic desert vegetation of Kumtorkala is distinctive of the other sand sites of the Caucasus rarely possessing the region species, some of which are shared with the Turan Lowland. *Calligonum aphyllum, Eremosparton aphyllum, Astragalus karakugensis, A. lehmannianus*, etc. are important here (Grossheim 1948). Some of these, rare and endemic species are entered in the Red Data Book of Russia (Zonn *et al.* 2010). It should be stressed that sand deserts and semi-deserts of the western part of the Circum-Caspian Lowland (eastern Caucasus), as previously pointed out, completely lack open woodland composed of saxaul (*Haloxylon* spp., fam. Amaranthaceae) and their psyllid species, i.e. consumer insects which are largely represented in countries of Eremian (Sahara-Sindian) and Irano-Turanian phytogeographic regions.

In the Caspian Sea coastal zone, typical sandy desert plant formations are formed by endemic *Calligonum* (phog) shrub species and C. *bakuense* (Agadjanov 1967; Prilipko & Agadjanov 1972; Soskov & Akhmed-Zade 1974; Atamov 2008). In the Aras River Gorge (Gorovan sands) another species, C. *polygonoides,* occurs (Prilipko 1945; Agadjanov 1962; Tadevosyan 2001, 2002). The coastal psammophitic flora consists of about 600 species, of which the great majority is represented by annual and perennial herbs (92%). In Azerbaijan, about 20 species may reach dominance in these communities (Agadjanov 1962). A very specific phytocoenosis is formed by *Convolvulus persicus* growing on shifting sands and dunes. On firm sands the psammophyllous *Artemisia scoparia* creates rather dense thickets. As other psammophites *Astragalus karakugensis*, the parasitic *Cistachne flava* and some common ephemers of the Poaceae family can be mentioned (Figure 53).

The rare species are richly represented in the coastal vegetation of Absheron Peninsula, and are mainly observed in spring, when they are biologically active, then they soon fade. According to Mailov (1989), sandy areas in Azerbaijan are about 117,650 ha of this, 24,150 ha is mobile sandy bodies located on the Caspian shores. As stated in Atamov (2008), the plant formation of Azerbaijan's Caspian shores belongs to the sandy-desert type vegetation of the Central (Middle) Asian variant, common in hardened sandy areas. The recent ecological and phytosociological investigations of the Caspian shores with a width of 100-150 m from Astara to Khachmas and Khudat, belong to Sakhsuvarov (1994) and Atamov (2008). Except for halophytic wet- and arid-deserts and semi- deserts, wet-swamp and forest, they outline dozens of formations and associations of sandy-desert vegetation types. In general, psammophyte plants are more common on the Caspian shores, especially leafless shrubs and semi-shrubs are characteristic in the rather narrow shore area. The authors mentioned indicate the following species as examples: *Calligonum bakuense*, *Ephedra distachya, Elaeagnus angustifolia, Nitraria schoberi, N. komarovii, Astragalus hyrcanus, A. caspius, Artemisia szowitsiana, A. arenaria,* and *Carex bordzilovski.* Psammophytic formations with *Convolvulus persicus - Argusia sibirica* and *Artemisia scoparia - Alhagi pseudalhagi - Linaria simplex* communities at Shirvan National Park are analyzed in detail by Schmidt (2007).

Figure 53. Sandy deserts of the sea-coastal area of eastern Azerbaijan. Communities consisting of psammophytes: parasitic *Cistanche salsa*; in the background: *Artemisia scoparoides* and *Calligonum bakuense* (1974).

Some lagoons of the Caspian Sea coast area, and particularly Kyzyl-Agach Bay are Important Bird Areas (IBA) for thousands of migrating and wintering water birds. The conservation value of these shelters and the shores, which do not underlie any further protection, is underlined in many conservationists, papers (Vereshchagin 1950; Feeny *et al.* 1968; Tegetmeyer *et al.* 2007).

Taxonomical composition of flora of the Dagestan coastal lowland, including sand semi-desert communities, was investigated by Soltanmuradova and Teymurov (2010).

The complicated history of the formation of the Caspian Sea and its repeated fluctuations during the Late Pleistocene and Holocene, have strongly influenced the formation of its coastal vegetation (Chapter 1.4.2.). The Caspian shore's vegetation underwent important succession due to the anthropogenic factors (herding, sand deportation for construction, etc.) (Atamov, 2008; Dimeyeva 2010).

2.2.1.2.3.2. 'Splendid Isolation' Patterns of Goravan Sands (Armenia) Using the Example of Psylloidea Fauna

Among sandy deserts of Transcaucasia significant attention is attracted by the Goravan Sanctuary Sands. This is one of the most recognized natural heritages of arid landscapes in the Caucasus, namely in the Aras River Valley (southern part of the South Caucasus), bordering on Iran. This tiny area – a clearly expressed insular system - supports many life forms encompassing a rare species of psammophilous flora and fauna (Gegechkori 1978, 2016, 2019 (in press); Tadevosyan 2001, 2002, 2006).

Species of *Calligonum* play a key role in the stability of desert and gravelly desert ecosystem, stabilizing the shifting sand dunes, as it consists of a dense network of roots in different arid climate countries. Species of the genus *Calligonum* are one of the economically important resources of the sandy deserts and play a significant role in the life of the deserts local nomadic population. (Soskov 1988; Dhief *et al.* 2011).

The Genus *Calligonum* L. (Polygonaceae) is a perennial shrub and sometimes grows as a small tree (e.g., *C. eriopodum*). Branches are articulated and fragile. Usually leafless shrubs (leaves may be reduced to short-scale) from 1mm to 35mm at the tip of jointed stems (Soskov 1988).

Plant species growing in the sand deserts of the Saharo-Gobian region are highly adapted to their peculiar environment. Psammophyte shrubs have evolved numerous peculiarities to increase their chance of survival. They induce photosynthetic and drought-deciduous stems, flexuous and joint woody branches, a long root system (with root suckers, which are able to explore and store comparatively high amounts of ground water stored from 1.5m depth in sandy soil and deeper) and double flowering ability of some species (Nechaeva *et al.* 1973). Annual green stems serve as assimilation organs (Korovin 1961, 1962; Serebrijakov 1962, 1985; Nikitin 1968; Nechaeva *et al.* 1973 and others cited in Soskov 1988).

It occurs throughout North Africa, South Europe, West and Central Asia. The genus is distinguished by unusual anatomical features including striking intra-individual polymorphism (Ford, 1940; Shepard, 1970; Mayr, 1974; Elenevskii, 1980 and others cited in Soskov 1988).

Taxonomic composition and phylogenetic relationships still remain problematic (Popov 1925; Botchancev & Lishpic 1955; Bobrov 1953 and others cited in Soskov 1989). According to Soskov (1988), quite a controversial view is outlined. For example, the described species of phogs in the 20th century, ranged from 174 to 90 (most frequently) and 35 species. Pavlov

(1936) in flora of the USSR recognized 117 species, Nabiev (1971) (in Soskov 1989) described from the Middle (Central) Asia 45 species, etc. As details of genetic relations were clarified, in addition, distribution, biology and new finding made in anatomy, Soskov (1988) revealed a total of 49 taxa of *Calligonum*, where 28 species, 5 subspecies and 16 natural hybrids were established. The results reveal the existence of three floristic provinces: (1) Irano-Turanian with 24 endemic species; (2) Sahara-Arabian (Sahara-Sindian) and Mediterranean with 1 native species; (3) the regions of the former USSR (Middle Asia, Kazakhstan and the Caucasus) enclose 19 species and all hybrids.

There are four sections within the monophyletic genus *Calligonum*: *Calliphysa* (with 1 species), *Medusa* (13 species), *Pterococcus* (5 species), *Calligonum* (9 species) (Soskov 1988).

Goravan Sands desert, as mentioned, is of particular interest first of all by its insularity and also by its rich and rare biota which differ significantly from other arid sites of the South Caucasus (Gegechkori 1978, 1984a, 2016; Tadevosyan 2001, 2002, 2006).

Goravan Sands Sanctuary was established as the Goravan State Nature Reservation ('*Goravanskie Peski*' in Russian) in 1959, after an old name of the nearest village Goravan. The total area of the reserve is nearly 200 ha and spreads in an elevation range of 850-1,050m a.s.l. (Tadevosyan 2001; Saghatelian 2006). The Goravan Reservation lies $39^{0}50'$ northern latitude and $44^{0}43'$ eastern longitude. It is situated within the arid climate zone (average annual precipitation is 200-300mm, average annual air temperature is $+12^{0}C$). Summer is dry and hot, winter is cold (absolute minimum is $-25^{0}C$). It is characterized by gentle sloping topography with, as mentioned, shallow sandy soil (Bagdasaryan *et al.* 1973).

The local 'sand-island' habitat is a home supporting about 160 species of vascular plants and 36 species of vertebrate animals (Bagdasaryan *et al.* 1973; Amiryan, 1999; Tadevosyan, 2001, 2002, 2007; Saghatelian 2006; Aslanyan *et al.* 2014), as well as many invertebrates (Tadevosyan, 2001, 2002, 2007).

Figure 54. Stand of the phog (*Calligonum polygonoides*), Goravan Sands, South Armenia (1972).

From local, mainly very specific psamophilic organisms, 12 species of plants and 9 species of vertebrate animals are included in the Red Book Plants (2010) and Red Book Animals (2010) of the Republic of Armenia. The gems of the sandy habitat are small populations of phog

(*Calligonum polygonoides*) and milfoil (*Achillea tenuifolia*), then *Astragalus paradoxus, Rhynopetalum gibbosum, Ceratocarpus arenarius, Allium akaka* and other psammophyte species. The sand body of Goravan is the only restricted place in Armenia, which harbours relict phog and *Rh. gibbosum* (Tadevosyan 2002, 2007) (Figure 54).

Among mammals Dahl's jird (*Meriones dahli*) fossorial psammophiles (Red List Category – EN) belong to endemics of the Aras River Valley, and the Goravan Sands is confirmed as one of the last harbouring sites (Tadevosyan 2007). As well as the Red Book Animals of the Republic of Armenia, this species is included in the 'Caucasus 50 animal species, having high priority', offered by CEPF (2005) and published in 2007 (Gegechkori 2007). The reservation itself is listed as an area of National Priority in 'Ecoregion Conservation Plan for the Caucasus' (2012) (Zazanashvili *et al.* (Eds.) 2012).

Among striking fauna of the sand desert two stenotopic species of lizards (*Rhynocephalus persicus* and *Eremias pleskei*) are extremely threatened on a world-wide scale and listed in the IUCN Red List of Threatened species as CR, including the Red Book of Armenia (Tadevosyan, 2006a,b; Bagdasaryan *et al.;* 1973, Aslanyan *et al.* 2014). Like the lizards, biotopically closely linked with some psammophite plants dominated by *C. polygonoides*, one of the endemic insect (beetle) species of Armenia is *Pharaonus caucasicus* (Aghasyan 2014).

As we see, the Goravan Sands area is a restricted refugium of stenochoric plant and animal species and likely to be sensitive to climate change and anthropogenic press.

In the framework of a comprehensive work project on the psyllids of the Aras River valley, fauna of Pachypsylloidini was investigated during three trips (March, April, 1973, April-May, 1974, May, 2005). Over 200 specimens belonging to 11 species of psyllids feeding on different host plants of the sandy patch were collected by me and my collaborator (Figure 55).

Figure 55. Collecting psyllids (Psyllodea) associated with *C. polygonoides*. Goravan Sands, South Armenia (1972).

Apart from the Caucasus, faunistic exploration trips were conducted with scientists of ZIN (Leningrad) in Middle (Central) Asia or Turan (1985-1989) (Gegechkori 1989, 2000), and single trips were implemented in the Repetek Biosphere State Reserve (East Karakum Desert, Turkmenistan (early springs of 1987-1988) (Gegechkori & Loginova 1990).

Among entomocomplexes represented in sandy deserts with the dominancy of *Calligonum* spp. and associated by feeding habits with the phogs there are host specific groups of insects which are restricted exclusively to the phogs. Among the latter groups should be mentioned psyllids of the tribe Pachypsylloidini (Loginova 1969; Gegechkori 1985; Burckhardt & Ouvrard 2012).

Pachypsylloidini is very specific and intimately tied to its hosts, psamophitic genus *Calligonum* (Polygonaceae) – phog (Loginova, 1974b, 1976).

The range of the extant species of the tribe, extends exclusively in sandy deserts of the Palaearctic realm (Loginova, 1970, 1971, 1976) from North Africa to the Mongolian Gobi, but the richest fauna is strictly represented in the sandy deserts of Middle (Central) Asia, including the former Soviet Union's five Republics of Turan: Kazakhstan (sometimes discussed independently from other countries of Middle Asia), Kyrgyzstan, Tajikistan, Turkmenistan, and Uzbekistan (Loginova 1960, 1969a,b; Baeva 1985; Gueguetchkori 1998).

Goravan Sands is the only place in the South Caucasus where the psammophilic fauna of sandy desert habitat is represented by 3 genera and 4 species of psyllids of the tribe Pachypsylloidini: *Shaerqia turkestanica* (Löw 1881) (new for the target area), *Pachypsylloides cornutus* (Log. 1974), *Eremopsylloides amirabilis* (Log. 1964) and *E. fedtschenkoi* (Löw 1881) (new for the target communities) (Gegechkori 1978; Gegechkori & Loginova 1990).

Using the example of scorpions, a four-pointed scale can be adopted in sandy deserts: (1) ultrapsammophiles (species found only in sand habitats with little vegetation); (2) psammophiles (species found in vegetative sandy areas); (3) fossorial psammophiles (species that spend almost all their time in burrows); (4) facultative psammophiles (species that inhabit sand and other substrates) (Newlands 1972; Fet *et al.* 1998).

Ecological demands and the life-modes of psylloids in different habitats are being tolerated via their host plants (Loginova 1974a, b). Therefore, unlike substrate inhabiting animals (e.g., scorpions), they are finding some relaxation in such a harsh ecological environment as the sandy deserts in the canopy of their host-plants. However, psyllids as upon host plant dependent insects – narrow oligophagous arthropods which feed on species of *Calligonum* should belong to the psammophilic group (categories 2).

It is worth stressing that the endemic phog shrubs in the Caspian Sea shore sands of Azerbaijan, to the north-eastern part of the Goravan Sands, possess *Calligonum bakuense* which, is represented as far more abundant in dense thickets and occupies an incomparably large area. However, these shrubs are not populated by the specific psammophilic species of psyllids, while after disjunction within this country, some sporadic distributions of two species (*E. amirabilis* and *Sh. turkestanica*) are also reported from the sandy formations of the NE Caucasus (Stavropol Krai and Dagestan). (Gegechkori 1984a; Gegechkori & Loginova 1990).

Absence of psammophilic psyllids in the sandy beaches and dunes of Azerbaijan is most properly associated with two circumstances: (1) biological particularity (host-plant chemistry) and (2) multiple transgressions of the Caspian Sea (Gegechkori 2016).

As already mentioned, the primary consumers of the phloem are sap-sucking Sternorryncha group (psyllids in general), which are mainly feeding on phogs and are characterized as in extremely close associations with their hosts (Loginova 1969a, 1976; Hodkinson 2009). However, according to the chemical composition, not all phog species saps are consumed equally, i.e., some potentially suitable host plants, based on the nutritional requirement of psylloids are unacceptable, hence selecting other species that would provide suitable food (host-seeking behaviour) (Loginova 1976).

Psyllids, associated with psammophytic species of *Calligonum* in desert ecosystems frequently display some form of survival capacity against harsh environmental conditions. Here it should be pointed out their very close dependency upon host plants life cycle adaptations to plant phenology and other life history and behavioural adaptations. For instance, species of Pachypsylloidini, according to their trophic relationship, are specific and intimately tied to their hosts (host range for all three genera tied to phogs are oligophagous and monophagous). It means that biotic – plant host-parasite interaction in sandy deserts of the Palaearctic has a long history (from the beginning of formations of extant sand deserts in the Palaearctic) (Loginova 1969; pers. observations).

Depending on the nutritive ratio and competition patterns on hosts, two genera (*Pachypsylloides* and *Eremopsylloides*) are known to have not less than 2-3 generations per annum, but *Sureaca* is univoltine (Loginova 1969a, 1970, 1971; Gegechkori 1978; Gegechkori & Loginova 1990).

One of the wonderful adaptations of an extreme environment of deserts is causing galls on host plants (Loginova 1974b).

Among phog-infested psyllids there are gall-inducing species. Larval stage species of *Pachypsylloides* are initiated mainly close galls on photosynthetic and vegetative stems and young twigs (*vs* non-gall midges habit of species of another two genera - *Shaerqia* (except *S. turkestanica* and *Eremopsylloides*). Some galls, particularly those affecting young stems, can be very damaging to plants (Loginova 1970, 1974b).

To escape high egg and first instars mortality caused by lethal high temperatures combined with low humidity (i.e., high Saturation Deficit Index), psyllids on *Calligonum* in sandy deserts of Middle Asia show a close phenological synchrony of development with that of the phenology of host plants.

In extended periods of drought seasons of the sandy deserts in Middle Asia, species of *Shaerqia* provide diapauses in both seasons – in summer (estivation) and in winter (hibernation). Both diapauses occur in larval instars. Estivation takes place in the first instars under the scales of dormant plant's buds and stems (leaves of the shrub are inconspicuous or sometimes reduced to scale) and ranges from mid-May to September-October. Hibernation of second and third instar immatures extended from October till mid-March and early April. Overwintering nymphs may be found in the cracks of the host plant's trunk (Loginova 1969a 1970, 1976; pers. Observations). In this case diapause of insects serves two goals: (1) to survive particularly harsh-dry and cold periods in sandy deserts habitats, when there are exceptional unsuitable environmental conditions, the host plant becomes unacceptable for further development of psyllids life history; (2) to synchronize the propagation with the short spring optimum period on the host. The first generation of winged psyllids in optimal abiotic conditions appear in May, i.e., in the period of oviposition. Then, by the end of May, the adults have disappeared completely (Loginova 1974b, 1976). Therefore, species of psammophilic psylloids trophically linked with phogs make aspects of the ephemeral group of invertebrate animals in hyper-arid (or close to these conditions) deserts. Obviously, such a behavioural tendency of sand dweller psyllids is one of the most powerful illustrations of adaptation tendency during the long-term, evolution development pathway in harsh conditions of deserts (Gegechkori 1985).

As underlined the total fauna of tribe Pachypsylloidini (fam. Liviidae) consists of 31 species belonging to three genera. The genus *Pachypsylloides* includes 13 species (42%), two other related genera *Eremopsylloides* – 10 (32%) and *Shaerqia* – 8 species (28%), respectively.

As regards the psammophilic psylloid of western Palaearctic, according to biogeographical investigation of Gegechkori (Gegechkori 2019, in press), at present the striking rich fauna from both a quantitative and qualitative viewpoint is found in Turan: three genera consist of 19 species, with 8 endemics (i.e., 61%/42%). Accordingly, after Turan the vast majority of psyllids trophically tied to *Calligonum* occur in Kazakhstan where the total number of specific species is equal to 12/1 (38%/8) (by its harsher climate than that of other four countries of Turan this latter country has been reviewed independently).

A comparatively small desert area of Israel consists of two endemic species, *Shaerqia loginovae* and *S. negevensis* (Spodek *et al.* 2017) closely related to other species of the same genus from Turan, at the same time represented as an allopatrically dispersed species group. So, with a large probability, the mentioned indigenous (autochtonic) species make the territory of Israel a unique biogeographic transitional area in the geologic past. Certain indexes for Israel are as follows – total number 5/2 (16%/40%). The third region represented with all three genera is the Caucasus with a total number – 5/0 (16%/0%).

All other five regions are restricted by only two or one genera of Pachypsylloidini. Central Asia (Mongolia) has total number of psylloids: 6/3 (19%/50%).

Today psyllids of the Chinese part of Central Asia (the Gobi Desert) are very low – 2/2 (6%/100%) (Gegechkori 2019 and references therein, in press).

Psyllids of the countries in the Middle (East) Asia, apart from Israel are far less investigated and Iran (1,648,000km^2) with 95 species is no exception. This number is certainly 50% less than the number of existing species (Burckhardt & Lauterer 1993; Zendedel et al. 2016). It should be said regarding Pachypsylloidini of this country, total number is 2/0 (6%/0%).

The same indexes for the Arabian Peninsula and North (Mediterranean) Africa are as follows – 3/0 (9%/0%) and 2/0 (6%/0%) (Gegechkori 2019 and references therein, in press).

So, on the base of the presented comparative analysis the following should finally be pointed out. The range of the tribe Pachypsylloidini encompasses the Saharo-Gobian (Sethyan, sensu Emelianov 1974) desert belt coincides with major boundaries within the Ancient Mediterranean area (Kryzhanovski 1965) with the center of current biodiversity, as we see, in the Irano-Turanian region (subregion) and significantly in sandy deserts of the Turan Lowland (The Turan Depression).

It is interesting, that from the genus *Calligonum comosum* in Sahara-Sindian deserts, on the one hand, and *C. roborovskii* (Kunlun Mts, 2,100-2,600m a.s.l., NW China) (Zhu et al. 2010), on the other, these are a good example of a species with an extreme westernmost-easternmost distribution pattern, occurring along the Sahara-Gobian deserts (Gegechkori 2019, in press). Today psylloid fauna of this large geographic area, as already mentioned, have been investigated unequally.

Finally, at the end of characterization of regional faunas, special attention should be paid to the fact that, as already revealed, all three taxa of overspecies level of Pachypsylloidini, similar to Middle Asia (Turan) and Kazakhstan are found in Transcaucasia. This is a minute site as in the Aras River Valley, namely, the Goravan Sands. Such kind of insularity on a tiny area is one of the most striking features because of the absence of analogues within the biota of the Caucasus (Gegechkori 2019, in press).

It is necessary to highlight that in Transcaucasia not all isolated patches with characteristic desert plant communities (which usually are hosts of psyllids) are populated by species of Psylloidea. Special research (Gegechkori 1984a) highlighted some interaction mechanisms between folivore arthropods (i. e. psyllids) and their hosts in the margins of the Irano-Turanian deserts. The infiltration pattern of deserts' biota of the Irano-Turanian biogeographical region within the Caucasus, as mentioned, shows clearly, the following regularity. Distribution patterns of host plants is broader than their monophagous or oligophagous consumer insects – psyllids. In other words, in the margins of the range of the Irano-Turanian deserts host plants occur ecologically more plastic and widespread than their consumer insects. For instance, dominant arboreal plants of Irano-Turanian xeric habitats lost their specific psylloid fauna in the most marginal western area, in East Georgia. Mention should be made of species of *Egeirotrioza* Boselli (1931), which are not found on *Populus* (*Turanga*) *euphratica* (a striking disjunct stand of this ancient poplar is found only in Vashlovani NP, East Georgia); species of *Brachystetha* Loginova (1964), do not link on *Nitraria schoberi* (in drylands of eastern Caucasus the population of nitre-bush shows a patchy, i.e., discontinuous distribution pattern); *Cyamophila oshanini* (Loginova 1964) – does not link on *Halymodendron halodendron* (outstanding, restricted stand of this shrub occurs in David Gareja Desert ecotone), etc. The rare exclusion from this regularity, it should once more be pointed out, is a species of Pachypsylloidini in the Goravan Sands.

From a historical point of view, it is highly likely that the sand desert patches of the Aras River Valley were refugia for the mentioned taxa of the tribe Pachypsylloidini during the Plio-Pleistocene time. Current distribution of the desert's hot climate tolerant species linked with *Calligonum polygonoides* is widely disjunct – there is interrupted distribution from their main ranges, deserts and semi-deserts of the Irano-Turanian phytogeographical region.

These and other (Gegechkori 1989, 2004, 2019, in press) investigations highlighted that the Turan floristic/faunistic region can serve as a 'donor' of dry climate tolerant taxa to 'recipient' neighbouring regions, including the Caucasus.

2.2.1.2.3.3. Pattern of Dispersion of Insect (Psyllids) - Host Plant in Zonal and Azonal Habitats of Deserts and Semi-Deserts (Middle (Central) Asia, the Caucasus)

Psyllids resemblance to other insects is as ectotherms, hence they are heavily dependent on abiotic factors. As oligophagous insects, strongly dependent on the chemistry of the host plant, psyllids, except for some exclusions, are closely associated to their hosts, flowering plants (Gegechkori 1968, 1984a; Loginova 1968; Hodkinson 1974, 2009).

Herbivorous insects and their host plants are rarely distributed randomly on either a local or a large geographical scale (Stanton, 1983). To quote this author: 'The mechanisms generating nonrandom herbivore distributions are not only important keys to understanding the evolution of plant-insect associations, but are also of great economic interest'. This point of view is far more valid for psyllids as generally narrow host specific insects.

In the margins of semi-arid and arid dweller psyllid - host associations distribution ranges (East Georgia) are gradually changing. Climatic conditions lead to decrease of consumer's voltinism, population density, alter foraging behavior, etc. (Gegechkori & Ginturi 2005; Gegechkori 2010c).

Psyllids are plant juice sucking insects, therefore host plants impact on their populations directly (life-history, nutritional conditions, etc.) (Loginova 1968; Hodkinson 2009). Despite this, in margins of ranges represented in semi-desert and desert regions of Middle (Central) Asia and the Caucasus, due to alteration of seasonal average temperature and total precipitation, the climate aridity began to decrease. These climatic peculiarities on the edge of the host-plant area cause changes of the synchrony in distribution ratio and interactions between herbivore-host associations. Some key taxa of psyllids do not remain attached to their hosts despite a large host range and its abundance in food.

Among others (see Marques *et al.* 2000 and references therein) one possible interpretation of the mentioned mismatches in psyllids-host plants chorology should be explained with more sensitivity of psyllids to abiotic factors than theirs hosts.

Bioclimatic envelopes (i.e., conditions, under which arid landscapes' psyllids life-history is going well) will shift to the margins of their ranges and higher in elevation. Hence, the geographic distribution of many desert psyllid species is more limited than their host distribution. The only exclusion of this regularity is the arid landscapes with psyllids intimately tied to their hosts in riparian woodlands. Changes in abiotic factors between bioclimatic zones and azonal riparian forests is a single opportunity shaping large-scale migration of psyllid species to inhabit arid bioclimate zones even outside of arid and semi-arid biomes.

One of the most common host plants on which psyllids develop in arid climate areas is the genus *Tamarix*.

It should be stressed that hosts psyllid's (tamarisks) range records, both from arid and humid regions of the Caucasus, are different. In the eastern Caucasus communities with tamarisks shrubland, except riparian habitat strip, largely occupy adjacent valley-bottom area, whereas in West Georgia the same shrubs are quite strictly restricted to riparian habitats.

Within these host plants psyllid species of the genus *Colposcenia* penetrate from eastern arid/semi-arid regions of the eastern Caucasus into western Georgia. They are found in riversides of lowland's Abkhazeti (Bzyb River and its tributaries) and in middle altitudes of Racha (Rioni River, and its tributaries), and Adjara (Adjaristskali River) regions. Among 7 species of psyllids on tamarisks collected in the Caucasus (Gegechkori 1984a), in lowland of Abkhazeti three species were collected: *Colposcenia aliena* (Löw), *C. kiritschenkoi* (Loginova) and *C. jakowleffi* (Scott). It is noteworthy to underline the fact that the latter 'most desert' dweller two species do not occur elsewhere in the arid eastern Caucasus, or frequently are found at a lower percentage of samples. The other two regions of West Georgia are poor of psyllid species feeding on tamarisk. In the riparian forests of Adjara *C. osmanica* (Vondraček) was collected (which is rarely found in the eastern arid and semi-arid Caucasus) and in Racha, the single species *C. aliena* was collected. It is the most common species, widely distributed in Palaearctic and outside this zoogeographical region and characterized by equally abundant populations in the Caucasus.

Based on the observed ecological patterns the following conclusion can be attempted. Some species of the arid climate-tolerant genus *Colposcenia* tend to disperse in mild climate regions of Western Georgia and find the most suitable habitats in the area close to sub-Mediterranean climate, i.e., 'climate of *Arbutus andrachne*' in the Abkhazeti and Shuakhevi regions of Adjara, the driest place of the Adjaristskali Gorge (the range of *C. osmanica* found in Shuakhevi, apart from Central Asia and the Caucasus and includes Bulgaria, Turkey, and Israel) (Gegechkori & Loginova 1990).

According to the above reviewed materials the following conclusions should be stated:

- Depending on the species, inhabiting riparian forests of arid and other landscapes may promote increased dispersal ability across environmental barriers involving high dispersal distances.

- Such a large scale (arid deserts to almost humid subtropics) range extension of exotic species of psyllids on tamarisks from the arid center of species diversity suggests a higher level of dispersion through riparian azonal communities (corridor). Such peculiarities of psyllid dispersion have already not been observed in other groups of insects inhabited tugai forests and floodplain forests with restricted host plant relationship in the Caucasus.

2.2.1.3. Fauna of Deserts and Semi-Deserts (a Brief Review of the Major Faunal Groups)

Animal composition in deserts and semi-deserts generally depends on edafic factors giving rise to many specialized species like psammophiles, halophiles and petrophiles. This certainly accounts for invertebrate animals (Gegechkori 1987 a, b, 1979; Gueguechkori 1998). On the other hand, the most arid zone of the Caucasus is the terminal stage of the xerophilization of psyllid fauna. There are many examples of expression of the highest morphological and remarkable life mode specialization of this group to the aridity of the climate and other extreme factors (species of *Agonoscena* Enderlein, *Shaerqia kemal & Koçak, Pachypsylloides* de Bergevin, *Eremophsylloides* Loginova, *Caillardia* de Bergevin and others) (Gegechkori 1985). Some species, however, and endothermic species in particular, have a very wide range of distribution and broad spectrum of adaptations (Vereshchagin 1958a). Typical invertebrate animals are the Lesser Asian scorpion (*Buthus eupeus*) (Figure 60), the South-Russian tarantula (*Lycosa singoriensis),* web spinning and soil inhabiting spiders, and various insects like grasshoppers and beetles. The amphibians are typically represented by the European green toad (*Bufo viridis).* Numerous are the reptiles with the following representative species: Caspian gecko (*Cytropodion caspium*), Caucasian agama lizard (*Paralaudakia caucasica* (= *Agama caucasica*) (Figure 58), Persian roundhead (*Phrynocephalus helioscopus horvathi*) (Figure 56), and various other lizards, the sand boa (*Eryx jaculus*), various whip (water) snakes, the Levant viper (*Macrovipera lebetina*) - the most venomous snake in the Caucasus (Figure 59) and others. Among the widely distributed subspecies of tortoises are the spur-thighed tortoise (*Testudo graeca)* and the Caspian turtle (*Clemmys caspica).* Typical desert birds are the big, usually walking species - little bustard (*Tetrax tetrax orientalis)* and great bustard (*Otis tarda tarda).* Raptors include *Falco columbarius, Aquila heliaca, A. rapax orientalis*, *Circaetos gallicus*, and *Gyps fulvus.* Conspicuous small and mid-size birds are bee-eaters (*Merops apiaster, M. superciliosus),* black-bellied sandgrouse (*Pterocles orientalis),* hoopoe (*Upupa epops),* redfronted serin *(Serinus pusillus),* nightjar (*Caprimulgus europaeus),* tree-sparrow (*Passer montanus),* skylark (*Alauda arvensis*), and other larks. The riverside (floodplain) forests are populated by Black francolin (*Francolinus francolinus)* and penduline tit (*Remiz pendulinus);* natural waters and reservoirs are inhabited by water birds (Voronov *et al.* 1985; Babaev *et al.* 1986).

Very typical are the mammals such as the rodents that include jerboas (*Allactaga* spp.), hamsters (*Mesocricetus* spp.), jirds (*Meriones* spp.), hares (*Lepus europaeus*), hedgehogs (*Hemiechinus auritus, Erinaceus europaeus*); the previously mentioned two characteristic species of ungulates: goitered gazelle (*Gazella subgutturosa subgutturosa*) and saiga antelope (*Saiga tatarica*); the predators: Caspian Sea wolf (*Canis lupus cubanensis*), jackal (*Canis aureus*), steppe fox (*Vulpes vulpes alpherakyi*), striped hyena (*Hyaena hyaena*) (Figure 57), and the jungle cat *(Felis chaus chaus*) (Voronov *et al.* 1985; Babaev *et al.* 1986).

The fauna of the Kura-Aras Lowland has been modified by human impact, especially through livestock grazing, which - according to the neighbouring, previously mentioned Gobustan's (Azerbaijan) and Early Holocene and Late Pleistocene rock art, has been practised here for centuries. In order to protect the domestic herds, the carnivores were exterminated, destroying the natural equilibrium in local ecosystem.

Figure 56. Nestling of snake-eating eagle (*Circaetus gallicus*) with its prey in its beak. Marneuli district environs, E. Georgia (1973).

Figure 57. Striped Hyena (*Hyaena hyaena*) – scavenger, but also skilled hunter. Currently very small population possibly survives in the middle reaches of the Aras River (Azerbaijan). Yerevan zoo (2005).

Figure 58. Caucasian agama (*Paralaudakia caucasica*) is a typical Caucasus desert's stony habitat dweller. Lizard have a moulting skin (which generally falls of in pieces). Uplistsikhe cave complex – the oldest (millennium BC) settlement site known in Georgia (2014).

Figure 59. The blant-nose viper (*Macrovipera lebetina*), one of the most dangerous poisonous snake of the eastern Caucasus (1962).

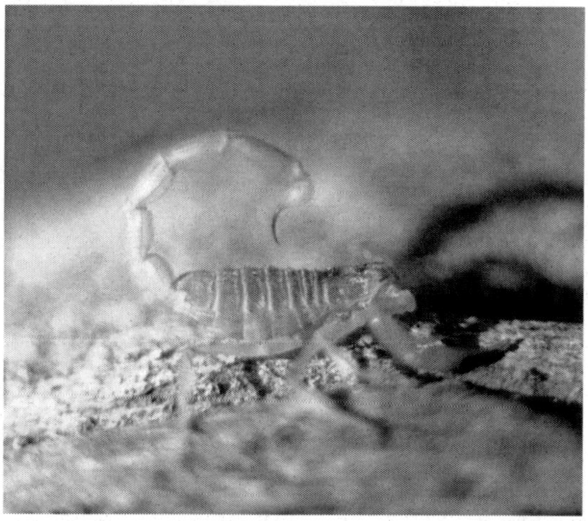

Figure 60. Scorpions (*Buthus eupeus*) are fusually ound in rocky, arid areas of the Caucasus. Armenia, the environs of the Ereburni State Reserve (2005).

In the Riverside (Tugai) forests animals also occur: European-Caucasian boar (*Sus scrofa atilla*); Caucasian deer (*Cervus elaphus maral*); lynx (*Lynx lynx*); badger (*Meles meles*); wild cat (*Felis silvestris caucasica*) and Colchian pheasant (*Phasianus colchicus*) (Voronov *et al.* 1985, Babaev *et al.* 1986).

Conclusion

- Among biomes of the Caucasus from the viewpoint of importance and extant range two biomes - deserts and semi-deserts and the forests (decidiuous, coniferous) are the most important.
- In the Northern Hemisphere is found a great arid and hyper arid belt extending from part of the Sahara Desert to the Gobi Desert. Deserts and semi-deserts of the Caucasus are a part of the Turanian Province and Irano-Turanian Subregion within the borders of the Ancient Mediterranean (Saharo-Gobian or Tethian) Subkingdom of the Holarctic Kingdom.

- Among botanists there are quite controversial views with regard to this biome of the Caucasus. According to one group of scholars, the driest places of this region belong to the deserts proper. On the other hand, another group of botanists refer vegetation community of the driest habitats of the same region as semi-deserts. This is in accordance with our study of psylloids and their food plants, on the one hand. However, some sources of scientific literature, including the author of the present work, on the other hand, are the best variant appearing to be used of both terms in complex way – 'deserts and semi-deserts'.
- The annual average precipitation in different parts of the region varies from 200 to 400 mm. The Caucasus deserts and semi-deserts hold characteristic zonal soils, including grey soils, solonchaks, rock and stone deserts (hammada, especially in the middle reaches of the Aras River Gorge), sand deserts with dunes, clay deserts, clay-takyr and clay-badland soils deserts.
- Unlike hot deserts of tropical-subtropical latitudes of the Earth, deserts and semi-deserts of the Caucasus belong to the cold-winter deserts of the temperate latitudes of the Northern Hemisphere.
- By botanists and zoologists three main sites of the Caucasian deserts and semi-deserts are highlighted: Ciscaucasian (western, restricted part of the Volga River Delta), Eastern Transcaucasian (the Kura-Aras Lowland), and South Transcaucasian (Aras River Valley).
- The flat circum-Caspian Lowland, on both sides of the Caspian Sea, i.e., Middle Asia (Turan) on one side and the Kura-Aras Lowland on the other, is represented by saline (saline-sodic), solonetzic and dry takyr-like soils. In such drylands, traditional drought-resistant agriculture crops are yielding in irrigation areas.
- Deserts and semi-deserts of a salinized soils with halophitic vegetation of dwarf shrub and semi-shrub halophytes (Amaranthaceae, in particular) occupy a vast area of the Kura-Aras Lowland in Azerbaijan and in Ciscaucasia (Northern Caspian Region). They are more diverse than the Artemisieto communities and exhibit polidominant vegetation.
- Vegetation of salt affected land in relation to soil salinity and their biological structures and ecological peculiarities can be divided into four groups: hyperhalophytes, euhalophytes, mesohalophytes (haloglycophytes) and hydrohalophytes.
- Hyperhalophytes are characterized by succulent assimilative organs, at the same time they are microphyllous or aphyllous (e.g., species of *Salsola*, *Nitratia*, *Limonium*, *Petrosimonia*, *Kalidium*, *Halostachys*, and *Halocnemum*). They develop in conditions of raised mineralization of soil solution.
- Euhalophytic communities are also well adapted to strongly saline soils, but prevail in phytocoenoses with lower salinity of the substrate. Between shrub and semi-shrub communities should be mentioned species of *Salsola* (*S. ericoides, S. nodulosa*), *Anabasis aphylla, Atriplex cana*, etc.
- Meso- or haloglycophytes, which broadly are known as plant communities adapted to clay (-semi) deserts are composed of species of wormwood, *Artemisia* spp., significantly *A. fragrnas*, as well as less saline-tolerant species of *Salsola, Kochia prostrata*, some ephemers and ephemeroids, some species of Poaceae, etc.

- Hydrohalophytes are growing in standing water with brackish water of saline (hemi-saline) small lakes, marshes, etc. These plant communities include species of *Phragmites, Arundo*, etc.
- Clay (semi-) deserts with the dominancy of wormwood and wormwood-saltwood, used by nomadic pastoralists of the eastern Caucasus for grazing livestock animals are, therefore, classified as rangeland and have been used since time immemorial. As a rangeland, a solonchak polydominant community has less importance.
- Among halophytic semi-brushes and brushes are found species of C_4 plants with Cranz anatomy structures (e.g., many Amaranthaceae species: *Atriplex, Camphorosma, Kochia, Suaeda, Salsola*, etc.).
- Shirvan and Ag-Gel National Parks in Azerbaijan are a perfect reflection of wildlife protection on flat deserts and semi-deserts with halophytic vegetations of the Caucasus (e.g., goitered gazelle, saiga, wintering and breeding sites of the rarest waterfowl birds).
- Patterns of extraordinary disjunction distribution of grey-purple-reddish hill deserts - peculiar formations of the Tertiary gypsum-bearing red clays of eastern Georgia - attract attention via traditional biogeographical framework.
- They form a sort of ecotone - isolated 'islands in the sky' against a 'sea' of mainland – arid, open woodland, steppe and semi-deserts, i.e., surrounding a quite different type of habitat. If taken into consideration that species richness of patches is determined by their size and spatial isolation, the typical example of such an 'island' found in eastern Georgia is Gareja Hollow. According to habitat diversity, apart from slightly or moderately saline-tolerant plant communities (e.g., shibliak) local reddish hills are very sparsely populated by species – pure representatives of hyper- and euhalophytes species.
- Patches of painted desert hills are discontinuously extended from east (Georgian-Azerbaijanian border) to south-west (southern Georgia adjacent to Turkey). Three relict patches, desertified hills are considered to be of ancient Mio-Pliocene origin. Thanks to topography, exclusively peculiar-terrain forms (absence of fertile soils), painted hills are home to drought-climate and poor soil-tolerant xerothermophillous species of relict plants and animals. Therefore, they should be classified as micro-refugia of the desert living species, mostly the Irano-Turanian center of origin. They managed to survive past climate deterioration and found their shelters in these, extraordinary refugial areas. This mosaic of natural patches disharmonious to the Kura-Aras Lowland region should be viewed as the net of archetypal ecotones favored allopatric speciation events and pattern of poor but unusual biodiversity characterized to each patch.
- The most significant species of flora of reddish hills is the presence of *Halimodendron halodendron* (Gareji Hollow – the most complex refugium, which is the only site in Georgia where this monotypic genus of legume occurs); *Nitraria schoberi* found in three - four discrete desert's habitat patches, apart from East Georgia gypsum, containing white and reddish rolling hills of Meskheti, south-eastern Georgia. Among other salt-tolerant xerohalophytes and halo-gypsophytes species with patchy distribution should be mentioned *Atriplex cana, Reaumuria alternifolia (=R. kuznetzovii), Gamanthus pilosus, Ceratocarpus arenarius* and others.

- The remoteness from vast, continuously extended deserts, even the smallest remnants of the rarest living organisms of isolated patches provide significant importance. Unfortunately, as mentioned, no sanctuaries or natural monuments exist even in the Gareji Hollow with climax formations of native halophytes, semi-halophytes and associated fauna. On the contrary, the latter site is a hardly disturbed area grazed by cattle, and other livestock, and needs immediate protection. The conservation status of such a micro-refugia is strongly insufficient. They sheltered the rarest for the extreme western edge of typical Turanian (Turan Basin) types of vegetation and with great probability some genera and species of animals as well.
- Riparian, azonal vegetation (tugai) of Middle (Central) Asia (in a broad sense), where a species-rich, tree-life form of *Populus* (section *Turanga*) forms extensive stands, within the marginal area of Georgia has been found as a small monodominant stand (Vashlovani NP), and as a far larger stand in the middle reaches of the Aras River Gorge. Other micro-habitats of hydrohalophyte vegetations are found either in, or around small saline lakes (e.g., similar to Sakhare Tba). All of them are small, closed basins with a high evaporation rate. Coastal and even surface areas of lakes usually support hyper- or euhalophytic vegetations (e.g., *Salsola crassa, Atriplex cana*, etc.).
- Some indicator groups of invertebrate animals, for example, Psylloidea, in saline deserts and semi-deserts of the eastern Caucasus are far more restricted in their ranges than that of their hosts.
- Sand deserts and semi-deserts stretch out in the Caspian Sea coastal area (Azerbaijan, North Caucasus) and extend in the most internal arid area of Azerbaijan. Among coastal psammophitic flora leafless *Calligonum bakuense* and *Nitraria schoberi* are the most typical species.
- The Caspian shore's vegetation was subjected to repeated fluctuations of the Caspian Sea during the Plio-Holocene time due to the anthropogenic press.
- Sarikhum Barchan (part of the Sulak Lowland) area is a sandy dune in the north-western Caucasus. With the other large sand site, the Kumtorkala sands, both constitute the largest dunes in Russia and Europe in general. Xeropsammophytic desert vegetation of Kumtorkala is distinctive of the other sand sites of the Caucasus possessing the species found very seldom in the region, some of which are shared with the Turan Lowland. Those which should be mentioned are *Calligonum aphyllum, Eremosparton aphyllum* and *Astragalus karakughensis*. It should be stressed that sand deserts and semi-deserts of the western part of the Circum-Caspian Lowland (eastern Caucasus) completely lack open woodland of species of saxaul (*Haloxylon* spp., fam. Amaranthaceae), largely represented in countries of the Sahara-Sindian and Irano-Turanian phytogeographic regions.
- By contrast, all edificator and co-edificator vegetations of solonchak deserts are still the dominant elements (i.e., habitat generalists) presented in the Kura-Aras Depression, especially in its eastern part within the borders of Azerbaijan.
- The desert and semi-deserts of the middle reaches of the Aras River Valley (in the border of South Armenia and Nakhichevan AR, Azerbaijan) are considered as a biome deserts and semi-deserts, but of the southern type. Such a type of arid landscape is the most northern outpost of the plant communitys characteristic of the countries of the eastern Mediterranean and Asian Interior. These kind of deserts and semi-deserts are

nutrient-poor represented on the pebble-gravel, pebble-sand and sandstone soil-grounds.

- In the Aras River Valley significant attention of naturalists is given to the Goravan Sands (Armenia). This disjunct relict patch confined as the harbouring site of rarest species of psamophillous flora and fauna, disjunction of which occurs in the huge sandy deserts stretching from the eastern Caspian coast. Many organisms of the Goravan Sands are stenochoric. For instance, another species of phog – *Calligonum polygonoides*, lizards – *Phrynocephalus helioscopus* (=*Ph. persicus*), *Eremias pleskei*, rarest to the Caucasus psylloid fauna (tribe Pachypsylloidini) associated with the native species of phog.

2.3. The Steppes (Temperate Grasslands)

2.3.1. General Features of Grasslands (as a Biome with an Extraordinary History and Remarkable Ability to Adapt to Various Types of Abiotic and Biotic Factors)

There is a wide variety of definitions of grasslands. A synthesized definition is given by Gibson (2009) and Dixon *et al.* (2014). Grassland biomes are areas where the vegetation is dominated by grasses as their naturally dominant vegetation. The grasslands typically occur in the interior of large continents, or occupy a central place among terrestrial ecosystems. They are represented in areas where there is not enough regular rainfall to support the growth of a forest, but not so little as to form a desert. Hence, in fact most grasslands are located both geographically and climatologically between the arid environments of deserts and semi-deserts, and humid landscapes of the temperate/boreal forests (Grossheim 1948; Walter 1975; Brown & Lomolino 1998; Sala *et al.* 2001; Gibson 2009), in other words this biome is a physiognomic continuum, transition area between two latter biomes (Dixon et al. 2014). About 40% of the Earth's land is covered with grasslands (Gibson 2009; Blair *et al.* 2014). They may occur naturally or as the result of human activity using them as croplands, livestock pastures and haymaking (anthropogenic grasslands) (Suttie et al. (Eds.) 2005 and sources cited therein). According to Gibson 2009 (cited in Strömberg 2011) '. . . grassland is extended by grazing and/or fire to form a plagioclimax in many areas that were previously forested' (Suttie *et al.* (Eds.) 2005).

Grasslands as open community are associated with one of several types of open land, areas with fairly flat to gently rolling (hilly) topographic relief (Anderson 2006). Historically, grasslands have been defined as land, where in the natural vegetation community occur predominantly grasses (either tall, medium, or short sizes). 'Although grasses provide the matrix in which other plant species co-occur, grasslands include other plant life forms, such as annual and perennial forbs (non-graminoid, nonwoody plants, shrubs and trees). The matrix-forming species in most of the world's major grasslands are perennial grasses. . .' (Blair *et al.* 2014). Usually they are mixed with herbs –forbs (broad-leaved herbaceous plants) and sometimes with shrubs, but trees and large shrubs are largely absent; few trees are found along streams - in areas near rivers (riparian communities) and lake-shore habitats (Lavrenko *et al.* 1991, Woodward 2003). In landscapes receiving >650 mm of precipitation, the vegetation changes from woodland to grassland. In areas of low precipitation (less than 300 mm), grasslands degrade into desert communities (Anderson 2012). Temperate grasslands are defined also as a non-wetland type of biome (Dixon *et al.* 2014).

Open habitat – grasslands – naturally occur on all continents except Antarctica (a grass species, Antarctic hairgrass, *Deschampsia antarctica* occurs on the Antarctic Peninsula and surrounding islands) (Blair *et al.* 2014). Grasslands in their wider sense, as a distinct biotic and ecological unit are divided into two main types: (1) temperate grasslands and (2) tropical grasslands, commonly called savannas (Dixon *et al.* 2014). Savannas experience warm weather all year long while temperate grasslands have four clearly defined seasons (seasonal changes happen every few months) with warm (or hot) summer and cold (very cold) winter (Smith 1973; Gibson 2009).

Savannas are generally dotted by arboreal plants, i.e., that can have trees and shrubs (of less than 40% or less, but not zero) scattered throughout over a homogenous grassy ground layer. Savannas are restricted to the wet/dry tropics, with rains in the summer months. Having a different root system and a different water economy, the two vegetation layers can be in an ecological equilibrium or not. Savannas are represented in Africa, Australia (Paleotropical savannas), Central and South Americas (Llannos and Central Brazilian cerrado), and southern south-east Asia. The term savanna is derived from a native American (Carib) word, denoting an area covered with a grass layer and lacking trees: the word first appeared in Europe in Spanish (*sabana*) in 1535 and then in English (1555) (Woodward 2003; Gottsberger & Silberbauer-Gottsberger 2006; Gegechkori 2008).

According to the definition of the majority of authors, temperate grasslands exclude tropical and subtropical savannas (e.g., Walter 1975; Lavrenko *et al.* 1991; Sala *et al.* 2001). The major manifestations of temperate grasslands are the steppes of Eurasia (and, its enclave–puszta of Hungary), the prairies (tallgrass – about 2m tall – or true prairie and shortgrass prairie) of central North America, the pampas of Argentina and Uruguay and velds of South Africa. One form of open grasslands should be mentioned, a tussock**,** which is common in many temperate southern hemisphere regions, notably in New Zealand, also in Australia, some sub Antarctic islands, etc. Some of these grasslands appear in more arid (desertified steppes) or in more mesic (meadows) places, while the others occur in the rain shadows or highlands of major mountain ranges (Vtorov & Drozdov 2001; Sala *et al.* 2001; Woodward 2003; Gegechkori 2008; Blair *et al.* 2014).

As well as the above mentioned two main types of grasslands, some authors, recognize flooded (mostly in the tropics) and montane (high altitude) grasslands (*https://en.wikipedia.org/wiki/Tussock-Grassland*).

2.3.2. Climate

The main driving force of formation and occurrence of grasslands across the world is climate. As mentioned already, grasslands are commonly found in the interior regions of the continents. In addition, they occur in disturbance-dependent ('disturbance-rich') ecosystems characterized by a climate that exhibits disjunct seasonal patterns both with respect to rainfall and temperature (Blair *et al.* 2014). Moderate annual average precipitation varies between 250 to 750 mm (Mordkovich 2014), extreme data fluctuate from 150 to 1200 mm (Sala *et al.* 2001).

Comparisons with other biomes indicate that grasslands are more responsive to variation in rainfall amounts than are most other biomes (Blair *et al.* 2014). Rainfall is irregular, but usually occurs in late spring and early summer. Potential evaporation is greater than precipitation. Mean annual temperatures vary from near 0^0C to around 26^0C (Blair *et al.* 2014). Native plant communities evolved during one or two severe droughts. Grasslands as open landscapes are windy, which in semi-arid (or semi-humid) conditions promote evaporation (Mordkovich 2014). Grassland ecosystems, with periodic drought, contribute to the accumulation of highly flammable overground plant detritus and the occurrence of periodic fires (Blair *et al.* 2014; Mordkovich 2014).

Grasslands, naturally, maybe lesser than other biomes of the world (i.e., savannas, heatlands, cerrado, Mediterranean chaparral, temperate and boreal coniferous forests), but belong to fire - dependent (or fire-controlled) ecosystems. In such ecosystems, wildfire as an obligatory abiotic distribution factor is essential and the species have evolved adaptations to respond positively to regular burning conditions of fire. They are often called fire-adapted or fire-maintained ecosystems (Bond & Keeley 2005a; Myers 2006). Grasses are able to survive fires because they grow from the bottom instead of the top. Their stems can grow again after being burned off (*environment.nationalgeographic.com/.../**gr***). Fire becomes particularly important in the grass-forest ecotones (Sala *et al.* 2001).

Such climatic conditions with frequent fires and trampling grazers (see below) kill seedling trees (but not grasses). These abiotic and biotic factors are major driving forces that prevent colonization of temperate grasslands by trees (in general arborescent forests) (Chibilev 1990).

2.3.3. Soils

Edaphic grasslands arise on particular soil types. Usually grasslands soils are characterized by various grades of fertility, especially black soil (cherozem) with a thick layer of humus containing large quantities of nutrients. Humus is accumulated in the upper portion of soil making it dark (Figure 61). Black soil with increasing climate aridization turns into gray and chestnut-brown kastanozems (see Gibson 2009, 158p.) or colored soils. High organic content and a relatively high concentration of calcium and potassium make them the most fertile soils of the world. The amount of precipitations in grasslands is not enough to leach salts (Blair *et al.* 2014; Mordkovich 2014).

Grasslands produce larger amount of new growth, but repeated disturbance caused usually by grazing and fire, prevents the accumulation of a thick layer of dead leaf litter. Occurring across a wide range of climatic and geologic conditions, grasslands are associated with many different types of soils. The nature of grass litter and its patterns of decomposition commonly result in the development of a dark, organically rich upper soil layer that can reach 300mm below the surface. This layer is different from the surface layer of rotting leaf litter typical of forest soils (Gibson, 2009; Mordkovich 2014).

Creation of the stable humus layer is closely linked with limited life existence and slow decomposition (i.e., slow mineralization) of grasses. The matted roots of the turf-forming layer and increased humus content cause the better utilization of soil moisture and soil ability to control the above mentioned nutrients leaching into water resources (Tomaškin 2007). The high proportion of organic material is attributable to two factors: (1) invertebrate species (e.g., earthworms or Oligochaeta) and vertebrates (particularly rodents) which leave behind them large amounts of excrements and burrows; (2) bacterial and decomposition processes (Müller 1986).

Figure 61. Black soil (chernozem) of steppe biome. Dedophlistskaro district, E. Georgia, 1998.

2.3.4. Characteristic Vegetation

Vegetation of grasslands is mainly formed by grass, forbs and small annual plants. The dominant group is grasses.

Grasses or graminoids, i.e., true grasses themselves contain members of the family Poaceae (formerly Gramineae). The grasslands complex of historic climax plant communities comprises the herbaceous monocot, hemicryptophytic life form, typically have long, narrow leaves (Gibson 2009). Except for the primary members of Poaceae, grasses may also include Cyperaceae, Restionaceae and other narrow leaved monocots (Walter 1975; Lavrenko *et al.* 1991; Mordkovich 2014).

Perennial grasses – the dominant growth forms –include both sod forming grasses (prevailing within more humid grasslands) that grow by their underground stems or rhizomes and bunch grasses (predominate in drier parts of the biome) that are reproduced by seeds and without rhizomes.

Grasses make up one of the largest families (Poaceae) of flowering plants, a family, which plays the most important role increasing biodiversity in grassland ecosystems, their ecological functions, and in man's economic activity providing multifunctional agricultural landscapes (Gibson 2009). This life-form of vascular plants includes about 10,000-11,000 species. Only three families surpass grasses in the number of species: Asteraceae (about 20,000 species), Orchidaceae (about 20,000 species) and Fabaceae (about 17,000 species). All other families are much smaller in number of species, and genera as well (Tzvelev 2006).

Grasses – a nearly ubiquitous family – are rather uniformly distributed on all continents and in all climatic zones. They have adapted to conditions in lush and humid rain forests, dry deserts and cold highlands (subalpine and alpine zones). They appear as dominant vegetation in some biomes and many habitats, including tropical savannas, temperate steppes, salt marshes, and reed swamps (Gibson 2009).

The most important food plants of mankind are such cereals as wheat, rice, maize, barley and millet. Domestication of these cereal crops lies at the foundation of sedimentary living and civilization around the world is very important (Gibson 2009 and reference therein).

There is a long and complex relationship between grasses, grasslands and humans. According to paleontologists, genetic studies of humans and pre-humans, i.e., subtribe Hominine within the Hominine taxonomical tribe (humans and biped ancestors) originated in the open savannas of Africa (Tzvelev 1989; Gibson, 2009; Blair *et al.* 2014). Poaceae still constitute the most economically important plant family nowadays, providing hay, forage, fiber, fuel and building materials (many bamboos and some other tall grasses) (Tzvelev 1989; Gibson 2009; Blair *et al.* 2014).

Therefore, it is clear that grasses and grasslands have played an important role in the history of humans and will continue to do so in the future (Blair *et al.* 2014).

'*Agrostology* is the science of classification and is essential to the study of grassland' (Gould 1955 cited in Gibson 2009).

The family Poaceae consist of two subfamilies, (1) Bambusoideae and (2) Pooideae or true grasses with numerous tribes (14 and 41, respectively), genera and species (Tzvelev 2006).

Above ground and below ground grasslands consist of three strata – herbaceous layer, ground (or mulch) layer and root layer. The herbaceous layer, as mentioned, is manifested mainly by short and tall grasses and forbs. Among herbaceous vegetation of grasslands, forbs (especially belong to families Asteraceae and Fabaceae) are often mixed within graminoids or patchily distributed among grasses. Within more continental regions with seasonal climatic aridification the same mixes occur by herbs and shrubs. Grasses versus forbs or shrubs are highly variable. They change their appearance and compete with each other throughout the latitudes and relief, largely depending on chemical composition of soils, their depth, season of the year, climate, importantly the average annual precipitation (nearby or far from ocean to get moisture), daily and seasonal temperature fluctuations, frequently of the fires, etc. (Walter 1975; Tzvelev 1989; Dixon et al. 2014 and references cited therein).

The ground layer consists of residues of herbs in various stages of decomposition. The herb remains form a thick layer of mulch or detritus. Usually grazing and fire reduce the mulch (Mordkovich 2014).

Narrow, leafy shoots or tillers are held upright to reduce heating in summer. Large fibrous root systems effectively trap moisture and nutrients from the upper horizonts of the soil, and the growth buds are at or just below the surface, where they are protected both from the fires that sweep through the canopy and from grazing animals (Woodward 2003). In temperate grasslands with sharp continental climates plant dormancy is a mechanism for surviving low winter temperature (Blair *et al.* 2014; Mordkovich 2014).

The success of the grasses results from their tolerance of grazing herbivores and fire, linked to their reproductive strategies, and their versatility in photosynthesis (Edwards et al. 2010 and references cited in).

The root systems of perennial grasses form complex mats that hold the soil in place. They are highly branched (fibrous) and do not have a well-defined central taproot.

According to Tomaškin (2007), the root production of plants represents 50-80% of total plant biomass (a high root to shoot ratio). 'This pattern of biomass allocation coupled with slow decomposition and weathering rates leads to significant accumulations of soil organic matter and often highly fertile soil' (Blair *et al.* 2014).

Root/shoot ratios vary from 1:1 in mesic to 1:5 in xeric environments (Mordkovich 2014, *www.gis.usu.edu/.../24&27-Feb-2006.Hndout.**pdf***).

Reproduction. Unlike many plants the growing points or meristems of grasses are located near the base of the plant or below the ground, rather than at the tips of the plant, hence can quickly recover from cropping at the top. Many grasses spread horizontally through the production of below ground stems called rhizomes or 'sod-forming' grasses (horizontal below ground stems meristematic tissue or belowground buds that send shoots above ground), and prostrate stems aboveground known as stolons (horizontal shoots that may produce vertical shoots). Thus, new grass shoots can emerge from either rhizomes or stolons.

Grasses produce seeds through cross-pollination between plants, self-fertilization and asexual reproduction. Many grasses reproduce clonally through vegetative parts. The most common means of such spreading involves rhizomes and stolons (Mordkovich 2014).

Grasses – generally anemophily display a wide variety of adaptations for dispersal and establishment of seeds. Their hairs (sometimes perform like parachutes), spines, barbs on the spisclets, etc. readily cling to animal fur. Large grazing and other animals eat grasses and disperse the seeds that pass through their digestive tract (Mordkovich 2014).

2.3.5. Fauna of Grasslands

2.3.5.1. Aboveground Diversity of Animals

Despite grasslands large area, they have only a few species of mammals and birds endemic to that region, moreover, faunal diversity is lower than in many other biomes. The number of birds and mammalian species that are found primarily in grasslands are estimated to be 477 and 245 respectively, representing only 5% of the world's species for each taxonomic group (Groombridge 1992 in Sala *et al.* 2001).

One of the striking features of temperate grasslands is the presence of large herbivores, i.e., ungulates (hoofed mammals). After a dominant position of grasses in the second half of the Cenozoic, they played a key and selective role in the evolution of grass-consuming animals (Jakobs *et al.* 1999 cited in Strömberg 2011). Open space of grasslands has led bovids to be open grasslands inhabitants. Herbivores obtain food from many species of nutrient-rich grasses and herbs in the diverse steppe vegetational associations (Formozov 1966). As it seems landscapes without closed canopy of forests, due to adaptation measures of antipredator defense, performed a crucial role in the emergence of evolution of gregariousness, i.e., in the formation of gregarious behavior in local ungulates (McNaughton 1984). So, in grasslands – 'grazing land' – wild herbivory is recognized as a key biotic factor to be responsible for the establishment, growth and reproduction success of this grassland as a biome (McNaughton 1984 and references therein). Many grasslands supporting a diverse animal community, include some of the most species-rich grazing food webs on the planet (Blair *et al.* 2014). Grass-eaters are able to take full advantage of the various grasses which occurred there. They have a specially adaptive digestive system to process the grasses (Blair *et al.* 2014). In temperate grasslands with a harsh continental climate many native animals become dormant or migrate to avoid harsh winter conditions (Blair *et al.* 2014; Mordkovich 2014).

The mammalian grazers come in two life forms – cursorial types (the fastest running animals on the Earth) such as bison (megaherbivore) and pronghorn in North America, asses and saiga antelope in Eurasia, a variety of antelopes and zebras in southern Africa and kangaroos in Australia. To the second life forms belong semi-fossorial burrowing mammalians: 'Through herbivory and ecosystem engineering they create distinctive and important habitats for many other species, thereby increasing biodiversity and habitat heterogeneity across the landscape' (Davidson *et al.* 2012).

For example, pocket gophers, should be mentioned, for example California ground squirrel, black-tailed and prairie dogs from North America, marmots, European ground squirrel, pika from Eurasia, degu, mara, vizcacha – from South America, springhare, ice rat, Cape ground squirrel – from South Africa and burrowing bettong, hairy-nosed wombat – from Australia (Davidson *et al.* 2012). Burrowing mammals dig underground tunnels, which provide in dens, shelter for many animals (Müller 1986, Davidson *et al.* 2012).

Herbivores serve as important prey for many carnivores such are lion, leopard, hyena, wolf, coyote, badger, foxes, etc. In the steppes of Eurasia and the prairies of North America - ecologically similar ecosystems - should be mentioned mammalian fauna. To shared genera belong *Microtus, Spermophilus, Mustela, Vulpes*, etc. In spite of each grasslands have their specific forms. For example, geophilous pocket gophers are absent in Eurasian steppes, and, by contrast, some Eurasian species have no exact ecological equivalent in North America (e.g., *Ochotona pallasii, Equus hemionus*) (Formozov 1966). Despite distant relationships these and other semi-fossorial herbivorous mammals play important functional roles in grasslands around the world (Davidson *et al.* 2012).

The characteristic birds of grasslands are strong fliers. For example, grassland's chickens, meadow larks, grouses, quails, sparrows, hawks and owls. A number of species of lizards and snakes are found in grasslands. The invertebrate fauna is rich and occupies all strata of grasslands. Common arthropods of steppes include spiders, grasshoppers, locusts, leaf hoppers, beetles, bees, wasps, ants, flies, etc. (Vtorov & Drozdov 2001).

Acridids (locusts and grasshoppers) are widely distributed in various ecosystems including grasslands. According to Lockwood *et al.* (Eds.) 2000, they range in wide spectrum: 'some are threatened with extinction and become subjects of intensive conservation efforts, while others are devastating pests and become the object of massive control'.

Figure 62. Italian locust (*Calliptamus italicus*) in East Georgia (Gardabani district), 2011.

Most of these orthopteran insects are abundant and inhabit natural and anthropogenic habitats. They stimulate plant growth, participate in nutrient cycling, and play an important role in food chains. On the other hand, when their populations grow to catastrophic dimension, according to their strong migratory habits, they become primary pests of agriculture crops.

Among native herbivores, grasshoppers rank among the most conspicuous and important consumers of aboveground vegetation in natural grasslands, often rivaling large vertebrate ungulates in impact. For example, a swarm of *Schistocerca* spp. sometimes includes ten thousand million insects, a flaying mass of 50,000 or 10,000 tonnes. A swarm can fly 3,000 miles overland. Hence, the total biomass of invertebrates in grasslands typically exceeds that of the much larger and more conspicuous vertebrates, except in regions containing large numbers of domestic stock. The species diversity of grasshoppers, for instance, in the Great Plains (North America), are similar to bird species diversity. These orthopterians belong to polyphagous insects, hence do not specialize on particular host plants (*plains humanities. unl.edu/encyclopedia/doc/egp.pe.031*).

In this context should be mentioned the following ground-dwelling species of grasslands and deserts: *Schistocerca gregaria, Locusta migratoria, Calliptamus italicus, Dociostaurus maroccanus* and others. To combat the acridids swarms sometimes demands transitional operations, which cost over half a billion US dollars to the world community (Latchininsky et al. 2011 and references therein) (Figure 62).

2.3.5.2. Belowground Diversity of Different Organisms

As it is well established apart from plants (i.e., producents being the primary producers), other consument organisms play key roles in terrestrial ecosystems. Terrestrial ecosystems are divided into aboveground and belowground compartments, which are tightly linked by plants, in other words this is an ecological model of plant - mediated linkages between aboveground and belowground food webs (Wardle *et al.* 2002, 2004).

Detritus – dead particulate organic material - typically includes organic wastes, for example, the bodies or fragments of dead organisms as well as fecal material. Usually detritus colonized by communities of microorganisms decomposes (i.e., is mineralized) the material (Swift *et al.* 1979). In terrestrial ecosystems, including grasslands, it is encountered as plant tissue (e.g., leaf litter) and other organic matter (animal tissue, feces, etc.) intermixed with soil. Such a combination of two materials is referred to as humus. The energy contained in detritus is not lost to the ecosystem as a whole, but it serves as a source of energy for a group of organisms – detritivores. They have their own food chain, i.e., detrital (detritus) food web (decomposer pathway), which are separated from the grazing food web (grazer pathway) organisms. In some ecosystems (e.g., upper layer of soils – humus), considerably more energy flows through the detritus food chain (aboveground biota) (Moore *et al.* 2004).

Detrital food webs (or soil food webs) include a lot of organisms which are distinguished by extreme heterogeneity. The following groups belonging to this food web should be mentioned: algae, bacteria, actinomycetes, fungi, protozoa, insects and their larvae, other arthropods (crustaceae, mites), molluscs, annelid worm (Olygochaeta), round worms (nematodes), millipedes and some vertebrates (mainly rodents).

As already mentioned, grasslands are very rich (or rich) with humus, therefore they are distinguished by tremendous representatives of soil biota. For example, soil can contain among other thing, between 5,000 and 10,000 species of microorganisms (Torsvik *et al.,* 1990 cited in Wardle *et al.* 2004).

As mentioned, soil-dwelling detritivores vary widely in size from macrofauna >10mm in length (earthworms, oribatid mites, dipteran larvae, millipedes, etc.) to micro- and mesofauna <10mm in length (protozoa, nematodes, collembolans) to microscopic forms (bacteria, saprophytic fungi). On the other hand, soil biotas are divided into major groups: herbivores (root feeding fauna - insects (weevil larvae, cabbage maggot, etc.), pathogens (bacterial, fungal), mutualists (mycorrhizal fungi, nitrogen - fixing bacteria) and predators (predatory mites, some millipedes, insect larvae, arachnids (tarantula), mole and other burrowing soil-inhabiting animals (Giljarov 1965).

2.3.5.3. Intimate Interaction Between Plants – Aboveground and Belowground Subsystems

In grasslands (belowground) the soil biota component of an ecosystem generally equals or exceeds that of the aboveground biomass (Paul *et al.* 1979 cited in Sala *et al.* 2001).

In shortgrass steppe the exclusion of mentioned patterns are arthropods, which take advantage of the large amount of belowground primary production (Lauenroth & Milchunas 1992 cited in Sala *et al.* 2001).

The recent understanding of the intimate interactions between two subsystems – aboveground biota and belowground communities are discussed in some works (e.g., Wardle 2002; Wardle *et al.* 2004, etc.). In both compartments, heterotrophic organisms are involved in numerous interactions with plants. At the same time the both compartments can be powerful mutual drivers (top-down forces and bottom-up forces), with both positive and negative feedbacks (Wardle *et al.* 2004).

The interactions between the above and belowground biotas in the fertile, productive ecosystems are positive supporting high soil food webs with a high density of micro- and macrofauna with a high density of earthworms and bacteria-based energy channel. On the contrary, the same interactions between two compartments in the infertile ecosystems are negative, support plants with low litter quantity and quality tend to support soil food webs dominated by fungi and arthropods (mites, collembolans) (Wardle 2002).

Thus, a balance of four major processes influences the ecology and development of grasslands on the Earth: climatic (semi-humid/semi-dry with more or less display drought seasons), pyrogenic (fires, natural or manmade in recent geological time), edaphic (mainly fertile humus-rich, black) and biotic (grazing by wild herbivores and in Holocene time domestic cattle, preventing spread of trees). The relative importance of fire in structuring grassland communities tends to be the most important (Blair *et al.* 2014).

As we see, in sustainable grasslands systems ecological linkage between above- and belowground biota soil ecosystem becomes increasingly important. The primary grasslands keep self-regulating processes. Although, among 14 major terrestrial biomes that have been most significantly altered globally by human activity are the temperate grasslands. More than two thirds of the area of mentioned grasslands (with Mediterranean forests) had been converted, primarily to agriculture (Reid *et al.* 2005). An important focus for ecologists is to provide the basic steps for developing and optimizing sustainable grasslands habitat similar, for example, to Askania Nova in Ukraine.

2.3.6. Photosynthetic Pathways – C_3, C_4 and CAM

Grasslands are a very large carbon sink at world level (Suttie et al. (Eds.) 2005. According to Minahi *et al.* (1993) (cited in Suttie *et al.* (Eds.) 2005), 'grasslands are almost as important as forests in the recycling of greenhouse gasses, and that soil organic matter under grassland is of the same magnitude as in tree biomes'.

Characterizing the ability to store carbon, in more than two-third of grasslands carbon is stored belowground. The large quantity of carbon is stored in grassland soils and could provide some emissions of carbon dioxide that drives global warming (CCSP–US Climate Change Science Program, 2008 cited in *Grassland Ecosystems*: *www.wildlifeadaptationstrategy.gov/.../Gras*).

As is broadly known photosynthesis is the fundamental biological process that transforms solar energy into the chemical fuel for life by generating sugars from water and CO_2 (Edwards *et al.* 2010).

The most significant variation in the internal structure of grass leaves involves anatomical differences associated with several (three) photosynthetic pathways. Depending upon the actual climatic conditions of a given region, photosynthetic pathways – C_3, C_4 and CAM are found in different biomes of the Earth and various green plant groups. These designations of mentioned metabolic pathways refer to different efficiencies of carbon fixation by which carbon dioxide is taken from the atmosphere and converted into carbohydrates during photosynthesis (Wataru et al. 2013, *agridr.in/tnauEAgri/eagri50/PPHY261/lec11.pdf*).

Plants, which use only the Calvin cycle for fixing carbon dioxide from the air are known as C_3 plants. C_4 carbon fixation is the next biochemical processes used to fix carbon. Finally, there is third pathway, in which intermediates between C_3 and C_4 photosynthesis are formed and this is known as a CAM (Crassulacean Acid Metabolism) photosynthesis. Explanations of these gradients has traditionally been focused on fundamental anatomical and physiological differences between those mentioned and considerably between the C_3 and C_4 pathways of photosynthesis (Edwards *et al.* 2010).

Plants that use solely C_3 fixation tend to settled in the areas where sunlight intensity and temperatures are moderate, CO_2 concentrations are rather high, 500 ppm (parts per million by volume) or higher, and groundwater is plentiful. C_3 plants lose 97% of the water taken up through their roots by transpiration. C_3 photosynthesis is typical of most green plants: about 83% of plant species and 95% of the Earth's plant biomass are C_3 plants. They include cereal grains (rice, barley) and most trees (Sage 2004; Edwards *et al.* 2010).

C_4 carbon fixation is the second of three biochemical processes used to fix carbon. The C_4 pathway of photosynthesis in plants with its series of anatomical and biochemical modifications shows some unique adaptations to certain environments that allow the plants to survive in rather harsh abiotic conditions. These conditions include high temperature drought, high salinity, waterlogging and nitrogen and CO_2 limitation. These adaptations allow the plants to overcome photorespiration and carbon shortage inhibiting plant growth. Avoiding this disadvantage, they use a two-stage strategy of photosynthesis, first via a 4-carbon organic acid, and second by the carboxylating enzyme Rubisco (these processes take place within C_4 plants with a characteristic leaf anatomy called Kranz anatomy, after the German word for wreath) (Sage 2004).

Therefore, the C_4 pathway reduces the loss of energy that takes place in C_3 plants as a consequence of the oxygen activity rates observed in C_4 plants under some conditions. Thus, the C_4 cycle is also a good example of how regulation of a metabolic pathway may be achieved. This biochemical peculiarity keeps CO_2 high and oxygen low in the chloroplast, where the Calvin cycle operates (*https://en.wikipedia.org/wiki/**C3**_carbon_fixation*).

All of the above becomes clearer from the following quote: 'C_4 photosynthesis is a coordinated system of anatomical and physiological traits that concentrate CO_2 around the C_3 photosynthetic machinery, through the use of a solar-powered biochemical cycle' (Edwards *et al.* 2010).

This different pathway for CO_2 fixation was explained by Hatch and Slack in 1966, hence it is sometimes called the Hatch-Slack pathway (Iglesias *et al.* 1986).

2.3.6.1. Climate Controls of Plesiomorphic C_3 vs Apomorphic C_4 Photosynthesis

Plants with the C_4 photosynthetic pathway dominate today's tropical savannas and warm-climate grasslands, and account for some 30% of global terrestrial carbon fixation (Still *et al.* 2003 cited in Edwards *et al.* 2010). Within these biomes with their high rates of foliage production are sustained the Earth's highest levels of herbivore consumption (Heckathron *et al.* 1999 cited in Edwards et al. 2010). C_4 carbon fixation has evolved on up to 45 independent evolutionary groups, occurring in different families of plants, making it one of the most striking examples of convergent evolution in plants (Sage 2004), 'and may have evolved up to 11 times' (Hattersley 1986 cited in Gibson 2009). About 7,600 angiosperm plant species use C_4 carbon fixation, which represents about 5% of the Earth's plant biomass and 3% of all terrestrial species (Hatch 1999; Bond *et al.* 2005b). With this negligible number of vascular plant species, flowering plants with C_4 photosynthesis have transformed the biosphere (Edwards *et al.* 2010). The C_4 pathway is less common in dicots than in monocots (4.5% and 40%, respectively). Despite this, only three families of monocots utilize C_4 carbon fixation compared to 15 dicot families. Of the monocot clades containing C_4 plants the grass (Poaceae) species use the C_4 photosynthetic pathway. Most, about, 50% of all grass species use C_4. They are responsible for about 25% of the Earth's terrestrial primary production (Ehlerlinger *et al.* 2002; Still *et al.* 2003; Sage 2004). Plants utilizing this pathway rank among the world's tremendously important crops: maize, sugarcane, millet and sorghum. It should be stressed that corn is a C_4 crop which can be grown under dry conditions (Osborne & Beerling 2006; Gibson 2009 cited in Stromberg & McInerney 2011).

Of the dicot clades containing C_4 species, the order Caryophyllales contains 33 families, 692 genera and 11,155 species. Major families in the order include Caryophyllaceae (2,200 species), Amaranthaceae (2,050-2,500 species), Aizoaceae (2,020 species), Cactaceae (1,500 species), Polygonaceae (1,100 species) etc. (Kadereit *et al.* 2003 cited in Osborne & Beerling 2006).

These advantages are manifested in the current distribution of grasses with C_4 photosynthesis in the tropics and subtropics (lowland areas) where the high air temperature contributes to higher possible levels of oxygenase activity by the enzyme Rubisco. This enzyme increases rates of photorespiration in C_3 plants (Sage & Monson (Eds.) 1999, Blair *et al.* 2014). Woodward (2003) pointed out that C_4 grasses are associated with temperate grassland as well, but with those parts which are characterized by hot summer. They are also referred to as warm-season grasses, because they do not grow until early summer, and peak growth occurs when temperatures reach about 35^0C. The examples of warm-season grasses are represented in the tallgrass prairie or 'true prairie' (625-1200mm precipitation per year) within the central grassland of North America (Anderson 2006).

On the contrary, grasses and arboreal plants with the plesiomorphic C_3 pathway are favored at high latitudes and altitudes with cool (or cooler) temperatures and under high CO_2, where photorespiration losses are low (Hartley 1973 cited in Strömberg & McInerney 2011). This specificity predicts that the C_3 vs C_4 crossover temperature is approximately 22^0C (Ehleringer *et al.* 1997; Collatz *et al.* 1998 cited in von Fischer *et al.* 2008). Today the knowledge of this climate (temperature) and biology (net carbon assimilation rates) relationship remains incomplete (Fischer *et al.* 2008).

CAM photosynthesis is also known as Crassulacean Acid Metabolism. This type of carbon fixation pathway is most common in desert plants. It evolved as an adaptation to increased efficiency in the use of water. In CAM photosynthesis, CO_2 is taken up only at night and is stored in vacuoles. The stoma is open only at night, when it is relatively cool and humid. Many CAM plants have photosynthetic cells (which are not mesophyll) on the stem. Photorespiration is not characteristic, unlike C_3 and C_4 (comparatively little) plants.

The metabolism was first studied in plants of the Crassulaceae family, which mainly include succulent forms (Ranson & Thomas 1960, *https://en.wikipedia.org./wiki/Crassulaceae _acid_metabolism*).

So, various photosynthetic pathways are adapted in different ecological conditions. C_3 photosynthetic pathway is best under moist conditions, C_4 – under warm, sunny, dry conditions, and CAM under desert conditions.

2.3.7. The Expansion of Grassland Ecosystems on the Earth in Relation to the Complexity of Abiotic and Biotic Events of the Cenozoic Era

2.3.7.1. General Review

According to such an important role of grasslands (Poaceae) for the biosphere and rapidly growing human population, the origin and subsequent evolution of grasses and grassland ecosystems continue to be a major target in evolutionary biology and paleontology (Strömberg 2011 and references therein). 'To appreciate how this prominent biome functions today –and how it will function in the future – it is vital to understand how it is formed' (Strömberg & McInerney 2011).

In the past decades, according to fossil records, much progress has been seen in researches about the understanding of grasses and grassland ecosystem origins and further evolutionary history (Edwards *et al.* 2010; Strömberg & McInerney 2011). In the late 20th and early 21st-century, grassland research has focused on the co-evolution of grasses and fire-grazing interactions (Anderson 2006; Janssen & Bremer 2004 cited in Gibson 2009).

Following current biological and geological knowledge, there are some hypotheses about the origin and ecological dispersion of grasslands.

Grasslands in common with other major biomes, especially Mediterranean communities, after the data of their earliest appearance, experienced the effects of major global changes in environmental factors and anthropogenic press. Depending on the severity or frequency of these factors biomes with herbs profoundly changed from one vegetation type to another. This historical patterns of ecological successions onto the grass family Poaceae, it seems, took place from mid Cretaceous (Gibson 2009).

The Cenozoic history of extension and contraction, when the Earth after the Paleocene-Eocene thermal maximum had begun a drying and cooling trend culminating in the glaciations of the Pleistocene Epoch, has shaped one of the most characteristic biomes of the Earth – grasslands. There are some hypothesis and scenarios regarding the major drivers of change. On the basis of compilation of the main scenarios all try to stress what were the main trends of evolution of grasses and grasslands? To shed light on this problem – the origin of Poaceae and formation of grassland ecosystem – today both direct and indirect records are generally used. The direct record includes macrofossils (fossilized parts of grasses) and microfossils (pollen, phytoliths, e.g., plant silica bodies, fossil record of stable carbon isotopes). The indirect one includes grazer mammalian anatomy (tooth morphology, long legs and others), which coincided in response to changing of diet style from browsing (folivore) to grazing (herbivore) (Strömberg & McInerney 2011; Dixon *et al.* 2014).

The ancient (Plesiomorphic or Calvin-Benson) pathway (C_3 photosynthesis) evolved in a CO_2 rich atmosphere more than 2,800 million years age (Edwards *et al.* 2010). Therefore, from this period 'photosynthetic CO_2 – fixation has provided the carbon for life on the Earth for at least the past 2.7 billion of years. More than 99% of the history of this ancient process has been dominated by C_3 photosynthesis' (Osborne & Beerling 2006).

This crucial event within the Archean Earth's history is associated with a common ancestor of extant cyanobacteria (Cardona *et al.* 2015).

Based on analyses of a wide range of fossil materials it is confirmed that plants colonized continental surfaces of the Earth sometime before 470 million year ago. Later, the Carboniferous period (about 300 mya) resulted from the rise of vascular land plants and the spread of rooted forests (Berner 2005 in Osborne & Beerling 2006). From this period evolutionary trends towards increased vegetation activity largely attracted both atmospheric composition and climate (e.g., Permo-Carboniferous - 300-250 mya) the longest and most severe glaciations of the Phanerozoic Eon. Therefore, a striking decrease in atmospheric CO_2 and an increase in O_2 during the Late Paleozoic occurred (Beerling 2002; Osborne & Beerling 2006).

As mentioned, the vast majority of grass diversity falls into two clades, in BEP (Bambusoideae, Ehrhartoideae, Pooideae) and PACMAD (Strömberg 2011).

The evolutionary history of grasses can be broken down into several distinct stages: (1) pre–Cenozoic appearance of C_3 grasses, as it seems in a marginal part of the closed canopy forests habitat (mainly BEP); (2) Cenozoic (Paleogene) appearance of C_3 and C_4 open – habitat grasses. Within the latter event: (2.1.) the Paleogene appearance of C_4 open-habitat grasses, (2.2.) the mid-late Cenozoic emergence of open, C_3 grasses-dominated habitats, and, finally, (2.3.) the late Neogene explosive evolution and dispersion of C_4 – dominated, grass-dominated habitats (Strömberg 2011 with some modification).

2.3.7.2. The Story of Pre-Cenozoic Evolution of Poaceae

To highlight significant gaps in our knowledge in the mentioned field of paleobotany we need answers to the following question – 'where and in what context did grasses evolve'? (Strömberg 2011)

On the base of recent molecular study and pollen and macrofossil records grasses were established in northern Gondwana in the latest Cretaceous and Paleocene (Tzvelev 1989;

Bremer 2002; Bouchenak-Khelladi *et al.* 2010). The fossil materials include several major modern Poaceae clades, for example members of the modern subclades PACMAD and BEP (Strömberg 2011). According to Prasad *et al.* (2005), cuticles with phytoliths of Oryzeae (including rice (*Oryza sativa*) and its relatives) have been found on the Indian subcontinent (Deccan Plateau) in the Late Cretaceous, 'pushing back the time of the origin of Poaceae as a whole'. The authors described microscopic silica bodies of grasses within the coprolites of titanosaurian sauropods. The same fossilized dinosaur coprolites have been found containing phytoliths of grasses related to modern rice and bamboo (Piperno & Suec 2005).

The second essential conclusion suggested that by this time the PACMAD and BEP lineages had split while Africa and India were still in contact (India broke away from the other fragments of Gondwana 100 mya ago) (Prasad *et al.* 2011).

Later, after India made contact with the South Asian margin in the Early Eocene period (~50 mya), dispersal of the major grass lineages would have taken place (Prasad *et al.* 2005).

Apart from the mentioned important evolutionary innovation, Gondwanatherian mammals, from the Late Cretaceous of South America, Madagascar, and India and the Paleogene of South America, and Antarctica may represent the earliest evolution of grazers (Prasad *et al.* 2005).

Grasses became widespread toward the end of the Cretaceous Period. However, the current simulations have suffered from a lack of direct evidence about the immediate ancestors and geographic center of the origin of the grasses. All these and similar questions regarding these evolutionary innovations of the plant kingdom still remain mysterious (Gibson 2009 and references therein).

It is suggested that grasses evolution occurred within or on the margin of tropical forests, adapted to forest shade habitats (Osborne & Beerling 2006). As Bambusoideae (as a subfamily of the Poaceae) generally still grow in forests and retain pleziomoprhic features in their flowers, they were possibly the first grasses (*Encyclopedia Britanica:http://www.britanica.com /plant/Poaceae*).

According to Tzvelev (1989) general climatic deterioration at the end of the Cretaceous period led to the extinction of the majority of Mesozoic plants of the lowlands, foothills and lower elevations. The free ecological niches were settled by flowering plants including grasses that descended from the upper limits of the forests (with seasonal climates) and appeared in high mountains during the time of orogeny. They, as already mentioned, occupied layers underneath the trees, i.e., began to grow in the shade of forests, and here, as Tzvelev suggests, the primitive bamboo grasses originated. Later their trend to gigantism via lignification of the stem occurred. The primitive species of 'arundinoid' and hydrophylous 'oryzoid' clades originated by adaptation of the first grasses to more humid habitats with sufficient moisture from open drainages to wetland ponds. Some of them (e.g., 'arundinoid' grasses) are more closely connected to the 'festucoid' grasses.

The fundamental feature in reproduction of grasses in new, more open habitats was shown by the change of the obligate entomophily into anemophily (Tzvelev 1989).

However according to a new study, a striking exception to the mentioned to-be-recognized viewpoint is an extraordinary discovery in the former Laurasian area – Myanmar, Burma ($26^0$20'N, $96^0$36'E). Namely, a fossil specimen completely resembling grass infected by a fungus (*Claviceps purpurea*), which consisted of an amber (fossilized in araucarian tree). Amber-bearing strata in the in the mine of Mayanmar dates from the Early Cretaceous (about 100 mya). So, this finding appears to contain the Earth's oldest grass fossil, about 50 mya older than any fossil grasses previously found (Poinar *et al.* 2015 and references therein).

2.3.7.3. The Peculiarities of the Cenozoic Evolution of Grasslands

The growing material of paleontological evidence and advances in molecular biology promise a current improvement in dating key events in grass evolutionary history, which finally is manifested by explosive emergence and ecological expansion of C_4 grasslands biomes in Neogene (the Late Miocene-to-Pliocene) (Gibson 2009 and references therein).

Poaceae subclades derived from the Cretaceous first grasses, as mentioned, began a nearly worldwide distribution (Strömberg 2011). At the same time, 'Pooideae and PACMAD last shared a common ancestor >50mya, allowing both lineages at a later period via selection-influenced, adaptive radiation including potential C_3 and C_4 photosynthetic pathways that played a major role in the origins of grasslands ecosystems in the Neogene' (Edwards *et al.* 2010).

The global climate deterioration improved scholars understanding of the further evolutionary history of grasses and grasslands. The most crucial abiotic event during the end of the Eocene of the Earth was the covering of the Antarctic by a continental-scale ice sheet. The latter geographic phenomenon is the most prominent feature of the mainland which covers over 99% of the continent (Drews 2015). In the Eocene/Oligocene boundary (about 34 mya), the rapid widespread scale glaciation of Antarctica occurred and the associated establishment of the Earth's cryosphere, i.e., global climate oscillations. The formation of such a large continental ice sheet forced by a gradual decline of atmospheric CO_2 is one of the most fundamental reorganizations of global climate known in the geologic record (DeConto & Pollard 2003; Pollard & DeConto 2015).

2.3.7.3.1. Carbon Dioxide and the Rise of C_4 Plants

Carbon dioxide (CO_2) – a potent greenhouse gas – currently constitutes about 0.04% (400ppm - parts per million by volume) of the atmosphere (Sage et al. 2011, Ballantyne et al. 2012). Study patterns of correlation of carbon dioxide and temperature over the Phanerozoic (about 570 mya) typically find that periods of low CO_2 indicate a strong coupling with long-lived, extensive continental glaciations. CO_2 concentration in the atmosphere has varied, ranging from as high as 7,000 ppm during the Early Paleozoic to as low as 180 ppm during the Pleistocene multiple glaciations (Royer 2006 and references therein).

Geochemical evidence indicates very high CO_2 volume concentrations between 600 and 400 million years ago (in the Cambrian and in the Ordovician periods of the Paleozoic Era) of about 7,000 ppm. The next rise to a high value of CO_2 occurred between 200 and 150 million years ago (on the boundary of the Permian and Triassic periods) of over 3,000 ppm; The Mesozoic Era is characterized by a well-described, long duration global warm event. During the Mesozoic, in general, CO_2 volume concentrations were thought to range between 1,000-3,000ppm (Royer 2006).

The Cretaceous period resulted in a further decrease in CO_2 concentration (about 1,000 ppm). The last geological, the Cenozoic Era (last 65 million years), as mentioned, began with the Paleocene-Eocene Thermal Maximum (PETM). This is a time of massive carbon injection, i.e., volcanic outgassing due to North Atmospheric rifting (Pearson & Palmer 2000). Atmospheric carbon dioxide values were at 700-800 ppm, but over periods of time of less than 1 million years the volumes of CO_2 rise to over 2,000 ppm (Royer *et al.* 2001). So, CO_2 forcing was the only direct impact of the PETM event, which caused global temperature increasing by

5-8^{0}C.Therefore, the planet was essentially ice-free (little to no ice) with a smaller difference in temperature from the equator to the poles (McInerney & Wing 2011).

Importantly, a sharp climate change in the Eocene-Oligocene transition, about 34 million years ago, is the geological time when two striking events took place: (1) the Eocene-Oligocene extinction and (2) when the above mentioned Antarctic ice sheet started towards its current form, in other words 'the Earth experienced a mysterious cooling trend' (Pearson *et al.* 2009).

The development of the Himalayas and the Tibetan Plateau may have been a trigger for a global reduction of CO_2 (Raymo & Ruddiman 1992). In the Eocene-Oligocene transition CO_2 is found to have been about 760ppm. Apparently, these events presented a critical test for the Paleogene abiotic-biotic paradigm in the Cenozoic history of the Earth (Pearson *et al.* 2009).

The appearance of herbaceous and annual life-forms including the C_4 photosynthesis pathway in the Oligocene was a key biotic factor, because the C_4 pathway, as stressed above, is very successful in these groups, particularly in arid and semi-arid ecosystems (Sage 2001). During the Oligocene the atmospheric CO_2 levels declined below (500 ppm) - the high levels of the Early Paleogene, hence, the global climate became drier, which triggered the formation of open landscapes in most tropical and subtropical climates favored by evolution of C_4 plants with a drought-adapted feature (Ehleringer *et al.* 1997; Wolfe 1997).

There is geochemical data that carbon dioxide concentrations were less than 300ppm by about 20 million years ago (Royer 2006). In more recent geological time regarding further decrease in CO_2 (data from ice core) climates deteriorated further, most notably during the Early Pleistocene. This next cooling led to the glacial-interglacial oscillation (Zachos *et al.* 2001).

The above-mentioned Antarctic icing with its continental ice-sheet undoubtedly caused a global chilling. These key climatological changes greatly influenced formation of open habitats. Habitats lacking tree layers with closed canopy strongly coincided with almost worldwide distribution of grasses in the Early Eocene. The Middle Eocene – earliest Oligocene member of more derived Poaceae subclades began to appear, including pooids and potential chloridois, bamboos with affinity to the genus *Chusquea* and stipoid pooids, and the first C_4 PACMADs in various regions of the Earth (Strömberg 2011 and references therein).

CO_2 reduction to <500 ppm during the Miocene-Pliocene transition (8-6 mya ago) was the primary agent forcing Antarctic glaciations (Pagani *et al.* 2011). Low CO_2 concentrations may have been a major selective force for an explosive evolution, ecological and geographical distributions of grasses possessing C_4 photosynthesis and forbs (Cerling *et al.* 1997, Ehleringer *et al.* 1991, 1997).

The expansion of plants using the C_4 photosynthetic pathway over those using the C_3 pathway occurred during and since the late Miocene (about 25 mya), after their first evolutionary radiation (Strömberg 2011). They increased in abundance and began wide-ranging successful expansion in some regions and latitudes of the Earth on the boundary of the Miocene and the Pliocene Epochs (7 and 5 million years ago), and later during full-glacial and interglacial periods (the Pleistocene) (Ehleringer *et al.* 1991; Osborne & Beerling 2006). This biotic event constitutes one of the most important ecosystem transformations of the Cenozoic (Flanagan *et al.* (Eds.) 2005).

2.3.7.3.2. The Expansion of the C_4 Photosynthetic Pathway (The CO_2 – Starvation Hypothesis) and Formation of Grassland Biomes (The Neogene Biogeography of C_4 Evolution)

Osborne and Beerling (2006) suggest that explaining the Late Miocene expansion of C_4 grassland requires a mechanism other than just CO_2, and, therefore, its causal explanation. For reconstructing the evolution of the Late Miocene to Pliocene geographical expansion of C_4 grasses and formation of grasslands biome, Edwards and co-authors (2010) directed their attention to the largely recognized biogeographical method of comparison of the same biomes of modern and remote past (paleo-grasslands). Edwards *et al.* (2010) pointed out that 'If we assume that paleo-grasslands had a similar diversity structure as their modern analogues, then understanding the Miocene expansion of C_4-dominated ecosystems hinges on the question 'How did particular grassland dominants (as opposed to C_4 grasses in general) come to occupy such a large fraction of the land surface?' The same viewpoint is manifested in Osborne and Beerling (2006).

Modern investigations highlighted the essence of the question, which is very complex.

So, 'when and how did key changes occur?' 'What was the main pattern of evolution of grasses and grasslands biome?' Recent advances in the understanding of C_4 plant's pathway of evolution are integrating in modern achievements across palaeontological (geology) and neontological (biogeography, ecology, physiology and molecular biology) sciences (Osborne & Beerling 2006, Osborne 2008). Below, on the base of compilation of various literature sources are given the main trends and scenarios in this field of natural sciences.

As previously noted, the Late-Oligocene CO_2 reduction below the critical threshold was the environmental pre-condition for C_4 photosynthesis which was repeatedly aroused in multiple grass lineages around 30 million years ago (Osborne & Beerling 2006). According to this viewpoint, global climates became cooler and drier in the past 40 million years, influencing the rise of arid-climate-adapted vegetation types (Sage 2001). However, both molecular evidence and fossil evidence suggest that large-scale expansion of C_4 plants and the original C_4–dominated biome occurred long after the C_4 grasses evolved (Scheiter *et al.* 2012; Hoetzel *et al.* 2013). Actually, as already pointed out, the dominant status of the C_4 photosynthetic pathway was reached only recently, during the Late Miocene and the Early Pliocene (Ehleringer *et al.* 1997, 2002; Osborne 2008; Gibson 2009). Declining atmospheric CO_2 has long been considered the key driver of this latter event. However, new geological evidence including an Ocean Drilling Program (Hoetzel et al. 2013), casts doubt on this simulation. Certainly, a C_4 carbon – concentrating mechanism evolving as an adaptation to CO_2 starvation was a plausible selective agent during the Late Paleogene, but a time gap, which remains between the origin and global expansion of grasses of around 20 mya becomes the subject of debate in this field (Osborne 2008). Recently proposed drivers of grasslands-dominated biome formation include complex, abiotic and biotic factors. For example, climate (increasing seasonality), increased aridity (periodic droughts), creation of a dynamic mosaic and then large open (of high light) habitats, frequent burning (fires), and large ungulate herbivores caused grassland ecosystems ecological generalization and adaptive radiation (Sage 2001; Osborne & Beerling 2006; Osborne 2008; Gibson 2009). To express the same ideas more briefly, grassland plants evolved under the influence of the periodic drought, frequent fires and grazing megafauna (Gibson 2009; Anderson 2012). All three factors play a crucial role to displace mesic forests in favor of grasslands. Among these ecological patterns, hypotheses of the Late Miocene-Pliocene C_4

grasslands expansions significantly strengthen ecological disturbance patterns including exotic, large, hooved mammals, and considerably profound changes (raised) in a fire regime (Osborne 2008; Hoetzel *et al.* 2013).

All of these viewpoints are largely based on fossil evidence. The latter materials indicate an abundance of woody C_3 plants throughout the Miocene, forming forests and other arboreal plant (trees) communities, which in some regions gradually became C_4 grasslands (Osborne 2008).

Attempt to bridge the knowledge gap between ecology of forests, on the one hand, and open biomes – savannas and grasslands, on the other hand, are given in some articles (e.g., Baudena *et al.* 2015). Thus, fully open grasslands with C_3 and C_4 grasses were likely a Late Miocene-to-Pliocene phenomenon (Osborne & Beerling 2006; Osborne 2008; Gibson 2009; Strömberg 2011).

The further paleogeographic event of the Paleogene and the Neogene boundary caused transgression of the sea and uplift of mountain systems and intensified the seasonal Indian monsoon climate with lighting strikes, significantly manifested in south Asia (Zhisheng *et al.* 2001; Zhang *et al.* 2007). Hence, the seasonal climate and increasing aridity brought on by the Asian monsoon favored grasses over trees and later mesic forests (Osborne & Beerling 2006).

As previously stressed, there are many morphological and physiological features that allow C_4 grasses to tolerance of various abiotic stress: (1) the occurrence of bulliform (bundle sheath) cells in leaves, which cause them to enroll, when they loss water, i.e., stomatal sensitivity to water loss, and they can grow under conditions of low-water potential; (2) utilization of the C_4 or Hatch photosynthetic pathway that adapts plants to extreme abiotic conditions and nitrogen-use efficiencies; (3) adaptation to high temperatures and a high level of solar radiation via well-protected buds and storage reserves; (4) high wet-season growth rates and slow rates of leaf decomposition result in large fuel accumulation inducing fire (Anderson 2006, 2012; Osborne & Beerling 2006; Edwards *et al.* 2010).

Significant change in seasonality, as previously noted, is closely linked with tectonic events, which caused mountain building such as the uplift of the Himalayas, the Tibetan Plateau, the reduction of the Paratethys, and, also the African Rift system (Keeley & Rundel 2005; Osborne 2008; Scheiter *et al.* 2012; Hoetzel *et al.* 2013).

Palaeontological evidence (pollen records, phytolith assemblages, mammalian tooth and bone morphology and flightless birds' bone and egg shell) suggest that a transition from close-canopy forest to open-woody vegetation resemblance to nowadays savannas or arid open woodlands vegetation with a significant C_3 (the Oligocene-Early Miocene) and later (the Late Miocene–Early Pliocene) C_4 grass components. Therefore, initially, grasses occupied open ground as grass-dominated patches within heterogenous ecosystems enriched with forest trees (Osborne & Beerling 2006; Osborne, 2008).

Increasing seasonality during the Late Neogene period with a hot growing season (creating fire regime) can reduce tree cover and convert closed-canopy forests into grassland-to-forest mosaic of woodlands with more open grass-dominated habitats at low rainfall. In other words, a major selective advantage over C_3 species in open woodland, savanna or grassland took place in seasonally arid climate habitats. Key factors which should be mentioned are water shortage and disturbance by fire (Bond *et al.* 2005b; Osborne 2008; Scheiter *et al.* 2012). Evidence for a heterogenous landscape is available from the Middle Miocene (e.g., Kenya) (Strömberg 2011). Experimental work has demonstrated that a climate with a hot season, and as a consequence creating a fire regime, is a strong predictor of the global distribution of the savanna biome (Staver & Levin 2012 and references therein).

2.3.8. Two Key Drivers - Disturbance Agents of Grasslands Biome: Fire and Large Ungulate Herbivores

2.3.8.1. A Brief History of Fire. Fire Triangle or Creating Combustion

The Earth is a flammable planet. Fire-prone ecosystems cover 40% of the Earth's present land surface (Chapin *et al.* 2012 in Belcher *et al.* 2010). Today fires (frequently disastrous) have occurred on every vegetated continent (Bowman *et al.* 2009). 'All terrestrial vegetation has the potential to be fuel' (Glasspool *et al.* 2015). Historically, different fire regimes for different geological times, as already discussed, created in grasslands, savannas, Mediterranean shrubland and boreal forests (Bond & Keeley 2005) their biotas (Bowman *et al.* 2009). Each year, global wildfires burn roughly 865 million acres of land (Howard 2014) and have high (multibillion US) damages (Bowman *et al.* 2009).

However until relatively recently, the classical view was that plant and animal species distributions can be explained solely by climate and soil. Therefore, the role of fire in global ecosystems and the role of plant-fire linkages and their evolutionary potential had not been identified and recognized. Bond and Keeley (2005) comment: 'It is difficult to find references to fire in general textbooks on ecology, conservation biology or biogeography, in spite of the fact that large parts of the world burn on a regular basis.' Pausas & Keeley (2009) are of the same opinion. To quote these authors: 'Ecologists, biogeographers, and paleobotanists have long ignored fire as a key ecosystem process', and further 'well before the emergence of humans on the Earth, fire played a key role in the origins of plant adaptations as well as in the distribution of ecosystems.

Nowadays fire is widely recognized as an important abiotic component of the Earth system and plays a key role on carbon and energy balances (biogeochemical cycles) and climate change. Fire is also a significant evolutionary force, helping to shape global biome distribution and ecosystem composition (Bond *et al.* 2005a; Bond & Keeley 2005; Bowman *et al.* 2009, 2011; Whitlock *et al.* 2010).

Describing the drivers of fire, the fire-fundamentals triangle (or combustion triangle) concept represents fire as a physiochemical process made up of three necessary ingredients: oxidizing agent (usually oxygen to sustain combustion), enough heat to raise the material to its ignition temperature, and some sort of fuel or combustible material. Oxygen, heat and fuel are referred to as the 'fire triangle' (Whitlock *et al.* 2010; Pyne 2001; Bowman *et al.* 2011). Usually wildfires occur when the ingredients are present and combined in the right mixture. It means that fire is actually an episodic event in nature, rather than thing (*https://en.wikipedia.org/wiki/fire_Triangle*). Variation in the Earth's atmospheric oxygen levels and changes in climatic and vegetation correlate with fire activity for hundreds of million years (Scott *et al.* 2000 in Marlon *et al.* 2013). As it is currently broadly known the fire-event triangle links with elements of climate regimes, fuel (broad scale, small-scale, i.e., fuel load) and topography (Whitlock *et al.* 2010). 'Flammability can interact in complex ways, sometimes producing positive or negative feedbacks on fire behavior' (Marlon *et al.* 2013 and literature cited therein).

Global fire occurrence patterns, landscape fires affecting biodiversity, carbon balance, human health, etc. are investigated by new discipline of pyrogeography (Bowman *et al.* 2009, 2013).

Information on past fire activity comes from a variety of sources, including geochemistry, tree-rings, charcoal, archeology, and written documents (Whitlock *et al.* 2010; Iglesias *et al.* 2015). Sedimentary charcoal (or its particles) records as an indicator of fire and vegetation history can provide information about past changes in fire regime, even into locations modified by recent human impacts (Marlon *et al.* 2013). It is well recognized in charcoal-based fire history reconstructions to understand the long-term relationship between fire, climate, and human activity (Scott *et al.* 2000; Pyne 2001; Bowman *et al.* 2009; Pausas & Keeley 2009; Whitlock *et al.* 2010).

As mentioned, three elements of fire are needed to ignite (fuel, oxygen and heat). Regarding the third element, heat sources, the vast majority of fossil wildfires are considered to have been initiated by lightning strikes, but less predictably from volcanic activity, sparks from rock falls or meteor strikes (Cope & Chaloner 1980; Bowman *et al.* 2009, 2011; Pausas & Keeley 2009).

To understand the evolution of the Earth system in deep time, the role of ancient fires associated with ancient plant communities and O_2 – forced climate thresholds during the Phanerozoic (Robinson 1989) needs to be taken into account (Scott *et al.* 2014; Belcher *et al.* 2010).

But before plants first invaded the land between about 480 and 360 mya ago, in the mid-Paleozoic Era (Kenrick & Crane 1997), cyanobacteria or blue-green algae appeared as aquatic photosynthetic organisms, i.e., oxygenic photoautotrophs (Bjőrn & Govindjee 2009). Fossil evidence of such prokaryotes falls into three categories: (1) microbial fossils, (2) stromatolite fossils and (3) chemical fossils. The oldest records of microbial biosphere were described within the Early Archean (~3.5 billion years ago) terrestrial rocks in Western Australia and South Africa (Awramik 1992).

Around 1.2 billion years ago the first photosynthesizers, on land algal scum as representatives of the microbial terrestrial biota, began to colonize the land (Raven & Edwards 2001).

The atmosphere of our planet in the pre-Phanerozoic (the pre-Cambrian) lacked both sufficient oxygen and also the land lacked terrestrial plants, i.e. fuels. It means that fire did not exist on Earth (Pausas & Keeley 2009). Globally there must have been a source of fuel roughly 400 mya ago through to the present (Scott *et al.* 2014). Research in this field has focused exclusively on the fossil record of vascular plants (Banks 1975; Chaloner & Sheerin 1979 cited in Kenrick & Crane 1997).

2.3.8.2. Oxygen, Forests and Paleofires Over the Phanerozoic Time (the Paleozoic, Mesozoic and Cenozoic Eras)

Given a fuel and ignition source, oxygen-enriched atmosphere leads to a greatly-increased fire hazard. It is the most ubiquitous natural terrestrial disturbance (Belcher *et al.* 2010). The amount of oxygen in the Earth's atmosphere and its significant fluctuations are determined by a wide range of physical and biological processes. Among others, the invasion of the land by plants ('terrestrialization') was one of the most important evolutionary events in the history of life on the Earth (Silvestro *et al.* 2015 and references therein).

Over the past 550 million years, the level of the Earth's atmospheric oxygen has varied considerably, roughly between 15% and 30% (or 35%) (the Great Oxygenation Event). Fossil charcoal, as evidence of paleofires has been found for all times that arboreal plants have populated the land (Berner *et al.* 2003). Belcher *et al.* (2010) revealed that fire activity on the Earth system would have been high during the Carboniferous (345-280 mya) and the Cretaceous (136-65 mya) periods; intermediate in the Permian (280-225 mya), Late Triassic (275-190 mya), and Jurassic (190-136 mya) and very low to lacking (despite the above mentioned information regarding charcoal) in the Early-Middle Triassic Period between 225-215 mya (the sequence of the GTS (Phanerozoic) periods of the Earth as quoted by Eicher & McAlester 1980). Hence, the carboniferous and Cretaceous are important periods of the Earth's geological timescale with regard to a range of feedback mechanisms that may have an impact on the evolution of life and biodiversity of our planet (Belcher *et al.* 2010; Brown *et al.* 2012).

2.3.8.2.1. The Paleozoic Era

The first evidence of a land flora of liverwort-like plants emerged and became well established from fossil spores of the mid-Ordovician age (470 mya) (Gray *et al.* 1985; Kenrick & Crane 1997). However, it seems that terrestrial fuels were sufficient to permit the possibility of wildfire.

Evidence of increasing land plant diversity and vegetational change is provided in the Early Silurian, around 425 mya ago. In other words, the start data for pyrogeography is the Silurian (Bowman *et al.* 2013). Once plants (and animals) became established on land they required new strategies for survival and contributed to changes in the nature of the physical and chemical processes in terrestrial environment: obtaining nutrients and water, avoiding desiccation, carrying out gas exchange and reproduction through the rhizoids and rigid vertical stem. Among representatives of flora which firstly colonized the land it should be mentioned bryophytes – hornworts, some mosses, and tracheophytes – early vascular plants, e.g., *Cooksonia, Rhynia* (Gray 1993; Kenrick & Crane 1997). Palaeobotanical evidence shows several major lineages of vascular plants had already evolved by the mid-Silurian. The first vascular plants to exist most probably lacked true leaves and roots, forming low vegetation mats similar in size (about 10cm) to modern-day mosses although some of the fossil plants of terrestrial ecosystems could reach one meter in height (e.g., genus *Cooksonia* with very simple branching patterns) (Kenrick & Crane 1997).

It is believed that fires in shallow wetlands of the Silurian must have been limited in amplitude (Pyne 2001). Accordingly, wildfire disturbance regimes were found on the basis of charred remains of low-growing vegetation of the stromata-bearing plants (the Early Silurian, 440mya); also charred coprolites, indicating a low-temperature surface fire (Glasspool *et al.* 2004 cited in Pausas & Keeley 2009).

The minimum oxygen requirement level permitting the possibility of wildfires should be 13% (Scott 2000) or 15% (Belcher & McElwain 2008). Hence, the atmospheric oxygen level in the Ordovician was too low for combustion, and as the Silurian vascular plant communities must have been rather restricted in distribution (Kenrick & Crane 1997; Scott *et al.* 2000), there was perhaps not sufficient fuel accumulated for fires to become widespread until the Devonian (Scott 2000).

For increased land plant diversification four fundamental geological phases were identified: the Devonian, the Carboniferous-Permian, the Triasssic-Jurassic, and the Early Cretaceous (Niklas *et al.* 1983 cited in Silvestro *et al.* 2015).

The subsequent fire history over geological time linked, as highlighted already, with the atmospheric oxygen concentration on Earth. Its balance modeling, suggests that there were periods of clearly high and low fire activity (Igamberdiev & Lea 2006). As mentioned, it was tied to changes in O_2 levels. In other words, throughout the different times of the Phanerozoic, O_2 differed significantly from the present atmospheric oxygen level (PAL) (Glasspool *et al.* 2015). The fire activity pattern is also tied to the fluctuations (origination-extinction) ratio of the tracheophyta across the Phanerozoic.

The Late Silurian and Devonian periods are considered as the critical interval, during which the initial diversification of early vascular plants (embryophytes) occurred and most of their basic features appeared (Banks 1975; Chaloner & Sheerin cited in Kenrick & Crane 1997).

The Devonian Period (395-345 mya) experienced the first significant adaptive radiation of terrestrial life. Kenrick & Strullu-Derrien (2014) comment: 'The Devonian witnessed the evolution of forest ecosystems from an earlier diminutive herbaceous vegetation of small leafless plants with simpler rhizoid-based rooting systems (RBRSs)'. Vascular plants encompassed the continents and there was a greening of various land masses. By the end of the Devonian (~345 mya) several free-sporing vascular plants had evolved leaves and true roots (e.g., Archaeopteris) (*broom02.revolvy.com/main/index.php?s=Archeopteris&item_type...*).

The widespread appearance of large (megaphyll) leaves, with their branched veins and planate form, root and vascular systems appeared in the Late Devonian (about 330 may ago). This anatomical novation caused global linkage between plant and atmospheric evolution (for example, atmospheric CO_2 levels 90%) dropped significantly in the Late Devonian). This important event took place about 40 mya after simple leafless vascular plants first colonized the land (Beerling *et al.* 2001, Osborne *et al.* 2004) and 'would have quickly propagated throughout terrestrial ecosystems by conferring significant selective benefits in competition for light and space' (Osborne *et al.* 2004). Beerling (2005) also comments: 'Megaphylls altered the evolutionary trajectory of terrestrial plant and animal life, the biogeochemical cycling of nutrients, water and carbon dioxide and the exchange of energy between the land surface and the atmosphere'.

Roots were essential to the evolution of large plants with the further essential functions: absorption of water and inorganic nutrients, anchorage, storage of food and nutrients, and vegetative reproduction. Root evolution indicates the broader impact on the geochemical carbon cycle in biosphere (Kenrick & Strullu-Derrien 2014). Leaves in higher plants are keys not only to plant life, but to all terrestrial life. This part of plants is the best way to collect solar energy for photosynthesis. But leaves do more than photosynthesize, e.g., evaporative cooling – transpiration (Beerling 2005).

From the Devonian Period has been observed the co-evolution of mycorrhized fungi (endophytic) and plants roots. For example, early symbiotic relationships with fungi in which the fungal network of filaments increases the efficiency of the plant root system and ectomycorrhizal mutualism evolved during the Cretaceous, during the period of rapid angiosperm radiation (Cairney 2000 and references therein).

Wang *et al.* 1999 (cited in Brundrett 2002) pointed out that fungi diverged from other living organisms deep in the pre-Cambrian, the Proterozoic, i.e., 'earlier life' (2.5 billion years ago to 570 mya). Fungi developed symbiotic partnership with photoautotrophs long before the evolution of vascular plants and the first terrestrial fungi in early land ecosystems appeared long before plants colonized the land. Soil surface microbial communities in the coastal environment containing well-integrated fungi and algae were probably the first terrestrial associations between highly co-evolved fungi and photosynthetic organisms. The fossil record for lichens is poor. The oldest fossil lichens date to the Paleozoic, Early Devonian, about 400 mya old, and independent lichenization events have occurred multiple times. By the end of the period the first seed-bearing plants and tetrapod animals appeared (Brundrett 2002 and references therein, *https:/en.wikipedia.org/wiki/Lichen*).

In the ***Carboniferous Period (345-280 mya)*** the continents were coming together (pre-Pangaean Earth). During the Late Paleozoic Era the remaining Late Devonian trees converted into giant woody plants. To the new groups belonged tree like horse-tails (the only representatives living today of the Late Paleozoic plants), scale trees (e.g., *Lepidodendron*), cycads, cordites (extinct members of the gymnosperms and precursors to the conifers), another groups of 'seed ferns' (actually gymnosperms) and others.

During the Carboniferous Period the land biomass became so high as to cause a fall in atmospheric carbon dioxide, triggering the longest period of the Phanerozoic glaciations. On the contrary, higher oxygen concentrations present in the atmosphere may have played a role enabling the earliest ancestors of modern dragonflies to possess wingspans of about 78 cm. Part of the land (Europe and North America) in the Carboniferous was covered by shallow seas and swamps. Organic material of the dead plants of forests did not completely decay and accumulated as peat. Later, covered by the sea and sediments heat and pressure transformed these organic remains into coal (which can be up to 11 to 12 meters thick) (Eicher & McAlester 1980; Monroe & Wicander 2008).

Thus, fossils indicate that land plants begin to appear in the fossil record as early as the Late Ordovician, plants vascular tissues (but spore-bearing ones) diversified during the Silurian. By the end of the Devonian Period ferns, horse-tails, and non-flowering seed plants populated the landscapes, giving rise to trees and the first forests, and the Carboniferous boost of massive, widespread and luxuriant forests, where major forest-building species were still represented by non-flowering seed plants. Amongst others the earliest relatives of the conifers appeared. Conifers used the massive windblown sporing strategy (Eicher & McAlester 1980, *sci.waikato.ac.nz/evolution/Evolution Of Life.schml*). A trend towards aridity and increase in terrestrial habitat led to the increasing importance of the amniotic egg – the greatest evolutionary innovations of the Late Paleozoic (Laurin 2004).

Extensive true charcoal deposits of the Late Devonian and the Carboniferous indicate the emergence of tree-like stature. These 'seedy ferns', true ferns, gymnosperms and other progymnosperms, were the first plants to develop into true wood ('lignophyte' clade). In the mid-Devonian trees reached heights of 8m, in the Late Devonian – 30m, and in the Late Devonian and the Carboniferous – 50m (Chaloner & Sheerin 1979). All of them branched multiple times, and the giant trees had true wood (Davis & Kenrick 2004).

Globally, fire-fundamental triangle – an exothermic oxidation reaction dependent on the rapid combination of fuel and oxygen in the presence of heat – was created from the Late Silurian (about 419 mya) and appears to the present (Glasspool *et al.* 2015). Clear evidence of charcoal appears in the Devonian and then more extensive deposits appear in the Carboniferous (Scott 2000; Glasspool *et al.* 2015), triggering the rapid rise in atmospheric oxygen from the Late Devonian to the Late Carboniferous (Pausas & Keeley 2009). Because of the high-oxygen (to over 30%) and high biomass during the Mid- and the Late Phanerozoic, compared with the current PAL (~21%), 'it would have greatly facilitated combustion even if the moisture level was high' (Pausas & Keeley 2009). It was at this time that both plants and animals became very adapted to a high fire environment that has continued to this day (Pausas & Keley 2009).

In the ***Permian Period (280-225 mya)*** the Earth was dominated by the single supercontinent Pangaea surrounded by the global ocean – the Panthalassa. In the Permian the extensive rainforest of the previous Carboniferous had collapsed and left behind vast regions of deserts within the continental interior. There were extensive glaciations in the southern parts of the landmass. The Permian ended with the largest mass extinction recorded in palaeontology and known as the Permian-Triassic extinction or 'The Great Dying' event. It affected many groups of organisms, which caused biosphere impoverishment (90-95% of marine and about 70% of land species became extinct). On the land the extinction created the way for other forms to dominate, for example, it led to the 'Age of Dinosaurs' (Raup 1986, Courtillot 2003). Palaeontological records highlighted five events of major extinctions in the life history on the Earth: Ordovician/Silurian, Late Devonian, Permian/Triassic, Triassic/Jurassic, and Cretaceous/Tertiary (Courtillot 2003).

Numerous possible explanations have been formulated to explain the events of the mass extinction (the formation of Pangaea, volcanism, glaciations, and impact with an asteroid).

Very complex forests were present across Pangaea with a diversity of seed ferns. Among seed ferns Glossopteridales with the best known and the largest genus *Glossopteris* were presented exclusively in the southern Hemisphere (across the intra-Gondwanian part of the Pangaea) and it coincided with the stratigraphic marker for the Permian throughout all the Gondwana continents (Chatterjee *et al.* 2014). It is noteworthy that the Austrian geologist Eduard Suess named this large land mass Gondwanaland after the district in India where the fossil plant *Glossopteris* was found. These plants went on to become the dominant elements of the southern biota through the Permian but disappeared in almost all places at the end of the Permian (McLoughin 2001, *https://en.wikipedia.org/wiki/Glossopteris*).

The radiation of many conifer groups occurred including the ancestors of many modern-day families. Cycads still flourished and the ginkgos (order Ginkgoales) evolved during this period. Swamp-loving vegetations became marginal elements. Modelling suggests a fall in oxygen at the end of the Permian (around 250 mya) to below the PAL. Therefore, as mentioned, fire activity in the Permian was rather low (Berner 2006 cited in Scott 2009).

Recent evidence suggests that atmospheric oxygen levels were suppressed worldwide between the Permian and Triassic (~250 mya ago) and the Triassic and Jurassic periods (~200 mya ago). The oxygen level of the surface fell to between 10 and 11% (in general about half of today's 21%), the lower being far below the PAL. These data are comparable to levels of oxygen found now at high altitudes (4,200 meters) (Ward 2006). 'However, models that suggest very low oxygen levels in the Jurassic, below 13%, are not in agreement with the charcoal record' (Scott 2009). But a rise in CO_2 levels and appearance of hot greenhouse conditions due to high volcanic activity was coupled with a drop in atmospheric oxygen. It is worth drawing attention to the fact that dinosaurs flourished in low oxygen levels.

2.3.8.2.2. The Mesozoic Era

The Triassic Period (225-190 mya). The world's landmasses were still combined in the Pangaea. By the end of the period (190 mya), tectonic forces had gradually begun to split the Pangaea in two separate landmasses, Laurasia to the north and Gondwana in the south. The Triassic climate was generally hot and dry, even the poles were ice-free. After the largest extinction at the end of the Permian, the survivors of that event spread and radiated into the niches left vacant (Eicher & McAlaster 1980).

Seed plants came to dominate the terrestrial flora. In the northern Hemisphere conifers, ginkgoes, cycads and Bennettitales became important. The Mesozoic ferns (including Osmundaceae), in some areas, inhabited damp and shaded habitats. By the end of the Triassic, both hemispheres gave way to conifer and cycad vegetations (Behrensmeyer *et al.* 1992; https://en.wikipedia.org./wiki/Triassic).

The Jurassic Period (190-136 mya). During this period, the continental drift caused the supercontinent Pangaea to split apart into Laurasia and Gondwana. This occurred over the following geological time of the Phanerozoic and significantly the second half of the Cenozoic, fragments of landmasses repositioned themselves close to modern continents. This rifting of mainlands, along with generally warmer global temperatures, allowed for diversification and dominance of the archosaurs. Dinosaurs were the chief land animals.

By the beginning of the Jurassic, non-vascular plants – mosses and liverworts - were confined to swampy areas. The dominant position was held by ferns, cycads, ginkgoes and glossopterids. Seed ferns, which were diverse and abundant during the Triassic-Jurassic time, declined in the Cretaceous accordingly with the radiation of angiosperms (McLoughlin *et al.* 2008 cited in Silvestro *et al.* 2015). During the Jurassic, cone-bearing conifers took possession of bisexual reproduction and by the end of the Jurassic, due to successful fertilization the gymnosperms were widespread and constituted the majority of large trees. The Jurassic conifers included Araucariaceae, Cephalotaxaceae, Pinaceae, Podocarpaceae, Taxaceae, and Taxodiceae. Many of them (cone-bearing cycads and southern conifers) are the most ancient of living seed plants little changed from their ancestors (e.g., *Agathis*, the Kauri pines from southern Hemisphere) (Behrensmeyer *et al.* (Eds.) 1992, *https://en.wikipedia.org/wiki/ Timeliene_of-plant-evolution*).

In the Late Triassic and Jurassic forests flammability was rather low, and in the Early-Middle Triassic Period the ecosystems, due to super low O_2 (<15% O_2), the full combustion was very limited (Belcher *et al.* 2010).

The Cretaceous Period (136-65 mya). Flowering seed plants (angiosperms), which include hardwood trees and grasses, originated then globally spread and diversified during the Early Cretaceous (they first appeared in fossil record about 130 mya). Comparative studies of living species suggest that angiosperms evolved from the extinct cycadeoides (seed ferns as its ancestors). Thereafter, to the end of the period they became dominant over gymnosperms. Today, angiosperms comprise the greatest terrestrial plant diversity – about 90 percent of the Kingdom plantae, estimated at over 250,000 species and at least 12,000 genera (Paton *et al.* 2008). A flower can be considered a powerful evolutionary innovation, because its presence (together with fruits and seeds) allowed the plant world to access new means and mechanisms for reproduction. They are completely different from non-flowering plants (Scott *et al.* 1960; Stebbins 1971; Taylor & Hickey (Eds.) 1996; Takhtajan 1969, 1997).

In addition to reproduction of flowers a vital role is played by the mutually beneficial animal-plant relationships. It has traditionally been viewed as a high degree of co-adaptive-mutualistic specialization in both angiosperm species and their biotic vector (mainly insects) for pollination and dispersal that is carrying pollen from one plant to another. Earliest pollinating insects were probably beetles, because of Cycadeoids were already specialized for pollination by beetles long before the appearance of angiosperms. Insects, as pollinators appeared after the origin of specialized lepidopterans (butterflies and moths) and hymenopterans (bees and bumble bees) during the Late Cretaceous – Early Tertiary periods. It's clear that animals deliver pollen more efficiently than wind. Such a trend of relationships is nowhere better developed than among the mentioned autotrophs and their consumers. Such a kind of mutualism is considered as a good example of co-evolution (Proctor *et al.* 1996, 2003; Crepet & Niklas 2009).

Box. 4. Erect Posture of Dinosaurs and the Case for High, Constant Body Temperature – Pattern of a Quasi-Evolution Event within Reptilians

One of the most debatable issues of the present-day paleontological research is the physiology of dinosaurs, particularly about thermoregulations of this group of reptiles. Many dinosaurs possessed comparatively long and erect limbs. Limbs resemble those of mammals and birds held under their bodies rather than sprawling out to the sides like those of reptiles and amphibians. Many groups of quadruped dinosaurs, including even tyrannosaurus, according to their skeletal features (beyond the erect posture) are clearly built as cursorial animals adapted to frequent sustained running. Hence, they were as fast as some quadrupedal mammalian carnivores and herbivores with highly aerobic metabolism (with two-part circulations driven by four-chambered hearts) (Bakker 1986; Reid 1997; *www.miketaylor.org.uk/dino/faq/s-misk/metabl/*). Erect posture with gracile leg bones were possessed by bipedal dinosaurs as well.

Robert Reid (1997) suggests that 'Dinosaurs were around for about 150 million years, so it is very likely that different groups evolved different metabolisms and thermoregulatory regimes, and that some developed different physiologies from the first dinosaurs. According to this suggestion, they were: (1) cold-blooded (poikilothermic); (2) warm-blooded (homeothermic); (3) their metabolism was intermediate between those of modern cold-blooded and warm-blooded animals; (4) they included taxa with two or three of these types of metabolism.

As dinosaurs appeared during a long period of low oxygen levels in the Permian-Triassic (P-Tr) lower oxygen event, therefore, as Dr. Peter Ward theorizes (Ward 2006), they developed highly efficient metabolism – an avian respiratory system. This anatomical feature, allowed them to survive the P-Tr extinction event. On the other hand, regarding ultra-low oxygen, it triggered mass extinct of terrestrial vertebrates including many mammals and mammal-like reptiles. Ward suggests that Saurischian dinosaurs had the extremely unusual, efficient breathing systems of birds, i.e., similar physiology (fairly high metabolic rates). In other words, Saurischian dinosaurs have bird-like lung and the extensive system of air sacs that made their skeleton lighter. Air sacs were extended throughout many of the bones (air sacs in many birds and dinosaurs often leave characteristic traces on the fossilized bones, making them hollow) (Sereno et al. 2008). According to this speculation the giant forms of Saurischians that include the gigantic brontosaurus 'rather than having a diaphragm to force air in and out of lungs, the mighty forms of reptiles had lungs attached to a series of thin-walled air sacs that may have acted something like bellows to move air through the body'.

No evidence of air sacs has been found in ornithischian dinosaurs, but they are mentioned and other scholars suggest that they should have had metabolic rates comparable to those of mammals. However, recent studies of bone histology and other anatomical features data seem to indicate that as a minimum some dinosaurs may have been endothermic (Reid 1997; Ruben & Jones 2000).

The apparently sudden appearance of nearly modern flowers in the fossil records, their mostly cryptic evolution during the Early Cretaceous and their subsequent rise to dominance in all terrestrial biomes and habitats during the Late Cretaceous remain one of the most puzzling mysteries in evolutionary biology. These events, initially posed as such a problem to the theory of evolution, were often quoted as an 'abominable mystery' by Charles Darwin (Davies *et al.* 2004). Some scientists believe that the answer 'lies in a confluence of features that collectively make the angiosperms unique among the land plants' (Crepet & Niklas 2009).

In recent time, on the bias of nuclear genome sequencing the family Amborellaceae (a male flower of *Amborella trichopoda*) from New Caledonia is regarded as being the sister clade to all other living flowering plants. ‘It is an early-diverging flowering plant and as such, it represents the equivalent of the duck-billed platypus in mammals’ (Adams 2013).

According to Brown *et al.* (2012), the Cretaceous can be considered a ‘high fire’ world (including all continents), when O_2 increased the present-day level: elevated O_2 concentrations predicted as in excess of 21% throughout this period and reached their highest (25%) level during some stages (e.g., Albian-Cenomanian transition).

2.3.8.2.3. The Cenozoic Era

Although there is rapid diversification of flowering plants beginning in the Late Cretaceous, it is not until the Cenozoic that angiosperms increased origination rates, contributing to their current ecological and evolutionary dominance (Wing *et al.* 1993), with some contributions from pteropsid ferns and ‘gymnosperms’, mainly pinopsids (Silvestro *et al.* 2015 and references therein).

During the Cretaceous-Paleogene event the almost complete downfall of seed ferns and the loss of primitive forms of pinopsids and ginkgoopsids, indisputably caused a major taxonomic shift in floristic diversity. This impoverishment of pre-Cretaceous taxa, together with the environmental changes following the Chicxulub impact event occurred in the Yucatan Peninsula in Mexico, appears to be tightly linked to the success of flowering plants, which after the mentioned events experienced a second burst of diversification (Silvestro *et al.* 2015).
The evolution event of the Poaceae during the Cenozoic Era is highlighted in previous pages.

2.3.8.2.4. Primary Ecological Trigger During the Cenozoic Era: The Role of Fire in the C_4 Photosynthesis Pathway Expansion

Fire has been mentioned among different abiotic and biotic phenomena as a part of the explanation for the Miocene-to-Pliocene expansion of C_4 grasslands, but frequently as a secondary factor following changes in aridity and CO_2 (Sage 2001; Bond *et al.* 2003 cited in Keeley & Rundel 2005).

But the coevolution of C_4 ecosystems and fire as general driver has recently been strengthened not only by neonatological, but also by palaeontological evidences. For example, materials from the Indian sub-continental (Himalayan foothills and Ganges floodplain) and charcoal abundance in Pacific Ocean sediments from the Early Miocene through the Pliocene offer complimentary evidence of fire frequency and the expansion of ecosystems dominated by C_4 plants (Osborne & Beerling 2006).

It should be mentioned about pollen charcoal (including microscopic charcoal), and the stable isotopic composition of plant mixes including a marine sediment core (Hoetzel *et al.* 2013).

The expansion of C_4 grasses occurred alongside Late Neogene’s climate changes via its seasonality and increasing aridity. As a sequence, ecological disturbances by fire were an essential feedback mechanism leading to the establishments of a grasslands biome in the Miocene and Pliocene. One of the key factors within the marked seasonality was changes of anatomical feature of the grasses through the seasons. In the warm moist season abundant biomass of grass production is promoted followed by a severe dry season capable of converting this primary fresh grass production into dead biomass, which after lighting as an ignition source represents highly flammable fuels that would readily carry intense fires (Cochrane 2003; Keeley & Rundel 2005; Scheiter *et al.* 2012).

The accumulation of highly flammable plant litter coinciding with more or less long periods of drought, open or relatively open landscapes, and a windy environment are highly combustible fuels to large-scale fires (Cochrane 2003).

During the Middle and the Early Miocene, when C_4 grasses have the possibility to hold a dominant position in the vegetation mosaic consisting of patches between woody plants or woodland-grassland ecotone, the hot season (summer heat) is thought to be the result of burning C_4 grasslands creating in this way a fire regime. It means the beginning of a new era of the Neogene's environment ecology, when ecosystem fire regimes show an important interaction with a changed paleoclimate. In the light of such a scenario, in the Late Miocene-Pliocene epochs during progressive aridization given sufficient periods of dry weather, fire should play a role of pre-conditioning , or environmental facilitator with further expansion of C_4 savanna and grassland biomes to displace mesic forests through sharp transitions in favor of distinctly different vegetation types – herbs with further expansion of C_4 savanna and grassland biomes (Osborne 2008). This all supports fire's phenomenon as a selective factor. Bond *et al.* (2005) (cited in Osborne 2008) pointed out that today fire reduces forest cover by ~50% at the global scale. This, fire-adapted mechanism is primarily acceptable to African Andropogoneae, which quickly accumulates biomass between fires (Edwards *et al.* 2010).

Disturbance by fire is 'especially common in monsoonal climates, where the dry season is followed by a high incidence of highlighting strikes accompanying convective rainfall. These favour C_4 grasses with their ability to grow more rapidly than C_3 grasses in hot, open conditions after a recent burn' (Osborne & Beerling 2006 and references cited therein). The same should be said regarding the fire-driven grasslands of African monsoonal climatic regions (Edwards *et al.* 2010).

The following examples should be noted for imagining the heavy effects of fire with ecosystems. Anderson noted that North American tallgrass prairie fires depending on topographic features can reach temperatures of 83^{0}C to 680^{0}C on the surface of the soil, and fire temperatures themselves range from 166^{0}C to 343^{0}C. Therefore, frequent fire and periodic droughts largely prevent woody plant invasion into grasslands (Gibson 2009; Anderson 2012 and references therein).

2.3.8.2.5. The Control of Fire by Early Humans

After manufacture of stone tools, control of fire 'was a turning point in human evolution' (Berna *et al.* 2012), in its anatomy, behavior and cognition. It provided hominids with an adaptive advantage over other nonhuman and human primates in terms of, for example, cooked food (proteins) compared to raw foods, more digestible foods might trigger brain expansion, and obtain warmth, activity was no longer restricted to the daylight hours, venture into harsh climates of high (temperate) latitudes, protected (with smoke) from predators and insects (James 1989; Wrangham & Rachel 2010).

Humans cultures have a long evolutionary history of an obligate dependence on fire (Wrangham & Rachel 2010). Humans and fire may have coevolved. However, hominins through controlled fire caused substantial ecological impact effects and anthropogenic global environment changes. On the other hand, fires are currently more or less under human control, especially following the Industrial Revolution (1780), therefore 'the combustion of carbon-rich fuels by humans is a defining feature of the Anthropogene, thereby linking this concept to 'pyro-geography' (Bowman *et al.* 2009, 2013).

The earliest definitive evidence for the fire adaptation of a cooked diet by *Homo erectus* occurred as early as approximately 1.9 million years ago (the Pleistocene Epoch or Paleolithic Era) (James 1989, Organ *et al.* 2011).

However, the evidence for controlled use of fire in early hominin sites is dated after 0.4 mya associating primarily with *Homo sapiens* and Neanderthals (Roebroeks & Villa 2011).

About 1.5-2 mya, groups of Hominid descendants of *Homo habilis/H. erectus,* and *H. ergaster* began leaving Africa and expanded their range into southern Europe and Asia. In Southern Eurasia - Caucasus (East Georgia) was occupied by *H. ergaster/H. georgicus* (divergent subgroup of *H. erectus*) about 1.8 mya (Vekua & Lordkipanidze 2010), and South-East Asia (northern China) was settled by *H. erectus* at least 1.66 mya (Wanpo *et al.* 1995).

It should be stressed that *Homo georgicus* from the Georgian Caucasus according to paleontological and experimental evidence does not support the hypothesis of using fire. Unlike African populations, there are no charcoal deposits found in the Dmanisi site (David Lordkipanidze, Maia Bukhsianidze, pers. comments), although Olduwan-style tools have been found at sites in Dmanisi. Instead, early humans needed to eat more meat, which increased the food supply options in seasonally arid and temperate environments (Vekua & Lordkipanidae 2010).

By the end of the Lower Paleolithic members of *Homo* extended across large areas of Eurasia, but northward expansion may have been limited (winter's temperatures at times dropped below freezing) by the lack of control of fire prior to 300,000-400,000 y ago. It is clear that fire became a significant part of the hominin technological repertoire (Roebroeks & Villa 2011).

The anatomically modern *Homo sapiens* arrived on the evolutionary scene in East Africa about 300,000 y ago, left Africa ('out of Africa') around 50,000 y ago and expanded throughout the planet entirely replacing all of the Neanderthals and other late archaic humans (Stringer & Andrews 1988). According to human genetics mitochondrial DNA 'Mitochondrial Eve' is the matrilineal most recent common ancestor (unbroken line on the one woman – mother's side) of all currently living humans (Cann *et al.* 1987). Studies concentrating on the mt DNA give support to the Stringer-Andrews model (1988). However, in the last decade, with regard to the Out of Africa model and interbreeding (hybridization-replacement) event of *Homo sapiens* with other archaic humans, 'fundamental assumptions underpinning the model have been violated' (Kendrick 2014).

Currently many indigenous people still use fire to farm local environments. For example, fire-stick farming of the aboriginal Australians, who regularly used fire to burn vegetation to change the composition of plant and animal species in different sites (Jones 2012).

Pre-Columbian native American Indians purposefully used fire as the most powerful tool to modify ecosystems, for example, clearing of forests to change forest plant communities into grasslands, savannas and open woodlands (Denevan 1992).

2.3.8.3. Grazing Mammals, Grasslands and Co-Evolution

As already stated, grassland plants evolved under the influence of periodic droughts, frequent burning, and grazing animals and adapted to all three of them (Anderson 2012). However, the post-Miocene expansion of open grasslands and savannas worldwide was associated with the adaptive radiation of large mammals with the evolution of more cursorial and saltational body forms adapted to grazing (Coughenour 1985 cited in Gibson 2009, McNaughton 1985 and references therein; Anderson 2006, 2012 and literature cited in).

Herbivore ecologists, according to diet selection, divide herbivore large mammals (ruminants) into three groups: (1) grazers (herbivores feeding purely on grass), (2) browsers (diets containing foliage, stems and fruits of tree and shrubs, i.e., entirely on green parts of dicotyledonous plants), and (3) intermediate or mixed feeders, facultative grazers that consume both herbs and browses (Hofmann & Stewart 1972; Hofmann 1968, 1989).

So, large mammalian herbivores tend to have two possible feeding strategies or dietary specializations: grazing and browsing. But there are fundamental differences between grasses and browses in architecture and physical structure (Searle & Shipley 2008)

Grazing is a form of herbivory in which most of the plant (both aboveground and belowground, including the roots) are consumed by herbivores. The consumers of plants consist of a wide variety of animal herbivores from microscopic invertebrates, small rodents to the large mammalian megafauna (Blair *et al.* 2014).

Although large grazing herbivores are an important component of biodiversity in grassland ecosystems, arthropods (specifically insect grazers) – the most species-rich eukaryotic group on Earth take a central place in all terrestrial food webs, including grasslands (Van Klink *et al.* 2015). Among insects and grasshoppers, periodically, can have a tremendous impact on grasslands biome through their effects on energy flow and nutrient cycling (Blair et al. 2014 and references cited therein). As it was already pointed out, nematodes and other root-feeding invertebrates, burrowing small mammals are actually the most significant consumers of underground plant biomass (Giljarov 1965).

Grasses (monocots) and forbs (herbaceous dicot plants) are essential herb vegetations, which are consumed by large ungulate (hoofed) herbivores (Shipley 1999; Blair *et al.* 2014).

'Grazing is, or was historically, an important process in nearly all grasslands and is considered a key factor affecting species composition and biomes production in grassland ecosystems' (Blair *et al.* 2014). Vertebrate herbivory has existed on land for about 300 mya (MacFadden 1997). But the Poaceae and mammals with hypsodont dentition first appeared in the Eocene (McNaughton 1985). The grazing adaptation (grassland/grazing interaction) principally developed in mammals much later, during the mid-Cenozoic ('grassland and grazers are relatively recent arrivals on the global ecological landscape', MacFadden 1997), i.e., underwent gradual adaptive radiations from the Miocene to the Pleistocene (McNaughton 1985 and references therein).

2.3.8.3.1. Large Ungulate Grazers

Unlike the non-avian dinosaurs, large grazing (graminivore) mammals are still roaming in current non-anthropogenic grasslands and other biomes. More than that, in protected areas of African savannas large (>350 kg with body mass) herbivore fauna in both key indicators – ungulate species diversity and proportion sometimes are described as having a pre-human, pre-Pleistocene character grazer/browser ratio (Bigalke 1978 cited in Bobe 2006).

In other words, large generalist herbivores with well-expressed grazer-grassland interactions are still occurring and, sometimes, they are abundant and important in protected grassland ecosystems (van Klink *et al.* 2015). According to McNaughton (1985 and references therein), 'Northern Hemisphere tundra, grassland, and semi-arid shrublands are exploited by Bovidae and Cervidae; Australian grasslands and shrublands supporting Macropodidae; and African and Asian grasslands and savannas utilized by Artiodactyla, Perissodactyla, and

Proboscidae'. Then: 'grazing ecosystems with large mammalian herbivores were characteristic of one-third to one-half of the Earth's terrestrial surface until the last century'. Usually, the most striking large grazers are still supported by tropical grasslands like the African savannas. The prairies of the North America are home of the American bison (*Bison bison*) and pronghorn antelope (*Antilocapra americana*) (*www.encyclopedia.com/doc/1G2-3408000157.html*).

The Great Plains of the pre-Columbian era were maybe not pristine, but little modified by Paleo-Indians grassland biome, may seem to be in general a vast natural area. After colonization of North America's grasslands by Europeans much of the Great Plains were increasingly integrated into arable land and rangeland, where cattle roamed free. Grazing of domestic cattle caused different stages of vegetational succession (the story of these changes has been reviewed many times, so it does not need to be repeated here).

In Asian steppe and deserts and semi-deserts there are found the remaining scanty and fragmented Asian (the Turkmen) wild ass – kulan (*Equus hemionus kulan*), wild horse (*Equus ferus przewalskii*); goitered gazelle (*Gazella subgutturosa*) and saiga antelope (*Saiga tatarica tatatica*). These subsist fragmented and discontinuous, sometimes occurring at an edge of a former habitats area. Until 18th they had a far wider range, but these (steppe) areas are evidently well suited to farming or herding today. Nowadays all mentioned charismatic species are preserved by intensive conservation efforts (Michel 2008).

Therefore, as livestock gradually replaced wild ungulates, outside of Africa, grasslands around the world turned into grazing and agriculture areas, whilst during the Pleistocene (100,000 to 10,000 years ago) rich grasslands of all five continents supported frequently unrelated or distantly related organisms (convergent evolution) – a very impressive diversity of comparatively large mammals (exceeded 45kg).

2.3.8.3.1.1. Feeding Ecology of Grazing (Hoofed) Mammals

As underlined in the latest Miocene and Pliocene Epochs of the Cenozoic, which spanned a period of about 25 million years, occurred the turning point when ancient forests declined and grasslands became widespread. In many parts of the world losing their homogeneity, complex landscapes with mosaics of flammable grasslands, forests and transition ecosystems – savannas and arid open woodlands became widely appreciated. Open habitats with seasonality, increased aridity, adaptive response to changes in atmospheric CO_2 concentration has had a profound effect on the global terrestrial biosphere including the emergence of the C_4 photosynthesis pathway in herbs.

As already stressed, expansion of Poaceae with warm-season C_4 grasses and transforming ecosystems across large regions and broadly separated parts of the globe has been widely documented around the world. Although, suggestion that: (1) C_4 plants gradually evolved from cool-season C_3 plants through various stages of C_3-C_4 intermediates or (2) some aspects of driving mechanisms of becoming dominant the grasses with Hatch–Slack pathway of photosynthesis still remain debatable.

As a long-term evolutionary response to predation, hence avoiding being (or restricted) eaten, among other features incorporate the ability to protect themselves by secreting microscopic plant silica (phytoliths) crystals (SiO_2), presented in their leaves (epidermal cells) (Anderson 2012).

The post-Miocene expansion of grasslands and savannas worldwide was associated with the long evolutionary histories of adaptive radiation of large ungulates familiarized to grazing (Stebbins 1981; Axelrod 1985). Even though feedback mechanisms between grasses and grazing animals are not well developed in systems of evolutionary history of grazing (Milchunas *et al.* 1988).

Shipley (1999) comments: 'Characteristics of food are one of the primary forces that shape animal behavior, physiology and anatomy,' and then: 'In mammalian herbivores diet choices correspond with morphological specialization within the gastro-intestinal tract, including the mouth and teeth, which may confer increased efficiency for extracting nutrients from grasses and browses'.

Essential differences exist between grasses (monocots) and browses (dicots). For example, grasses predominate in open landscapes and tend to have an anti-herbivory strategy. They include a thicker cell wall than browses, predominate in forests, at the same time including a greater proportion of this slowly-digestible plant fiber, while browse has lignified fiber. In addition, higher silica concentration, which cause leaf blades to have sharp and rigid edges, is harder than tooth enamel and thus increases tooth wear more than brows does, therefore, reducing the ability of herbivores to digest fiber. Within the grasslands the grinding of C_4 grasses needs distinctively more force than that of C_3 grasses. Grasses, which typically grow close to the ground provide a relatively homogenous food source for ruminant herbivores (Shipley 1999; Searle & Shipley 2008 and references therein).

In this work digestive tract physiology of large ungulates (grazers and browsers) (e.g., Hofmann 1989) is not reviewed, but we turn our attention to morphology, considerably tooth morphology. Adaptations to browse or grass – the main components of ruminant's diets and, accordingly, herbivore foraging behavior (Searle & Shipley 2007) influence crown height: hypsodont dentition of grazers and brachydont dentition of browsers. For instance, browsers tend to have short or lower-crowned molars (i.e., brachydont teeth) with prominent cusps for dissection of leaves, stems and fruits of dicotyledons. They inhabit closed-canopied forests or woodlands and feed upon green parts and fruits of trees and shrubs (Hofmann 1989; MacFadden 1997). In contrast, grazers generally exhibit finely ridged surfaces on high-crowned (or ever-growing) molars, i.e., hypsodont teeth for grinding the fine and fibrous leaves typical of graminoids. They are predominantly grazers, living in more open biomes and feeding predominantly upon abrasive grasses (Estes 1991; MacFadden 1997).

On average, according to Clauss *et al.* (2008) and references therein, grazing ruminants are larger than browsing ruminants. However, there is no strict correlation between body mass and natural diet of ruminants. For example, the largest extent ruminant, the giraffe, is a browser and the largest known terrestrial mammalian herbivore ever, the *Paraceratherium* (=*Indricotherium*) (length about 7.40 meters, weight – 15-20 tonnes) lived in the Oligocene (34-23 mya ago) was also a browser (has low-crowned teeth and fed chiefly on C_3 trees and shrubs). As was pointed out above, extensive grasslands did not exist when species of three or four of the mentioned giant herbivore genera inhabited the Earth (Clauss *et al.* 2008; *https://en.wikipedia.org/wiki/Paraceratherium*).

MacFadden (1997) and references therein) pointed out that 'of the 34 orders of fossil and modern mammals, about two-thirds (24) contain species adapted to herbivory, and 13 are exclusively herbivorous. Many of these orders have representatives that evolved high-crowned teeth, interpreted to be a grazing adaptation', and, 'the greatest diversification of high-crowned mammals has occurred during the second half of the Cenozoic, during the past 30-35 mya'.

Among the three subfamilies mentioned, Pooideae evolved around 50 million years ago and suggested its probable origin on the northern land mass – in temperate Eurasia (Hartley 1973 cited in Stebbins 1981). Some largest genera (*Stipa, Oryzopsis*) of the tribe Stipeae have centers of diversity in Central Asia (Komarov 1934 cited in Stebbins 1981). Consequently, pre-anthropological, intercontinental migration of grasses is the result of expansion of these herbs via Greenland (between northern Europe and (to) North America as ancient (Paleogene) root of range spread. Later the same expansion occurred (Plio-Pleistocene) via Bering Bridge between Asia and North America (Beard 2012), and in mid-Pliocene - via Panama Bridge between the two Americas (Stebbins 1981; Beard 2012). The physiological adaptations to non-tropical, strongly seasonal and partly cool climate environments were one of the most significant processes in the early evolution of the subfamily. The taxonomic splitting of the C_3 Pooideae group appears in Europe (the Oligocene ~33.5-26 mya) (Beard 2012). C_4 Pooideae apparently emerged in a thermically favorable climate of this group of subfamilies in Africa (~33-30 mya), with further migration from the center of diversity into west Gondwana (South America) and to East Gondwana (India and Australia) (Beard 2012).

With regard to grass consuming large mammals, it is worth presenting the following citation: 'In a classic monograph published in 1873, the Russian palaeontologist Vladimir Kovalevsky believed that the evolution of high-crowned teeth and elongated limbs in fossil horses was an adaptive response to the spread of grasslands during the Miocene'; and then pointed out: 'This idea has since been accepted as the model to explain a time of major global change in ancient terrestrial ecosystems, which George Gaylord Simpson called the 'Great Transformation' (MacFadden 1997).

Due to an almost-complete fossil record found in North American sedimentary deposits from the Early Eocene to the present, the horse (Equidae, Perissodactyla) provides one of the best examples of above-mentioned definitions (quotations) of famous scholars. In other words, an excellent confirmation of interaction between grasses and herbivorous mammals is the progression of the horse (Chapman 1996 cited in Gibson 2009). Transformation of the Eocene's tropical forest-dwelling *Eohippus*, through the long evolutionary sequence, into the modern horse involved at least 12 genera and several hundred species. Fossilized plants found in different strata shed light on patterns of speciation. It shows that the marshy forested environments in which *Eohippus* lived became gradually drier, harder ground covered with mainly grassland vegetations. The environmental changes triggered off modification in stature and behavior of ancient horses. The major trends may be summarized as follows: (1) increase in size (from 0.4m to 1.5m), (2) lengthening of limbs and feet, reduction of lateral digits, increase in length and thickness of the third digit (the replacement of the splayed-out foot by the hoofed foot), the feature by which horse(s) with long legs becoming far swifter runners for better speed out in the open space to escape from predators); (3) increases in tooth length, crown height of molars equipped with cement layer, i.e., developing tougher, 'hypsodont' teeth suited for chewing abrasive grass (due to the silica in the leaves) instead of soft vegetation (Romer 1966; Simpson 1961 cited in Stebbins 1981; *https://en.wikipedia/org/wiki/*).

From the Early Miocene (~25 mya ago) to the Early Pliocene (~5 mya ago), a species of horses occupied niches to which they had completely adapted (including grazing the dominating grasslands).

The genus *Equus* with 15 or so species of horses appears in mid-Pliocene (~4 mya ago). *Equus* occupied North America for the entire Pleistocene Epoch (from 2.5 mya ago until their extinction). Therefore, the record of coevolution between grasses and herbivores included horses spanning most of the Tertiary and Quaternary periods (Stebbins 1981; Gunter 2005).

The boundary of the Late Pliocene-Early Pleistocene (~2.6 mya ago), coincided with a global cooling event, species of *Equus* crossed to the Old World via the large Bering Land bridge that once connected Alaska and Siberia and spread into Asia, across the Middle East, Europe and Africa. The Old World provides a suitable home and, evolving into the true horse, *Equus caballus*, onagers, wild asses and zebras (all of them, Asiatic and African equids belong to the genus *Equus*). In North America during the Pleistocene extinction event, like many other large mammals (including camels) on the continent had been largely demised before disappearing completely (about 7,600 years ago). These profound changes in native biota have been attributed to climate deterioration, altering vegetation shift from grasslands to tundra, resulting in reduction in the animals' food supply and the impact of the arrival from Asia of human beings – Paleo-Indians (about 13,000-13,500 years ago), who have used some artifacts (stone tools) to hunt horse (*www.horsetalk.co.nz>Post>Features*), and possibly vulnerable to new fire regimes instituted by Paleo-Indians (Gunter 2005).

Across the same Bering Land bridge in North America bison arrived (*Bison bison* and now extinct *Bison latifrons*) around 40,000 years ago. Prehistoric bison-invasive species – quickly filled grazing niches of megafauna emptied by the wave of Quaternary extinction event. Until the great slaughter of American bison, even in 16th century North America an estimated 30 to 60 million bison living in North America. 'Present ecological communities in North America do not function appropriately in the absence of megafauna, because much of the native flora and fauna evolved under the influence of large mammals' (Martin 2005).

Summarizing the main ideas reviewed above, Stebbins's (1981) opinion is worth citing: 'The combined study of these grass fossils and the mammalian remains associated with them provides the best opportunity for analyzing coevolution of terrestrial plants and animals over a long period of time that exists anywhere in the world', and, finally, 'the record of coevolution between grasses and herbivores spans most of the Tertiary and Quaternary Periods'.

Today the Serengeti Volcanic Grasslands ecoregion has the most diverse and abundant grazing-browsing fauna in the world, the last remnants of the great Pleistocene herds of megafauna (McNaughton 1985). Wooded savanna (its fundamental architecture) of Africa nowadays is probably the best analog of past landscapes, i.e., its extant habitats are similar and reflect the rich history of the Plio-Pleistocene palaeoenvironment, its entire food web and the nature of plant-herbivore relations in our planet's deep history. Finally, the complex investigations of the East Africa savannas can be used to answer questions of patterns and processes in the evolution of both grasslands and their large consumers−mammalian herbivores (McNaughton, 1985).

Grasslands as natural pasture were used by human – pastoralist as rangelands from the beginning of the Holocene (about 12,000 years ago), when people started to domesticate animals and plants (Grice & Hodgkinson 2002 cited in Gibson 2009). Extremely fertile soils provide much grasslands perfect for cropping and pasturing. Continuous, thousand years of agricultural, intensive introduction of domesticated grazing animals, exclusively expansion of cropland and forage production development substantially altered the temperate grassland. Millions of hectares were sacrificed to cropping in the 1950s and 1960s in the former USSR. So, today the steppe and its analogues are the most-altered and least protected biome on Earth. Through the successions, reconstruction of the former natural environment is almost impossible (Mordkovich *at al.* 1997). Programmes for assessing rangeland health are currently being developed since 2001 by the Sustainable Rangeland Roundtable (SRR) (Gibson, 2009, 227p.).

2.3.9. The Pontic-Caspian Steppe Land as an Important Part of the Eurasian Steppes

Among the temperate grasslands biome, the Eurasian steppe is the largest grassland in the world, extending from Hungary to Mongolia and adjacent China (Lavrenko 1940, 1954; Lavrenko *et al.* 1991; Archibald 1995). According to Lavrenko *et al.* (1991), the Eurasian steppe, due to the difference in degree of continentality of the climate, is divided into two subregions: Black Sea – Kazakhstan and Central Asian subregions.

The Pontic-Caspian steppe forms a part of the Black Sea–Kazakhstan or western European steppe. This steppe land stretching from the northern shores of the Black Sea as far east as the Caspian Sea (extends roughly from the Dniepr River to the Ural River or 30^0 to 55^0 east longitude, and from the Black Sea and the Caucasus in the south to the temperate forests and taiga (*https:en.wikipedia.org/wiki/Pontic-Caspian_Steppe*) (Figure 63).

Figure 63. Map of the Pontic-Caspian steppe (https:en.wikipedia.org.wiki/Fire:Pontic_ Caspian-_Climate. Png).

The Eurasian steppe in ancient times represented a unique region, which has played a significant role in the formation of cultural (equestrian culture) – economic systems of nomadic, horse-warrior people and their empires. Here should be mentioned Scythia, Cimmeria, Sarmatia, the Hunic Empire, Mongols, etc. (Mordkovich *et al.* 1997).

The flat expanse of the Eurasian steppe provides an ideal route between Asia and Europe. The most famous trade route on the Eurasian steppe is the Silk Road – trade and cultural networks (notably in the area of religions), connecting China, India and Europe (established around 200 BCE and existed up to 1450s CE) (Wood 2004). From the Mongolian traditions in the east to the Cossack traditions of western Russia, these cultures have relied on horses for travel, trade, and conquest on the vast steppe. (*nationalgeographic.org/encyclopedia/steppe/*).

Archaeological and genetic research has identified the region across the Pontic-Caspian steppe, and probably far larger area from southern Ukraine (east of the Dniepr River) to northern Kazakhstan as the most probable place, where horses were first domesticated (Kosintsev 2006 cited in Warmuth 2011).

The Eurasian steppe, especially its Black Sea–Kazakhstan part, has been settled by ancient civilizations practically uninterrupted since the Paleolithic. Hence, the interaction between humans and the natural environment across the entire steppe subregion - fertile black soil (chernozem) has very ancient roots. 'A primary pastoral and partly nomadic, economy was replaced over time by settled arable farming. The waves of settlers ploughed up the fertile, virgin black soil and planted cereals, converting pastures, meadow and 'natural' steppe to arable fields' (Moon 2008).

Today the Biosphere Reserve Askania Nova (southern Ukraine, Kherson region) as one of the last 'islands' of the virgin Eurasian steppe providing conservation of the largest (total area of 825 km^2) area of feather grass steppe of relict flora-faunistic complex in Europe. At the end of the 19th it was created by German Baron Friedrich Pfalz-Fein. Later Askania Nova was declared as a State Steppe Reserve in 1921, and finally was reorganized into a Biosphere Reserve in 1983. Its autochthonous flora is represented by 505 flowering species of plants, the fauna of care area is presented by 1,154 species. In the zoological park, 'Askania Nova' are protected 75 species of birds and 41 species of mammals (among others *Equus ferus przewalskii* and *E. hemionus kulan*). Wild, introduced animals are found in the wild or in semi-free conditions (zebra, antelopes, bison, etc.).

2.3.9.1. A Brief Review of the Holocene History of Grasslands of the Javakheti Plateau (South Caucasus); Pattern of Abiotic and Human Impacts

2.3.9.1.1. Javakheti Lakes and Wetlands as a Potential Ramsar Site

The Javakheti Highland represents the northernmost part of the mountain plateaus of the Near East. The total area covered by the central plateau of the Javakheti region is approximately 700km^2. The plateau is composed of a flat table and softly undulating terrain. It is intersected by the two ridges of Samsari, and Javakheti.

Geologically Javakheti Highland is produced by volcanic activity (Maruashvili 1964). The volcanic plateau of Javakheti plateau is composed of basalts and minor differentiated material, which was built in successive period of volcanism between 3.8 and 1.0 mya. Volcanic landforms are heavily marked by cryogenic activity associated with the most recent periods and those many volcanic products such as obsidians were, according to the ancient location of hominin settlements, used as raw materials for tools during the Paleolithic period (Nomade *et al.* 2013).

Javakheti upland with its mountainous, semi-arid to subalpine bioclimate zone is the coldest place in Georgia. It distinguished with its harsh, predominantly dry continental climate and considered as the most extreme in all of Georgia and adjacent plateau area of Armenia. Average temperature of the warmest month is about 15^{0}C and winter temperatures are below freezing - 8^{0}C-10^{0}C. The average annual precipitation is 500-700mm (Javakhishvili 1972 cited in Shetekauri & Chelidze 2016). Javakheti Upland (within the borders of South Georgia, is under the influence of the severe climate of the landlocked plateau system of the northern sector of the Middle East. From west to east these highlands are represented by the eastern Anatolian and Armenian plateaus (between which occurs the Javakheti Highland) (Maruashvili 1964).

The highest point of Javakheti is Great Abuli Mount (3,300m a.s.l.), the second peak after Aragats (4,095 m) on the Lesser Caucasus (Armenia) (Maruashvili 1964).

Javakheti is a woodless region. There are artificial plantations of Caucasian pine (*Pinus kochiana*) stands. The relict remnants of former forests consist of *Quercus macranthera, Populus tremula, Betula pubescens litwinowii, Pinus kochiana* with small fragments of dwarf rhododendron (*Rhododendron caucasicum*) thickets in understory. Large areas of the region are covered with a mountain steppe and subalpine and alpine meadows. The Samsari and Javakheti Ridges are used as pastureland, hayfields and forage (Ketskhoveli 1960; Gulisashvili *et al.* 1975; Connor & Kvavadze 2008; Shetekauri & Chelidze 2016).

Various coniferous, mixed and deciduous forests are thought to have existed on parts of the plateau in the historic and prehistoric past (Grossheim 1948; Ketskhoveli 1960).

In specific conditions in the subalpine zone, rarely but occasionally, occur stands of Oriental beech (*Fagus orientalis*). Primeval stands of beech continue to be discovered in the somewhat differences in climate, soils, and overall habitat conditions. In recent time, a new, krummholz-form beech wood has been discovered. This is a small patch (1600 m^2) of *Fagus orientalis* woodland on the Javakheti Upland (NE slope of Mt Tavkvetili, northern tip of the Samsari Range, about 2.580 m a.s.l.) (Arabuli *et al.* 2008). The authors suggest that the beech wood patch (Fageta nuda) of Mt. Tavkvetili has existed for at least 2,000 years (the Late Holocene) as relict and is a remnant of more extensive beech forest on the mentioned area.

During the Holocene time (the last 10,000 y) pollen-based, microfossil and macrofossil charcoal investigations of the Lesser Caucasus were carried out by Connor (2011) (South Georgia) and by Joannin *et al.* 2014 (Armenian part of the Javakheti Plateau).

The Javakheti Plateau extends across north-western Armenia, Southern Georgia and north-eastern Turkey. The main volcanic activity during the Pleistocene ended in the Late Pleistocene (Joannin *et al.* 2014 and references therein). A subalpine to mountain semi-arid bioclimate zone characterizes the Javakheti Plateau, the major vegetation types are steppe and pine woods (Connor & Kvavadze 2008).

Across the Armenian part of Javakheti Plateau, is Zarishat fen, the site above the tree line, where dominant vegetation consists of sedge species (Cyperaceae). During last millennia of the Pleistocene (Holocene) this site dramatically changed in climate, vegetation and fire regime. Abrupt post-glacial shifts in vegetation (cold steppe, steppe-forest, and forest composition), vegetation structure and fire regime also coincided with human arrival (e.g., remains of cultivated cereals, charcoals and obsidian tools) (Connor 2011; Joannin *et al.* 2014).

According to Messager *et al.* (2013) cited in Joannin *et al.* 2014), deforestation on the Javakheti Plateau started at 2000-3000 BP, driven by deterioration of the climate and human impact.

Joannin *et al.* (2014) pointed out that the Early Holocene is distinguished by a more humid and warmer climate with lower seasonality, but the Mid-Late Holocene favored higher, drier phases of approximately 6400, 5300-4900, 3000, 2200-1500, and 400 BP. These observations highlighted the long-term decreasing trends of the presence of conifers (*Picea, Abies*) and deciduous trees (*Fagus, Carpinus* and *Quercus*) in the internal plateau's area (Messager *et al.* 2013 cited in Joannin *et al.* 2014).

Alongside the climate fluctuations, the arboreal plants shift on the Javakheti Upland, to have resulted from the fire activity being controlled by climate and human actions during the Neolithic agricultural expansion. The Holocene fire history with a low fire frequency at 6200-5800 BP based on microcharcoals, is determined in southern Georgia by Connor (2011) (cited Joannin *et al.* 2014).

The depression of Javakheti Plateau supports a rainfall accumulation forming a system of permanent standing waterbodies: numerous lakes, ponds and wetlands. There are six large fresh lakes within the plateau's area: Kartsakhi, Khanchali (shared with Turkey), Madatafa, Saghamo, Paravani and Tabatskuri (the largest one in Georgia is Lake Paravani located on 2.073 a.s.l.).

As well as the large lakes more than 60 smaller lakes are scattered over the plateau area. All lakes were formed in depressions created as a result of volcanic activity. In winter Javakheti lakes freeze for a long time. Wetlands occupy large territories, ranging from 1,500 to 3,300m.

The unique ecosystems of lakes and wetlands are characterized by a high level of endemism and a large diversity of terrestrial and aquatic species. The local ecosystem is recognized as one of the most important migratory corridors for water birds, most of which are included in the IUCN Red List.

Being a high-mountain specific ecosystem with its rare flora and fauna, the local system of lakes has ensured ecological balance of the adjacent extensive area. The high biodiversity importance of local terrestrial (mountains steppe) and marine (lakes and wetlands) value of these areas is underlined in 'An Ecoregional Conservation Plan for the Caucasus' (2006).

The lakes are threatened by many problems, including invasive invertebrate animal species. Local, mostly poor people in this rural area of Javakheti, which has few natural resources after the breakup of the USSR, exist on a disorderly development of the local economic structure. It should be mentioned that pressures include harvesting artificial pine wood, overgrazing, random cultivation, hunting and fishing.

In recent time (from the beginning of 21th century) within the lakes of Javakheti the narrow-clawed crayfish (*Astacus leptodactylus*) from neighboring Armenia was introduced. This species of crayfish is an indigenous species in Armenia (freshwater habitats of Ararat Plain). In the 1980s it was introduced into Lake Sevan. There is a large-scale fishery of the crayfish in the lakes as it is actively consumed in the internal market of Armenia (over 300 tons) and partially abroad in the late 1990s (some countries of West Europe, from 20 to 1120 tons). During 2008-2012 the main exported crayfish stock came from Lake Sevan (Aghababyan et al. 2015). *A. leptodactylus* as other species of crayfishes, are polytrophic organisms, i.e., consumers of the detritus and the submerged vegetation. At the same time, it is a main predator of benthos animals. It had rapidly occupied the ecological niche filled by the Sevan Khramulya (*Capoeta capoeta sevangi*), another detritivore, the abundance of which has declined due to overfishing (Aghababyan *et al.* 2015 and references therein).

Crayfish introduced from Armenia to Javakheti lakes was a new step to organize some business depending on crayfish farming. This event has negatively impacted resulting in disturbance to local, Javakhetian aquatic biocoenosis via predation and an overlapping feeding habitat of local benthos animals due to its prolific breeding ability. Later it began to decline in the quantity of dominant local species of fishes (*Coregonus* spp.) (competition with the introduced benthos animals' consumer) and hence has been considered a pest (pers. comment of local rural fishermen).

Agriculture is the main occupation of the human population in the lake system area. Due to an inadequate mechanism of waste and wastewater management, the hydro ecosystems in the mentioned region have been endangered by pollution. All the mentioned deteriorations require increasing effects for their conservation.

Javakheti lakes and wetlands are a potential Ramsar site (second after Colchis or Kolkheti) as a unique azonal wetland ecosystems.

Among hygro- and hydrophyllous herb species are worth mentioning these species of grassy plants (fam. Cyperaceae) – sedges (*Carex* spp., 9 species), among which *C. juncella* is considered as a relict species; species of *Calamagrostis* (fam. Poaceae, 2 species), *Sagittaria sagittifolia* (Fam. Alismataceae), growing in water, *Utricularia vulgaris* (fam. Lentibulariaceae), aquatic species, etc. The peaty hills and hummocky habitats include *Carex acutiformis* and *C. acuta* (Kimeridze 1966, 1975 and references therein).

The Javakheti wetlands project, as a potential Ramsar site, will contribute to its designation to the List of Wetlands of International Importance, which would help conservation and improvement of the biota of local water bodies.

Due to its uniqueness the ecosystem of Javakheti plateau was selected as a priority conservation area.

Javakheti became a protected area in 2011. About 16.000 ha of land developed into the Javakheti National Park, also into Kartsakhi Managed Reserve, Sulda Managed Reserve, Khanchali Managed Reserve, Bugdasheni Managed Reserve and Madatafa Managed Reserve.

The Ramsar convention gives special attention to Javakheti wetlands to be included in the Ramsar List of Wetlands of International Inspiration, although the area has no official conservation status.

Today the international donor community in particular has embraced the concept of transboundary conservation. One of the main goals of this effort is the application of joint land use management policies associated with sustainable economic development. The creation of a tri-national transboundary biosphere reserve involving the countries of Georgia, Armenia and Turkey in the Caucasus Ecoregion is expected to benefit aimed at the protection of border zones as trans frontier conservation areas (Schuerholz 2004).

Box. 5. Fire in Mythology: Prometheus and his Georgian Analogue – Amirani

From the beginning, as their essential, starting point, humankind has two basic phenomena: hostility and humanness, war and peace, good and evil. Among these two controversial phenomena which have never been in equilibrium as though 'we are born sinful, rebellious, and basically despicable' (Doctrine of Humanity: *https://peacetheology/net.doctrina/07-doctrine-of-humanity/*). The roots of hostility vs tolerance go back to primordial time. The mentioned controversy is still a vital, fundamental problem for humankind, we are still looking for the peace as order and the love as kindly feelings. Because without the basic need of peace and kindness, we are 'vulnerable and fragile, easily damaged' (Figure 64).

From early civilizations, in their mythology, side-by-side when love and trust died among humans resulting in total disharmony there appeared heroes who made something striking for their people or for mankind in general. In early cultures, in every nation it was as essential, even more so, as air and water, like the love and kind relationship of our biological species, a part of a human being looked for this spiritual phenomenon, beyond physical and social needs. This is an expression of super kindness, altruism. As it is broadly known, Greece was among the first Great Western Civilizations. Many created myths originated in Greece, one of the most known of which comes from the writings of Ancient Greek peoples; the main god was a powerful, thunder-bearing god, Zeus. Among Olympian (Mount Olympus) gods was the Titan Prometheus, who was the creator of mankind and its greatest benefactor. According to Greek stories he made a crucial step.

Figure 64. Monument to Prometheus near Abastumani town, S. Georgia (2003).

As stated in Greek mythology, Prometheus defied the gods, stole the sacred fire that was reserved for the gods and he was also suffering at the hands of Zeus. Hence, he is also sometimes known as the god of fire or fire bringer (pyrphoros).

Prometheus in his eternal punishment, was chained to a rock in the Caucasus, Kazbek Mountain or Mountain of Khvamli, where his liver is eaten daily by an eagle, only to be regenerated by night, due to his immortality (the eagle is a symbol of Zeus himself). Years later, the Greek hero Heracles (Hercules) slays the eagle and frees Prometheus from the eagle's torment.

In Georgian mythology, Amirani is a hero, who challenged the chief god, and like Prometheus was chained on the Caucasus mountains, where raptors would eat his organs. 'This aspect of the myth had a significant influence of the Greek imagination' (Braund 1994).

Writing in late antiquity of the fourth and fifth centuries, the influence of the myth of Prometheus extends well into all periods from antiquity up to nowadays. The myth is a symbol of completeness encircling all the civilizing arts, such as artists, writers, poetry, philosophers of many nations and civilizations.

2.3.10. The Steppe Biome of the Caucasus

2.3.10.1. General Review

Distribution of the steppe biome in the North Caucasus is zonal, especially at lower altitudes in the western and central parts of this region, where the primary steppes usually coincide with the presence of chernozem (black) soils. They were largely used as arable land

(Gulisashvili *et al.* 1975), therefore their typological composition is considerably impoverished. In the South Caucasus, also in the more arid regions, there are no primary steppes. All the present steppes of Transcaucasia are secondary and have developed after the depletion of the forests. These secondary steppes are thus intrazonal, but they look like primary steppes as their vegetation and soils have changed. The largest secondary steppe is found in the mountains of Armenia and South Georgia (Javakhetian plateau), where locally poor fragments of primary forests survive. In the past, they must have covered considerable areas (Zedelmeyer 1933; Grossheim 1948; Ketskhoveli 1960; Gulisashvili *et al.* 1975).

In the western and central parts of the North Caucasus the climate corresponds to that of the winter cold steppes of Asia with hot and dry summers (mean July temperatures between, (20 – 25°) cold winters (mean January temperatures as low as -30^{0}) and a persistent snow cover. Mean annual precipitation varies between 300-650 mm with maximum precipitation falling in May and summer and minimum precipitation, as snow, in winter (Gulisashvili 1964).

In the South Caucasus the climate is dry subtropical with continental tendencies. The summer is dry and also the winter is rather dry with a thin and non-persistent snow cover. Average annual precipitation ranges between 200-300 mm (Figurovski 1919; Grossheim 1948; Gulisashvili 1948; Gulisashvili *et al.* 1975).

The dominant soil type of the steppes in the North Caucasus is chernozem (black soil), in the western part of the South Caucasus castanozem (chestnut soil), in its eastern part brown, grey-brown soils (Zakharov 1939; Gulisashvili 1964).

2.3.10.2. Steppes of the North Caucasus (Ciscaucasia)

In the Caucasus the main xerophilous turf-producing grass species are feathergrasses (*Stipa pennata* and *Phleum phleoides*). On the foothills of the Greater Caucasus *Stipa pulcherrima is* dominant (Flerov & Balandin 1931; Keller (Ed.) 1931; Tanfiljev & Kononov 1972).

In the western and central parts of the North Caucasus meadow-steppes have developed that are composed of meso-xerophytic grasses like *Festuca pratensis, F. sulcata* and *Bromopsis (= Bromus) riparia* and herbs like *Filipendula, Myosotis* and *Coronilla*.

Figure 65. Steppe, grassland. Ciscaucasia, Krasnodar Kray (Russia) (1972).

Some phytocoenoses on humus-rich chernozem consist of xerophilous feathergrass species like *Stipa tirsa, S. lessingiana, S. ucrainica* and *S. capillata* mixed with *Festuca sulcata,*

Koeleria gracilis, Paeonia tenuifolia, Adonis vernalis, Carex humilis, Achillea millefolium, Thymus sp. and other species (Tanfiljev, 1971 a,b, 1977, 1980, 1987; Tanfiljev & Kononov1972; Lisitskaya 2005; Levantsova 2009). Because of the more arid climate the steppe landscape of the eastern part of the North Caucasus is mainly shaped by *Festuca* species, but in excessively dry areas the leading position is occupied by feathergrass-wormwood communities (Grossheim 1948; Gulisashvili *et al.* 1975; Khasanova 2003) (Figure 65).

2.3.10.3. Steppes of the South Caucasus (Transcaucasia)

Also, the most distinguished grassland region in East Georgia, the Shiraki Plateau (Figure 66), is a secondary steppe formed by the destruction of arid open forests, primarily hemi-xerophilous shrubland, mainly of Christ's thorn (*Paliurus spina-christi)* and deciduous forests in the foothills (Zedelmeyer 1933; Khachidze 1975, 1985; Lachashvili *et al.* 2007). As good arable land they are under heavy anthropogenic pressure. Within East Georgia (Kakheti district) the same can be said with respect to the steppes in the gorges of the rivers Alazani, Kura and Iori. Fragmentally, primary steppe communities have survived in Garedji, around the Tbilisi Sea (Nakaidze & Sokhadze 1979) and in eastern Azerbaijan, Gobustan (Peper 2010). The Garedji steppe is well investigated. The leading species is *Stipa tirsa* that is mixed with *Bothriochloa ischaemum* and *Festuca sulcata* and species of *Filipendula, Potentilla, Galium, Medicago* and *Dianthus.* A characteristic xerophytic species is *Salvia garedji*, an endemic restricted to this area and described by Troitski 1930 cited in Grossheim 1948; Ketskhoveli 1960; Sokhadze 1970).

Figure 66. Vast grassland hills with *Botrichloa ischaemum* dominancy. Dedophlistkaro district, E. Georgia (2017).

Wild sorghum (*Bothriochloa ischaemum*) (Figure 66) a representative of a widespread subtropical-tropical genus in both hemispheres of the world, is one of the leading species in the steppe, but it is also widely distributed in arid open, park-like forests, with the grass species *Poa bulbosa, Phleum phleoides* and *Festuca sulcata* and herbs like *Galium verum, Achillea setacea, Glycyrrhiza glabra and Salvia nemorosa* (Sokhadze & Sokhadze 1952; Sokhadze 1977). Some botanists (e.g., Ketskhoveli 1960) are inclined to attribute *Bothriochloa* to semi-steppes, other scientists (e.g., Sakhokia 1958) prefer the term 'savannoid vegetation'. To support his viewpoint, the latter points to the presence of savanna grasses like *Imperata cylindrica* and *Erianthus ravennae* and to the fact that *Bothriochloa* communities share features like the origin of their flora, phenology and groundcover with savanna. Lachashvili *et al.* (2007) considers it more appropriate to attribute these grass communities to the steppe biome (Bothrichloeta xeroherbata). Anyhow, these communities hold several Caucasus endemics, such as *Polygala transcaucasica, Aster ibericus, Cephalaria media* and *Thymus tiflisiensis.* Sometimes a mix of ephemers can be found (*Botrichloeta ephemerosa*) together with the geophytes: *Alyssum campestre, Sideritis montana, Callipeltis cucullaria, Ziziphora capitata, Allium atroviolaceum*, etc. The above-mentioned communities mainly occur on chestnut soils and rarely on chernozems (Sokhadze 1977; Khachidze 1985). Feathergrasses (*Stipa* spp.) were investigated in the dry steppes of Nakhichevan by Novruzova *et al.* (1978) and Sadikhov (1981).

2.3.10.4. Mountain Steppes of the South Caucasus

The mountain steppe in the southeastern and southern parts of the South Caucasus, namely in Javakheti volcanic mountain plateau (E. Georgia), the Lori steppe in the Bezobdal Range and on the slopes of the Sevan Basin (Armenia), has another species composition (Mirimanian 1959; Nakhutsrishvili, Sh. 1966; Fajvush 1990, 1991, 1992, 1996). Many upland territories, however, have been ploughed up or are overgrazed by domestic animals (Berg 1952).

Grossheim (1948) recognized two types of mountain steppes: feathergrass-fescue and mountainous-feathergrass steppes. The first type – the most characteristic species of grass-herbal polydominant steppes - occurs from 1,800 to 2,100 m a.s.l. and extends as a more or less continuous belt. The main turf producing species are *Stipa capillata* and *Festuca sulcata* (Figure 67) with an admixture of *Koeleria gracilis* or – far more rarely – *Zerna variegata.* Common herbs include *Artemisia absinthium, Thymus* spp., *Medicago grandiflora, Galium cruciatum, Polygala anatolica*, etc. In total, 81 species of plants contribute to the formation of this type of steppe.

Figure 67. Mountainous steppe-hayfield, forbs mixed with grasses (species of *Stipa* and *Festuca*). In background: lake Pharavani (2,073m), Ninotsminda district, Javakheti plateau, S. Georgia (2014).

Mountainous steppes occur higher and occupy altitudes from 1,800 m to 2,500 m.a.s.l. The dominant species is *Stipa stenophilla*, co-dominants are *Koeleria gracilis* and *Festuca sulcata*. In total 65 species of higher plants constitute these communities, including species of *Galium, Thymus, Dianthus* and *Onobrychis*.

The flora of this region is characterized by the presence of several rare and endemic species. Here should be mentioned *Scorzonera kozlowskyi, Campanula crispa, Achillea sedelmeyeriana, Erysimum krynitzkii and Asphodeline taurica* (Magakian 1941; Grossheim 1948).

Figure 68. Highmountain steppe composed of feather grasses (*Stipa capillata, S. tirsa*) embodies into a matrix of subalpine meadow herbs. Tsalka town environs, 1,950 m a.s.l., S. Georgia (2014).

Figure 69. Mountain steppe of South Georgia (Adigeni district, 1,450m a.s.l.) (2018).

In addition, secondary meadows can also be found here that are developed on the areas owing to intensive logging once occupied by primary broad-leaved forests (Ketskhoveli 1960). Resembling primary communities, such meadows are created for the variants of polydominant herbs including species of *Agrostis, Bromopsis, Calamagrostris, Alchemilla, Centaurea, Lotus, Trifolium,* etc. From monodominant meadows can be mentioned: *Nardus stricta, Agrostis planifolia, Anemone fasciculata, Bromopsis variegata,* etc. (Akhalkatsi & Tarkhnishvili 2012 and references therein).

The dry slopes of the Javakheti upland are densely populated by sedge (*Carex humilis*) dominated communities. This species is one example of a plant species widely distributed in the other postglacial period. *C. humilis* and some other species (*Festuca valesiaca, F. ovina, Filipendula vulgaris, Stipa tirsa,* etc.) of meadows of *Meskhet*-Javakheti turned into steppes (Akhalkatsi & Tarkhnishvili 2012) (Figures 68, 69).

Magakian (1941) distinguished four types of grasslands in mountainous Armenia: (1) secondary wild dry steppe vegetation with a predominance of *Zerna riparia fibrosa, Festuca sulcata, Koeleria cristata, Xeranthemum squarrosum, Artemisia armeniaca, Triticum boeticum,* and *T. araraticum*; (2) feathergrass steppe vegetation. In the southern part of the South Caucasus 12 species of feathergrasses occur, of which many are components of steppe and meadow-steppe vegetation. On the slopes and the plateau, the most important grasses are *Stipa tirsa,* S. *lessingiana, S. pontica* and *S. pennata*. Large areas of feathergrass steppe with *S. stenophylla* as the dominant grass occur on the slopes of Mt. Aragats, in the Sevan Lake Basin and on the Javakheti Plateau. Co-dominants are *Phleum phleoides, Koeleria caucasica, Festuca sulcata, Poa bulbosa, Potentilla recta* and *Carex buschiorum*; (3) fescue steppe vegetation where the dominant grass is *Festuca sulcata,* which in the southern part of the South Caucasus occurs from the semi-deserts to the highlands (Magakian 1941). Usually it covers hilly habitats or plains in the most arid climatic conditions. The species composition of this vegetation type is very diverse. Next to fescue, species like *Koeleria cristata, Stipa tirsa, Poa densa, Galium verum, Carex buschiorum* and *Scutellaria orientalis* occur in these dry steppes. Fescue is very resistant to trampling and under heavy grazing outcompetes other grasses, leading to a secondary *Festuca* dominated steppe phytocoenosis; (4) steppe communities with a dominance of *Agropyron cristatum* are very rare and only found on the slopes of Mt. Alagez (Armenia) (Magakian 1941). The diverse herb steppe on the south slopes of the mountains of Armenia includes as common species: *Thymus kotschyanus, Teucrium chamaedrys, Galium verum, Scabiosa bipinnata, Achillea setacea* and *Trifolium alpestre* (Magakian 1941).

2.3.10.5. Steppe Fauna of the Caucasus (a Brief Review)

The steppe fauna is characterized by a great diversity (Chernov 1975) because next to typical species other animal species of deserts and semi-deserts floodplains and summer-green forests are found, including saiga antelope (*Saiga tatarica*), jungle cat (*Felis chaus*), wild cat (*F. silvestris caucasica*), wolf (*Canis lupus*), foxes (*Vulpes vulpes caucasica, V. corsak*), marbled polecat (*Vormela peregusna*), weasel (*Mustela nivalis caucasica*), wild boar (*Sus scrofa*), European hedgehog (*Erinaceus europaeus*), long-eared hedgehog (*Hemiechinus auritus*), European hare (*Lepus europaeus*), rodents like *Cricetus cricetus, Mesocricetus auratus, M. brandti, Cricetulus migratorius, Allactaga jaculus, Meriones tristrami, Microtus socialis, M. arvalis,* and the Caucasian mole (*Talpa caucasica*). The bird fauna is represented by both bustard species (*Tetrax tetrax, Otis tarda*), lesser kestrel (*Falco naumanni*), pallid harrier (*Circus macrourus*), hen harrier (*C. cyaneus*), long-legged buzzard (*Buteo rufinus*), demoiselle crane (*Anthropoides virgo*), steppe eagle (*Aquila nipalensis*), quail (*Coturnix coturnix*), hoopoe (*Upupa epops*) and skylarks (*Alauda arvensis, Galerida cristata,* etc.) (Zhordania *et al.* 1999; Abuladze 2013). Many reptile species are shared with the deserts. Water habitats are populated by frogs and toads including *Rana camerani* (highland's steppe of the Lesser Caucasus), *R. macrocnemys*, *Bufo viridis* and *Pelobates syriacus* (Tarkhnishvili, 1996); typical reptile species are Mediterranean tortoise (*Testudo graeca*), some species of racers (*Dolichophis* (*Coluber*) *schmidti, Elaphe sauromates*), lizards (*Lacerta strigata, L. media*) (Tarkhnishvili & Kikodze (Eds.) 1996). Insects are represented mainly by numerous species of the order Orthoptera (grasshoppers, crickets, locusts) and other arthropods such as spiders, tarantulas and scorpions (Gulisashvili 1964) (Figure 70).

Figure 70. The great bustard (*Otis tarda*). Dedophlistskaro district, Vashlovani NP environs, E. Georgia (1978).

Conclusion

- Grasslands, generally open and continuous biome, are found on every continent except Antarctica. Grasslands are transitional land situated between forests and deserts and semi-deserts.
- There are two main kinds of grasslands: (1) tropical and subtropical grasslands, shrublands and open canopy woodlands – savannas – located between the Tropic of Cancer and the Tropic of Capricorn. Savannas represent a transition from tropical evergreen rain forests to hot deserts and semi-deserts. Hence, savannas have no cold seasons; (2) temperate grasslands found toward the interior of continents in mid-latitudes between temperate summer green broadleaf forests and cold winter deserts and semi-deserts.
- Temperate grasslands can be further subdivided into: (1) with tall grasses (North America's prairies and South America's pampas (tend to get more precipitation) and (2) biome with short grasses (temperate Eurasia's steppes).
- In the western part of the vast Eurasian Steppe Belt occurs the Pontic-Caspian steppe ranged from the mouth of the Danube to Ural Mts. The southern part of this grassland region is home of the Black Sea – Caspian steppe encompassing the western half of the northern Caucasian plain.

- Determined by temperature, rainfall, soil conditions, fire and grazers, the major manifestation of a grasslands biome is having grasses as the dominant, climax vegetation life form. The vegetation is restricted to a single layer. The flora is dominated by graminoid grasses, but usually with participation of herbaceous flowering plants (forbs); and geophytes (bulbs) are frequently abundant.
- The grasses die back to the ground annually (each winter), while the roots and the new buds found on significantly belowground stems are protected by the upper layer of soils. Therefore, they are able to survive from temperature fluctuations (winter's cold and summer's dry conditions) and from fire and grazing animals.
- Over thousands of years the evolution and spread of grasses in interior areas occurred through the adaptations to severe abiotic and biotic factors. Among many striking adaptation features of grasses resulted from their ability to grow is a major part (more than 60%) of biomass belowground, in the extensive root systems.
- Temperate grasslands are, in most cases, associated with semi-arid – dry summer and cold winter continental climates. This biome, depending on latitude, receives about 400 to 800 mm of rainfall per year, but it tends to decrease at low precipitation level, 400 mm, along the Caspian Sea. Rain usually occurs in the late spring and early summer (plant growing season), but much of the precipitations falls as snow in winter. Temperatures vary greatly between summer (can be as high as 38^0C) and winter (can be as low as -40^0C) (dormant season). Such a kind of climate has originated, because there are no large water bodies (e.g., ocean) nearby to obtain moisture.
- Seasonal drought causes periodic, spontaneous wildfires. It is often set off by lightning but currently human-induced fire is the major reason for ignition of grassland fire. However, their effects aren't dramatic to native grassland because, as mentioned, the plants are adapted (using a wide variety of strategies) to survive fires. Fires help to rid the ground off the old grasses; hence they are very important to local biodiversity. In temperate grasslands drought, fire and ungulates prevent large forests with tall trees from growing.
- Temperate grasslands have some of the darkest (chernozem), deepest (with fertile upper layer) and nutrient-richest soils in the world.
- Rich with grasses, temperate grasslands make an ideal environment for use as pasture lands for animals. The fauna include species of grazing herbivorous wild and domestic (livestock) ungulates and their predators. Temperate grasslands have representatives of mammalian grazers including megafauna, but do not have species richness and abundance in populations compare to the savannas. Small mammals are represented by (-semi) fossorial rodents and lagomorphs that mainly live in burrows. Today large wild animals mainly survive in natural reserves. This biome is a home for numerous arthropods (grasshoppers and other insects, spiders, etc.).
- Fossil evidence had suggested that grasses evolved on the Cretaceous, about 100 mya ago. The decrease of atmospheric carbon dioxide over the Cenozoic Era (across the last 65 mya) accompanied the 'CO_2 starvation', led to the wide expansion of plants with C_4 photosynthesic metabolism. The latter pathway of flowering plants, which is responsible for converting light energy into chemical energy is considered as one of the most convergent evolutionary phenomena within the Kingdom Plantae, especially in the family Poaceae (which has evolved in up to 45 independent evolutionary

groups). C_4 grasslands further development and spread in a given environment is often viewed as from the Miocene global climate oscillations which resulted from the consecutive dry season on the inland area. Such change of climate has been triggered by the formation of open landscapes and, on the other hand, a gradual decline of atmospheric CO_2. Later, during pre-glacial and glacial climate deterioration, as an important ecological factor, caused an increase in seasonality. As a result, the C_3 forests largely retreated (C_4 *versus* C_3).

- From the Pliocene epoch grasslands – this highly adaptable plant community of open landscapes often coincides with interaction (coevolution) with fire regimes and wild ungulate's grazing, trampling and migratory patterns.
- Grassland biome plays an important role of not only a proper ecological equilibrium, but also it is essential for humankind's survival. On the other hand, since time immemorial, when humankind changed from a hunter-gatherer into agriculture, currently truly virgin steppe with climax vegetation has become a rarity. Only small isolated islands of steppe similar to the Askaniya Nova in Ukraine, were preserved.
- In the Caucasus primary steppe, as mentioned, it envelopes its north-western part (Ciscaucasia). Grasslands of Transcaucasia represented in both landscapes – lowlands and highlands are of secondary origin ('semi-natural'), which historically was largely influenced by anthropogenic activities.

2.4. Biomes of the Arid Open Woodland (Wooded Grasslands)

2.4.1. General Characteristics

In arid and semi-arid countries, a park-like open woodland biome is found. Walter (1975) distinguishes two types of open woodland: firstly, the forest-steppe of regions with a moderate climate in the northern latitudes, formed by the intermixing of summer-green broadleaf forests and grassland vegetations of the steppe zone. Secondly the savannas, a rolling grassland scattered with shrubs and isolated trees between the tropical rainforest and the desert biomes of the southern latitudes.

In open woodlands, trees are so small or widely spaced that the canopy does not close and grassland can persist in the herb layer. In the forest-steppe tree densities are locally higher (Mil'kov 1950; Rikhter & Mil'kov (Eds.) 1956; Golubev 1965; Avanesova 2006). With respect to their physiognomy the open woodlands of the Caucasus show an extraordinary situation: within the same moderate climate zone both types of open woodland are represented. In the north-western part of the North Caucasus, open woodland exists as a transition of deciduous forests to the herb-rich meadow-steppes of the steppe zone. In the eastern part of the South Caucasus, on the other hand, open woodland has originated from the intermixing of summer-green broadleaf forest with vegetation of winter cold deserts and semi-deserts (Grossheim 1948; Gulisashvili 1964, 1980; Gulisashvili *et al.* 1975; Gegechkori 2008, 2010b).

It is beyond discussion that the North Caucasus forest-steppe belongs to the open forests of the moderate climate of the northern latitudes (Gulisashvili *et al.* 1975). The viewpoints of specialists on the open woodlands of the eastern part of the South Caucasus are, however, quite controversial. One group of botanists regards the arid open woodlands of eastern Transcaucasia to be the northern version of (sub-) tropical savanna (Gulisashvili 1980), other specialists consider the arid open formations as quite independent communities (savannoid vegetation) that, except for physiognomic similarity, have nothing in common with savanna from countries with year-round warm climate (Ketskhoveli 1958, 1960; Gegechkori 2008, 2010b).

What types of arid and semi-arid open woodlands are represented in the Caucasus? And does some version of tropical-subtropical savanna exist in this region? To answer these questions, it is necessary to analyze the different types of (semi-) arid open woodlands in more detail.

2.4.2. The Forest-Steppe of the Western North Caucasus (Ciscaucasia)

The forest-steppe ecotone occurs between steppes and forests in the Northern Hemisphere. Forest steppes have a more humid climate than steppes and in their natural condition represent a mosaic of meadow steppe vegetation and woodlands. The forest-steppe serves as a demarcation line between humid and arid environments, and here annual precipitation and evaporation are approximately equal. The forest-steppe extends as an uninterrupted zone from the Carpathian foothills to the Altay, including southeastern Europe, south Russia, southern Siberia, and North China (Nesterov & Fedotov 2005). Steppes and forest-steppes have been transformed by human activities more than in any other part of Northern Eurasia, hence once a virgin land and natural environments were turned into arable fields (Gulisashvili *et al.* 1975).

Figure 71. Forest-steppe biome of lowland and the foothills of Ciscaucasia, transition area between dense deciduous broadleaf woodland and open grassland of primary origin (1973).

The forest-steppe of the western part of the North Caucasus is found between 45° and 47° latitude. Annual precipitation in this area ranges between 500 and 600 mm, the winter is cold, the summer rather warm, and most rainfall occurs in May and during the summer months. The soils consist of chernozem and grey soils.

This forest-steppe (Figure 71) is located where the deciduous forest (with a dominance of local pedunculate oak (*Quercus robur*) on the foothills of the north slopes of the Greater Caucasus begins to mix with the (primary and secondary) steppes of the lowland. Hence, the forest-steppe communities are situated at higher elevations than the steppe grasslands (Gulisashvili *et al.* 1975).

The forest component usually consists of dense stands of trees with a well-developed canopy ('forest islands') that interchange with areas of steppe. Grossheim (1948) already pointed out that the forest-steppe of this part of the Caucasus represents a complex alternation of typical forest and typical steppe in regions, where both biomes border on each other. Because of this origin, the vegetation structure differs substantially from that of arid open woodland, where grassland species and trees are mixed in a much more diffusive way. As we shall see below, deciduous broadleaf forest species are unusual in open woodlands.

The oak-hornbeam forest associations of the North Caucasus open woodland are closely related to the forests of Eastern Europe. They are composed of *Quercus petraea, Q. robur, Carpinus betulus* (=*C. caucasica*), *Fraxinus excelsior, Acer campestre, Tilia platyphyllos, T. begoniifolia* (= *T. caucasica*), *Pyrus communis, Ulmus minor* and *Sorbus torminalis*. The undergrowth consists of *Crataegus kyrtostyla, Cornus mas, Mespilus germanica*, and *Corylus avellana*. Liana (vine) species include *Humulus lupulus* and *Vitis vinifera sylvestris* (Gulisashvili 1964).

The steppe sections of the forest-steppe biome of the western part ('meadow-steppes') are more mesic than those of the eastern part of the North Caucasus and are composed of yellow oat grass (*Trisetum pratense*) as the most conspicuous species and *Elymus repens, Festuca pratensis, Phleum pratense, Poa pratensis, Achillea millefolium, Cerastium arvense, Geranium ruprechtii* and *Verbascum orientale* (Shiffers 1953). The animal composition of the biome has noticeably changed due to the excessive cultivation of local arable land. In the historical past the forest-steppe was inhabited by widely distributed forest and steppe species typical of moderate climates of Europe.

Many xeric invertebrate animals, even from the adjacent open arid woodland of the eastern part of the North Caucasus, do not penetrate into the more mesic open landscape of the western part (Gegechkori 1984a).

Though a more arid climate is characteristic for eastern parts of the North and South Caucasus, it is remarkable that foothills of the Gelendzhik Black Sea coast area of Krasnodar Krai in the North-West Caucasus is populated by arid open woodland, composed of the same pistachio-juniper communities (Shanina 1987; Lipka 2006 a, b).

2.4.3. Arid Open Woodlands of the Eastern North Caucasus (Ciscaucasia)

The arid open woodlands of the eastern part of the North Caucasus occupy a region of the Caucasus that is characterized by a far more extreme arid climate than the previous region, by very steep slopes, and by intense agricultural activities that have strongly damaged the woodland ecosystem.

The main deciduous trees of these open woodlands, *Pistacia* and *Celtis*, were in prehistoric times absent in the North Caucasus. On the limestone slopes the only arboreal plants are shrubs: *Rhamnus pallasii*, *Spiraea crenata, Ephedra procera* and *Cotoneaster racemiflorus*. On non-limestone slopes this phytocoenosis is enriched by *Juniperus oblonga, Lonicera iberica* and *Paliurus spina-christi*. The herb layer includes species of *Salvia, Scabiosa* and *Convolvulus* (Gulisashvili *et al.* 1975).

The pine-juniper open woodland of Lower Predgorny Dagestan was investigated by Azimov & Teymurov (2003) and Azimov (2005). From six florocoenotypes of arid woodland, the floristic composition exceeds 700 species of vascular plants.

2.4.4. Arid Open Woodlands of the South Caucasus (Transcaucasia)

2.4.4.1. General Characteristics

The arid open woodland of the South Caucasus is sometimes called 'light forest' (Sosnovski 1928; Ketskhoveli 1960) to denote that the open canopy allows sufficient light for a continuous xerophytic herbaceous ground layer. Indeed, due to the scattered occurrence of trees, many open places occur in the arid woodland, but the term 'forest' is invalid, because in the open woodland no forest communities, with a closed canopy and shade enduring undergrowth occur (Gegechkori 2008, 2010b). Physiognomically similar to anthropogenic parks, the open arid woodlands consist of sparse stands of drought-resistant trees and shrubs with a ground layer consisting of xerophytic, light-demanding grasses and herbs (Gulisashvili 1942; Ketskhoveli 1960).

Apart from the Caucasus, arid open woodland biocoenoses are known from adjacent regions in Southern Europe, from the Asian Interior and from Middle (Central) Asia.

Arid open woodland covers territories in the eastern (E. Georgia, Azerbaijan) and southern parts (Armenia, Azerbaijan) of the South Caucasus. Depending on orography and climate, the biome occurs at various altitudinal levels, e.g., on low-mountain slopes (400-500 m a.s.l.) in the environs of Mtskheta (near Tbilisi), whereas in the more continental southern part of Transcaucasia its vertical distribution ranges from 800 to 1,500 m a.s.l. (Ketskhoveli 1958, 1959; Chincharauli 1968; Kebadze 1969; Khachidze 1984; Ivanov *et al.* 2000; Lachashvili *et al.* 2004; Aleksanyan 2011 a, b).

Arid open woodland mainly developed on grey-cinnamonic soils as well as on chestnut, loam slightly saline grey and cherhozem soils (Sabashvili 1948, 1965; Urushadze & Blum 2014). The climate in the area of the biome varies between dry subtropical and warm temperate with very hot summers. The average annual precipitation ranges from 400 to 500 mm, with a minimum of 250 mm and a maximum of 560 mm. Maximum precipitation falls in May-June, whereas July and August are very dry.

Two main types of arid open woodland can be distinguished: woodland with a dominance of mastic trees (*Pistacia atlantica* (= *P. mutica*) and woodland with a predominance of juniper trees (*Juniperus* spp.). Next to these, there are arid light forests with a dominance of nettle tree (*Celtis caucasica*), complex of willow leaf pear (e.g., *Pyrus salicifolia*), Georgian almond (*Amygdalus georgica*), Georgian maple (*Acer ibericum*) and/or Eldar's pine *Pinus brutia eldarica* (=*P. eldarica*). In the lowlands, wormwood (*Artemisia fragrans*) and yellow acacia (*Caragana frutex*) are the most characteristic and common species among the xeric herbs and shrubs (Ketskhoveli 1960, Lachashvili *et al.* 2004, 2007). Under trees, as a rule, mesic forest shrubs and herbs are absent (Grossheim 1948).

Arid open woodlands of the South Caucasus historically have been dramatically reduced. Natives as well as invader tribes greatly reduced open woodlands cover, most dramatically in the 19th after overpopulation of semi-arid lowland and foothill areas of the region. Pistachio-mastic and nettle trees were extensively harvested for firewood and charcoal production (especially in Tbilisi, capital of Georgia, and its surrounding areas). Widespread destruction of open woodlands to clear land for intensive agriculture, cattle grazing, urban development, and quarries remained after mined limestone ('Heidelberg Cement') have eliminated open woodlands from much of their former range. Unfortunately, this practice continues to the present day. Some remains of open woodlands mainly on badland hillsides like Vashlovani National Park provide the only hints of the extent of former transitional biome of Transcaucasia.

2.4.4.2. Communities Dominated by Mastic (Pistachio) Tree

Communities dominated by mastic (*Pistacia atlantica*) are present on cinnamon and chernozem soils (Figure 73). These formations belong to the west-Iranian vegetation type. Pistachio communities prefer the lower parts of the slopes and gentle forms of relief. The undergrowth of this open woodland includes a rich variety of shrubs such as *Paliurus spina-christi, Rhamnus pallasii, Lonicera iberica, Caragana grandiflora, Colutea orientalis, Cotinus coggygria, Prunus* (*Cerasus*) *incana, Pyrus salicifolia, Acer ibericum, Tamarix* spp., *Ulmus minor, Amygdalus georgica, Punica granatum, Rhus coriaria, Ephedra procera*, and *Juniperus* spp. The ground layer is formed by steppe grasses: *Agropyron, Stipagrostis (=Aristida), Bromopsis, Bothriochloa, Festuca, Stipa, Poa* and herbs and semi-herbs like *Kochia prostrata, Salsola ericoides, S. glauca, Peganum harmala, Xeranthemum squarrosum, Achillea nobilis* and *Artemisia fragrans* (Lachashvili *et al.* 2004, 2007).

2.4.4.2.1. Vashlovani National Park (VNP) as an Ecotone: The Paradigm of Charismatic Patterns of Mosaic Habitats

Sometimes in arid open woodland, the tree layer consists entirely of mastic or pistachio (*Pistacia atlatinca*). Such a site of open woodland, which escaped destruction, is protected in the Vashlovani National Park (southeastern Georgia, Kakheti region, close to the Azerbaijan border). It is located on the Iori Plateau, 150-750m a.s.l. altitude. The amount of annual precipitation is 400mm (Lachashvili *et al.* 2004). It was founded as a State Reserve in 1935 (about 4,000 ha), In its current boundaries, in 2003 the Strict Nature Reserve area was expanded to 10,143 ha, and the bigger area of Vashlovani Protected Areas (VPA) (about 36 ha) was established around it. VPA includes the Vashlovani Reserve, the Vashlovani National Park (24,598 ha), and three Natural Monuments: The Eagle Canyon (100.4 ha), Takhti-Tafa (9.7 ha), and Kaklisyure Alazani floodplain forest (204.4 ha) (Ministry of Environment and Natural Resources Protection of Georgia, 2006). Plans are also to establish protected landscapes on the Iori Uplands (IUCN category V, area 173,000 ha), although this has not yet been realized

(Fourth National Report to the United Nations Convention on Biological Diversity: Georgia). The territory of the reserve is characterized by a dry subtropical climate (Lachashvili *et al.* 2004, Gegechkori 1997, 2008). In this large reserve – the nucleus of the National Park - 581 vascular plant species have been recorded, including 33 Caucasus endemics and two endemics for Georgia (Lachashvili *et al.* 2004) (Figure 72).

A great diversity of flora and habitats in the Vashlovani Reserve have been determined by the heterogeneity of physico-geographic conditions: gorges, plains, hills, badlands, plateau, and cliffs-of-the canyons (the latter is known as 'Sharp Walls'), geological history and location in a crossroads of different (arid, semi-arid, temperate deciduous forests) biomes (Ketskhoveli 1960, 1980).

This open arid woodland is associated with foothills and plains among adjacent drought-resistant grass cover (steppe), on the one hand, and drought-tolerant desert—semi-desert vegetations, on the other hand, i.e., a central area of transition. There is a sharp environmental change from two neighbouring types of biomes toward the open woodland-grassland from the west and deserts and semi-deserts from the east. This is a crucial large-scale mosaic topographical event why this 'island' of semi-xeric woodland is referred to as a typical edge effects, i.e., ecotone (Senft 2009 and references therein).

Figure 72. Arid open woodland ecotone enclosing an area between deciduous and semi-deserts. A panoramic view over Vashlovani NP consists of *Pistacia atlantica* and *Juniperus* spp. (1968).

The reserve is rich with ecosystems. In Vashlovani the following ecological communities coincide: arid forests (deciduous and coniferous), phryganoid vegetation, deserts and semi-deserts (mainly foothill and partly plain deserts), and riparian forest fragments (Ketskhoveli 1960; Lachashvili *et al.* 2004; Nakhutsrishvili 1999, 2013).

The Vashlovani National Park is the only place in the Caucasus, where *Pistacia atlantica* most frequently reaches its maximum size–growing to 10-15 meters tall, with a domed crown approximately the same – 10-15 meters wide. The trees are widely spaced, scattered, therefore the canopy does not close. According to the physiognomy of Vashlovani NP, the term *Vashlovani* is believed to be derived from the Georgian word *vashli* (an apple), i.e., community of mastic tree in central part of NP with its wide spacing (not less 15m) canopy happens to look similar to the apple's orchard at first sight. Therefore, 'Vashlovani' means 'garden of apple trees' (Ketskhoveli 1980; Gegechkori 1982).

Figure 73. Pistachio (*Pistacia atlantica*) is the archetypal plant of eastern Transcaucasia, significantly in Vashlovani NP. In the foreground: grassland with a dominancy of *Botrichloa ischaemum* (1968).

Figure 74. Fruits of the pistachio tree are greatly appreciated by wildlife (invertebrate and vertebrate animals), Vashlovani NP (2012).

Vashlovani NP represents one of the few, if not the only, western Eurasian areas, where savanna-like (savanoid) forests are still present. The most striking communities are represented by open woodland with pure mastic woodlands, i.e., edificator (dominant) tree is represented by *P. atlantica* with admixture of single trees and shrubs (e.g., *Ulmus minor, Pyrus salicifolia*, etc.) (Nakhutsrishvili 2013). To quote Ketszkhoveli (1960): 'The arid open woodland consisting of mastic-trees is one of the few and possibly only woodland in the USSR, where this tree is proposed to be preserved in such a good way'.

Best known as the pistachio, species of *Pistacia* trees are the terebinth and mastic (Ketskhoveli 1960, 1980) (Figures 73, 74).

Placement of *Pistacia* in the family Anacardiaceae has never been questioned among taxonomists. The level of speciation and relationship within the genus is still unclear and controversial. Comprehensive, phylogenetic study of the genus is given by Partiff & Badenes (1997), Al-Saghir *et al.* (2006) and Kozhoridze *et al.* (2015). It largely depends on either the plant's anatomy (leaf morphology) and geographical distribution or recent molecular genetic method (variability of chloroplast DNA) (Parfitt & Badenes 1997 cited in Al-Saghir & Porter 2012). Linnaeus (1753) recognized 6 species, Zohari (1952) – 11 species, Partiff and Badenes (1997) – 11 species, Yi *et al.* (2008) – 12 species, Al-saghir & Porter (2012) – 9 species and 5 subspecies. In general, in the Northern Hemisphere 11 disjunct distributed species are recognized (Partiff & Badenes 19970).

Seven species are derived from the Mediterranean basin to Central Asia (P. atlantica, *P. integerrima, P. khinjuk, P. lentiscus, P. palaestina, P. terebinthus,* and *P. vera*), two species – in eastern Asia (*P. chinensis, P. weinmannifolia*), and two species from the southwestern United States to Central America (*P. mexicana* and *P. texana*). *P. chinensis* extends into tropical Asia as far as Myanmar and the Phillippines (Zohary 1958 cited in Yi *et al.* 2008).

In all classifications *Pistacia mutica* Fisch. & C. A. Mey is considered to be either a subspecies of *P. atlantica* Desf. or treated as it is synonymous or, finally, not recognized as an independent taxon (species/subspecies). On the contrary, modern Caucasian and former Soviet Union's botanists recognize mastic as *Pistacia mutica* (Grossheim 1932b (vol. VI), 1948; Gagnidze 2005; Lachashvili & Kachidze 2005), Czerepanov 2007 and so on). From adjacent to the Caucasus countries the same opinion is perceived by Esmail-pour (1998) for Iran and Atli et al. (1998) for Turkey.

According to the latter author, in Turkey '*P. atlantica* is recorded in the same areas, where *P. mutica* is also found. Although *P. mutica* has a rounded tree shape, different from that of *P. atlantica*, the latter is often accepted as synonymous with *P. mutica*'.

The genus is divided into two monophyletic groups – sections – *Terebinths* and *Lentiscus* (Partiff & Badenes 1997). The former – deciduous one - contains three sister groups. Among others the first group includes *P. atlantica, P. mutica* and *P. eury*carpa (Al-Saghir 2006 and references therein).

P. atlantica including *P. mutica* as a subspecies or without it ranges from Central Asia to the Middle East and all along North Africa and Canary Islands (Zohari 1952; Browicz 1988; Sheibani 1994 cited in Kozhoridze *et al.* 2015).

The range of *P. mutica*, if it is recognized as an independent species, encompasses the Eastern Mediterranean region, Balkan Peninsula, Asia Minor, Iran (western part), Afghanistan, the Crimea, Western Caucasus, eastern and southern Transcaucasia (*www.agroatlas ru/ru/content/related/Pistacia_mutica*).

Al-Saghir (2006) in the list 'specimens examined' dedicated to *P. atlantica* (p.93), among other countries are named Azerbaijan and Georgia (Vashlovani Reserve), but in 'General distribution' (p.94) the same species is missing in the 'Caucasus' as a region, as well as countries of South Caucasus, unlike, for example, neighbouring countries (Turkey, Russia, Iran). Unfortunately, this fault is repeated in the map (Map. 1) dedicated to the range of the same *Pistacia* (page 31) (an area between the Black and Caspian Seas is not dotted). In later work G. Al-Saghir co-authoring with D.M. Porter (2012), for the same *P. atlantica* from the South Caucasus mentioned also two countries (Armenia, Azerbaijan), but in this case Georgia (with Vashlovani Reserve) is lacking (p.15).

Species of the genus cover a wide and disjunct area located in arid and semi-arid regions. In the Old World two centers of species diversity are known. One consists of the Mediterranean region of Southern Europe, Northern Africa, and the Middle East. A major center of speciation also developed in West and Central Asia. Naturally, as mentioned, two species occur in the Western Hemisphere, North America.

Pistacia plants are shrubs or small trees (growing to 5-15m tall) and can be either evergreen or deciduous. *Pistacia* is an ancient genus. It is estimated to be from the Late Cretaceous Period (about 80 mya old). A common ancestor originated in Central Asia (Zohary 1952; Parfitt & Badenes 1997; Al-Saghir & Porter 2012).

Two centers of diversity have been recorded: (1) the Mediterranean region of Europe, Northern Africa, and the Middle East, (2) West and Central Asia (Kafkas et al. 2006). The species of the genus occurs naturally from North Africa to the Philippines and in the New World in Honduras and Mexico to Texas (Al-Saghir 2006). *Pistacia* exhibits a disjunct distribution in Mediterranean Eurasia and adjacent North Africa, eastern Asia, and North to Central America. A vicariance event has occurred mainly from the Late to Early Miocene (Xie *et al.* 2014).

Species of *Pistacia* easily form interspecific hybrids, suggesting very close relationships (Zohary, 1952; Parfitt & Badenes 1997). In other words, some species of *Pistacia* appear not to be genealogical species, and evidence for reticulate evolution is presented (e.g., *P. saportae* was shown to be a hybrid with *P. lentiscus* (maternal) and *P. terebinthus* (parental) as the parental taxa (Yi *et a*l. 2008).

Pistacia is believed to have originated in the boreal forest and its migration pathways might have been linked in relation to climate change (Kozhoridze *et al.* 2015 and references therein): '. . . *Pistacia*, as it is distributed today, most probably originated in Central Asia, speciated in the Northern Hemisphere and reached the corresponding North American part of Pangaea, but not South America, . . . while four species have been recorded in East Africa, along the rift valley' and then 'the current global distribution of *Pistacia* in a narrow belt between latitude 10^0 North and 45^0 North was changed and shaped by continuous cycles of cooling and warming since the continental drift' (Kozhoridze *et al.* 2015 and references therein). New World species suggested are a result of historic distribution pattern (Parfitt & Badenes 1997).

The most primitive species in the genus *Pistacia* is recognized as *P. vera* (Al-Saghir 2006) or *P. vera* and *P. khinjuk* (Partiff & Badens 1997). This is manifested in the Central Asian center of diversity of the genus. Both are the only *Pistacia* species with large edible nuts. The trend from large to small seed is connected to reproductive strategy from seed distribution via zoochory (bird - or mammal – mediated distribution) (Parfitt & Badenes 1997).

Except for nuts, an evolutionary trend outlines also leaves, when few and large leaflets turn into many and small leaflets (Parfitt & Badenes 1997).

Species of *Pistacia* are adapted to dry continental climate areas or summer drought typical of Mediterranean or sub-Mediterranean climates, therefore have a high tolerance to saline soil and a harsh climate. Thus, deciduous trees (about 7 species) can survive in temperatures ranging from -26^{0}C in winter to 46^{0}C in summer (especially Central Asian *P. khiniuk* and *P. vera*), but evergreen (about 4) species adapted to the temperature ranging from -8^{0}C to 41^{0}C and prefer a well-drained soil. Species of *Pistacia* grow and survive in degraded and rocky land areas, where many other shrub and tree species have been eliminated (Kozhoridze *et al.* 2015).

Pistachio (*P. vera*) is economically the most important species in the genus, edible seeds of which have large commercial value. It is indigenous to Central Asia. It has been cultivated for 3000-4000 years in Iran and was introduced into Mediterranean Europe by the Romans at the beginning of the Christian era (Crane 1978 cited in Al-Saghir 2006). For seeds the pistachio was introduced into the United States (significantly in California) and into Australia (Al-Saghir & Porter (2012) and references therein). The pistachio seeds were common food as early as 6570 BC (excavations at Jarmo in Iraq (*https://en.wikipedia.org/wiki/Pistacia*). Iran, the United States, Turkey and Syria are the major producers of pistachios (90% of the world production). Seeds are consumed fresh, roasted and salted. Processed uses include candy, baked goods, ice cream, paste and butter. They also have traditional, medicinal uses. The resin is used as a blood-clotting agent, while husks are used for dying and tanning. Other species of terebinth resin were widely used as a preservative in ancient wine (Iran) (Al-Saghi & Porter (2012) and references therein).

P. atlantica (=*P. mutica*) produces a resin called mastic. Many of them are planted as ornamental trees due to their attractiveness (Ketskhoveli 1958, 1980, 1982; Abashidze 1959).

The resin is used to manufacture alcohol and lacquer. In the Caucasus and Iran, the dry resin named 'keva' is used for chewing. From this term was derived the Russian 'Keva tree' (*кевовое дерево*) and Georgian '*Kevis khe*' (khe (ხე) – tree in Georgian). Essential oils are used in the perfume industry. From a single tree, up to 200g of resin can be extracted (Esmail-pour 1998). Canopy, fruit and autumn foliage (especially *P. chinesins*) are used as an understock for grafting *P. vera*. The woods are used for production of furniture. The leaves are rich in tannins (up to 20%) in the galls caused by an aphid parasite, used as the raw material for tanning in the leather industry (Abashidze 1959; Ketskhoveli 1960; Wulf & Maleeva 1969).

Except aphids, among gall-producing species of insects should be mentioned psyllids (Hemiptera, Psylloidea). *Megagonoscena viridis* in the South Caucasus sometimes are abundant, causing visible damage to the host-plant through galling (roll leaf galls) of leaf tissue. The galls apparently do not cause significant harm to the tree. Leaves are not galled by another species – *Agonoscena pistaceae*, co-existing with *M. viridis* on the leaves of mastic tree (Gegechkori 1980, 1984a; Gegechkori & Loginova 1990).

Except communities dominated by *P. atlantica*, there is a unique relict and virgin forest of junipers (*Juniperus foetidissima, J. polycarpos*) in Vashlovani NP (Lachashvili *et al.* 2004; Lachashvili & Kakhidze 2005).

Phryganoid vegetation, according to Lachashvili *et al.* (2004), is represented by shrub communities including *Caragana grandiflora, Atraphaxis spinosa* and *Reaumuria alternifolia* (=*R. kuznetzovii*). This community is connected with clayey and clayey-sandy (argillaceous and argillo-arenaceous) slopes, but on a more stable substrate. On comparatively small areas are found thorny cushion (tragacanth) plant (*Astragalus* spp.) phytocoenoses.

Desert and semi-desert plant communities are mainly found on the slopes of unstable and nutrient-poor soils – badlands and more rarely on plane sites. The presence of *Artemisia fragrans, Salsola nodulosa* and *Gamanthus pilosus* (Lachashvili *et al.* 2004) should be mentioned.

The fact that *Bothriochloa* communities are combined with arid open forests, which in Pliocene times had zonal characteristics in the Caucasus, indicates that these communities with their key species *B. ischaemum* florogenetically are closely related to true savannas (Sakhokia 1958 cited in Nakhutsrishvili 2013), but not savannas *sensu* Gulisashvili (1980).

Under the influence of animal grazing, whereas the community gets floristically poorer, *Bothrichloa* becomes a co-edificator of the species of *Stipa* (Nakhutsrishvili 2013).

There is a restricted area with saline-poor grounds covered by fragments of grasslands consisting of *B. ischaemum* and secondary origin *Stipa lessingiana, S. capillata.* Apart from grass herbs stabilized patches are dominated with hemixerophillous shrubs of the shibliak community (*Paliurus spina-christi, Rhamnus pallasii*). Geophyte bulbous plants consist of rare charismatic species - *Iris iberica* and *Tulipa eichleri* listed in 'Red List of Georgia' (Figures 75, 76).

In the Vashlovani NP, on banks of dry ravines of Pantishara Gorge (*panta* (პანტა – wild pear in Georgian) the groundwater holds a peculiar riparian forest fragment (phytocoenosis) composed of *Punica granatum, Vitis vinifera silvestris, Ulmus minor, Hippophae rhamnoides, and Elaeagnus angustifolia.* The presence of the Pliocene relict grasses - *Imperata cylindrica* and *Erianthus ravennae* in these communities clearly shows that a tropical savanna was present here in the geological past. The former existence of savanna biocoenoses similar to those of the present-day African savanna is confirmed by megafauna fossils found in East Georgia (e.g., in the Vashlovani NP and adjacent area, including remnants, as previously mentioned, of the Southern elephant (*Mammuthus primigenius* (= *Archidiskodon meridionalis taribarensis*), giraffe and rhinoceros (Gabunia & Vekua 1963) (Figure 77).

Figure 75. Awesome beauty of the Georgian iris (*Iris iberica*), endemic to E. Georgia and adjacent Azerbaijan (1984).

Figure 76. Eichler's tulip (*Tulipa eichleri*), the showiest plant growing in mosses ground cover, endemic to E. Georgia and adjacent Azerbaijan (1984).

Figure 77. Reconstructed skeleton of the Taribana elephant (*Mammuthus primigenius* (= *Archidiskodon meridionalis taribarensis*) from the Early Pleistocene (~2 mya), discovered in 1956; nowadays steppe biome, Kakheti Province, E. Georgia. Georgian National Museum (2006).

According to fossil plants from the Late Pliocene (around 3 mya old) discovered from this area and described by Kolakovski & Ratiani (1967), these fossil plant complexes mostly belong to Mediterranean elements in the Late Tertiary of western Eurasia. A high number of living Mediterranean elements of the Vashlovani Reserve (species of *Juniperus, Jasminum, Punica, Cotinus* and others), and comparing the modern and the fossil vegetation, give the opportunity to scientists to express the following idea. The mentioned site of the eastern Georgia is an ideal area for studying the climatic development in eastern Transcaucasia from the Pliocene until today.

The Pantishara valley is famous for its temporary or seasonal brooks or small rivers unlike other dry creeks of the National Park, which rarely fill and flow furiously after sufficient rain. The latter is known as 'khevkhmeli' (ხევხმელი) ('dry gorge' in Georgian like the Arabic and Hebrew term wadi), therefore, there are a lot of tracks of boar, lynx and especially bears and their cubs. The bachelor students study the ichnological peculiarities of the gorges (locations of bears or their tracks, population estimation, number of cubs associated with their mothers, track sizes, measurements of front paw width, etc. (Gegechkori & Khochiashvili; pers. observation).

The Datviskhevi Gorge ('The Gorge of a Bear' or 'Bear Canyon' – Datvis Khevi (დათვის ხევი) in Georgian) in the Vashlovani NP, another gorge is adjacent to the Pantishara valley with discharging groundwater.

Figure 78. The showy and unique Pliocene relict, river bank's poplar (*Populus* (*Turanga*) *euphratica*) grove in Vashlovani NP (Datvis Khevi (Bear's Gorge). In background: *Juniperus* spp., E. Georgia (1982).

This place has been the bottom of the sea (Sarmatian or Paratethys in a broader sense) since ancient, Pliocene times. The evidence of this event is the shells of marine molluscs, which are still kept in the sediments of the clearly expressed limestone layers on the naked slopes of local hills (Gegechkori 1982, p.188) (Figure 83). This gorge also displays well-preserved Taribana elephant's fossilized remains deposited today in the Dedoplistskaro natural history museum (Figure 77).

Datvis Khevi is the single location, as mentioned, where *Populus* (*Turanga*) *euphratica* occurs in Georgia (although it is more widely represented in the extreme south part of southern Transcaucasia). This poplar is a relict of the Tertiary Period as well and a remnant of warm-winter tropical-subtropical deserts (Kebadze 1965) (Figure 78).

P. (*T.*) *euphratica* is one of the most important species in the riparian (tugai) forests of the arid and semi-arid regions of West and Central Asia to the north of Africa. This species of poplar is a 'living fossil', i.e., is the most ancient, primitive tree of the Salicaceae family that, as mentioned, dates back to the Tertiary period and has existed possibly, for more than 60 mya. Fossils of this kind of poplar have been discovered in the strata of the Oligocene in China (Shiji et al. 1996). *P.* (*T.*) *euphratica* has extraordinary characteristics that make its population tolerable to both chilling winters and droughty summers-growing in a salinized desert belt enabling them to withstand waterlogging and high saline-alkali concentrations (Chen *et al.* 2004).

Except for a surprising ability to great abiotic stress resistance, this species of poplar has characteristic anatomical adaptive significance. The latter is expressed in heteromorphism of the leaves, i.e., leaf shape variation with the growing of the tree. The leaves lanceolate in the early growth period, and then change into different shapes such as oval, extravagant oval, broad triangular and dentate broad-ovate leaves (Li & Zheng 2005; Calagari *et al.* 2006; pers. observation). It has an extensive root net with strong ability to absorb water and withstand salt (Chen *et al.* 2004). Because of human interference, poplar riparian forests in many countries, including the Caucasus region have been destroyed, leaving only sparse and degraded forests. Currently, the largest populations, of the *P.* (*T.*) *euphratica* in the world are concentrated in China (Tarin River drainage area) and some mid-Asian countries (Middle Asia, Kazakhstan) (National Commision of the People's Republic of China for UNESCO, 2010). Materials concerning distribution and conservation of the *P.* (*T.*) *euphratica* in the Caucasus is given in *Peper et al.* (2010).

Thuia (*Biota orientalis*) has been planted in another site of NP, in the Natural Monument Eagle Canyon's territory by Shiraki (Kakheti) forester and became secondary naturalized (Akhalkatsi 2015).

So, nowadays Vashlovani NP is represented as the last outpost of some deserts and semi-deserts organisms in the eastern South Caucasus (for example, *Populus* (*T.*) *euphratica, Hyaena hyaena, Gazella subgutturosa* and others; the two latter species have become extinct in recent times). In broad meaning, several plant and animal species except those of Caucasian, Mediterranean, Irano-Turanian and Asia Minor origin have their eastern, western or northern limit of distribution in the Vashlovani NP ecotone (Gegechkori 1984 a, b; Gagnidze 2000; Gagnidze & Davitadze 2000; Lachashvili *et al.* 2004). Vashlovani NP is a natural edge, where complexes of biomes meet. In other words, here the convergence amongst different biomes, as underlined, grassland with open woodlands, deserts and semi-deserts with open woodland – the mosaic approach of nature is well expressed.

Figure 79. Black stork (*Ciconia nigra*) with chicks in the nest. Artsivis Kheoba (an eagle's gorge), part of Vashlovani NP (photo by Kochiashvili, 2013).

NP has long been well-known for its richness in large mammals. This site is a home to different species of animals from two types of biomes – temperate deciduous/mixed forests (migrated formerly from the southern slope of the Greater Caucasus Range) and arid open woodlands (settled by living forms from open woodlands proper and neighboring deserts and semi-deserts). They can survive side by side in the same community, living in a particular 'island' area. For example, brown bear, wolf, hyena, lynx, gazelle, and birds (*Aquila heliaca, Aegypus monachus, Gyps fulvus, Otis tetrax* and rarest *Cicconia nigra*, etc.) (Figure 79).

Since 2003, the reserve has been inhabited by a male leopard (*Panthera pardus ciscaucasica*), again a unique fact for the arid lowland and foothills area of Georgia. It seems that the animal migrated to the Vashlovani area from the extreme southern part of South Transcaucasia (Butkhuzi 2004; Gegechkori 2007). The leopard population has evidently gone through a bottleneck several times in Vashlovani open woodland's history (Gavashelishvili & Lukarevsky 2008) (Figure 80).

Figure 80. Leopard (*Panthera pardus ciscacuasia*). Area of open woodland of Vashlovani NP which has been inhabited by asingle male predator during 2004 -? 2008. Phototrap picture has been detected by staff of the NACRES (2014).

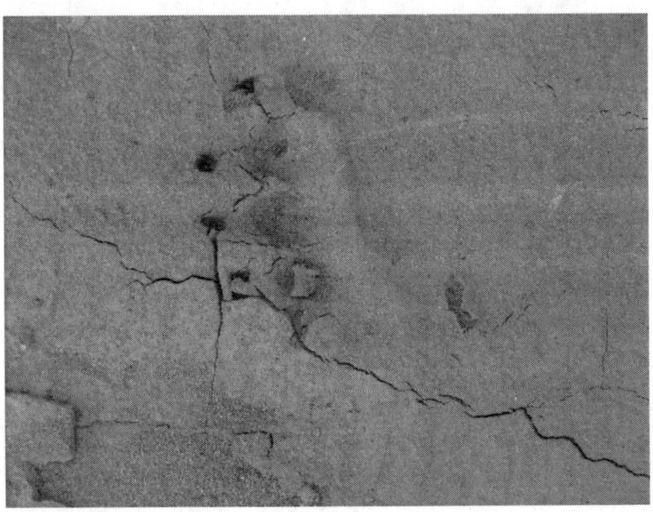

Figure 81. The tracks of brown bear (*Ursus arctos*) roaming in Vashlovani NP. (2016).

The Vashlovani NP also holds various endemic invertebrate species. The psyllid species - *Heterotrioza bathiashvilii* (as already pointed out), which feeds on *Kochia prostrata* (Amaranthaceae), has never been observed outside the territory of the reserve, despite the presence of its hostover a wide range within the Caucasus (Gegechkori 1968, 1984a).

Today Vashlovani NP is officially recognized as a natural and cultural heritage for its high-aesthetic ecological, scientific, educational, recreational and historical value, including both historical and significantly natural monuments.

The Turianchai Reserve in Azerbaijan, adjacent to Vashlovani, also holds arid woodland with a dominance of pistachio, but here the biocoenosis is much poorer.

Apart from mesophilous forests of the Greater Caucasus and Lesser Caucasus Ranges, the Eurasian brown bear (*Ursus arctos arctos*) – one of the largest extant terrestrial carnivores, occurs in the extreme south-eastern part of Georgia – in Iori Upland. Within the region it inhabits much of the arid ecosystems (Dinnik 1914; Satunin 1915 cited in Lordkipanidze 2003). The brown bear's range in the arid and semi-arid areas of East Georgia until 1930s of 20th century include unforested hills, fringes of deserts and semi-deserts. Thus, its range extended to the village of Udabno (desert - in English), not far from Tbilisi (capital of Georgia) (Markov 1934 in Lordkipanidze 2003). But in the 1960s the bear's range shrank and it didn't occur even in David Gareji Hollow and its surrounding, which is one of the wildest area in East Georgia. It was found only in the eastern part of the Iori Upland (Arabuli 1987 in Lordkipanidze 2003). In arid and semi-arid landscapes of Georgia, the species' main requirements areas include arid open woodlands and riparian forests of the Alazani and Iori Rivers.

Today the brown bear is mostly found in VNP. In 1993-2015 the bear's population ranged from 4-5 to 10-15 individuals (Lordkipanidze 2003, *https://matsne.gov.ge/ka/document/view/62824*) (Figure 81).

In Vashlovani Protected Areas the brown bear avoids areas occupied by shepherds. Its principle range includes VNP. Despite the fact that they are not highly territorial, several adult bears gather closely and roam over the same site, primarily in the most remote area far from humans in the central part of the park. The bear's main requirements are areas with sparsely, but narrow forested gorges with paths – a corridor, which would provide a passage for an animal occuringin in another location – another gorge. In this way it is less vulnerable and can easily escape from poachers guns. During the summer, when shepherds lead sheep to the high alpine pastures, free from anthropogenic press, free-roaming bears are seen in almost all gorges of the park (Lordkipanidze 2003).

The brown bear is omnivorous, its diet in the VNP includes various vegetations – from grass and shoots in the spring to fruits of the mastic tree, wild pomegranate, wild plum (*Prunus spinosa*), shrub roses in the summer and autumn. In summer it mainly consumes in the fields of maize and sunflower, foods which have the highest nutritional value. This forage habit of the animal became the reason for human-bear negative relationships, which can result in a bear's death – animals damaging property can be shot (in 2015 a bear was poached collared by biologists in a sunflower field, near VNP). Notably on the base of a questionnaire shepherds from arid pastures of Kakheti, unlike mountainous area, never confirm that bears kill and eat farm animals despite the fact that they can kill even a cow with one blow.

Since 2000s a team of biologists from NACRES headed by B. Lordkipanidze has been conducting a brown bear study in Georgia and adjacent Azerbaijan including the Iori Upland. From 2015 methods included satellite technology and telemetry (GPS telemetry study). Biologists have been able to reliably estimate the VPA population and determine the population's level of risk. Brown bear immigration/emigration in VNP is minimal and there is an increase in population from adjacent sites in the park's territory – over much of their historic range in arid and semi-arid ecosystems (*www.nacres.org/g_tkve_datvi.htm*).

According to restricted home range of arid open woodland, where food sources are sparse, the female generally lays down one, seldomly two offspring (Khochiashvili, pers. communication). Hibernation is very short or usually may not occur at all, i.e., in arid open woodland it does not become very lethargic, although the bears in the park have a lot of protected spots to make a den (caves, crevices, etc.).

Depending on the season, as its broadly known, brown bears may travel hundreds of kilometers, especially in autumn to locate food supplies. Some researchers (Kudatkin & Chestin 1993 in Lordkipanidze 2003) pointed out that such migration sometimes has occurred in East Georgia between Lagodekhi Protected Areas (typical site of the Caucasian broad-leaved deciduous forests) and VNP, i.e., as though this migration route was available to connect two populations: one from the forests of the Greater Caucasus Range and the second – arid open woodlands of the Iori Upland. On the contrary, E. Markov (Markov 1938 in Lordkipanidze 2003) pointed out that in Lagodekhi Protected Areas except primarily vertical, seasonal migration of bear he did not find another, large scale travel migration route of the animal. Nowadays in VNP there is no information of such seasonal or other type of migration of brown bear, and, as it seems, such pathways from arid woodland to another habitat have no place in the recent past. A rising population of humans during the last centuries between two sites can be an important isolating mechanism, which kept the two populations of brown bear separated (Lordkipanidze 2003). In other words, once diverged from main populations, migrants of open arid woodlands find 'no place back home'.

In this context it is noteworthy to underline that even 'bear populations from the Greater Caucasus and the Lesser Caucasus Mountains, separated by the densely populated Transcaucasian Depression, have been matrilineally isolated since the Early Holocene Epoch, i.e., after permanent human settlements appeared throughout the area' (Murtskhvaladze *et al.* 2010 cited in 'brown bear'. *htts://en.wikipedia.org/wiki/Brown_bear*).

Summarizing the above reviewed context, the following should be stressed.

- In many sites of drylands this large carnivore, as a keystone species was hunted virtually to extinction in the beginning of 20th century, and part of its population is highly endangered and its remaining range is still under threat.
- Today VNP, the bear's principle range, includes in its central part, a brown bear population estimated at about 10 specimens. It is believed that such a tiny population could soon be extinct due to threat from poachers and bear-human (shepherds) interactions.
- Due to its ability to survive in arid climates, the brown bear, once native to much of arid territory of East Georgia and adjacent Azerbaijan including even desert edges, has greatly decreased today.
- Recently, in the VNP, as the bear's fragmented and relatively stable population – the largest in Iori Upland – became an isolated, insular refugium. Variable scanty numbers of bear, which still occur in adjacent small patches of Vashlovani Protected Areas, tending to concentrate in VNP. This fact indicates that among other arid ecosystems, arid open woodlands became for brown bears the most hopeful, year-round shelter.

- Natural selection in insular situations, over the centuries, has forced them to adapt to life on the large arid woodland ecosystem. At the same time it led local populations of this species of carnivore to elaborate different behavioral and structural adaptations: to live in maturing woodland, but with gaps in the canopy, limited water supply - standing open water, even more, water stress is associated with summer drought; another composition of native food web with low species diversity, first of all, low primary productions (edible plant resources) than that, which was represented in initial, historical habitat - forests with closed canopy; co-existence (biotic interactions) with other competitors such as wolf, leopard and other stressful conditions. This complex of adaptations includes also philopatry (i.e., 'home-like') well-adapted to open woodland composed of mastic-junipers unlike their deciduous or coniferous forest-dwelling cousins in the Caucasus. This standpoint clearly expressed the subsequent divergence event of these two, actually allopatric populations, which nowadays are geographically isolated from each other. In this case, divergence is facilitated by the absence of gene flow across two arid and mesic ecosystems.
- The basic purpose of this review concerning the brown bear's settled arid ecosystem is to once more emphasize the following two main points: (1) the combination of increased tourist visiting, advances in hunting technology, and after breakup of the former USSR, changes in livestock husbandry, brown bear's unique populations range with other umbrella inhabitants (e.g., leopard, lynx, wolf) in the arid environment of Georgia has shrunk. Hence, the local relict population within today's small range faces a high risk of extinction; (2) following, far more careful preservation of VNP's current ecosystem has long been recognized as a unique ecotone initially by botanists, but the park with its richness in animals, other organisms, and finally genetic diversity, has a unique position. Extraordinary, complex communities of VNP is a truly natural heritage represented *in situ*, an insular laboratory of rare and endangered species.
- From the conservation point of view, the brown bear's population in VNP appears to be a key link with other, very scanty and scattered populations of this species in adjacent, mosaic patches of the Iori Upland. Accordingly, saving the brown bear's population within arid ecosystems in strictly protected territory and outside the VNP represents contemporary challenges in conservation biology of Georgia with WWF of the Caucasus, integrated it into a coherent framework for predicting wildlife population trends.

Box. 6. Vashlovani – the Woodland of Kindness

Georgia . . . if one sees this place, he will never forget its natural resources: its magnificent and haughty mountains, its fine and mild seaside and impressive forests between.

We are used to imagine Georgia's nature like this, but in this magical laboratory of the nature there is one completely different part.

There is woodland there, but it is not dense deciduous or coniferous forests very characteristic of this country, or dense relict forests of Colchis (Kolkheti). It is woodland too, but arid, sparse community embodied with completely different 'laws' and 'strategy', another environmental structure, competition rules, resource quality and with different life secrets and wildlife features.

Figure 82. Crucial large-scale mosaic of the grassland, desert and arid open woodland, color patterns of ecotons (2016).

It is open (light) arid woodlands edged with a hot rim of deserts and semi-deserts. Yes, Vashlovani National Park in the far east of Georgia is somewhat like an African savanna.

Inordinate is the color palette of the plains, deserts and semi-deserts, and open (light) arid woodlands of this place, its color conglomeration: grey, yellow, red, brick-red, rusty and their intermediate colors (Gegechkori 1982) (Figure 82).

2.4.4.2.2. Transhumance and Nomadic Pastoralism vs Vashlovani National Park's Conservation Goals and Principles

Semi-arid habitats, for example, steppes, semi-deserts and arid open woodland provide man with an attractive rangeland for his livestock. Parts of VNP and adjacent area are used as winter pasture for livestock belonging to sheep farmers of East Georgia. It has a seasonal transhumant form. There is a total of 45 farm buildings within the national park and 17 in close areas. After the breakup of the Soviet Union (1990s) and at the same time loss of the trailing livestock in winter pastures of the Caspian Sea coastal zone of Dagestan (the Russian Federation), livestock grazing is allowed in the traditional usage zone of VNP.

Even though grazing is prohibited in strict protection, managed protection, restoration, administration and visitor zones through recently received protected areas legislation (*Law on the System of Protected Areas*, 1993, 1996). Although, there are no unified regulations for pasture use and grazing norms (Centre for Biodiversity Conservation and Research–NACRES 2013).

Unsustainable farming presents the greatest threat to species and ecosystems. Indeed, one of the WWF's high priority sites, VNP is negatively impacted by livestock grazing. This circumstance evidently causes an internationally recognized intricate web: 'local farmers vs environmental equilibrium'. Taking into account vulnerability of local shepherds by means of low economic and social status, regardless of their being no attempts to escape from the harsh transhumance pastoralism working conditions, most of them, being ecologically marginalized people, it seems they do not care for nature protection.

Sheep farming through traditional transhumant form, legislated during centuries, became part of lifestyle of some native communities. Therefore, negative interactions have been formed between shepherd's pastoralism and National Park's staff principal goal – defense of natural (ecosystem) balance. Such differences in motives of primary interests led to conflict of interest (COI).

Unfavorable socio-economic conditions are the reason for extensive seasonal grazing within a nomadic land-use system, which is obviously negatively attracted to remarkable landscapes and habitats of the national park. Over-intensification of the winter pasture by livestock and increased human population density with farming infrastructure in adjacent territory of the VNP, i.e., human's interference into extensive and intensive utilization of semi-arid lands for pasture and farming may entail irreparable damage to the functioning of the arid open woodland. Human exploitation via overgrazing of semi-arid habitats means local ecosystem productivity decreases – palatable and nutritious forbs are replaced by unpalatable thorny plants.

It should be taken into account that transhumant shepherds' psychology towards the nature (or pasture) is quite controversial compared to those of high mountain indigenous peoples of the Great Caucasus Range, for example, the Khevsurians (eastern part of mountainous Georgia). They, as a rule, recognize the historical interactions between local rural communities and surrounding environment that they are dependent on. Traditional knowledge and practices have been established through many generations in 'caring for their traditional territories, and for the harvesting of wild food and animals, medicines, water, and other basic needs' (Langton *et al.* 2005 in Piermattei 2013). During my numerous field trips, I was eyewitness to how the native hunters of the Great Range usually shot only one specimen of native big game goat (tur) either for adoration of their shrines with prey's horns or for their nutrient requirements in wild animal's meat or both. This is a good argument of the mentioned opinion.

Figure 83. Fossilized seashells found in exposed rocks of Vashlovani NP, Datvis Khevi (Bear's Gorge) from the the Early Pliocene (~6 mya). Fossil materials (marine molluscs) have indicated that in ancient times the Paratethis sea spread over the land of East Transcaucasia (2014).

Large scale deforestation in the Iori Upland region is leading to impoverishment of the native communities of mastic tree and junipers. Anthropogenic activities can alter riparian habitats of the VNP with a striking complex of plant diversity including wild progenitors of domesticated plants (e.g., wild forms of pomegranate, fig tree, and wild grape) or a single population of *Populus* (*T.*) *euphratica* in Georgia. The latter – relict of desert's riverside communities – is particularly vulnerable because of settlers in NP on specific geologic formations, soils, marginal gorge, i.e., microhabitat conditions. Such a type of insularization makes the population vulnerable to any abiotic, biotic and anthropogenic forces that can lead to elimination of one of the valued biological heritages of this country (Figure 83).

Further increase of livestock grazing within NP and rangeland near its boundaries will have the potential or real impact to herbaceous and semi-herbaceous plant biodiversity. For example, such exclusively rare endemic species as are *Iris iberica, Tulipa eichleri* may be affected by grazing and trampling, which can result in damage to the species populations. In general, overgrazing of the VNP's natural pasture or rangelands, despite the above-mentioned legislation, prohibits an increase in livestock unit in the park and will cause degradation or loss of suitable habitats altering vegetation composition, introducing non-native plant species causing harm to environmental heterogeneity (see below).

Today on the base of the Government's Orders, well trained and equipped staff of VNP carefully monitor and protect the local natural, cultural and historical resources. On the other hand, shepherds are worried about protection of wildlife from hunters' guns, because the wild boars have multiplied quickly, which on the other hand becomes an important source of food for predators – lynx, bears, and especially, wolves, which in their turn support an increase in predator numbers.

Illegal hunting expansion of the national park's wildlife has had a negative impact on umbrella species or caused their migration being confined to remote areas outside of VNP (the result of such a way of stress toward the wildlife, it seems, is missing a single male leopard from NP).

Figure 84. Flock of sheep and goats. The goats had made a browse line in the taller shrubs: In the picture, the almond tree is being consumed by goats. Environs of Vashlovani NP (2004).

Another 'harsh' problem includes extensive sheep grazing which can also stimulate severe competition for pastures between herds of domestic livestock and recently reintroduced goitered gazelle for forage. Some shepherds are willing to hunt this ungulate species which, for whatever reason, is illegal (Figure 84).

The fact is that shepherds and poachers override the park's resources, its environmental heterogeneity and values including plants and animals are listed under the International and Regional Red Books and Red Lists. Grazing domestic animals via impoverishment or altering habitats indirectly impact on the local communities of birds of prey: imperial eagle (*Aquila heliaca*), griffon vulture (*Gyps fulvus*), harrier (*Circus cyaneus*), buzzard (*Buteo rufinus*), short-toed snake eagle (*Circaetus gallicus*) and Egyptian vulture (*Neophron percnopterus*) (in general 26 species of raptors).

Scientist predicted that other bird species that depend on the open woodland's understory and ground story vegetations would be most negatively affected by livestock grazing. It could cause changes in bird species composition, density, and relative abundance across riparian and non-riparian habitat types. To this group of avian communities belong little bustard (*Tetrax tetrax*), great bustard (*Otis tarda*) (the latter rarely inhabited the park's area), black francolin (*Francolinus francolinus*), common pheasant (*Phasianus colchicus*), chuker (*Alectoris chukar*), black stork (*Ciconia nigra*), etc.

Human interference into the VNP can be manifested in another way, i.e., tourism along established routes, which is becoming widespread and growing in VNP.

2.4.4.2.3. Invasive Alien Species with Considerable Attention to *Yucca gloriosa* in Arid and Semi-Arid Landscapes

Recent global warming is among the important environmental events throughout the current century.

Further deterioration of the arid and semi-arid biomes and among them arid open woodland ecosystems are closely associated with changes in climate. Global warming could affect most terrestrial ecoregions and therein ecosystems.

Climate deterioration including droughts is becoming more intense in arid and semi-arid regions. It is expected to become the main driver of biodiversity loss in the future, affecting physiology, phenology, and species distribution (Dai 2011). Semi-arid ecosystems - one of the environmental units – is most vulnerable to the impacts of climate change.

Ecological deterioration should be involved in changes in plant cover (e.g., to reduce the abundance of native vegetation in favor of non-native plants and drought-resistant competitors) with consequent modification of wildlife taxonomic structure, which will have a significant impact on soil erosion rates, increase salinisation and loss of normal physiognomy of landscapes (Dai 2010 and references therein).

The largest centers and areas of strong warming, with the mean annual air temperature rising at the speed of more than 1.2^{0}C have been observed in eastern Georgia. Namely, among others should be mentioned the southern part of Kakheti Ridge and in its driest part including the Iori Upland (Gare Kakheti region) (Maruashvili 1970).

The calculations conducted with climate models demonstrated that compared to the average of 2010, the average annual temperature in Kakheti will be increased by 1.1^{0}C by 2050 and by 3.5^{0}C by 2100 (Climate Change and Agriculture in Kakheti Region, 2014).

During the last decade Dedoplistskaro district in East Georgia (Gare Kakheti Region) is already experiencing fragmented drought episodes via increasing frequency of drought in both rural and natural habitats including primarily the driest part of Georgia - arid and semi-arid Iori Plateau and Shirak-Eldari Lowlands (Maruashvili 1970).

Grazing lands (pastures) are concentrated on the drought-impacted mentioned areas. Today most of them are weedy, non-irrigated and depleted due to drought increase and overgrazing and need restoration with drought-resistant vegetations (Lachashvili *et al.* 2004).

Side by side of the flora of natural habitats, the flora of human or near-human settlements, i.e., alien species has received considerable attention in the last decades (Even *et al.* 2012). 'Stimulated by the rapid global expansion of trade, transport and travel, invasive species and their costs to society are increasing at an alarming rate' (National Invasive Species Council, 2001).

With respect to the plant introduction, researchers distinguished two categories of flora: (1) *native* (indigenous) and (2) *alien* (introduced) (Pyšek 1998).

In other words, species are considered to be alien if they have been assisted in reaching the country actively or passively, as the result of human activities, i.e., man-accompanying flora and fauna (Elven *et al.* 2012 in Gederaas *et al.* (Eds.) 2007). 'Those alien species that become established in a new environment, then proliferate and spread in ways that are destructive to human interests and are considered 'invasive alien species' (IAS) (McNeely *et al.* (Eds.) 2001).

If the alien taxon is unable to form a self-replacing population, it is considered as casual (not established). On the contrary, alien taxa, which are capable of reproducing self-sustaining populations for a long time (without direct human assistance in the area where it exists is considered as naturalized (established) alien taxa. The latter taxa have the potential to spread over a considerable area from parent plants (Lambdon *et al.* 2008 cited in Galanos 2015).

A subset of naturalized plants that produce reproductive offspring, are often in very large numbers and at considerable distances from the parent plants and thus have the potential to spread over a long area. These are considered as invasive (invading) plants. Introduced species are most problematic for biodiversity and humankind.

Aliens can be divided further into two groups: (3) archaephytes and neophytes. The first group includes alien plant species introduced before the year 1492 (i.e., before the discovery of America by European colonizers. This date is conservationally rounded off to 1500). Neophytes occur as a result of later introductions after the year 1492. Neophytes are usually escaped plants from the cultivation of crops and ornamental plants (Pyšek 1998).

All mentioned impacts increase the risk of expanding the range of invasive alien species either on purpose or accidentally. Globalization processes have created a new pathway for the introduction of alien species elsewhere. IAS are a threat to biodiversity, agriculture and human health. Some of them (plants, animals, other organisms) are characterized by the ability to out-compete native species and taking over their niches. These may have extensive impacts. The damages can be devastating. Researchers of IAS biological harm in the USA and India show that the economic cost of IAS in these countries amount to approximately 130$ billion per year.

Among other major taxonomic groups 10% of the world's 300,000 or so vascular plants, have the potential to invade other ecosystems and affect native biota (McNeely *et al.* (Eds.) 2001). According to the DASIE Database, 6,658 terrestrial plant species are at present classified as aliens in Europe, at the same time six new alien taxa are penetrating each year in the Europian mainland and are considered to become established (Pyšek *et al.* 2009 cited in Galanos (2015).

According to Kikodze *et al.* (2010), from 4,200 species of vascular plants of Georgia, 368 species (8.6%) are alien. Out of the alien species 129 species (35%) have become naturalized and 16 species have become invasive (5.5%). The remaining species are either adventive or of unknown (sub spontaneous) status. Among others *Ambrosia artemisifolia*, *Phytolacca americana, Amorpha fruticosa*, and *Robinia pseudoacacia* are believed to occupy the widest range.

In arid and semi-arid regions such a type of threat includes thermophilic exotic species. This is also true in Georgia, where alien species have recently received considerable attention (Kikodze *et al.* 2010). Among nonnative plant species our considerable attention is focused on the Spanish dagger (glorious yucca), *Yucca gloriosa* (Agavaceae).

Yucca L. comprises of about 40 species and dozens of cultivars. It is a native of the New World and is distributed in southern USA, Mexico and the West Indies (Ji & Meerow 2000 cited in Kanwal *et al.* 2012). *Y. gloriosa* is a rosette-shaped evergreen shrub. High salt tolerance makes Spanish dagger ideal for seaside plantings (*http://edis.ifas.ufl.edu*). It is frost (about -20^{0}C) and drought tolerant (Cullen 1986). In some countries *Yucca* is often considered to be an invasive ornamental weed shrub. For example, among 49 invasive species catalogued in Spain (Valencia), is listed *Y. gloriosa* (Merino & Donat 2011). It has been found as the secondary wild plant in Belgium, France, Italy, Netherland, etc. distributed (as invaders) in coastal dunes, 'possibly as a result of global warming' (*Yucca*. Manual of the Alien Plants of Belgium 2015).

Glorious yucca is wonderful with its abundant blossoming bell-shaped white flowers, the plant is widely cultivated in urban landscapes worldwide and in arid-loving in East Georgia as well. The climate area suited for glorious yucca cultivation is restricted to the open landscapes of the Eldari Lowlands. This full sun tolerant plant occurs in areas with high, dry heat, high wind and clayey, poorly draining soils. It is frost resistant in winter. For example, *Y. gloriosa* was planted in the northern part of Eldari Lowlands due to its number of medicinal uses. Industrial plantations of this alien plant were founded on about 190 hectares. Upland in order to provide tigogenin production from raw materials - the leaves; another medicinal component (furo- and spirostanols) were isolated from the strong root system-rhizomes; metabolism bioregulator – the steroid saponins have been found in blossoming flowers. Previously treatment of seeds or spraying of seedlings with water solutions of Alexin increases wheat, barley, soya, potatoes and tomatoes yields by 20-55%. The research staff of the Institute of Pharmacochemistry (Tbilisi) considered that *Y. gloriosa* cultivated in Georgia, can produce an opportunity for the development of a number of fields of pharmacochemical, agricultural and even textile industry of the country. The institute of Tbilisi had collaborated with the institutions of the same profile of the city of Odessa in Ukraine. The Research Institute of Odessa bought the raw materials yielded in Kakheti (Kemertelidze 2007).

It should be stressed that industrial plantations of glorious yucca have grown in the margin area of the Eldari grasslands, close spacing (about 100-150 meters) to Vashlovani National Park, the complex relief of which is found in the borders of VNP, i.e., in the Vashlovani Hollow of the Iori Upland. It is notable that establishment of glorious Yucca, out of plantations, within the VNP did not occur hitherto (Figure 85).

Figure 85. Yucca (*Yucca gloriosa*) plantation being close to the edge of the Vashlovani National Park (2010); background: The GCR.

Today a patch of Eldari Lowland settled by plantations of glorious yucca has profoundly altered views of the local steppe's landscape. Recently (the last ten years) this patch is completely abandoned without any research and commercial interests. Therefore, nowadays non-native plant plantations are the next (if not primary) problem within a number of other problems threatening the Vashlovani National Park.

Y. gloriosa till nowadays is not included in a check-list of 'The alien flora of Georgia' (Kikodze *et al.* 2010). Despite this fact, we must take into consideration the climatologists viewpoint based on possible warming scenarios. In this case climate warming will impact the range and production of the vast majority of arid and semi-arid environment including not only land degradation, but also habitat loss and fragmentation, changes in plant community composition, and effect on dryland biodiversity from competition of alien species (Sala *et al.* 2000 cited in Safriel & Adeel 2005). In this situation introduced species can spread faster than native ones can. Therefore, taking a leading position in local environment the so called 'ecosystem engineer' species can be held (sensu Jones *et al.* 1994 cited in Safriel & Adeel 2005), which should modify the arid and semi-arid environment for the benefit of alien species. Among other alien species, for example, glorious yucca has a high potential to invade new areas – the Vashlovani Hollow and Alazani and Iori Rivers entire catchment, and with a high possibility to establish self-sustaining populations will cause large-scale changes or losses of the arid and semi-arid-specific keystone species, such as mastic and junipers. During biological invasion, arriving in new areas, the former casual species can be labeled as 'invasive alien species' of the flora of Georgia (the Caucasus, in general). According to this simulation, the center of radiation of glorious yucca will be the Kakheti region.

Altering ecosystem, alien species should be characterized by low biodiversity due to a near monoculture. At the same time dominant invasive species can support other biodiversity: another complex of shrubs and annuals will coincide in the matrix of change communities and soil crust, another complex of animal species will depend on it for food shelter, nest and burrow sites, etc., another complex of arthropod and other invertebrates species and unicellular organisms will be participating within the process of dead plant biomass decomposition, i.e., constituting the first link of nutrient cycling. Indirect threats include the loss of interest to VNP as the most exotic for Georgia recreation and ecotourism site.

2.4.4.3. Arid Open Woodland Composed of Junipers (Juniperus spp.)

The genus *Juniperus* - evergreen shrubs or trees. It is a major component of semi-arid and arid tree/shrub ecosystems throughout the Northern Hemisphere (Adams 2014; Farjon 2005 cited in Mao *et al.* 2010; Adams, 2014). The same should be said concerning the arid regions of the Caucasus (Gulisashvili *et al.* 1975; Lachashvili *et al.* 2004) (Figure 86).

Figure 86. In the open woodland, species of Juniper (*Juniperus* spp.) are a distinctive constituent of the steep slopes of foothills, Vashlovani NP (1968).

Genus *Juniperus* is believed to be monophyletic and includes three monophyletic sections: *Caryocedrus*, with one species, sect. *Juniperus*, with 10 species, and section *Sabina*, with 56 species. Therefore, it is estimated that total number of the junipers is 67 species. Another recent author, Dr. Adams (2014), has made estimates where sect. *Juniperus* includes 14 species, sect. *Sabina* – about 60 species, in total three sections encompass approximately 75 species. Caryocedrus and *Juniperus* are native to the Mediterranean, and *Sabina* - to North America, Asia and the Mediterranean (with an extended range in Africa and the Canary Islands). *Juniperus communis* has a circumboreal range. Altitudinal distribution of junipers ranges from sea level to above the timberline. Preferably with a limestone substrate, many species successfully colonize other substrate including sand dunes, granite, sandstone, etc. from drylands to bogs (Adams 2014). Among others ecologically one of the most acceptable habitats is open woodlands (Mao *et al.* 2010).

Today the genus *Juniperus* is one of the most advanced conifers (Rai *et al.* 2008 cited in Adams 2014).

Central Asia, the Mediterranean region and North America, i.e., former Laurasian mainlands are the three major centers of diversity for junipers. According to fossil records (sect. *Sabina*), the genus probably originated in the Eocene-Oligocene boundary, about 50-30 mya (Kraček 2002 in Mao *et al.* 2010). To reconstruct the biogeography pattern of *Juniperus*, the biogeographic intercontinental connections between the Old and New Worlds by estimating divergence time within the genus, indicate that the New World was colonized at least three times independently, in the Eocene-Oligocene, in the Miocene (~20-15 mya,) and once more recently (Mao *et al.* 2010 and references therein). Therefore, 'Junipers had become a component of the Madrean-Tethyan vegetation belt, on either side of the Atlantic' (Axelrod 1975 cited in Mao *et al.* 2010; Adams 2014).

In the South Caucasus the arid open woodland composed of junipers (*Juniperus* spp.) covers a vast territory, mainly the lower altitudes of south Karabakh and the Zangezur Mountains. Here, in the southern part of the South Caucasus, open juniper woodland sometimes also occurs in the highland, up to altitudes of 1,800 to 2,000 m a.s.l. On the south slope of Shakh-Dagh, near Lake Sevan (Armenia), for example, a typical open juniper woodland is found amidst mountain steppe communities. Similar phytocoenoses clothe the mountain slopes in the middle reaches of the Aras River (Nakhichevan and South Armenia). Here, in the Khosrov Reserve (1,400-1,700 m a.s.l.), as mentioned, in the Armenian part of the river valley, a unique refugium of juniper woodland forms the most important outpost of organisms of Asia Interior in the Caucasus (Gulisashvili *et al.* 1975).

Box 7. The Juniper Woodland of Armaziskhevi and the Armazian Bilingua

Georgia is a heterogeneous country, in part with respect to the mental spirit, behavior, ethnography and culture of its inhabitants (Bagrationi 1742-1745 in Kaukhcishvili (Ed.) 1955; Kartlis Tskhovreba (literally 'Life of Kartli', or A History of Georgia) Metreveli (Ed-in-chief) 2008). This certainly applies to East Georgia, or Iberia (Iveria) as it was known in the Graeco-Roman world (Lang 1962).

Iberia was encircled by the Greater and Lesser Caucasus, which are intersected by numerous, rich valleys. The Greek geographer Strabo (± 63 BC-24 AD) states that the plain dwellers practiced agriculture and, unlike the Iberians that inhabited the mountains, were inclined to peace (Braund 1994).

Around 1000 AD the political, economic and cultural centre of Iberia was Mtskheta (currently about 21 km west of the capital of Georgia, Tbilisi). Situated at the junction of the rivers Aragvi and Kura (Mtkvari) it inspired the Russian Romantic writer and poet Mikhail Lermontov to write his 'Mtsiri' (1839), one of the most famous poems in Russian literature.

Northwest of Mtskheta the Iberian plain and the Great Caucasus merge. In contrast to Mtskheta and its direct surroundings, the moderate semi-humid climate of this area allows the prevalence of broadleaf deciduous forests, consisting of Georgian oak *(Quercus petraea* spp.

iberica). In the gorges of the Pshavi Aragvi and Mtiuleti Aragvi Rivers the latter gradually turns into even more humidity demanding beech (*Fagus orientalis*) forests. In the drier, almost Mediterranean climate southwest of Mtskheta, the hardwood community gradually changes into an open woodland with juniper dominance that covers the low, canyon-like slopes of the Armazi Ridge. This part of the Lesser Caucasus has - fortunately - not been subject to deforestation up to the present day. The open juniper woodland for its part is replaced by semi-desert habitats with *Artemisia, Capparis, Colutea, Rhamnus, Ephedra* and shibliak (*Paliurus spina-Christi*). The very steep slopes are honeycombed by huge caves, a favorite habitat of scavengers (Gegechkori, pers. observation).

The dwellers of Mtskheta enjoyed all these natural riches, including the crystal-clear waters of the Aragvi river (that today provide most of the drinking water for the Capital of Georgia, Tbilisi), the rich soils, the pleasant climate, the forests and the diverse wildlife. But even more importantly: at the very beginning of the Christian Era, Mtskheta was positioned at the junction of some principal waterways, one from the north (where the Sarmatians reigned) and one that connected west and east (Apakidze 1968). This certainly helped the Iberians to change from a pastoral to a sedentary way of life in the first centuries BC.

Early in the first century AD Iberia, and especially Mtskheta, were known for their impressive prosperity that also Strabo described admiringly. In and around the city, grand buildings from the Hellenistic period, such as the columned hall at Bagineti and public baths, possibly inspired by contact with the Roman Empire and perhaps even built by Roman craftsmen still bear evidence of this wealth (Apakidze 1968).

In the first half of the 20th century Georgian archeologists discovered some burials adjacent to Armaziskhevi. They provided us with much information on the Iberian elite from the reign of Hadrian (117-138 AD) through the third century AD.

The main language of the elite seems to have been Greek, at least from the second century AD onwards. At the same time the local Iberian version of Aramaic also played an important role. From the first century BC, and probably earlier, a form of Aramaic script was in widespread use in Georgia, Armenia, Syria, and Northern Mesopotamia. The Transcaucasian version is often termed Armazian or Armazic (G. Tsereteli 1942, 1949; K. Tsereteli 1991). Its origin remains obscure, but a persuasive hypothesis sees it as a descendant of the last script from which Parsi and also, Pahlavi originated (Braund 1994).

The most striking finding during the excavations was that of 25 November 1940, when the tombstone of a young woman was discovered, on which next to Greek also Armazian script was inscribed. This bilingual epitaph allowed the decipherment of the script by Prof. G. Tsereteli (1942). The Greek text reads (Kaukhishvili, 1941): ‘Serapeitis (Serapita), daughter of Zevakh the Younger, pitiakhsh, wife of Iodmandagan, son of Publicius Agrippa, pitiakhsh (Iodmandagan), who won many battles as epitropos of the great king of the Iberians, Ksefarnug - died, younger than 21 years, who had inimitable beauty’. The Armazian text (after K. Tsereteli 1991) is similar, but not the same: ‘I am Serapita, daughter of Zevakh the Younger, pitiakhsh of Pharsman the King, and wife of lodmandagan the victorious, winner of many conquests, master of the court officer of Ksefarnug, the great king of the Iberians, and son of Agrippa, master of the court King Pharsman. Woe, woe, for the sake of her, who was not of full age, whose years were not completed, and so fine and beautiful that no one was like her in equal in excellence, and she died at the age of twenty-one’.

The Armazian bilingual epitaph indicates that during the first centuries AD Kartli (the former Iberia, a contemporary province of East Georgia) was headed by a king, who carried the title of King of Iberia. Within the royal society there were hereditary pitiax (pitiakhsh), whose rank approximated to that of viceroy or satrap. For official documentations, the Armazian alphabet, which relates to Georgian, was used (Figure 87).

Figure 87. Armazian bilingua discovered in Armaziskhevi (1940) (E. Georgia). This site is covered by extensive *Juniperus* spp. woodland (1962).

In the mountains of Nakhichevan, near Shakh-Buse, arid woodland is represented by wild almond (*Amygdalus fenzliana*).

An impressive massif of juniper woodland occupies the middle reaches of the Kura River near the town of Mtskheta, near Shio-Mghvime and Karsani, in southern Kiziki (E. Georgia), in the form of small stands, etc. The same conifer species as in Vashlovani NP populate the scraped mountain slopes: *Juniperus foetidissima* (widely distributed in the eastern part of the Mediterranean), *J. polycarpus* (a species with Asia Interior as centers of distribution), and *J. oxycedrus* (= *J. rufescens*), *J. oblonga*. On steeper slopes, the junipers grow far more densely, than on gentle slopes or on low mountains. In juniper woodland xeric shrubs (*Ephedra procera, Colutea orientalis, Rhamnus pallasii, Jasminum fruticans, Atraphaxis spinosa,* etc.) and steppe grassland prevail (Lachashvili *et al.* 2004; Lachashvili & Khachidze 2005; Togonidze 2011).

In the arid area of South Transcaucasia and Middle Asia, the juniper forests on mid elevations are generally replaced by coniferous forests of pine, spruce and fir (Gulisashvili 1964).

2.4.4.3.1. Khosrov Reserve (Armenia) – a Unique Site of the Organisms of Different Biogeographical Origins

Khosrov Reserve was founded in 334-338 by the order of the Armenian king Khosrov II Kotack to improve the natural-climatic conditions of territories adjacent to Artashat city – the capital of Armenia of the given period and the newly established Dwin city to ensure conservation and enrichment of flora and fauna species, served as a ground for royal hunting, military exercises and entertainments.

Over the centuries the Khosrov Reserve has been mentioned in the history of Armenia as a hunting-ground for the nobility and animals were brought and bred in the Reserve from various regions, especially Persia.

People perpetuate the name of the Khosrov King not only giving his name to the forest, but also to Khosrov settlement, Khosrov River and Khosrovasar Mountain (*en.wikipedia.org/wiki/Khosrov_State_Reserve*).

In 1958 the Khosrov forest consisting of 8 separate lots, was officially declared as a 'Khosrov Forest' State Reserve and its borders were widened in 1992 to create a single parcel of land (Bannikov (Ed.) 1969).

The Reserve is located in the Ararat Province (Marz) in the Gegam (Gueghama) Mountain Range (south-western Armenia), in the basins of the Azat and Vedi Rivers at altitudes of 1,400-2,250 meters a.s.l., and extends over an area of 23,213 ha, 9,000 of which are covered with forests. The importance of the reserve is to play a significant role in the entire ecology of the country; protection of the Azat River water resources, juniper and high mountain oak communities, arid mountain vegetation, rare animals and plants.

Atmogenic processes, the hot climate of the arid area and water create in the reserve very complicated relief with amazing landscapes, numerous extraordinary monuments – figures, caves, shady canyons, towers, pyramids, quaint cliffs, etc.

Figure 88. Khosrov State Reserve, Armenia. It is Asia Interior's living beings the most important outpost in the Caucasus. Background – open woodland composed of *Juniperus* spp. (1972).

According to Bannikov (Ed.) (1969) and Khanjyan (2004) Khosrov Strict Nature Reserve is one of the oldest protected areas in the world having a history of about 1700 years. The reserve contains unique plant and animal species, a large diversity of scenic landscapes, preserves ancient cultural amenities and historical-architectural monuments (Figure 88).

Due to the complicated mountainous relief and peculiarities of soil-climate conditions, here within the reserve and adjacent territory were formed landscape types from deserts and semi-deserts to mountainous and alpine meadows. Deserts and semi-deserts area spread at 900 up to 1,250m a.s.l., dry steppes - at 1,250 up to 2,500 m, arid open woodlands extend from 1,500 up to 2,500 m, and deciduous forests at an altitude between 1,600-2,500 m a.s.l.. Then the latter biomes are transferred into subalpine and alpine meadows in the high mountain zone (2,500-3,200m). Parts of the desert and semi-desert and highland habitats (e.g., meadows) lie outside of the reserve. The intrazonal water-swampy ecosystem is presented along the river banks.

Due to local mountains (e.g., Khosrovasar) mostly steep slopes, river valleys cut by deep ravines and rivers upper reaches, the larger part of reserve is distinguished by inaccessibility and absence of paths and roads. The peculiarities of the reserve's relief mentioned determine the complex and unique ecosystems in the Caucasus with very diverse species compositions, numerous endemic, relict and other rare plant and animal species.

In the reserve area grow 1849 species of vascular plants (more than half of the species growing in Armenia) 156 of them are considered rare and endangered species, 24 species are endemic. It is the only Caucasian reserve with such a diversity of climatic areas and plant types (Bannikov (Ed.) 1969).

Deserts and semi-deserts with prevailing *Artemisia fragrans* occupy large areas spreading over the foothills and lower mountain zones of the Reserve. In spring in the lower mountain zone's, some places are covered with bulbous ephemerals such as *Tulipa biflora, T. julia, Iris reticulata, I. elegantissima*, etc. and ephemeral annuals. In summer they are changed to blossoming perennial plants: *Capparis spinosa*, species of *Salvia, Centaurea, Verbascum* and others. In habitats with a halo-gypsophilous soil area, dominant position have species of *Salsola – S. ericoides* and *S. dendroides* (Figure 89a, b, c, d).

Rocky eroded bare slopes expand over the upper part of the deserts and semi-deserts zone. These skeletal rocks are the habitat for xerophilous Mediterranean flora. Mention should be made of phrigana represented by *Amygdalus fenzliana, Prunus* (*Cerasus*) *mahaleb, P.* (*C.*) *incana, Rhamnus pallasii, Spiraea hypericifolia*, species of wild pears, especially *Pyrus salicifolia,* sometimes with *Celtis glabrata, Pistacia atlantica, Rhus coriaria, Ephedra procera, Atraphaxis spinosa* and others. Trees never form a close canopy.

Barren rocks and slopes of some gorges and such characteristic for Reserve's canyons are clothed by tomillares: *Stachys lavandulifolia, Thymus kotschyanus, Ziziphora clinopodiodes*, etc.

In the middle mountain zone (1,400-2,200m a.s.l.), the mountainous steppe ecosystems are the most widely spread in Armenia. They occupy considerable spaces in the Reserve as well. Although grassland steppes consist of mainly *Stipa capillata, S. pulcherrima, S. pontica* which are represented in relatively small areas, where *Nepeta, Trifolium, Medicago* and other species dominate. Tragacanth thorn cushion formations with ordinary shrubs are dominant in the dense thickets of *Astragalus microcephalus, A. lagurus, Onobrychis cornuta, Acantholimon armenum* (tragacanthic and cushion sainform formations).

(a) (b) (c) (d)

Figure 89 (a,b,c,d). Khosrov Reserve (1,700m a.s.l.), along with the exclusive biodiversity is a peculiar natural (*in vivo*) laboratory. Here like Mendel's experiment on garden pea plants about genes hereditarianess, the nature made a similar experiment on wild tulips (red and yellow are typical colors of tulip, other ones-hybrid forms (1972).

Within the same middle mountain zone, the most typical landscape creates an open woodland. Here the juniper communities are more widely spread, where *J. polycarpos* dominate. The presence of *Acer ibericum, Quercus infectoria veneris* (=*Q. araxina*), *Pistacia atlantica, Celtis glabrata, Padus ramosissima, Amygdalus fenzliana, Viburnum lantana, Rosa spinosissima*, etc. as constant species is typical for these communities.

There are a number of endemic and rare species in the arid open woodlands and shibliaks of Armenia. Here should be mentioned, *Amygdalus nairica, Crataegus armena, C. zangezura, Colutea komarovii, Smyrniopsis armena, Cousinia fedorovii, C. takhtajanii, Isatis sevangensis, Polygala urartu, Eryngium wanaturii, Prangos arcis-romanae, Convolvulus calvertii, Iris grossheimii*, etc. (Bannikov (Ed.) 1969).

Here, next to harboring on exclusive biodiversity, the reserve is a natural genetical laboratory, where the nature *ex situ* repeats Mendel's experiment on the heredity of garden pea plants: in 1973 red, yellow, reddish-yellow and yellow-reddish forms of wild tulip (*Tulipa julia*) were growing side by side in the juniper woodland on some slopes of the Reserve (Gegechkori 2007) (Figure 89a,b,c,d).

The distribution area of 'highland oak' *Quercus macranthera* is confined to mid-mountain and high-mountain (1,600-2,300 m) zones. Similar to junipers, highland oak forms open woodland, but seldom large dense monodominant oak forests sometimes accompanied by *Fraxinus excelsior, Sorbus aucuparia, Acer ibericum*, species of *Pyrus, Rosa, Crataegus* and others.

Meadows steppes are spread at an altitude of 2,200-2,600m populated by species of *Phleum, Helictotrichon, Hordeum* and others.

Of the arboreal components of the river banks the following species should be mentioned: *Populus pyramidalis, Salix* spp., *Elaeagnus angustifolia, Tamarix* spp. and so on.

The vertebrate mammalian fauna is presented by *Ursus arctus, Canis lupus, Sus scrofa, Vulpes vulpes, Meles meles, Lepus europaeus, Martes foina*, species of Insectivora, Chiroptera and Rodentia. *Hystrix indica* and *Lynx lynx* occur rarely. The reserve is famous for populations of ungulates – *Capra aegragus* and *Ovis orientalis gmelini*. The most striking representative of the mentioned group of fauna is *Panthera pardus ciscaucasica*. According to WWF (2010) about 3-5 leopards live in southern Armenia, in Syunik marz, preferring juniper and oak forests.

In general, vertebrate species in the Khosrov Reserve are represented by 280 species (mammals - 41, birds – 192, reptiles – 33, amphibians – 5, and fish - 9). There are numerous species of invertebrates in Reserve (Bannikov (Ed.) 1969).

2.4.4.3.1.1. Diversity of Psylloidea at Khosrov Reserve

Khosrov Reserve contains a remarkable diversity of psyllid (Psylloidea) fauna in the Caucasus. Investigation of this fauna was carried out in the years of 1972-1973 during all vegetation seasons and in all biomes and altitudinal zones of the middle reaches of the Aras River valley, including 'Khosrov Forest' State Reserve. In the history of the mentioned region research on psyllids Loginova (1955, IV-VI) pioneered the scientific study of the group. Another collection during the second half of 20th has either random (Triapitzin, Rikhter, Ter-Minasyan, Logvinenko), or applied purposes (Arakelyan, Avetyan, Azaryan, Simonyan) and included 42 species for the Aras River valley and three species for Khosrov Reserve (Loginova 1968).

Taking into consideration the extraordinary importance of the Aras River valley in general and Khosrov Reserve in particular, during our field trip the checklist of psyllids of the region numbers 130 and Reserve 58 species, two of which remain unidentified due to insufficient material avialable (Gegechkori 1978). Both sets of data are the highest recorded among natural-historical regions of the Caucasus among all Strict Nature Reserves and National Parks of the same region (as it was mentioned from this region are known about 210 species) (Gegechkori 1984a; Gegechkori & Loginova 1990).

From 130 species 32 are new for Armenia, 1 – for Transcaucasia, 21 – for the Caucasus, 13 – for science, from which 8 species were described by the author.

Within the mentioned region Khosrov Reserve is the most distinguished natural monument by its rare psylloid fauna with restricted distribution. Within the Reserve's fauna, 3 species are considered as a local endemics: *Cacopsylla initialis* (Loginova) (on *Salix* spp.), *Bactericera nana* (Gegechkori*) (on *Valeriana* sp.), *Eryngiofaga armeniaca* Gegechkori* (host plant unknown). It is doubtful to present this species even in adjacent Nakhichevan.

Endemic to the Aras River valley, distributed only in Khosrov Reserve (Armenia) and Nakhichevan is *Craspedolepta araxica* Gegechkori (host-plant – *Artemisia fragrans*). A less restricted range has *Camarotoscena fulgidipennis* Loginova, associated with *Populus pyramidalis* and occuring on both sides of the Aras River (Nakhichevan, Armenia and Iran) (Gegechkori & Loginova 1990); *Egeirotrioza (Astutia) gegechkorii* (Loginova) (host - *Populus* sp.) was described based on the materials collected by the author in Khosrov Reserve, and since has been discovered in Greece (Burckhardt, 1988b). *Bactericera kantshavelii* (Gegechkori)* (host unknown) is limited in its distribution to South Georgia (Gegechkori & Loginova 1990) and adjacent Turkey (Burckardt & Onuçar 1993); *Aphalara grandicula* (Gegechkori)* (host plant *Rumex* spp.) was described based on the materials found in the Aras River valley and former Yugoslavia (Gegechkori 1981). This species is replaced in northeastern Iran by vicarious *A. loginovae* (Burckhardt & Lauterer) (1993); *Cacopsylla abdominalis* (Meyer-Dür) (on *Salix* spp.) is distributed in the Khosrov Reserve in the Caucasus and, on the other hand, in Middle Asia, Central Asia (Kazakhstan, Mongolia) and Europe; *C. fulgidipennis* is associated with *Populus pyramidalis*, except in the southern part of South Caucasus it occurs in adjacent Iran (Gegechkori & Loginova 1990; Burckhardt & Lauterer 1993).

Biogeographical analysis was performed on 53 species. It attracts a distribution of psyllid species through very contrasting habitats from deserts and semi-deserts up to the subalpine zone. The mesophilic complex of psyllids linked with mountainous deciduous forests and subalpine meadows as is presently known has a very wide range within Holarctic (e.g., *Psyllopsis fraxinicola* Förster), wide Palaearctic (10 species), and Europe (9 species). A single species (*Colposcenia aliena* (Löw) is multiregional (Palearctic-Oriental-Ethiopian) biogeographical element; one species (*Diaphorina chobauti* Puton) has an Ancient Mediterranean range of distribution. Xerophilic and mesoxerophilic psyllids are of Turanian (6 species) and Irano-Turanian (4 species) origins. They are distributed within deserts and semi-deserts desertificated steppes, steppes, mountain-xerophites, and arid woodlands. Part of the psyllid fauna, as is currently known, are endemics (4 species) and semi-endemics (4 species) and ecologically belong to three groups: xerophilic, 2 species: *Craspedolepta araxica**, *Cyamophila astragalicola* (Gegechkori)*; meso-xerophilic, 4 species: *C. fulgidipennis, A. grandicula*, B. kantshavelii*, B. nana**; mesophilic psyllids, associated with riparian forests are represented by 2 species: *C. initialis, E. (A.) gegechkorii.*

2.4.4.4. Arid Open Woodland with Xerophitic Willow-Leaved Pear Species (Pyrus spp.)

In Georgia the leading species are: *Pyrus salicifolia, P. georgica, P. demetrii* with the less characteristic *P. syriaca*. The other narrow-leaved xeric wild pears – endemics for East Georgia have quite restricted areas. Examples of such rare species are: *P. federovii, P. ketskhovelii, P. oxyprion, P. sakhokiana, P. salicifolia angustifolia* (Figures 90, 91).

Among other species of xerophytic pears *P. eldarica* hold a small area in Azerbaijan (Eliar-Ougi Ridge bordering East Georgia), and, *P. zangezura, P. takhtadziani* - in Armenia.

Figure 90. Solitary tree of xerophytic pear (*Pyrus salicifolia*). Iori Upland (E. Georgia). Ground layer composed of *Artemisia fragrans* providing an invaluable food for thousands of sheep in winter time (2016).

Figure 91. The fruits of *Pyrus salicifolia*. Iori Upland, E. Georgia (2016).

Among the other species accompanying wild pears are: *Paliurus spina-Christi, Punica granatum, Tamarix* spp., *Berberis vulgaris, Rosa* spp., *Celtis caucasica*, etc. (Kutateladze 1947, 1980; Akopian 2007; Asanidze *et al.* 2011).

2.4.4.4.1. Mountain Dryland Landscape of the South Caucasus as a Trigger for Reticulate (Net-Like) - Microevolutionary Processes Based on Sympatric Speciation on the Example of Xerophitic Pears

Species of pear (*Pyrus* L.) are extended in the temperate zone of the Northern Hemisphere (except North America), and penetrate into northwestern Africa (Browicz 1993). Notes on species numbers within the genus *Pyrus* are quite controversial. The reason should be looking for data in the following definition: '*Pyrus* L. is a taxonomically difficult genus mainly because of interspecific and intergeneric hybridization' (Zamani *et al.* 2015). There are botanists, who focus their interests on the sub generic (sections, subsections, series, races) taxonomy of this genus of rosaceous family (Koehne 1890 (cited in Zamani *et al.* 2015); Rubtsov 1930; Maleev 1939; Grossheim 1952a,b; Tuz 1972; Kutateladze 1980; Gladkova 1990a, b, etc.). They do tend to express the mentioned opinion in unison. According to the basic parameters of taxonomic viewpoint of each author, this number of species varies in a wide range, from 20 to 80 and more (Silva *et al.* 2014; Aydin & Dönmez 2015 and references therein). Browicz (1993) indicates not more than 41 species. Among recent, or existing wild pears about half of the species are found in Armenia (Browicz 1993).

In accordance with paleontological evidences, the origin of the *Pyrus* goes back to the Early Tertiary (65-50 mya) in the mountains of western and southwestern China (Komarov 1908, 1931 cited in Asanidze *et al.* 2011; Rubtsov 1944 cited in Zamani *et al.* 2015) or even in more ancient (the Cretaceous) time (Rubtsov 1944 cited in Zamani *et al.* 2015). Following the Alpine-Himalayan orogenesis in the Early Cenozoic, diversification of pears occurred in western Eurasia and eastern Asia, significantly in China. These two endemic centers have provided important clues to the evolutionary history with indisputable taxonomic diversity (radiations) in two distinct groups of species, western and eastern (Rubtsov 1944 cited in Aydin & Dönmez 2015). The western species are found in Europe, the Caucasus, northern Africa, in countries of the Near East and Central Asia, while the species of eastern area are found mainly in Eastern Asia (Silva *et al.* 2014; Aydin & Dönmez 2015 and references cited in).

Figure 92. The unique tree stand of *Pyrus georgica* consist of large shrubs to small trees (growing up to 8-10m tall). Akhaltsikhe Hollow, Adigeni district environs, 1,450m a.s.l. (2018).

According to Vavilov (1926, 1992), there are three centers of diversity for pears: China, Central Asia, and the Caucasus. Therefore, it is believed that the process of domestication took place in these regions including the mountains of the Caucasus (Figures 92, 93).

This suggestion is confirmed by the fossil remains of leaves of *P. theobroma*. These fossil records were found in western Europe (Austria) and in the Caucasus (Kakheti, East Georgia) in middle and lower-middle Sarmatian (~10-7 mya) within broad-leaved deciduous and evergreen laurisilva fossil plant assemblages (Rubtsov 1944 cited in Asanidze *et al.* 2011, Silva *et al.* 2014).

Table 4. Species Number of *Pyrus* L. Growing in the Three Republics of the South Caucasus*

#	Species of *Pyrus*	Republics of the South Caucasus			Sections of pears
		Armenia	Azerbaijan	Georgia	
1.	*P. acutiserrata* Gladk.	+	+	-	x
2.	*P. balansae* Decne	-	-	+	A
3.	*P. boisseriana* Buhse	-	+	-	P
4.	*P. browiczii* Mulk.	+	-	-	P
5.	*P. caucasica* Gladk.	+	+	+	A
6.	*P. chosrovica* Gladk.	+	+	-	a
7.	*P. complexa* Rubtz.	+	+	-	x
8.	*P. daralaghezi* Mulk.	+	-	-	x
9.	*P. demetrii* Kuth.	+	+	+	x
10.	*P. elata* Rubtz.	+	+	-	x
11.	*P. eldarica* Grossh.	-	+	+	x
12.	*P. fedorovii* Kuth.	+	+	+	a
13.	*P. georgica* Kuth.	+	+	+	a
14.	*P. gergerana* Gladk.	+	-	-	A
15.	*P. grossheimii* Fed.	+	+	-	A
16.	*P. hajastana* Mulk.	+	-	-	A
17.	*P. hyrcana* Fed.	+	+	-	A
18.	*P. ketzkhovelii* Kuth.	+	-	+	A
19.	*P. medvedevii* Rubtz.	+	+	-	a
20.	*P. megrica* Gladk.	+	+	-	a
21.	*P. nutans* Rubtz.	+	+	-	x
22.	*P. oxyprion* Woronow	+	+	+	a
23.	*P. pseudosyriaca* Gladk.	+	+	-	x
24.	*P. raddeana* Woronow	+	+	-	x
25.	*P. sachokiana* Kuth.	-	-	+	a
26.	*P. salicifolia* Pall.	+	+	+	a
27.	*P. sosnowskyi* Fed.	+	-	-	A
28.	*P. syriaca* Boiss.	+	+	+	x
29.	*P. takhtadziani* Fed.	+	-	+	a
30.	*P. tamamschjanae* Fed.	+	-	-	A
31.	*P. ? taochia* Woronow	?+	-	-	a
32.	*P. theodorovii* Rubtz.	+	-	-	a
33.	*P. turkomanica* Maleev.	+	-	-	A
34.	*P. voronovii* Rubtz.	+	+	-	x
35.	*P. vsevolodii* Heideman	+	+	-	A
36.	*P. zangezura* Maleev	+	+	-	x

P.S. *P. communis* L. as a cultivated form of pears in the list not included

- From the Caucasus Ecoregion *P. boisseriana* is found in Iran and *P. balansae* in Turkey as well.

A –sect. *Achras* Kochne emend. Fed. (*Pyrus* Tuz) (10 species);

a – *Argyromalon* Fed. ex Kuthatheladze (12 species);

P – *Pashia* Buch.-Ham. Ex D.Don (2 species);

x – *Xeropyrenia* Fed. Ex. Kuthatheladze (12 species).

* According to Akopian (2007) with some additions or corrections from the following sources: Gagnidze (2005); Aydin & Dönmez (2015); Zamani *et al.* (2015).

Figure 93. *Pyrus georgica* with extremely abudance of fruits. Adigeni district (2018).

2.4.4.4.1.1. Analyses of Diversification and Species Richness in the Subtribe Pyrinae in the South Caucasus and Adjacent Countries

According to Table 4, currently an estimated total number of 36 species of wild pears have been reported from the South Caucasus. A remarkable number, more than 88% (32 identified species with nine endemics) belong to Armenia (29,743 sq. km), 63% (23 species without endemics) belong to Azerbaijan (86,600 sq. km), and about 37% (12 species with one endemic) belong to Georgia (69,700 sq. km). Therefore, all three countries, if we take into account their small area of territories are surprisingly rich with pear species. Although, taking into consideration its smallest area, with the biggest concentration of wild pear species and exceptionally high rate of endemism, Armenia is the most striking country in Transcaucasia (Browicz 1993; Akopian 2007). It should be stressed there is same data for ecologically more or less identical, adjacent countries, territories which are hardly larger than all three countries together that comprise the South Caucasus. For example, genus *Pyrus* is represented by 23 species in Iran (1,648,195 sq. km) (Zamani *et al.* 2008, 2015), Turkey (783,562 sq. km) – by 12 species (Aydin & Dönmez 2015). It should be noted that in semi-arid habitats of Armenia this trend of major biodiversity patterns is confirmed by the presence of other wild relatives of domestic cultures. This country is considered a global center for the species of wild wheat, rye and barley (Vavilov, 1926, 1992).

Nakhichevan AR, exclave of Azerbaijan is distinguished by the most severe, continental climate in the South Caucasus. The present study of wild pears indicates that this other extremely southern region of the South Caucasus region is a home of *Pyrus* species abundance (17 (57%), without endemics). The latter, landlocked region, is completely surrounded by other landlocked countries or regions represented in middle reaches of the Aras River. It covers only 5,500 sq. km. (Talibov & Ibrahimov 2009).

13 species (40%) of pears are endemic to the South Caucasus.

According to successful field investigations in recent times (including adjacent Iran), it should be predicted that further taxa: new species, subspecies and hybrids of pears – wild genetic resources, and even new areas of a high species diversity will be described.

As for the taxonomic structure of the genus *Pyrus*, generations of botanists have been interested to clarify and improve the current sections of pears (Silva *et al.* 2014; Aydin & Dönmez 2015).

Gladkova (1990b) classified the genus *Pyrus* L. of South Caucasus in four sections (Table 4): *Achras, Argyromalon*, *Xeropyrenica* and *Pashia.*

More than 44% of the identified species belonged to the section *Xeropyrenia* (*P. acutiserrata, P. complexa, P. daralaghezi, P. demetrii, P. elata, P. eldarica, P. nutans, P. pseudosyriaca, P. raddeana, P. syriaca, P. voronovii, P. zangezura*, 12 species); 44% belonged to the section *Argyromalon* (*P. chosrovica, P. fedorovii, P. georgica, P. hajastana, P. medvedevii, P. megrica, P. oxyprion, P. sachokiana, P. salicifolia, P. takhtadzhiani, P. ? taochia, P. theodorovii*, 12 species); 27% belonged to the section *Achras* (*P. balansae, P. caucasica, P. gergerana, P. grossheimii, P. hyrcana, P. ketzkhovelii, P. sosnovskyi, P. tamamschjanae, P. turkomanica, P. vsevolodii,* 10 species); finally, 5% belonged to the section *Pashia* (*P. boisseriana, P. browiczii*, 2 species).

According to Akopian (2007), complexity of the four sections is the result of a strong evolutionary tendency which may be favored by natural selection. Major types of ancestral forms should be mesophilous pears (sect. *Pashisa*) occuring in milder climate of Colchis - Hyrcanian type. Later (from the Pliocene, ~ 7-2.5mya), according to the Late Neogene's more severe global climate deterioration, pears through the xeromesophilous forms (sect. *Achras*) converted into xerophitic descendants growing in harsh conditions (*Xeropyrenia, Argyromalon*) (Zamani *et al.* 2015; Akopian 2007). This evolutionary trend found its influence in pears anatomy and physiology (for example, from broad and oval, glossy dark green above leaves (sect. *Pashia*) to narrow, sharply lanceolate forms, with densely, silvery-hairy (pubescence) leaves (e.g., some species of *Xeropyrenia, Argyromalon*) (Akopian 2007).

P. pashia, from sect. *Pashia* found within the area from mountains in Yunnan Province (China) to the foothills of Hindukush in Afganistan, is classified traditionally as morphologically intermediate between Chinese proper and Central Asian pears (Zhukovsky 1964). Zheng *et al.* (2014), based on molecular investigations, studied the phylogenetic tree of *Pyrus* of the southern China. This region, where a large number of *P. pashia* populations are found 'is probably the most important diversification center of *Pyrus*' (Zheng *et al.* 2014).

In the Caucasus Ecoregion Hyrcanian forests in the north, Iran is the center of concentration of species of pears belonging to sect. *Pashia*. Hyrcan harbours four species of pears of this section (*P. cordifolia*, *P. boissieriana, P. longipedicellata*, *P. pashia*) (Zamani *et al.* 2008, 2015).

Ayidin & Dönmez (2015) suggest that the diversity center of *Pyrus* sect. *Achras* (*Pyrus*) appears to be mostly in Turkey (with West Georgia, pers. comment). The other two sections – *Xeropyrenia* and *Argyromalon* – comprise several narrow local microspecies. Most of them have a hybrid interspecific origin, generally from the South Caucasus (Akopian 2015, Aydin and Dönmez 2015). Such species of hybrid origin are characterized by species, intraspecies diversity and polymorphism. The anatomical and molecular study at the interspecific level (Akopian 2007, 2015; Zamani *et al.* 2008, 2015; Aydin & Dönmez 2015) has revealed architecture of such variation and, at the same time, the center where the hybrid origin species richness is primarily located. As pointed out above, such kind of event occurred exclusively in southern Transcaucasia and, it seems, the surrounding or adjoining core (-semi) arid regions of Iran and Turkey. This area, in general, can be considered as vital sites of reproductive migrations of pears. Although, this fact that 'the 'heart' of concentration of the narrow endemic pears is the southern Armenia was emphasized by Akopian (2007), by illustrating the following fact: 'even in only one gorge of the Elegis River (the Darelegiz Floristic Region) are growing 16 species of genus *Pyrus* including all four sections'.

This remarkable diversity of xerophytic species (and other, xeromesophitic taxa) of pears is highly correlated with different habitats commonly found in arid and semi-arid environments (open arid woodlands, dry (open) hills, dry slopes of mountains, forest remnants at altitudes from 800-900 up to 1,800-2,400 m a.s.l., also shibliak (scrub-lands) of lower mountains. In arid and semi-arid habitats narrow-leaf pears, trees or shrubs are key structural components, especially in some arid open woodlands and shibliak scrub communities. They grow mostly solitarily or in groups. There are sites, where two or three, sometimes, even more species, similar to the Elegis River gorge, coexist at the same time intensively overlapping ranges. It means that via differentiation their niches, numerous hybrids are able to co-habit (Akopian 2007, 2015).

In a woody plant community, including willow-leaf complex of pears, commonly co-occur species of *Quercus, Juniperus, Pistacia, Amygdalus, Prunus* (*Cerasus*), etc. (Akopian, 2007; Musayev *et al.* 2014).

To explain patterns of intensive speciation processes in the genus *Pyrus* on the territory of southern Armenia a number of hypothesis have been proposed by botanists: (1) isolation of species (other taxa) by means of strong fragmentation of relief; (2) the further global warming – changing environmental conditions; (3) intensive hybridization; (4) historical activity of humans for domestication of native wild plants (later some of cultivars promptly managed to escape to the wild nature) (Akopian 2007 and references therein).

2.4.4.4.1.2. Patterns of Reticulate Evolution

Interspecific hybridization among several genera of Pyrinae is usually a form of reproduction in the *Pyrus*. Browicz introduced 47 hybrid and feral taxa of pears (Browicz 1993 cited in Zamani *et al.* 2015). 'This circumstance has led to reticulated evolution of the members of the subtribe, notably *Pyrus*' (Zamani *et al.* 2015 and references therein).

Patterns of reticulate evolution are a major topic within angiosperms by means of natural hybridization. Hybridization can create new species (Arnold 1997; Sessa *et al.* 2012). It should be stressed that a lot of papers dedicated to the study of hybrid taxa of *Pyrus* in the Caucasus Ecoregion are not focused on a such theoretical aspects. Nevertheless, recent examples of such investigations, which arose out of the Caucasus isthmus, are a major topic with regard to the subtribe Pyrinae, significantly *Pyrus* (Zheng *et al.* 2008 cited in Zamani *et al.* 2015).

Until recent times the study of the reticulate evolution was not recognized for vertebrate animals. However, over the past decades a large number of investigations highlighted a network evolution between two ancestor lineages which leads to speciation among vertebrates. This new evolutionary framework in turn triggers a new conception of life (Dobzhansky 1937, Borkin & Darevski 1980; Rubtson, etc. cited in Arakelyan 2012; Spengenberg *et al.* 2017). For instance, similar to pears, the same evolutionary mechanism was pointed out by Arakelyan in reptiles living in Armenia, Nagorno-karabakh and the Aras River Valley (Nakhichevan) (Arakelyan 2012). It encompasses populations of four groups of reptiles: tortoise (*Testudo*), turtle (*Emys*), parthenogenetic and bisexual species of rock lizards (*Darevskia, Lacerta*). This study, similar to Pyrinae, revealed the relationship of close taxa by a phylogenetic network rather than a bifurcating pathway as one of the above pointed evolutionary mechanisms among two or more independent evolutionary lineages of reptiles. As the major reasons of development of microevolutionary processes in sympatric populations of reptiles of Armenia and adjacent territories are named by Arakelyan 2012 (and other herpetologists cited) the same complexity and a strong fragmentation of geographical relief of the Armenian highlands (Armenian upland and part of the Lesser Caucasus) with its highland's continental climate.

Thereby, chiefly, the Aras River valley and adjacent territories of the three landlocked uplands (plateaus) of the Near Middle East (Armenian, Iranian and Anatolian) are represented as the most active centers of microevolution among some groups of plants and animals, which are reproduced via hybridization.

2.4.4.5. Arid Open Woodland with Hackberry Species (Celtis spp.)

Fragments of these communities have remained in Georgia, the south slopes of the Karabakh Range, and the eastern part of North Caucasus (Dagestan). Xeric species of hackberry consist of *Celtis caucasica* and *C. australis*. These species are frequently associated with *Pistacia atlantica*, more infrequently with *Quercus infectoria veneris*, growing together with *Tamarix ramosissima, Elaeagnus angustifolia*, narrow-leaf pears (*Pyrus* spp.) and *Juniperus* spp., *Acer ibericum* in understory (Gulisashvili *et al.* 1975; Akhalkatsi & Tarkhnishvili 2012; Nakhutsishvili 1999, 2013).

Figure 94. Open arid woodland with a dominancy of *Celtis caucasica*. Near Tsalka town, S. Georgia (2015).

Open woodlands of *Celtis. caucasica* considerably extended its area in the lower reach of the Mashavera River (Kvemo Kartly region, East Georgia). Here, side by side with riparian forests, in foothills and mid-altitude (550-900 m a.s.l.), areas are covered by patches of arid open woodlands. Usually they spread on scree slopes, sandstone and sandy-gravelly river terraces with shallow soil in ravines. Being a remnant of a presumably larger area, today strongly degraded and modified by human activities, patches of natural semi-dry plants consist of *C. caucasica* with admixture of *Pistacia atlantica, Acer ibericum, A. hyrcanum, Ficus carica*, etc. The dominant zonal type of natural shrub vegetation is shrubland of *Rhamnus pallasii, Paliurus spina-christi, Spiraea hypericifolia, Cotinus coggigria*, etc. Treeless vegetation includes various types of steppes and semi-deserts (species of *Stipa, Festuca,*

Achillea, Artemisia, Salsola, Xeranthemum, etc.). In upper altitudes (700-900 m) hemixerophitic communities form admixture with broad-leaved deciduous trees, especially *Quercus petrae iberica* and *Carpinus betulus* (Ketskhoveli 1960; own research) (Figures 94, 95).

Figure 95. Fruits of *Celtis caucasica*. David-Gareja Monastery environs, E. Georgia (2015).

2.4.4.6. Arid Open Woodland with Eldari Pine (Pinus Brutia Eldarica)

The last Caucasus woodland to be discussed is the very impressive open woodland with Eldari pine (*P. brutia eldarica*). It survived as a remnant of the Pliocene maritime pine forest on the Eliar-Ougi Ridge (Azerbaijan, bordering Georgia). The Eldari pine, a relict from the Tertiary Period that once (Miocene Epoch, Sarmatian Age, 14,0-10,5 mya) covered the seaside territories of the ancient, almost landlocked Sarmatian Sea. This relict sea in the beginning was the straight descendent of the Tethis Ocean and then Paratethis which lately disintegrated into the Black, Azov and Caspian Seas (Chapter 1.4.1., 1.4.2.). Eldarian pine is a close relative of another maritime Bichvinta (=Pitsunda) pine species (*P. brutia var. pityusa*) (Schmidt, 2002), which occurs along the Black Sea coast. With regard to chorology, the geographical range of Eldari pine includes the Kura-Arasian desert and mountain steppe province of the Afro-Asian desert region (Nakhutsrishvili 2013). Within the natural habitat of Eldari pine usually other woody plants are confined, for example, species of juniper, pistachio tree, Christ's thorn and others (Ketskhoveli 1960) (Figure 96).

For the protection of Eldari pine, the genetic heritage and biological diversity of ecological systems of this unique forest the Eldari Shami (pine) State Nature Reserve was established in 2004. Before 2004 this territory was a branch of Gey-Gel reserve. The unique natural grove of Eldari pine trees on the Eldar (i) plain in the Samukh region has been protected since the end of the 19th century and this area can be referred to as the oldest specially protected site on the arid territory of the Caucasus. This is the only place where *P. b. eldarica* trees have been preserved in their natural form. The pine trees growing here are 100-120 years old and 2-6m high. The main part of the reserve is typical for semi-desert and arid open woodland flora (e.g., European dog-wood, medlar, pistachio and others are spread here).

Figure 96. Eliar Ougi Ridge (Azerbaijan) provides shelter for the unique, former (the Early Pliocene) maritime, now xerophilic Eldari pine. In the background: Desertified 'waves' of hills – bottomland of former Paratethis Sea (1975).

Box. 8. Arid Open Woodland of Middle (Central) Asia

The analysis of the arid woodland biome would be imperfect without discussing the arid open woodland of Middle Asia, which is well known and typically developed in the Badhiz Reserve in Turkmenistan along the borders with Iran and Afghanistan. Occupying latitudes from 35° to 36°, the woodland is located in the subtropical zone and its bio communities are characterized as subtropical semi-savannas (Bannikov (Ed.) 1969). Badhiz is a vast (86,000 km^2), hilly area with an average elevation of 800 m a.s.l. between the snow-covered Hindu Kush Range (south) and the Kara Kum desert (north). Average annual precipitation is 250 mm. The herb layer is dominated by *Poa* and *Carex*, when the soils become more salty and slightly saline by *Artemisia* and *Salsola*, and finally by the giant Apiaceae, *Ferula*. Spring is the most colorful time with many ephemeral and bulbous species (*Tulipa, Iris, Allium*) flowering. In the western part of the reserve the prevailing woody plant is *Pistacia vera*, a vicariant of *P. atlantica*, which is distributed west of the Caspian Sea (Figure 97, 98).

Figure 97. In Central (Middle) Asia *Pistacia atlantica* is replaced with *P. vera*; allochtonic distribution of two species, Uzbekistan, (1990).

Figure 98. The fruits of *Pistacia vera*. Uzbekistan (1990).

In contrast to the South Caucasus, the open woodland of southern Middle (Central) Asia is far richer in (sub-) tropical animal species. In the Badhiz Reserve next to a unique wild horse (*Equus hemionus*) species occur of porcupine (*Hystrix indica*), honeyeater (*Mellivora capensis*), leopard (*Panthera pardus*), and striped hyena (*Hyaena hyaena)*, cobra (*Naja naja*), grey varan (*Varanus griseus*), termites, etc. In the recent past the territory was inhabited by the Asiatic cheetah (*Acinonyx jubatus venaticus*), while this critically endangered predator is still represented in central Iran (Farhadinia 2004; Farhadinia & Hemami 2010).

2.4.5. Are There Savannas in the South Caucasus?

Different opinions exist on the relationship between the arid open landscapes of the South Caucasus and the tropical-subtropical savannas. Is the arid open woodland of moderate climate latitudes the northern version of the savannas?

A scholar who supported this opinion is the famous Georgian forester Prof. V. Gulisashvili (1961, 1980), who wrote in his book 'Savannas - open woodland' (1980): 'We assign the open woodland of the South Caucasus to savanna'. In an earlier monograph, co-authored with other botanists, he had not yet been so categorical: 'Open woodland has many features in common with savanna' (Gulisashvili *et al.* 1975). The father of Caucasian botany, Grossheim, took the opposite position. In spite of the physiognomic similarity of the two biomes, he st ressed the fundamental differences in biology and phenology between the two ecosystem types, not to mention their different floristic basis (Grossheim 1948). This position was supported by Prof. N. Ketskhoveli in his capital work 'Vegetation cover of Georgia' (1960). Our investigations of the arid open woodlands of the North Caucasus, the South Caucasus and Middle (Central) Asia (Gegechkori 1984a,b, 1985, 1997b), together with observations in the savannas of eastern tropical Africa (Gegechkori 2008, 2010b) and analysis of the literature sources support the opinion of Grossheim.

Except for the physiognomic differences between the arid open woodlands of the North and South Caucasus, noticeable differences exist in the tree, shrub and herb strata. In the arid open woodland, the dominant tree taxa *Pistacia, Juniperus*, *Celtis* and *Amygdalus* belong to other botanical families and genera and are drought-resistant, unlike the mesic trees and shrubs of the forest-steppe. The genus *Pyrus* contains only xeric species, like for example, *P. salicifolia*. Also, the wild grasses are, next to *Festuca* and *Agropyron*, represented by different genera. Therefore, meadow-steppe communities of the forest-steppe can easily be distinguished from the dry steppe communities of the arid open woodland, depending chiefly on soil mechanical and chemical composition. In the ground layer of the latter, apart from steppe grasses, a key position is occupied by deserts and semi-deserts plants.

The fauna of arid open woodland comprises a mixture of mesophilous (brown bear, wild boar, etc.) and xerophilous (hyena, gazelle, etc.) species, where the xerophilous group is evidently dominating. This in particular applies to the invertebrate fauna (Gegechkori 1984a, b). The arid open (light) woodland of the South Caucasus is thus an independent type among the arid open ecosystems. Similar communities are widely distributed in the neighboring interior Asia and Middle Asia, but not in regions at lower latitudes, where savanna prevails.

The savanna is a completely different biome for the following reasons:

- The most characteristic abiotic factor that fully determines life in the savanna is the seasonal distribution of annual precipitation: very long (3-9 months) dry winters alternating with wet (rainy) summers. Air temperature doesn't change a lot (20°-30° C). When it does, it is very gradual and not drastic;
- The dormancy of arboreal plants is determined by winter drought, not by low temperatures. This is in contrast to moderate climates;
- The long dry period with intense insolation, strong evaporation and irregular rain is decisive for the physiognomy of the plants. Many plants are equipped with features to limit transpiration losses. The various kinds of succulence are largely unknown from the arid (light) open forests, whereas on the other hand, spring geophytes and annual ephemerals that are so characteristic for moderate open woodland are absent in the savanna;
- Savanna vegetation is characterized by facultative and obligate pyrophytes, e.g., fireproof trees. Frequent fires have selected the savanna flora for fire resistance: thick barks, trunks that can store water, covered fruit, seeds that can tolerate temperatures of over 100°C without losing their germination capacity. All these features are unknown from open woodlands;
- The all year-round positive temperature with abundant food availability determines the exceptional richness of the fauna, especially the various ungulates and their predators, including the incomparable megafauna. The tropical grassland dotted with scattered trees is populated by wandering herds of wildlife that give the landscape its special character. Climate fluctuations never force animals into winter or summer hibernation, but stimulate seasonal migration on both sides of the equator.

In conclusion: between the arid open woodland of the South Caucasus and the savanna of tropical Africa indeed similarities exist, but these are mainly physiognomic. They share some grassland taxa, such as *Imperata, Erianthus, Bothriochloa, Cynodon* and *Eragrostis*, but these are common pantropical, xeric elements. They have also two big mammals in common, leopard and striped hyena, but these also have a general tropical-subtropical origin and distribution, from where their natural range extends into temperate regions. With respect to all other abiotic and biotic factors, there are noticeable differences between the savanna and the arid open woodland (Figure 99).

Thus, from our viewpoint, the conclusion of Prof. V.Z. Gulisashvili and his followers that the arid open woodlands of the South Caucasus are a northern version of savanna is not valid, not even with respect to the southern part of Central (Middle) Asia that is situated in the dry subtropics (e.g., the Badhiz Reserve in Turkmenistan).

Figure 99. Savanna, open woodland consists of *Acacia* sp. A herd of Thomson's gazelle (*Gazella thomsonii*) resting in the shade of acacia. Africa, Serengeti NP (photo by Dr. Gegechkori, 2010).

From palaeobiological data it is clear that savanna conditions, including the typical rhythmic summer peak rainfall, prevailed in the South Caucasus in the Late Miocene and Pliocene. The Pliocene fossil record, as previously stressed, shows the Southern elephant (*Mammathus primigenius* (= *Archidiskodon meridionalis taribarensis*), ostriches and antelopes. The arid and semi-arid flora were represented by species of the Boraginaceae family (*Anchusa, Lycopsis, Lithospermum*) and by *Celtis* as an open woodland dominant. In the Early Pleistocene, as previously mentioned, *Homo ergaster* dwelled in East Georgia (Gabunia & Vekua 1963; Gabunia *et al.* 2000a,b,c; Vekua & Lordkipanidze 2008). Fossil bones from the place of discovery, Dmanisi town, include the following species of open landscapes: *Mammuthus primigenius* (=*A.m. taribarensis*), *Hippopotamus georgicus, Equus sussenbornensis, E. stenosis, Dicerorhinus etruscus, Gazella borbonica, Giraffa paleotraginae*, swordtooth tiger (*Megantereon megantereon*), and *Struthio dmanisensis*. The occurrence of some species of mesic autoecology in the same area (*Ursus etruscus, Martes* sp., *Cervus perrieri*, etc.) shows that open landscapes bordered mountain gorges that were covered by forest with closed canopy. Pollen data indicate that the deciduous forests consisted of *Fagus, Castanea, Alnus, Tilia, Carpinus, Ulmus, Salix, Betula* and *Pyrus*, whereas coniferous communities with *Abies* and *Pinus* must also have been present (Klopotovskaya *et al.* 1989; Kvavadze 1997).

The Late Pleistocene climate deterioration caused the extinction of the tropical savanna fauna by changing the savanna biome into a savannoid open arid woodland that only by physiognomic convergence has similarity to the savanna. The latter cannot be said from the open woodland - forest-steppe communities of the North Caucasus.

The apparent resemblance of the three types of open woodlands is the same display of convergence (but on a landscape scale) as, for example, the occurrence of highland megaphytes of various botanical families on different continents (Africa – *Senecio*, South America – *Espeletia* and *Puya*), or the external similarity of marine vertebrates like dolphin (Mammalia), tunny fish (Pisces), and the extinct ichthyosaurs (Reptilia) (Gegechkori 2008).

Conclusion

- There are two types of transition zones or an ecotone between deciduous hardwood forests and open, woodless vegetation areas in the Caucasus.
- The ecotone between woodlands (oak-hornbeam) with closed canopy and steppe (with grasslands and herb-rich (forbs) meadows) is considered as forest-steppe. Such kind of transition zone is particularly well suited to environments of lowlands and significantly foothills of the Ciscaucasia. The essential ecological features of the mentioned transition zone are approximately equal in order to ratio of annual rainfall (annual average – 500-600mm) and its evaporation.
- In the forest-grassland transition zone, it is considered its sensitive dry edge of the closed forest belt, as an agriculturally important area, is largely depends on human activity. Currently, the forest-steppe biome of the North Caucasus is over-exploited land due to enlarging agriculturally improved grasslands (arable land). Natural summer green forests, grasslands and herb meadows become reduced and fragmented, therefore, remained patchy, surviving human-non-modified during millennia mini-sites. As 'eyewitness' of the past virgin landscapes, they need protection and restoration.
- Between the deciduous forests, on the one hand, and deserts and semi-deserts biome, on the other hand, appear woodlands without close canopy. By this feature, he latter one is certainly more distinctive than that of forest-steppe. In scientific literature this transitional biome is mentioned as 'arid (light) open woodland'.
- In Transcaucasia open woodland is one of the most typical ecosystems with complex habitats. They mainly developed on slightly saline soils; the average annual rainfall ranges from 400 to 500 mm.
- Effects of canopy opening do not reduce sunbeams to filter. Therefore, these natural parkland's undergrowth consists of xeromorphic shrubs and ground layer – grasses and herbs. Among sparse stands of drought-resistant trees and shrubs ground layer vegetations are growing in abundance, but in marginal areas of parklands sparse semi-desert vegetation (which is represented as arid rangeland) became dominant.
- The dominant keystone species of arid woodland is the pistachio tree (*Pistacia atlantica*). It develops in gentle forms of relief with a predominance of juniper trees (*Juniperus* spp.) preferring slopes of the hills and mid-elevations of mountains.
- In the Caucasus, communities of open woodlands dominated by hackberry species (*Celtis* spp.), willow-leaved xeric pears (*Pyrus* spp.), wild almonds (*Amygdalus* spp.), and Eldari pine (*Pinus brutia eldarica*) are represented as well.

- Vashlovani National Park (Georgia) and Khosrov Reserve (Armenia) as ecotones and insular, natural laboratories of rare and endangered species exemplify *ex situ* natural heritages of the Transcaucasia.
- The arguments of some scholars who considered arid open woodlands as a northern version of savannas are not driven by evidence of abiotic and biotic features (unlike the geological past, e.g., mid-Pliocene).

2.5. Biome of the Mountain-Xerophyte Vegetation (Sclerophylous Scrub)

2.5.1. General Characteristics

According to the majority of botanists (Takhtajan 1937; Grossheim 1948; Gulisashvili *et al.* 1975 and others) phytocoenosis like hammada is not considered in deserts and semi-deserts biomes. The same should be said with regard to some other xerophytes' complexes like phrygana, tragacathic, shibliak, tomillares and discussed in mountain xerophytic communities (sclerophyllous scrub).

Mountain-xerophyte vegetation is a very peculiar type of vegetation that is characteristic of the arid regions of the Earth. It is a zonal biome that is similar to other biomes like, for example, forests. Mountain-xerophyte ecosystems extend as a wide belt through all Mediterranean countries and through a vast area in Asia Interior and particularly in Middle (Central) Asia. It consists of drought-resistant shrub woods and occurs in its most distinctive form on mountain slopes with stony, quiches- and clay-stony soils. Because of lack of soil this type of community is also called 'vegetation of skeleton soils' (Takhtajan 1937, 1941; Grossheim 1948; Gulisashvili *et al.* 1975).

Figure 100. Aras river flowing in the southern part of the South Caucasus, Nakhichevan AR, general view (1971).

Figure 101. Zoological expedition in the middle reaches of the Aras River Gorge (1971).

A large variety of mountain-xerophyte vegetation types is distinguished on the basis of their floristic composition, structure and seasonal development rhythm that in their turn are determined by abiotic site factors like relief, soils, exposure steepness and height above sea level (Grossheim 1948). Most frequent are the floristically polydominant formations, but in some habitats also mono- and bidominant phytocoenoses are found.

In the Greater Caucasus the biome occupies a significant area, especially in mountainous Dagestan. In South Georgia a biome of mountain-xerophyte vegetation is well manifested in Javakheti Highland (Akhaltsikhe Hollow) and ranges from 900 up to 2,200m (Ivanishvili 1969; Khintibidze 1990). In the southern part of the South Caucasus, it forms a natural zone on local mountains from 1,200 to 1,600 m a.s.l. (Grossheim 1948; Gulisashvili *et al.* 1975) (Figures 100,101).

2.5.2. Hammada Communities

Hammada or hamada (arabic *hamm*āda) is a type of desert landscape consisting of high, largely barren, hard, rocky plateaus, with very little sand, because this has been removed by deflation (Gautier 1935 in Fairbridge 1968).

Hammada communities consist of xeric semi-shrubs, chiefly solonchak, that occupy cloddy and stony places absolutely devoid of soil. Regions of rocky highland desert are found on gypsum deposits in the South Caucasus and occur only in the southernmost part of the region, in the middle reaches of the Aras River Valley in Nakhichevan and southern Armenia.

A broad strip of this type of mountinous deserts runs through the Sahara, Palestine and Asia Interior. Sparse, poor vegetation cover is typical for rocky deserts, but the plant species composition is unique and there are few annual plants.

On gypsum containing soils along the middle reaches of the Aras River (around Julfa town, Azerbaijan) a special phytocoenosis with *Achillea tenuifolia* can be found in which *Neogallonia szovitsii*, *Leontice armeniaca* and *Stipagrostis* (=*Aristida*) *plumosa* are common. Frequently occurring species are *Acanthophyllum pungens A. squarrosum* and *Zygophyllum atriplicoides* (Grossheim 1948).

Figure 102. River Aras Gorge, Julpha Hollow, Julpha town approach area covered by hammada xerophytic vegetation (1972).

In the Aras River Valley, in lower latitudes (39^{0}N), hammada alternates on tiny areas with solonchak and sand biotopes with halophytes (*Salsola glauca, S. cana* and *Anabasis eugeniae*) and psammophytes (*Calligonum polygonoides*) respectively (Akhundov *et al.* 1978; Gegechkori 1978, 1984a; Ibragimov 2005; Ibadullayeva *et al.* 2011; Ibragimov & Nabiyeva 2015 and references therein). The floodplain forest of the Aras River and its tributaries consists of various species of willow (*Salix* spp.), oleaster (*Elaeagnus angustifolia*), tamarisk (*Tamarix* spp.) and, typical of desert riverside tugai-communities, the Euphrates poplar (*Populus* (=*Turanga*) *euphratica*) (Gegechkori 1984a) (Figure 102).

The hammada of the Aras River Gorge is considered to be a northern outpost of these communities (Grossheim 1948).

In the Caucasus the middle reaches of the Aras River are distinguished by its relief and unique biodiversity. The Aras River basin, including Nakhichevan and part of Turkey, like the other regions of the Caucasus Isthmus, suffers with human impact, causing further desertification of the region. Reason for desertification and measure of the fight against this process is the actual problem the native people are facing today (Imat & Bababeyli 2010).

2.5.3. Phrygana Communities

Phrygana ('dried out' in Greek) communities include low xeromorphic shrubs and semi-shrubs or scrubs with an admixture of therophytes and geophytes found around the

Mediterranean countries. In the Caucasus they are usually found as a xeric mountain formation (Grossheim, 1948). Some Caucasian botanists differentiate between local phrygana and the proper Mediterranean communities (phrygana – in Greece, matorral – in Spanish) and describe the former as 'phryganoidic' vegetation (Yaroshenko 1956; Gulisashvili *et al.* 1975).

Phrygana (phryganoidic) communities are frequently polydominant, more rarely they are dominated by one or few species. The most typical habitats are of stony, rock-stony with rock debris, loamy, clayey and clay-sandy slopes of mountains and hills.

The most peculiar phrygana formations are situated on slopes covered by limestone, as is typical of the Aras River Valley (Ganbarli 1973). In spring time these slopes are clothed by annual and perennial terrophytes and geophytes. In June they are replaced by rich vegetations with species of the Cruciferae, Umbelliferae and Labiatae families. Some species reach 1-1.5 m in height (*Ferula oopoda, Prangos ferulacea, Salvia limbata*).

Xeric plants are well adapted to the local severe environment: droughts in summer, low temperature and lack of snow cover in winter. Adaptations to these conditions include a cover of dense hair, leaf metamorphosis, cushion-shaped form, etc. Many species are also ether-bearing plants (Gulisashvili *et al.* 1975).

The most distinguished mountain-xerophyte vegetations with the richest flora are found in Nakhichevan (Azerbaijan), where, on a rather small territory (5.300km^2), 2,015 species of vascular plants are found. 807 species are represented at low altitude, whereas 1,761 species are found in mountainous Nakhichevan (some other species are common to both zones), the latter group being very rich in xeric species. From the phytogeographic viewpoint they belong to various groups: Asia Interior, Asia Minor, Armenian, Iranian, North-Iranian, transitive Armenian-Iranian, Irano-Turanian, etc. (Gulisashvili *et al.* 1975).

Within Nakhichevan various types of phrygana communities are distinguished. Grossheim (1948) recognized 4 types of these xeric formations, Gulisashvili *et al.* (1975) - 15 types. To the most characteristic phrygana phytocoenosis belong the following:

1. Phrygana on limestone boulder is populated by *Zygophyllum atriplicoides, Salsola glauca, Salvia dracocephaloides, Astragalus szovitsii* and many other perennials and semi-shrubs. This type of phrygana has only a few annual plant species (Figure 103);

Figure 103. River Aras Gorge, Julpha Hollow, Nakhichevan AR. Phrygana community with dominancy of *Zygophyllum atriplicoides* (1972).

2. Another type on clayey-rock debris slopes which has more specific plants: *Hedysarum formosum, H. atropatanum, Hypericum scabrum* and *Stachys fruticulosa*;
3. On stony and rock-stony habitats the most dominating species is *Stachys inflata*;
4. In some places in Nakhichevan (e.g., Ordubad district) the most distributed form of phrygana has a clear dominance of *Thymus kotschyanus* and *Satureja macrantha* that are sometimes accompanied by *S. inflata*. Grossheim (1948) relates this phytocoenosis to Mediterranean tomillares. 'Tomillo' is thyme in Spanish and include order Labiata species (Gratsianskii 1971 and references therein, Morales 1997). Unlike Takhtajan (1937), Grossheim (1948) does not see the two formations as completely identical. In the same mountains, phrygana is replaced by mountainous steppe communities at an altitude of about 1,500-1,700 m a.s.l. (Figure 104).

Figure 104. River Aras Gorge, Julpha Hollow, Nakhichevan AR. Environs of geizer populated by tomillares vegetation confined to mother rocks (1972).

5. In the southeastern part of the Caucasus (Azerbaijan, Lenkoran district) in the Diabar Depression the luxurious deciduous Talysh forests completely turn into xeric (phrygana, particularly tragacanthic) vegetation communities above 1,000–1,600 m a.s.l.

2.5.4. Shibliak Communities

In lowland and partiularly in mountain steppe and semi-desert landscapes an intrazonal drought-resistant thorny shrub wood occurs that in geobotanical literature is named 'shibliak' and has *Paliurus spina-christi, Berberis vulgaris, Lonicera iberica, Rosa iberica, Juniperus oblonga, Cotinus coggygria, Punica granatum* and *Carpinus orientalis* as the main constituents. All these arboreal plants are represented in the Caucasus. Some Caucasian authors consider the drought-resistant communities in the Caucasus to be analogues of the Mediterranean shibliak (sensu Adamovich 1899, 1905, 1909; Rikli 1943-1948) and use the term to refer to a plant formation that in the Caucasus (in contrast to the evergreen communities of Mediterranean countries) consists of deciduous shrubs.

Figure 105. Shibliak communities with mainly thorny shrubs (*Rhamnus pallasii*). Tbilisi (Georgia) environs (2014).

Primary shibliak is widely distributed and composed of a rich variety of xerophytic shrub genera (Grossheim 1948). Secondary shibliak originated from the destruction of forests, first of all arid open woodlands (Sakhokia 1958). In east and south Transcaucasia, mainly in the areas under a strong anthropogenic pressure, shibliak formations consist of *Rhamnus pallasii, Atraphaxis spinosa, Cotoneaster racemiflorus, Amygdalus fenzliana, Spiraea crenata* and more frequently of *Caragana grandiflora, Colutea komaronii* and *Ephedra procera* (Figure 105).

Sakhokia (1958) distinguishes 3 types of shibliak:

1. Primary arid thorn shrublands untouched by human activities;
2. The derivatives of arid open woodland dominated by xeromorphic *Pyrus* and *Pistacia*;
3. Secondary shibliak, appearing after destruction of deciduous foothill forests.

The most xeric version of shibliak is confined to mother rocks and stony slopes. In regions with a continental climate phrygana clearly prevails over shibliak.

2.5.5. Tragacanthic Communities

Tragacanthic communities are composed of non-thorn cushion plants and thorny cushion plants. These oreoxerophytic plant species are almost completely absent from the proper Mediterranean countries and express the ancient connection with the flora of Asia Interior (Grossheim 1948; Sakhokia 1958; Gulisashvili *et al.* 1975). All species of these communities are characterized by their compact, low, cushion-shaped growth form and their xeromorphic leaves that usually end in thorn (Krivonogova 1960). They are dominated by milk vetch (species of *Astragalus*) and prickly thrift (species of *Acantholimon*), rarely by sainfoin (species of *Onobrychis*) that all form thorny (spiky) cushions. Thorn less cushion formations occur at higher elevation and are apparently more resistant (Figure 106).

Figure 106. Tragacanthic community with prickley thrift (*Acantholimon glumaceum*) in the highland (1,800m a.s.l.) of the Nakhichevan AR (Ordubad district) (1974).

Figure 107. Tragacanthic community with milk vetch (*Astragalus raddeanus*) in the Uraveli River Gorge, Meskhet-Javakheti province, S. Georgia (1976).

Lavrenko (1962) and Ivanishvili (1969) consider tragacanthic communities to be physiognomically, phytocoenologically, and in particular with respect to their floristical composition sufficiently different to classify them as a separate formation, whereas other specialists prefer to assign them to one of the three other types of mountain-xerophyte vegetations (Gulisashvili *et al.* 1975).

Tragacanthic formations have a vast distribution in arid regions of Iran, Afghanistan, Turkey, South Mongolia, Central (Middle) Asia, and partially in South Europe.

In the Caucasus they occupy the slopes of some gorges in the Greater Caucasus with an arid climate and particularly the southern part of South Caucasus (Nakhichevan, Armenia, South Karabakh, Diabar Hollow) and also in the north slopes of the Greater Caucasus, especially in Inner Dagestan (Ivanishvili 1969, 1973; Ochiauri 1965). Climatically the areas with tragacanthic formations are distinguished by a dry summer (4-5 months), low air humidity (with annual precipitation ranges from 300 to 400 mm) and significant daily temperature fluctuations. Furthermore, they are characterized by poor soils.

The Diabar Hollow in the Zuvand district of Azerbaijan is an area where tragacanthic communities are well investigated. The constituent species include *Astragalus caucasicus, A. aureus, A. caspius, A. microcephalus* and *A. erinaceus,* of which the thorny *A. microcephalus* and *A. marshallianus* are most frequent. Although thorny cushion formations are not particularly diverse, they harbour representatives of various families including the Asteraceae, Fabaceae, Rosaceae, Apiaceae, Lamiaceae and Caryophyllaceae. On the steep slopes of the Diabar Hollow the following species are found within the tragacanthic phytocoenosis: *Scutellaria orientalis, Veronica multifida, Agropyron tarbagaticum* (=*A. trichophorum*), *Polygala anatolica, Onobrychis cyri, O. iberica, Salvia verbascifolia, S. nemorosa* and *Thymus collinus* (Figure 107).

Tragacanthic communities occur mostly at mid- and higher altitudes of Nakhichevan. Local tragacanthic vegetation contains *Astracantha aurea, Astragalus lagurus, A. aznabjurticus, A. mesitis, Acanthophyllum mucronatum, A. pungens, Acantholimon karelini, A. araxanum, A. caryophylaceum, A. hohenackeri, A. sahendicum, A. quinquilobum* and several species of characteristic herbs, such as *Ceratocarpus arenarius, Gypsophila arctioides, Thymus collinus, Tulipa florenskyi, Prangos ferulaceae, Ziziphora biebersteiniana* (Ibragimov & Nabiyeva 2016).

In the central part of the Greater Caucasus (Pirikita Khevsureti and Khevi regions of Georgia, Chechnya, Ingushetia of the Russian Federation) tragacanth communities with *Astragalus denudatus* are dominating. The cushion shrub with its dominancy occurs in the subalpine zone, at the altitudinal 1,700-2,200m a.s.l., at the slopes of dry gorges. Typical tragacanth vegetation is manifested by *Astragalus raddeanus, A. kazbeki, A. arguricus, Ephedra procera, Juniperus hemisphaerica, Berberis vulgaris, Artemisia chamaemelifolia, A. splendens, A. marschalliana, Dianthus cretaceaus, Allium* spp. and others.

Figure 108. Rocky-scree habitat with sainform (*Onobrychis sasnowskyi*) in Khertvisi fortress environs, Meskhet-Djavakheti province, S. Georgia (2014).

Figure 109. Tigran's elderberry (*Sambucus tigranii*) belongs to the rarest species (endemic of South Caucasus – Armenia, S. Georgia (Khertvisi fortress environs). Inhabits in rocky habitats and other dry places (2014).

At middle elevations of South Georgia (Meskheti), within this community there are 199 species of vascular plants (Khintibidze 1990). They occupy a large distribution range in many gorges of this region, where *Astragalus raddeanus* and *A. arguricus,* coincide with that of *Onobrychis sosnowskyi*. In the borders between Meskheti and the Javakheti plateau the two following dominant species are represented - *Astragalus microcephalus* and *A. caucasicus* (Ivanishvili 1969, 1973; Khintibidze 1990). According to personal observation in South Georgia (environments of Khertvisi fortress) inhabits one of the rarest species, *Sambucus tigranii*) (Figures 108,109).

2.5.6. Cushion Sainforn formations

Cushion sainforn formations have the thorny sainfoin (*Onobrychis cornuta*) as the main constituent, a species that is higher (50-90 cm) and creates a larger cushion (100-150 cm in diameter) than the tragacathic species. The shrubs are closely connected to each other, forming a dense thicket. The altitudinal distribution of sainfoin cushions ranges from 600-700 m a.s.l. near Julfa town in Nakhichevan to sometimes the subalpine zone. They occur on stony slopes of various exposures. In May the shrub begins to blossom with copious violet-pink flowers that give it the appearance of an ornamental plant.

Armenia, Nakhichevan and the Diabar Hollow are the regions, where *Onobrychis* occurs most abundantly; some small spots are also found in the Greater Caucasus.

The floristic composition of the *Onobrychis* formations is significantly richer than those of *Astragalus* and - next to the above-mentioned species - includes mesoxerophytes and even mesophytes (Gulisashvili *et al.* 1975) (Figure 108). The following species have been registered: *Astragalus microcephalus* (sometimes producing bidominant phytoassociations with sainfoin), *Plantago saxatilis, Inula conyza* (=*I. vulgaris*)*, Astragalus johannis-howellii, Polygala anatolica, Euphorbia condylocarpa*.

Cushion sainform formations of Armenia (especially Zangakatum community) as well as *Onobrychis cornuta*, contains 22 perennial species of plants (Takhtajan & Fyodorov 1972; Ghukasyan 2012).

There are cushion plants in Nakhichevan Autonomous Republic of Azerbaijan – thorny ones (e.g., *O. cornuta*) and thornless (e.g., *Allochusa versicolar, Senecio taraxacifolia*). In total about 30 cushion plant species are identified belonging to 14 families and 26 genera (Ibragimov & Nabiyeva 2016).

2.5.7. Fauna of the Mountain-Xerophyte Biome

The biome of mountain-xerophyte vegetation is represented mainly by a mixture of the consumers of various arid landscapes. In the southern part of the South Caucasus till the 1970s, naturalists were impressed by the rather high populations of wild goat (*Capra aegagrus*) and sheep (*Ovis ammon gmelini* (NT). Individuals of the latter species shaped the local landscapes. Many more specimens of leopard (*Panthera pardus ciscaucasica*) and hyena (*Hyaena hyaena* (NT) were known from just the middle reaches of the river Aras. Therefore, in this region an ecological model of predator-prey interaction was expressed. Today (over the last 40-50 years) in the Caucasus, such circumstances of interactions among big game ungulates, on the one hand, and large beasts, on the other, were more and more wiped out (Gegechkori 1978, 2007).

Typical mammal species are *Canis aureus, Vulpes vulpes, Meles meles, Dryomys nitedula* (LC), *Apodemus witherby* (=*A. fulvipectus*) (LC) common birds species, including: *Alectoris chukar*, in shibliak type communities – *Otus scops* (LC) and dozens of songbird species. Dominating species of reptiles are: *Lacerta media* and *Testudo graeca*, *Malpolon monspessulanus* (Figure 110) (Akhalkatsi & Tarkhnishvili 2012 and references therein); and a high diversity of butterflies (Didmanidze 1979, 2005; Nekrutenko 1990; Tshikolovets & Nekrutenko 2012).

As mentioned, fauna of psyllids (Hemiptera, Psylloidea) of the mid-reaches of the Aras River with about 130 species is richer than any other natural-historical regions of the Caucasus. The richest and exclusively most interesting nucleus of fauna includes a biome of mountain-xerophyte vegetation (67 species, 51%). The same should be said concerning the psyllid species associated with mountain-xerophyte vegetation of other regions of Transcaucasia and Ciscaucasia. Mentioned should made of: *Diaphorina kopetdaghi* (Loginova) associated with *Zygophyllum atriplicoides, Eurotica alticola* (Loginova), *E. distincta* (Loginova) – with *Eurotia ceratoides, Cyamophila astragalicola* (Gegechkori)* - with *Astragalus* sp., *C. oshanini* (Loginova) – with *Halimodendron halodendron, Aphalara grandicula* (Gegechkori)* - *with Rumex scutatus*, etc. In riparian (tugai) riverside arboreal plants are populated by *Egeirotrioza gracilis* (Bajeva), *E. verrucifica* (Loginova), *Camarotoscena unicolor* (Loginova) (host plant – *Populus* (*T.*) *pruinosa*), *Craspedolepta georgica* (Gegechkori)* - with *Artemisia chamaemelifolia, Cacopsylla incerta ornatipennis* – with *Rhamnus pallasii, C. fasciata* – with *Spirea hypericifolia, Bactericera perrisii* – with *Artemisia splendens*, *B. magnisetosa* (Loginova) – with *Elaeagnus angustifolia, B. scotti* (Löw) – with *Berberis vulgaris*, etc. All mentioned species are known mainly on the both sides (South Caucasus, Iran) of the Aras River (Gegechkori 1973, 1975a, b, 1978, 1984a; Gegechkori & Loginova 1990).

Figure 110. Montpellier snake (*Malpolon monspessulanus*), large (up to 2m) snake, mainly feeds on lizards. In picture: dramatic event - snake eats Caucasian agama alive. Nakhichevan AR, S. Caucasus (1972).

Below is given a comparatively complete characterization of one of the umbrella species of the region - wild sheep.

2.5.7.1. Some Keystone Species of the Mountain-Xerophyte Biome

2.5.7.1.1. The Armenian (Gmelin's) Mouflon (*Ovis ammon gmelini*)

The Armenian (Gmelin's) mouflon (*Ovis ammon gmelini*) is a rare subspecies of wild sheep, the ancestor form of modern domestic sheep. The Transcaucasian population of Armenian mouflon occupies the north-westernmost borderlands of the species range, which occurs mainly in Iran. For example, only Kabutan island in the southwest part of the Urmia Lake is home to 2,000-2,500 Armenian mouflons. This population probably represents the largest population of this mouflon's distribution which is in western and northwestern Iran (Ebrahimi 2007). It is found in the Lesser Caucasus, mainly along the river in the Aras Gorge and adjacent territory to the north (Khosrov Reserve, part of Zangezur Range, etc.), therefore it occurs in Armenia and the Nakhichevan Autonomous Republic of Azerbaijan (Khorozyan et al. 2009). There are a few herds of mouflon in Turkey, as well as in some areas along the Turkish-Iranian border (Shackleton 1997). Total number of mouflon ranges from 350 to 400 individuals for both Armenia and Nakhichevan. It inhabits gentle slopes and plateaus of the mountain–steppe, including subalpine and alpine belts at altitudes from 800 to 3,000m a.s.l. Biotopes avoid intersecting and rocky cliff relief. By its general appearance, structure of horns and sizes *O. ammon* is extremely variable. Therefore, this wild sheep is not comparable to its 25 subspecies throughout the world (Gegechkori 2007).

The Armenian mouflon is similar in size to the Cyprus (Cypriot) wild sheep; it belongs to the smallest subspecies – the adult male will stand 65 – 75 cm of the shoulder, weighing around 25 -40 kg – a quarter of the size of the Pamirian wild sheep (the biggest wild sheep of the world 'on the top of the world'), weighing – 200 -220 kg. The Armenian mouflon's horns are not large and are slightly crooked. Females are smaller than males. Mountain sheep are very slender and handsome, their back and sides have monotonic yellowish color or brownish-grey with slight nuances (Gegechkori 2007) (Figures 111,112).

Figure 111. Armenian mouflon (*Ovis ammon gmelini*). Armenian part of the middle flow of the Aras River (photo by K. Manvelian).

Figure 112. Achamenid Persian aristocracy during hunting the mouflon (6th-5th BC).

Mature adult males live separate from their female counterparts during the summer months. They are seasonal and will migrate over great distances. Until recently, the sheep would migrate as far as Iran and Turkey. A herd of sheep is represented as a group that consists of 10 -15 animals, rarely – far more individuals. Mating takes place from late November. The female gives birth in late April-May, bearing 1 to 3 lambs, usually – 2. The sheep feed on herbage, preferring cereals and rarely eat leaves and branches of some xerophyte bushes and trees (Gegechkori 2007).

The Armenian mouflon is categorized as Near Threatened by the IUCN Red List (Harris & Reading 2008) and has entered the Red Data Books of the USSR, Armenia and Azerbaijan (Khorozyan *et al.* 2009; Talibov *et al.* 2009).

Conclusion

- The biome of mountain-xerophyte vegetation (sclerophyllous scrub communities) is represented by hammada, phrygana, shibliak and tragacanthic cushion plants. They encompass as 'vegetation of skeleton soils'.
- Mountain-xerophyte vegetations occupy a large area of the Caucasus. They are dominant on different kinds of rocky desert highlands.
- With striking plant species composition, they should be highlighted as independent biomes and, hence, are characterized independently from the above characterized deserts and semi-deserts and arid open woodlands, in which some variants of mountain-xerophyte vegetations are mixed.
- As the most typical example this biome can be found in the extreme southern part of the South Caucasus (the middle reaches of the Aras River Valley). At present, this incredible area is epitomized as an exclave of the Near East's mountain xerophytes.

2.6. Temperate Broadleaf Deciduous (Nemoral) and Coniferous (Boreal) Forests

Indicator plant species of two biomes, - according to Kaliyakin and co-authors (2016), - were sorted by two complexes: 'nemoral' – associated with formation of deciduous broadleaf trees, and 'boreal' – associated to formation of dark coniferous (evergreen) trees.

2.6.1. Biome of Temperate Broadleaf Deciduous (Summer-Green) Forests

2.6.1.1. General Characteristics

2.6.1.1.1. Distribution

This biome is listed as Temperate Broadleaf Deciduous Forest (TBDF) in worldwide ecological literature sources. These forests are also called the Arcto-Tertiary Geoflora (ATG) (*sensu* Cheney (1959); see below) as well as 'nemoral forests' (*nemoralis* is derived from Latin *Nemus* (*nemoralis*) meaning of woods, grows, or open-glades). The TBDF biome covering over 20% of the landmass (Willis 2009). This biome stretches round the Northern Hemisphere and borders to the north of taiga and to the south either moist subtropics or in semi-dry regions to forestless grassland ecosystems (Walter 1974; Vtorov & Drozdov 2001; Gegechkori 2008, p.320).

The northward (upper/northern) distribution of temperate deciduous forests is limited by thermal conditions (growing season warmth and coldness and mean temperature for the coldest month). The southward (lower/southern) border of distribution of this biome is restricted by climatic continentality (the moisture gradient, indicated by annual average precipitation), which is primarily an abiotic condition (Mordkovich 2005). A transition intermediate area (ecotone) between two biomes is occupied by mixed (deciduous/coniferous) forests (Walter 1974).

The summer-green deciduous forest is found between 40^0 and 60^0 latitude north and south of the equator. However, the majority of this biome is almost entirely limited to the mentioned middle-latitude landmass of the Northern Hemisphere, and lies between 40^0 and 45^0 (Manual & Molles 1999). In North America the northern border of the biome is located at latitudes of 50^0, in eastern Asia it extends to about 54^0. The forest reaches north to latitudes of 50^0 to 60^0 N in the northeast of China (Manchuria), the Kamchatka Peninsula, and central Japan (Woodward 2003). The reason for its northern extension towards higher latitudes of Europe (58^0) is a powerful current in the Atlantic Ocean – the Gulf Stream. It keeps part of Europe quite warm at relatively high latitude. Western Europe is especially favored in this way due to maritime climates, i.e., largely to the moderating effects of the Atlantic Ocean (Walter 1974).

TBDF biome arises 'in the region of intermediate temperate, precipitation and latitude' (Reich & Fredich 2002).

Geographically isolated from each other deciduous forests are reduced to three segments: eastern Asia, North America, and Europe. Currently, remnants of the Arcto-Tertiary forests occur largely in two main refugia, in eastern Asia and eastern North America (i.e., broadly recognized today in the eastern North America-eastern Asia disjunction of closely related flora and fauna) (Walter 1974; Willis 2009).

The impact of global climate change has been noticeable in eastern Europe. In eastern-central Europe during the last full-glacial period the thermophilic woody species of native deciduous broadleaf forests largely disappeared. The forests of eastern Europe presently occur completely isolated from TBDF of the Caucasus and partially of the Crimean Mountains by another biome – steppes and forest-steppes (Walter 1974; Woodward 2003; Mordkovich 2005; Gegechkori 2008; Willis 2009).

South Caucasus two refugial areas - Colchis and Talysh (Euxine-Hyrcanian province) are floristically very rich with numerous endemics and relicts of the Arcto-Tertiary origin (Walter 1974; Browicz 1989).

A vast area of the TBDF is found in the Northern Hemisphere, although restricted places of TBDF are familiar to inhabitants of southern New Zealand, South America, southeastern Australia, and north and west Tasmania (Walter 1974; Willis 2009). These forests are among the major biomes on Earth (Woodward 2003; Gegechkori 2008, p.320). During the Early Cenozoic (the Paleocene-Eocene periods) thermal maximum up to the early/mid-Oligocene, temperate broadleaf forests, as extensive and uninterrupted in distribution, extended across the Northern Hemisphere's high latitudes (circumpolar distribution). But, according to global climate deterioration and geologic (tectonic) events, today this biome exhibits interrupted distribution. According to reduced precipitation (areas receiving at least 800-1000 mm of rainfall per annum) deciduous forest is replaced by grasslands (Vtorov & Drozdov 2001).

From the beginning (e.g., Carboniferous Period of the Paleozoic) the sustainable conditions of forest ecosystems are fundamental phenomena to keep many principal environmental and protective functions: support the balance between two gaseous elements (CO_2/O_2) in the atmosphere, reduction of gas emission impacts, influence local air circulation, control of wind velocity (the forest canopy slows down the wind) and protection of fresh water resources from time immemorial. Forests decrease the speed of water dispersion, reducing flooding. It is frequently the principal supply source for drinking water. Forests play a key role in maintaining soil protection and its fertility (protect against wind and water erosion). Forests, except for storing of water, evaporate huge amount of water, therefore play an important role in mitigation of the regional climate (Barnes *et al.* 1998; Gigauri 2000).

Owing to its size and structural diversity, this biome offers a greater variety of habitats and food resources (fruits, nutrient-rich foliage) to fauna than any other non-forest terrestrial biomes. Forests support about 65% of the world's terrestrial taxa and are characterized by the highest diversity of many taxonomic groups of wildlife (vertebrates, especially birds and invertebrates) (Barnes *et al.* 1998; Ravanbakhsh & Amini 2013).

Forest ecosystems have a perfect, protective effect on adjacent human settlements and the productivity of cropland; today forests are broadly recognized in their invaluable role in the restoration of degraded fragile ecosystems of highlands and drylands (open woodlands, artificial forest stands) used by human as transhuman (summer and winter) pasturelands. They are important in the recreational and other social (cultural, spiritual) functions. All these functions of forest ecosystems, directly or indirectly are of tremendous importance in drylands (open woodlands, dry and harsh plateaus), high mountainous forest (krummholtz), also coastal, riverine and wetland forests, Mediterranean maquis type forest-poor fragile environments (Gigauri 2000; Woodward 2003, *www.fao.org/docrep/.../w6251e06.ht...*).

National parks and strict reserves are key sites for conserving forest diversity (from genetics to landscape scale) and preserving endemic genetic resources. However, protected areas alone are insufficient to high level conservation of forest ecosystems and their biodiversity, because 92% of the world's forests occur outside protected areas (Lindenmayer *et al.* 2006).

According to Woodward (2003), the seasonality of life processes is one of the key characteristics of the Temperate Broadleaf Deciduous Forest, where the great majority of woody plants are represented by deciduous species, falling of the leaves in autumn, forming an abundant litter layer.

Box & Fujiwara (Eds.) 2015) indicate 'the most typical temperate climate is expressed as a humid subcontinental climate with warm summers, cold winters with significant frost, and four characteristic seasons of roughly equal length'. Taking into account the large latitudinal spectrum (40^0 up to 60^0), the mentioned authors differentiate temperate deciduous forests of the Northern Hemisphere into two subtypes based on the relative warmth or coldness of the summers and winters; a cool-temperate climate, with cooler summers, i.e., nemoral, after Walter (1974) and a warm-temperate climate, with milder winters: warmer summers, and a longer growing season (Walter 1974). In the Caucasus are represented the first subtypes of climate with corresponding biota. However, in the Caucasus Ecoregion's two refugia, significantly in the Guria-Adjarian and adjacent Colchis part of Colchis Triangle (i.e., Western Lesser Caucasus) the region with its clearly expressed oceanic climate differs from 'typical temperate', which permits evergreen broad-leaved woody flora to be seen as extraordinary undergrowth layer. In North Colchis (Novorossyisk—Tuapse area and Abkhazeti) with more dry summer than that in South Colchis, as an intermediate climatic region between nemoral to Mediterranean coastal area it produces 'sub-Mediterranean' climatic conditions.

In the Euro-Siberian region, the indicator species in the canopy layer of temperate climate's deciduous forests are the deciduous species of oaks (*Quercus* spp.) (Wulff 1944; Yurtsev 1966; Emeljanov 1999; Box & Fujiwara (Eds.) 2015).

The forest biome covers 18.52% of the total area of the Caucasus Ecoregion and about 17% of the Caucasus proper. It is primarily found between altitudes of 500-2,000 m and mainly grows on steep slopes. This biome is the most important factor for biodiversity conservation in the Ecoregion. Most forests are broad-leaved, but a number of different types (needle-leaved, narrow-leaved) are found. In Georgia the total area covered by forests is about 38%, while the forest area in Azerbaijan amounts to 14-15%, in the Northern Caucasus to 14.4%, in Armenia 10%. Mountain forests, which make up the greater part of the forest biome in the Caucasus Ecoregion play a critical role to keep a sustainable environment (Ecoregional Conservation Plan for the Caucasus (ECPC 2012).

Forest vegetation predominates over all other vegetation types in the Caucasus. In the Western Caucasus forests are found from sea level up to the alpine belt. In the Eastern Caucasus low precipitation, low air humidity and high air temperatures prevent forests from occurring at low altitudes and they are only found altitudinally above steppes, semi-desert and desert ecosystems. The Javakheti Plateau (South Transcaucasia) is completely devoid of forests. In that area not only climatic and edaphic factors, but particularly deforestation has led to a replacement of the forest coenoses by secondary vegetation types, especially in the high-mountains, where forests found their natural altitudinal boundary. The altitudinal distribution of forest types in the Caucasus strongly depends on complex climatic and orographic conditions (Grossheim 1948; Gulisashvili *et al.* 1975; Safarov & Olisaiev 1991).

According to ECPC (2012), today, only about 14% of forests are preserved in protected areas. In the long term, an additional 10% of the forests in the Ecoregion should be given protection equivalent to IUCN categories I-IV. This would bring the area of protected forests to nearly a quarter of the Ecoregion's forested area.

2.6.1.1.2. Climate

Abiotic (climatic) conditions determining temperate forest ecosystems include a marked seasonality, i.e., inhabitants of this biome experience all four seasons (Woodward 2003, Willis 2009). The TBDF biome is well supplied with rainfall. It has the second greatest amount of precipitation (after the tropical rainforest). Typically, precipitation is distributed throughout the year, and average annual precipitation ranges between 750 and 1250mm to over 3000 mm (Manuel & Molles 1999; Woodward 2003). Seasonal variation is determined more by temperature than precipitations. Temperate climates are not considered to have extremes of hot or cold. The average annual temperature is 10^0C. Temperature ranges between - 30^0 (winter) to 30^0C (summer). The winter is rather cold (but mild), with snow and lasts from three to four months. Summer is warm moist, rarely hot. Cool winter, reduced water loss encourages the trees drop their leaves each year (abscission). That is why such ecosystems are often termed 'summer-green' or 'cold-deciduous' forests. Remaining leafless in the winter, trees exhibit dormancy, when air and soil temperatures drop below freezing and bloom again in the spring. That is why temperate forests have less net productivity than tropical forests. In autumn, among various biochemical changes, plants withdraw (break down) chlorophyll from leaves, hence lose the green color, allowing other pigments a brief but beautiful display of amazing colors of the autumn forests, before the leaves are shed completely (Willis 2009; Taiz & Zeiger 2010).

This feature, the abscission of leaves, 'has also provided an important marker in the geological record for the onset of deciduousness of the global distribution of the first temperature forests' (Willis 2009).

2.6.1.1.3. Soils

Brown (chestnut-colored) forest soils (alfisols, according to American soil taxonomy) develop under the deciduous broadleaf forests. The soil in temperate deciduous forests tends to be deep and rich in organic and inorganic nutrients. This is because it receives a new blanket of leaves every autumn, hence a thick layer of leaf litter on forest floors returns nutrients to the soil creating abundant dark humus (i.e., the organic layer contains nutrient-rich leaf litter). Under deciduous forests develops one of the most fertile (with steppes chernozem) soils. Resembling a northern biome - taiga, the pedogenic- soil-forming process of local pedosphere is podsolization, but moderate by the warm wet summer's decomposition occurs quickly. Deposited in autumn, it begins to decay in spring (Walter 1974). Such a kind of soil (alfisols), similar to steppe, has the perfect features for cultivated crops. Many have been under continuous cultivation since the Neolithic (Walter, 1974). Toward lower, subtropical latitudes and in warmer, oceanic climate countries (regions), for example in Colchis, less fertile ultisols, red and yellow (*terra rosa*) dominate. A high content of oxidized iron compounds as found in moist subtropics, give soils a reddish or yellowish color (Vtorov & Drozdov 2001; Woodward 2003).

2.6.1.1.4. Vegetation

Because temperate forests have moderate annual rainfall and temperatures, the dominant plants are the deciduous trees ('deciduous' means 'falling off' and 'temperate' means 'mild'). Due to enough rainfall annually, the main characteristics include large broadleaf trees (species of elm (*Ulmus*), alder (*Alnus*), beech (*Fagus*), chestnut (*Castanea*), maple (*Acer*), horse chestnut (*Aesculus*), etc. The temperate deciduous forest contains a variety of trees and other plants. Deciduous trees are also called hardwoods. They are usually slow growing and long lived. The dominant plants of temperate forests are the deciduous trees. Tops of the mature trees in the first layer – crowns, interlock and form an enclosure – a canopy, or roof.

Temperate forests show less diversity of tree species than do tropical and subtropical wet forests. Mostly, TBDF forests are represented by 1-3 dominant species of trees, but in the richest communities of summer-green forests of eastern Asia and eastern North America the same summer-green forests can be dominated by any of 10-20 or more species (Vtorov & Drozdov 2001).

Usually, temperate forests, towards altitudinal limits, may go through mixed stands with some evergreen conifers (Mordkovich 2005). For example, species of *Pinus, Picea* and *Abies* in the Caucasus (Dolukhanov 2010). Many coniferous trees as a canopy component grow very tall, sometimes taller than deciduous trees (Gigauri 2000).

In the Southern Hemisphere this biome is occupied by Acacia (*Acacia* spp.), native genera of deciduous evergreen southern beeches (*Nothofagus* spp.), eucalypts (*Eucalyptus* spp.) and also various species of conifers (e.g., *Podocarpus, Araucaria, Dacrydium*) (Manual & Molles 1999, Woodward 2003, Willis 2009) (Figure 113).

Figure 113. The author at a stand of southern beech *Nothofagus gunii* (Nothofagaceae). Tasmania (2000).

Deciduous forests are multi-layered, strongly stratified biomes. Five layers are recognized: (1) The uppermost layer is the canopy composed of tall trees ranging in a mature (climax) 0.5 to 1.5 meters in diameter and from 30 to 60m in height during their life span. Forming a leafy canopy, this layer is blocking the sun and casting dappled shadows over the ground (e.g., *Fagus orientalis* in the Caucasus); (2) small or comparatively small trees (or saplings), sometimes determined as a sub-canopy (e.g., *Pyrus caucasica*); (3) below the canopy (and sub-canopy) grows a shorter, shade-tolerant, shrub layer. There are two types of this layer: 3(1) evergreens under forest in humid cold-temperate regions (e.g., *Rhododendron ponticum*), 3(2) deciduous understory (e.g., *Quercus pontica*). In temperate deciduous forests, trees and tall shrubs are used by woody and herb lianas. For example, the species of *Clematis* and *Vitis* genera; (4) rich herb layer of perennial forbs wildflowers, including annual plants. In climax deciduous forests light may be scarce, but the sunlight reaches the forest floor. Perennial and considerably annual herbs appear in early spring, when full sun can reach them (spring ephemerals). For example, violas, primroses and others quickly bloom as the trees' leafy canopy begins to form overhead; (5) ground (forest floor) layer, i.e., the lowest layer consists of small plants that live on top of the soil. It should be mentioned species of mosses, shade-loving ferns and lichens, which at the same time can grow on the trunks of trees. Lichens tend to be less common in the ground cover than in boreal forests (Calow (Ed.) 1999). In addition, the ground layer includes fungi, algae and other organisms (other than plants including lichens). The decaying leaf litter and also dead plant stems, small twigs and fallen fruits in the autumn all provide plenty of material for decomposers: soil bacteria, worms, the majority of millipedes, and fungi. Debris from decaying organisms, i.e., decomposing leaf litter adds valuable nutrients to the soil and helps it to retain moisture. Karpachevsky (2006) comments: 'the annual fall of plant litter, organic horizons (mort mass), on the soil surface ranges from 1 to 30 ton/ha'. Among saprophages, necrophages and detritophages, the main component of living forms dwelling in litter and organic horizons is earthworms. The burrowing and feeding activity of earthworms has numerous benefits. According to the type of feeding, pointed out by Karpachevsky (2006), earthworms (Oligochaeta, Lumbricidae) are divided into two ecological groups: feeding on leaf litter and detritus on the ground and proper soil forms (feeding on raw humus). These Oligochaetes may accelerate litter decomposition by 2-3 times and at the same time 'earthworms produce coprolites – excrements rich in organic compounds' (Karpachevski 2006). In redistribution of nutrient microorganisms and nitrogen compounds within the pedosphere should be underlined the certain role of protozoans, nematodes, beetle larvae, etc. Manuel and Molles (1999) indicate that most important consumers of all are fungi and bacteria. 'Fungi decompose wood and litter, and accelerate carbon cycling in ecosystems' (Karpachevsky 2006). Moles discard the soil material from deeper horizons up the soil surface: the discarded material (totaling to 50 ton/ha) becomes involved in soil formation. A molehill is often the only evidence to a moles underground existence in the forest. Wild boar actively reallocates soil material, mixing the upper and lower layers of soil down to the depth of 20 cm (Karpachevsky 2006 and references therein) (*http:/www.radford.edu/~swoodwar/CLASSES/GEOG 235/biomes/intro.html*), with some additions regarding the present book).

Species of *Fagus, Carpinus, Acer, Alnus*, but especially oaks (*Quercus* spp.) are recognized as the most dominant (indicator) tree species of TBDF (Wulff 1944).

Box. 9. Temperate Rain Forests

TBDF is sometimes referred to as a temperate rainforest because so much rainfall is received (Walter 1974). These forests, frequently mixed with stands of coniferous trees are found in temperate zones predominantly in areas with warm summers and cool winters. Many of these needleleaf evergreen trees are extremely large and long-lived vs the much shorter lifespans of deciduous trees (Alaback 1991 and references therein). The essential abiotic feature of such a forest is milder winters than those typical of the deciduous forests. Consequently, they are characterized by abundant (over 1400 mm) precipitation, mostly of rain rather than snow. Mean annual temperature is between 4 and 12^{0}C. In such climatic conditions lush vegetation occurs, but predominantly by the above mentioned humid-tolerant coniferous trees (temperate coniferous forests). These conifers differ from those of the boreal forests proper species of needleleaf trees (Woodward 2003).

Amongst other regions should be stressed the Pacific Northwest (coastal range of northern California, the USA). This area is home to over 20 species of native conifers growing from sea-level to the timberline. For example, one should mention redwood, cedar, spruce, pine, fir, yellow cypress, etc. (Elman & Salisbury 2009).

As it generally known, the tallest of all trees that grow in a temperate rainforest is the coast redwood and the giant sequoia found in California (the USA). *Sequoia sempervirens* grows 115m in height, 8.9 m in diameter, and *Sequoiadendron giganteum* – 94.8 m in height and 8.2 m in diameter (both from Cypressaceae family). They are the largest living things by volume (Van Gelder 1982 cited in Woodward 2003). In Walter (1974) humid-tolerant coniferous forests are discussed in boreal forests (Figure 114).

Temperate rainforest regions are represented in North America (except Pacific temperate rainforests, should be mentioned Appalachian temperate rainforests), South America (Valdivian and Magellanic temperate rainforests), Europe (Atlantic oakwood forests), Colchis (Georgia and adjacent area of Turkey), Iberian Peninsula (Spain), Asia (Caspian or Hyrcanian forests in Iran and Azerbaijan). Some regions of Japan, Taiwan, Korea, Russian Far East), Oceanic/southeastern Australia, Tasmania, Southern and New Zealand (Mordkovich 2005).

Figure 114. Extensive stand of the coastal redwood (*Sequoia sempervirens*). California (USA) (2000).

There are different definitions for temperate rainforest in scientific publications in general and Colchis among them frequently causes intense public debate on both–definition and geographic range. For example, humid temperate forests (Denk *et al.* 2001), warm-temperate deciduous forests (Box & Fujiwara (Eds.) 2015), temperate rainforests (Dolukhanov 1980; Dolukhanov & Nakhutsrishvili 2003; Nakhutsrishvili 2013), temperate broadleaf deciduous forests (TBDF) or the Arcto-Tertiary Geoflora (Vtorov & Drozdov 2001, Gegechkori 2008, 2011, 2012a, 2012b), and temperate deciduous angiosperm forest (Graham 1999).

2.6.1.1.5. Animals

Temperate forests, resembling the tropical rainforests are vertically stratified. Accordingly, the native trees support a wide range of plant eating animals and other wildlife. For example, the majority of resident and migratory birds, mammals and arthropods make use of all layers. Small mammals and birds use the tree canopy. So, a large majority of invertebrates and vertebrates show adaptations to an arboreal life (among others should be mentioned woodpeckers). Other mammals (deer, roe-deer, bear, wolf, lynx) live on the forest floor. For the most part of temperate regions some species of large mammals have been exterminated (e.g., wolf (*Canis lupus* - top predator in deciduous forests), forest ruminants (*Cervus elaphus, Bison bison, B. bonasus*); and the forest-dwelling tarpan (*Equus caballus*) is extinct in the wild (Woodward 2003 and references therein). Some small mammals burrow into the rich forest soil. Temperate forests are home for many kinds of amphibians and reptiles.

In autumn some species of birds and mammals store a rich supply of food in their burrows and dens, which includes nuts, pulpy fruits and fungi.

Animals in temperate deciduous forests have developed special ways of adaptation – a state of inactivity (deep sleep) allows them to conserve energy, and survive the winter with very little or no food and water. Many eat extra food in the fall while it is plentiful. It is stored as body fat to be used later for energy. Others grow fur to protect themselves from winter temperatures. The thicker pelage and thick layer of fat help insulate mammals in a severe winter. Birds face so many challenges to survive the winter. Many species solve the cold season problem by migrating long distance into warm countries. For protection many use camouflage (Vtorov & Drozdov 2001; Voronov *et al.* 1985; Mordkovich 2005).

2.6.1.1.6. A Brief History of Origin

The composition of various floristic assemblages (and fauna closely associated with them) can be understood as a process of migration and isolation, further taxonomic differentiation and evolution through time and space. This complex event in the first stage of dramatic changes tightly coincides with key transformation in climate, palaeogeography (increasing continental dispersion and uplift event) and the environment (biome) as a whole (Komarov (1908) cited in Budantsev 1992).

TBDF supposedly descendent from evergreen subtropical forests (Woodward 2003 and references cited in), forming mixed mesophytic forest that comprised a mix of deciduous and evergreen trees, also newly originated herbs (Tiffney 1985a cited in Milne & Abbott 2002; Basinger *et al.* 1994).

Temperate deciduous (predominantly angiosperms) forest ecosystems are one of the most recent groups to evolve (Early Cretaceous, ~130 mya; Willis 2009). For comparison it should be underlined the original events of the first tracheophytes - green vascular plants – first-spore-producing trees and first seed-producing trees. Evidence for the appearance of the first angiosperm woody flora occurred ~90 mya, i.e., Mid/Late Cretaceous). The earliest angiosperm trees to evolve belong to members of the Ulmaceae (e.g., genus *Ulmus*), Juglandaceae (e.g., *Juglans*), Fagaceae (e.g., *Castanea*), Betulaceae (e.g., *Betula*), Nothofagaceae (e.g., *Nothofagus*) (Willis 2009).

Regardless of trees of flowering plants they did not form a recognizable TBDF until about the Early Paleogene (the Early Paleocene, 60 mya) (Willis 2009). Willis (2009) indicates: 'From 60 mya (the Early Tertiary) a polar broad-leaved deciduous forest became established at approximately latitudes 70^0N and 70^0S and above in both the Northern and Southern Hemispheres. The nearest living equivalent to this polar broad-leaved deciduous forest in terms of composition is the temperate forest biome'.

Regarding the close relationship between Cenozoic plant communities of the Arctic region and the modern northern temperate deciduous flora of northern, Europe and East Asia, Grimsson *et al.* (2015) stated that 'Engler (1882) coined the term 'Arcto Tertiary Element'. After this Cheney (1959) extended this concept into a broader context, establishing the term 'Arcto-Teriary Geoflora'. Grimsson *et al.* (2015) concerning ATG the rejected Cheney's viewpoint stating: 'they must all have had a common area of origin during Cretaceous and Early Tertiary time at high northern latitudes, where we find their earliest record of their occurrence'.

According to the Tertiary plant biogeography during the Paleogene time (65-15mya) moderate warm climate-tolerant plant genera had spread all around the Northern Hemisphere, i.e., as a large circumpolar plant community, which, as mentioned, nowadays is known as the Arcto-Tertiary Geoflora (ATG). The ATG concept currently accepted that the former higher-latitude Northern Hemisphere (Arcto) vegetation existed in the above indicated time of the Cenozoic (Tertiary).

The Arcto-Tertiary Geoflora was first introduced by the paleobotanists J.S. Gardner & C. Ettinghausen in 1879 (*https://en.wikipedia.org/wiki/Arcto-Tertiary_Geoflora*). The Arcto-Tertiary Geoflora is understood to be the Paleocene Greenland and the Eocene-Oligocene Turgayan (in Russian terminology) flora sensu of A.N. Krystofovich (1955), and the Paleogene's eastern Asian and eastern North American flora (Wang 1961 cited in Upchurch & Wolfe 1987).

The concept was intended concerning the disjunctive distribution of closely related plant species between eastern Asian and western North American members of the Arcto-Tertiary floras. This phenomenon has long been recognized (*https://en.wikipedia.org/wiki(Arcto-Tertiary_Geoflora*). The Arcto-Tertiary Geoflora of temperate deciduous broad-leaved and coniferous genera appeared to have occupied extensively high northern latitudes of the Northern Hemisphere in the Cretaceous and Early Tertiary time. Many of the genera were found as far north as the Palaeo-Arctic Circle, where fossils were recorded. These fossils were associated with certain older Tertiary rocks of Greenland, Spitzbergen, Alaska and arctic Canada, which has spread southward through Eurasia and North America during the Cenozoic (Chaney, 1947; Upchurch & Wolfe 1987). 'The pollen suggests remarkable diversity of the family Fagaceae in the Early Cenozoic Greenland' (Grimsson *et al.* 2015). Discussion about high-latitude biota should be taken into consideration with the abiotic factor of the Arctic Circle (66^0 latitude). For example, as Askin & Spicer (1995) indicate 'light and particularly temperature are the principle controlling factors for polar vegetation, whereas precipitation is generally the prime factor at lower latitudes' (Ziegler 1990 cited in Askin & Spicer 1995).

According to Akhmetiev (2010), the Early Eocene Raichikha Formation (Kamchatka Peninsula, Russian Far East), when a warmer episode took place has yielded rich fossil records of thermophilic (polar Arcto-Tertiary) plants (Sabal, Myrtaceae, and Lauraceae). This flora, probably of northward migration reached and existed near the recent polar circle. 'It is more similar, -comments Akhmetiev, - to the recent flora of south-west and south China'.

The Paleocene wetlands and peat swamps of the Tsagayan Formation (occurring in the same Kamchatka Peninsula) were dominated by woody plant communities consisting of *Taxodium, Nyssa* and *Mesocyparis* swamp forests and herbaceous plant communities including *Equisetum*, Ferns (*Osmunda, Woodwardia*). The Kivda wetlands were the home of *Taxodium, Metasequoia, Pseudolarix, Myrica,* etc. According to the mentioned plant communities 'the climate was humid warm-temperate to temperate' (Kodrul 2009).

Early Tertiary fossil plants belonging to other polar Arcto-Tertiary vegetation communities were found in the northernmost part of the Canadian Arctic Archipelago (Ellesmere and Axel Heiberg islands). This site occurs far more north latitudinal areas (within the Arctic Circle, 75-80^0N) (Basinger *et al.* 1994).

Basinger and co-authors (1994) pointed out that Early Tertiary forests of the Canadian Arctic Archipelago's frigid zone experienced the seasonal stress of polar light cycles – prolonged periods of continuous daylight in the summer and continuous darkness in winter. Despite this, the primarily deciduous vegetation, i.e., greenhouse world was clearly dominant.

According to the opposite point of view Wolfe (1977): "The existence of a broad-leaved evergreen forest in Alaska, such as that of the early Ravenian (late middle Eocene), indicates that extended periods of darkness may not have existed all that time. The Alaskan Paleogene flora as well as those from other regions, indicate that the Earth's axis of rotation may not have had as great an inclination in the past".

The ATG afterwards migrated southward to middle latitudes in Eurasia and North America. Over millions of years this undoubted representative of deciduous plants and some warmth-demanding conifers, i.e., mixed conifer-broad-leafed assemblages have been distributed across terrain toward southern latitudes of the Northern Hemisphere (Sinitsin 1965).

On the basis of current molecular and fossil records, according to dispersal-vicariance analysis (Milne & Abbot 2002 and references therein), such dispersal events took place via both the Atlantic and Pacific oceans. Few if any of them appear to have been capable of ocean crossings including the terrestrial animals (Flerov *et al.* 1974). Although, members of the Arcto-Tertiary forests flora, via stepping stones managed to cross the North Atlantic Land Bridge throughout much of the Tertiary (40-10 mya). In the Neogene between two continents the North Atlantic Ocean barrier, biogeographical filter developed. Today, this migration pattern is broadly known as a transatlantic Tertiary relict disjunct event.

Regarding contacts across Beringia, it may have occurred between 5 and 10 mya through far larger and complete terrain was the Bering Land Bridge 'before it was severed after about 5 mya, i.e., intercontinental change of living organisms was prevented by abiotic factors' (Sinistin 1965). An especially close relationship between eastern Asia and western North American members of the Arcto-Tertiary forests floras has been largely recognized (Xiang *et al.* 2004). 'During such movements in space and time, many genera became extinct and others became restricted to central and southern China and western and south-western North America. (Sakai 1971 and references therein)

Controversial issues are dedicated to the center of origin of the ATG. According to botanical-geographical literature (Wulff 1944; Lavrenko 1951; Takhtajan 1969 and others) it was considered that the center of origin and distribution of Temperate Deciduous Flora was East Asia. A contrary version was accepted by other scholars. Namely, Meyen (1981) suggests that East Asia is not the center of origin of the initial Arcto-Tertiary flora but actually it is a survival and modern diversity region of relict genera.

Other palaeontological studies provide an important framework to consider the former pathway of origin, and ecology (i.e., climatic requirement) of the Paleogene ATG relicts. Axelrod (1965, 1966, 1968) suggests that similar to later original boreal flora, nemoral communities of vegetation might have been present in high altitudinal zones of the Rocky Mountains of the Eocene.

Following this (mountainous origin) hypothesis, according to Fotjanova (1984), pre-Turgayan flora from western Kamchatka and eastern Sakhalin in the beginning of the Late Eocene, after the Paleocene-Eocene boundary's temperature maxima, the process of climatic cooling started. Climate deterioration as a key abiotic factor has triggered the creation of floristic assemblages of the Eocene of mentioned regions of East Asia. The Arcto-boreal flora almost completely disappeared and warm-temperate broadleaf deciduous flora and thermophilic conifers appeared. It should be mentioned genera of Fagaceae (*Fagus, Quercus, Castanea, Fagopsis*), species of *Ulmus, Acer* (*A. arcticum*), etc. and exotic conifers (*Pseudotsuga, Cryptomeria, Cunninghamia, Dacrydium*, etc., appeared for the first time. This complex of flora, it seems, began to originate as a mountain version of native biota across the time of climatic optimum. Later, when global cooling accelerated by early Late Eocene, the first mixed (broadleaf and needle leaf) temperate flora became widely distributed, and for the first time occupied large areas (Fotjanova 1984 cited in Gegechkori 1985, p. 90).

Long ago (up to the Late Miocene) Tertiary taxa became more widely distributed across the North Hemisphere, forming Mixed Mesophytic Forest communities (Xiang *et al.* 1998 and references therein). However, the Late Miocene temperature decline synchronized with the growth of Antarctic Ice-Sheets and closure of the epicontinental Turgai Strait (or West Siberian Sea) (Akhmetiev & Beniamovski 2009), led to fragmentation and modification of warmth-demanding temperate forests. Global climate deterioration of the Early Neogene led to the disappearance of warmth-demanding plants from mid-altitude forests (Woodward 2003).

In high latitude, once continuously distributed, the Arcto-Tertiary Geoflora in the Late Cenozoic was restricted and today remnants exist within three main refugial regions of the North Hemisphere: in the eastern United States, eastern Asia, and western Eurasia (with fewer surviving genera in the latter region) (Whittaker 1961; Woodward 2003).

The Late Tertiary orogenies and further (the Plio-Pleistocene) climate oscillations eliminated whole families of plants of the temperate deciduous angiosperm forest in Europe exposed to high level of taxonomic mortality in many regions of western Eurasia. But some genera survived the Quaternary glaciations in pockets (small refugia) with favorable conditions of the continents situated north of the equator. It should be mentioned the Russian Far East and Colchis and Hyrcania in the Caucasus Ecoregion with their outstanding endemics of the Arcto-Tertiary Geoflora (e.g., *Parrotia persica* in Hyrcan, *Betula megrelica* in Colchis, *Rhododendron dahuricum* in the Russian Far East).

2.6.1.1.7. Human Effects

Clean air, a rich variety of plants and animals, fertile soil and pure water have attracted humans since time immemorial. Clearly the fertile soils make this biome very cultivable for agriculture. Temperate forest biome is one of the most productive on the globe in terms of human occupation (Willis 2009). Therefore, up to today TBDF of all continents became a center of agricultural development. Forest resources support the energy needs of an ever-growing human population, especially in the temperate forest regions of Europe, Asia (Japan, China) and eastern North America. The history of the forests of these regions is characterized by heavy exploitation by man. Agriculture was practiced via deforestation and burning. This is a history of humans' effect on the temperate forests during the pre-industrial period, i.e., Neolithic Age – the period of man's sedentary settlement. Temperate deciduous forest provided a vital resource (wild animals and natural plant products - timber, charcoal, and non-timber forest products which were harvested). It is used as pasture for domestic animals and for recreation. Dramatically transformation of nemoral forest ecosystems has been ongoing for millennia: since the Agricultural Revolution in the Neolithic era (i.e., the period of sedentary settlement during five thousand years in Eurasia and during some centuries in North America). But biome's and within them habitat destruction has accelerate sharply over the last two-three centuries. After the Industrial Revolution (1760-1830) most of the original forests have been felled. Nearly 2 million tons of timber are cut from the forests each year (*ic.galegroup.com/.../ReferenceDetailsWindow?*). The global area of forests has been reduced by roughly half over the past three centuries (Laurance 2010). Hence, all kinds of mentioned human activities have caused large alternations of forest composition, structure and diversity and converted much of the forest to farmland, pasture, and human settlements. Hence, few remnants of primeval forest have survived (Reich & Frelich 2002). 'Thus, temperate forests have experienced more severe impacts over broader areas than any other large forest biome' (Willis 1989 cited in Reich & Frelich 2002; Fischer *et al.* 2013).

In 1993, by environmental and human right campaigners, forest defender organizations began to recognize an increasing mentality manifested following: 'support environmentally appropriate, socially beneficial and economically viable management of the world's forests'. Countries which have temperate forested areas, in their State Legislations are (or will be) focused on certain issues (including the Forest Code) the following truth: 'It is now vital that the criteria for sustainable forestry management that have been suggested are implemented in order for the forests of today to be available for the enjoyment and use of future generations' (Sustainable Management of Temperate Forests: *www. curruculimpress.co.uk*).

2.6.1.2. TBDF of Lowlands, Foothills and Lower Mountain Zones Enriched with Floristic Elements of the ATG, a Short Review

2.6.1.2.1. Global Importance of Lowland and Riverside (Riparian[3]) Forests of the Colchis Region

2.6.1.2.1.1. Lowland and Riverside (Riparian) Forests

2.6.1.2.1.1.1. General Characteristics

In the lowlands of the Caucasus, deciduous broad-leaf forests are typical for swamps, floodplains and riversides, where the availability of river and groundwater allow the development of forest vegetation (Grossheim 1948). Lowland forests are characterized by *Quercus robur, Q. robur pedunculiflora, Ulmus minor* and *Morus alba* with *Cornus mas* and *Crataegus* spp. as undergrowth. Typical for these forests, especially their margins, are lianas, including *Clematis vitalba, Humulus lupulus, Periploca graeca, Smilax excelsa* (Gulisashvili *et al.* 1975).

Some authors have considered riverside forests as alluvial forests. As Sharitz & Mitsch (1993) stated, 'It is a hardwood forest found on low levels, ridges and terraces within the floodplains of streams and rivers. This type of forest grows in areas that are slightly elevated above floodplain swamp and are usually flooded for a portion of the growing seasons. It is greatly influenced by the ongoing disturbance created by a fluctuating river bed which is both eroding and depositing substrates.

Riverside forests are developed both in the forest zone and areas outside the forests. In these places it grows as a narrow strip along the river-bed. In the forest zone the riparian forest is less distributed from the synusia of the bordering forest. Despite this, it has a characteristic complex of species (Gulisashvili *et al.* 1975 and references therein; Akhalkatsi & Tarkhnishvili 2012).

---

[3] Forested riparian areas have also been called 'streamside forests', 'river woods', and 'wet woods'. Natural riparian forests are recognized as a buffer zone, which is a stable area for moist tolerant vegetations and specific wildlife, ly in arid biomes (Palone & Todd (Eds.) (1998) and *Vitis vinifera silvestris* (Gulisashvili 1964). Riparian forests are considered to be azonal and will be treated separately.

2.6.1.2.1.1.1.1. Lowland and riverside forests of Trancaucasia

2.6.1.2.1.1.1.1.1. Lowland and riverside forests of NW Transcaucasia (Imereti Lowland), Russia

Imereti lowland is located on the southern slope of the Greater Caucasus mountain range and is bordered by the rivers Mzymta and Psou and typical mountain streams (Adler district of Sochi, Krasnodar Kray, 30km west from Sochi) (therefore, it should be differentiated the Imereti district of the Republic of Georgia from those of the Imereti Lowland of South Russia).

A considerable part of the lowland forests of NW Trancaucasia has been destroyed by timber logging. Remnants of these forests consist of penduculate oak (*Quercus robur*) with *Carpinus betulus*, *Acer campestre, Fraxinus excelsior, Malus orientalis, Pyrus caucasica* and a well-developed understory (e.g., *Prunus laurocerasus*) (Figure 115) and herb layer. Various types of *Q. robur* forests have been distinguished on the basis of associated species, such as those with *Corylus avellana*, *Molinia caerulea*, *Fraxinus excelsior*, *Euphorbia amygdaloides* and *Aristolochia clematitis* (Grudzinskaia 1953; Elagin 1953; Zelenko 1969). The remnants of riverside forest are represented by *Populus alba* and willow (*Salix*) species.

The Imereti Lowland's wetlands display a diverse vegetation with many endemic species and relicts from the glacial period. Including the Mediterranean, Mediterranean mountainous and Boreal species, this site's flora was enriched by adventive species (Akatov *et al.* 2009).

Since the beginning of historic time the Imereti Lowland was completely covered by luxuriant forests and mire vegetation. The forests were Colchic type consisting of many mesophylous arboreal species. They play a considerable part in landscape formation in lowland. The dominating position was occupied by alder (*Alnus barbata*) (Akatov *et al.* 2009).

Currently most of the Imereti Valley is occupied by farmland, the remaining natural vegetation close to a primal one which was represented here till the end of 19th century, is presented only in specific niches — depressed areas of relief.

Figure 115. *Prunus laurocerasus* blooms in spring.

Most articles in the first half of 20th century were dedicated to a remarkable area populated by royal fern (*Osmunda regalis*) which occurred abundantly in the ditches of the peatland. (Figure 116).

Figure 116. Royal fern (*Osmunda regalis*). A fossil rhizome of this genus was discovered from the Early Jurassic (1978).

From the Colchis type of vegetation in the same book are included: *Pterocarya fraxinifolia, Diospyros lotus* (very rare mostly in cultural landscapes), *Buxus colchica* (only in graveyards), *Ostrya carpiniflolia, Dioscorea caucasica* and *Staphylea colchica* (Tilba *et al.* 2006; Akatov *et al.* 2009) (Figures 117, 118).

Figure 117. Caucasian dioscorea (*Dioscorea caucasica*) (Dioscoreaceae), perennial herb liana, relict and endemic of north-western Transcaucasia (Abkhazeti, Adler district [Russia]); grown in the oak-hornbeem forests of the lower and middle belts of the Greater Caucasus (picture: Tbilisi Botanical Garden, 2015).

Figure 118. Caucasian bladdernut (*Staphylea colchica*) (Staphyleaceae); shrub, native of the Colchis region and Balkan Peninsula, relict of the Tertiary Period. Cooked flowers are common dish in Georgia (1986).

The fauna of invertebrates of Imereti Lowland includes many endemic, relict and other rare (for Russia) species (Gazaryan & Shevchenko 2014).

2.6.1.2.1.1.1.1.2. Biota of the Colchis wetlands: (focused attention on Western Georgia, Kolkheti National Park)

Characterization of riparian and humid temperate forests of West Georgia (Colchis Lowland) are given by Grossheim (1948), Kolakowski (1961), Gulisashvili *et al.* (1975), Ketsckhoveli (1960), more thoroughly – by Flerow (1951) (Figure 119).

Figure 119. Panoramic view of the Kolkheti NP, Ispani 2 wetland near Kobuleti town, Adjara (2008).

In the riparian forest of West Transcaucasia primary trees found include Arcto-Tertiary relict wing–nut (*Pterocaria fraxinicolia*), common alder (*Alnus barbata*), which is more typical in the marshy forest, white alder (*A. incana*) in mountains' large rivers and their tributaries, willow (*Salix* spp.), aspen (*Populus hybrida*) and persimmon (*Diospyros lotus*).

Using the example of riparian forests of the Paleostomi Lake and the Pichora River (which are seasonally flooded) consisting of relict trees: *Pterocaria fraxinicolia*, *Quercus hartwissiana, Q. dschorochensis*, *Fraxinus excelsior, Carpinus betulus* and *Ulmus carpinifolia*. Understory is represented by species of *Rhododendron, Ruscus, Vaccinium*, etc. This riparian forest is outstanding with its very rich lianas (species of *Smilax, Clematis, Hedera, Periploca*, etc. (Ketskhoveli 1960).

The Colchis Lowland Rioni River and especially its delta ae frequently made up of silt. The banks are covered with shrubs and small trees which consist of *Hippophea rhamnoides, Salix* spp., *Tamarix ramosissima, Rubus* spp., and many groundcover-herbal plants (Akhalkatsi & Tarkhnishvili 2012).

In some areas of the Greater Caucasus and the Lesser Caucasus the riparian forest areas deeply intrude into the mountain gorges.

Figure 120. TBDF of Colchis lowland, Adjarian part of the Kolkheti NP. *Quercus hartwissiana* in the overstory. In the foreground: A noxious invasive alien species (*Amorpha fruticosa*) (2015).

The Colchis Lowland in the historical region of Colchis in western Georgia is a young accumulative plain formed during the Pliocene-Quaternary. In this geological time the whole region experienced uplift (Gulisashvili 1964 and references therein). Many lagoons which lie on a coastal plain on the Black Sea developed into mires (Joosten *et al.* 2003). There is a dense network of rivers (mainly the tributaries of the Rioni River), lakes (the most famous one is the Paleostomi or Paliastomi lake) (Figure 120).

The Colchis mires display a diverse vegetation with many endemic and endemic-relict species. They are characterized by various origin roots: the Tertiary, last Glacial Period (Würm) and its maximum (LGM), sub-Mediterranean and boreal (Joosten *et al.* 2003). The largest and richest with vegetations are found in the down to sea level in the middle part of the Colchis

Lowland, encompassing the Odishi-Guria Lowland (it belongs to three provinces of Western Georgia: Adjara, Guria, Samegrelo).

In Guria and Samegrelo considerable attention of scholars is attracted to the outstanding Ispani 2 mires in Adjara — Ispani 1 and particularly Ispani 2 (near the town of Kobuleti, a famous marine resort). Four species of *Sphagnum* peatlands (*S. austrinii, S. papillosum, S. palustre, S. capillifolium*) make this landscape a living 'carpet'. In both peatlands other dominated communities are represented by sedges (*Carex* spp.), among other components of native vegetations should be mentioned purple moor grass (*Molinia litoralis*) and saw grass (*Cladium mariscus*). Many of them participate to build a domed-shape mire with clearly expressed peat cupolas known as a percolation bog (sensu Joosten *et al.* 2003). Local wetlands are characterized by the presence of the Tertiary relict – *Kosteletzkya pentacarpa*) (Joosten *et al.* 2003).

Mires of the Colchis are filled with reeds (*Phragmites australis*), and comparative poorly represented royal fern (*Osmunda regalis*). Boreal elements of mires flora include *Drosera rotundifolia* and *Calluna vulgaris*. A low herb layer composed of thickets dominated by *Polygonum thunbergii*, species of *Juncus, Iris pseudacorus* and others. Mainly on the edges of marshes grow *Alnus barbata* and *Salix rubens* (Denk *et al.* 2001) while close to the mire margin shrublands (*Frangula alnus, Rhododendron luteum, Rh. ponticum*) occur.

The constant species of riparian communities, which are rather rich with relict and endemic species, include *Quercus hartwissiana, Q. chorochensis, Pterocarya fraxinicolia, Fraxinus excelsior, Carpinus betulus* and *Ulmus carpinifolia* (Ketskhoveli 1960; Gagnidze & Davitadze 2000). Lianas are abundant (*Smilax excelsa, Periploca graeca, Hedera helix, H. colchica*), far more seldomly occurs the wild grapevine (*Vitis vinifera sylvestris*). *Amorpha fruticosa* (particularly) and *Andropogon capillipes* are noxious invasive species (Akhalkatsi & Tarkhnishvili 2012).

Swamp forests of Colchis Lowland are represented by approximately 160 plant species (Kolakovski 1961).

Marshes and their azonal plant communities are largely distributed in western Georgia and involve all altitudes from sea level up to the high mountainous area (Kimeridze 1966).

Wetlands of Colchis provide numerous habitats for fish and wildlife, among which are large or comparatively large mammals: *Canis aureus, Vulpes vulpes alticola, Felis silvestris, Lutra lutra, Sus scrofa* and *Capreolus capreolus*. They support large game birds - *Phasianus colchicus colchicus*; amphibians: *Hyla arborea, Pelophylax ridibundus, Triturus vulgaris* and *T. vittatus*; reptiles: *Natrix megalocephala, N. tesselata* and *Emys orbicularis* (Denk *et al.* 2001; Gokhelashvili & Kikodze 2006). In almost all pristine riparian and flood forests occur an immense variety of species of insects and other invertebrates; local wetland ecosystem habitats are impressive with countless large webs of the garden spider (*Argiope aurantia*), etc.

According to the Ramsar Convention (Georgia joined this Convention in 1996), Colchis wetlands play an important role supporting the conservation of hundreds of African-Eurasian migratory waterbirds (species of geese, ducks, swans, etc.) and birds of prey (species of hawks, eagles, buzzards, kites, kestrels, etc.).

For protection of Colchis unique biota with its incredible wetland biodiversity and food web, in 1947 a State Nature Reserve was established. Later, in 1998 came Kolkheti National Park (Gokhelashvili & Kikodze 2006). Consequently, today a wide range of environmental problems are solved – peat mining (as fertilizer for farmlands), *Sphagnum* peatlands drainage (for cultivation of citrus and tea plantations) (Georgia produced almost the entire citrus and tea crop for the entire Soviet Union) (Gegechkori 2008).

2.6.1.2.1.1.1.1.3. Forests enriched with the tertiary relict trees survived in the Imereti Lowlands (West Georgia)

Imeretian oak *Quercus robur imeretina* (*Q. imeretina*)[4] is a dominant species in large areas of less flooded or well drained Colchis lowland and foothills.

Imeretian Oak forests are confined to the left banks of the Rioni and Kvirila Rivers (Western Georgia, Imereti Region, from 80 up to 120-200m a.s.l.). This site belongs to one of the rarest refugial areas. Thus, this is the only place where forest building, two relict species – *Q. r. imeretina* and *Zelkova carpinicolia* are preserved as its almost only communities. In prehistoric times, these forests found in relatively dry habitats, were widely distributed in Colchis Lowland. Its current, isolated and disjunct population of mentioned relict species was affected by numerous agents (abiotic factors) such as geology, regional climate, soil types, historical biogeographical processes (significantly Messinian Salinity Crisis) and human activity (Bakhsoliani 1972 in Dolukhanov 2010).

Sometimes Imeretian oak forms monodominant forests, but occurring in mixed phytocoenosis, except the zelkova tree, mingled with other deciduous taxa such as *Carpinus betulus, Fraxinus excelsior* and, listed in the Red Data Book of Georgia, *Pterocarya fraxinicolia, Diospyros lotus, Quercus hartvissiana, Ruscus colchicus* and *Buxus colchica.* Deciduous species of sparse understory are represented by yellow azalea, medlar, rose, hawthorn, Colchis bladdernut (capers), etc. (Dolukhanov 2010).

Figure 121. Colchis lowland's forests consist of *Quercus robur imeretina* mixed with *Zelkova carpinifolia*, and *Ruscus ponticus* in the underground layer. Adjameti Reserve (1984).

In West Georgia forests of Imeretian oak with considerable admixture of the zelkova from 1946 are preserved in Adjameti Reserve (an area of 4,848ha, of which 4,738 ha is covered with forests). There are two major massifs in the reserve: an Adjameti oak forest (3,742 ha, between

[4] Some authors treated this subspecies as species – Q. imeretina (e.g., Firsov 1998) (Red List Category: Vulnerable).

the left tributaries of the river Rioni – Kvirila and Khanistskali Rivers) and Vartsikhe forested sector (the left bank of Khanistskali River (1,106ha) (Matikashvili 1948; Makhatadze & Urushadze 1962, 1964; Bakhsoliani 1972 in Dolukhanov 2010).

The pure stand of zelkova in reserve covers 1,5ha. The diameter of some trees reaches 105cm, the age of oaks in the forest's ranges between 120-200 (Kozlovski & Mamisashvili 1969 in Bannikov (Ed.) 1969; Gagnidze & Davitadze 2000; Gokhelashvili & Kikodze 2007).

Q.r. imeretina requires rather abundant precipitation (mean annual rainfall 1200-1300 mm) and an underground water supply.

Five types of *Q. r. imeretina* forests have been distinguished (Gulisashvili *et al.* 1975):

1. *Q. r. imeretina* forests with *Carpinus betulus, Quercus petrae iberica* and *Zelkova carpinifolia*. In contrast to the rich underwood, the herb layer is poorly developed (with *Viola, Primula, Fragaria, Poa*, etc.);
2. *Q. r. imeretina* forests with *Hypericum xylosteifolium* (= *H. inodorum*) in the ground layer. In this type *Quercus* is the leading species, other tree species hardly occur;
3. *Q. r. imeretina* forests with a well-developed moss cover and *Crataegus lagenaria* and *Ruscus ponticus* in the undergrowth;
4. *Q. r. imeretina* forests with grasses. Underwood is absent, the herb layer is rich in graminoids like *Poa nemoralis, Agrostis vulgaris, Briza media, Dactylis glomerata* and *Carex* sp.;
5. *Q. r. imeretina* forests with *Ruscus ponticus* in the ground layer. In this type the herb layer is very poorly developed (Figure 121).

2.6.1.2.1.1.1.1.4. Colchis Proper Plant Communities Exclave in Eastern Ciscaucasia (Samur River Delta)

In eastern ward Ciscaucasia, in the Caspian Sea coastal area, on the territory of the Samur River Delta is represented a unique, for this region, lowland and riparian forest enriched with representatives of relict Arcto-Tertiary flora. The rare floristic community is located in the Magaramkent and Derbent districts of Dagestan. Today (from 2009) on this territory is established the State Nature Zakaznik of Federal (Russian Federation) importance. The Zakaznik covers 11,200 ha. The territory includes floodplain deciduous liana forests of the Samur River Delta, dissected by numerous river arms and rivulets, also coastal lakes, lagoons, beaches and a 500–metre zone of the Caspian Sea shallows. Flora of Zakaznik has more than 1,000 species, among them many endemic and as mentioned relict forms, as well as rare and endangered species entered into the Red Data Books of Russia and Dagestan. For example, *Pterocaria fraxinicolia, Atropa belladona, Hedera pastuchowii, Ophrys sphegodes mammosa*, etc. In general, Samur forests represent a complex of poplar, alder, oak and hornbeam forests. Zakaznik holds wild boar, roe deer, jungle cat, wildcat, and other mammalians, it is stopovers and wintering sites of more than 300 bird species. There are recorded 51 rare and endangered species listed in the Red Data Books of Dagestan, Russia, and IUCN Red List (flamingo, Dalmatian pelican, the black stork, white-tailed eagle, purple swamp hen, etc.). The Samur River Mouth is recognized to be an Important Bird Area of global significance (Sokolov & Syroechkovski (Ed.) 1990).

Despite the official 'Task Force for the Protection of Federal Zakazniks' main protection goal there are problems. Due to general trends of resettlement patterns in Dagestan. In the southern part of the republic, in settlements along the Lower Samur River, and in particular in two sites mentioned (Magaramkent and Derbent), the focus is on immigration with respect to favorable local climatic conditions that are conducive to high agriculture yields (within the borders of Zakaznik there are occur 7 villages). Therefore, resettlement has accelerated the logging of unique liana forests of the Samur River Delta. According to Eldarov *et al.* (2007), under present rates of deforestation, that vegetation community will disappear in another 10 to 15 years.

However, despite the main ecological and environment protection goals of the officials of the Zakaznik, the main aim is supported by the Caspian Coastal Greenaway project (Boskhomdziev 2010).

2.6.1.2.1.2. The Peculiarities of Colchis Lower-Mountain, Broad-Leaved Forests

Mixed forests of Colchis occupy a considerable part of West Transcaucasia, namely in Colchis Lowland, foothills and lower mountain zone of the Greater Caucasus and the Adjaro-Imeretian Ridge (Gulisashvili *et al.* 1975).

Its altitudinal distribution ranges from 50m to 500-600 m a.s.l. In upper elevations it gradually changes to deciduous forests, poorer with relict species in the first stratum of canopy, and higher up replaced by coniferous forests (Ketskhoveli 1960, Gulisashvili 1964).

Soils are represented by yellow, brown and red colors with a well-developed profile (Sabashvili 1936; Urushadze & Blum 2014).

Grossheim (1948) indicated that in polydominant forests of Colchis, major forests building species include 5 species: chestnut (*Castanea sativa*), oak (*Quercus hartwissiana*), beech (*Fagus orientalis*), hornbeam (*Carpinus betulus*), and alder (*Alnus barbata*).

According to the mentioned associations, Albov (1899) proposed the following definition: "Colchic mixed forests of broadleaved deciduous species". Some Caucasian botanists suppose that the same oligo-and polydominant forests belong to mixed (moist) subtropics (Grossheim 1948, Gulisashvili 1964, Gulisashvili *et al.* 1975).

Besides the above-mentioned leading representatives of forests' first stratum, the local canopy consists of the following subdominants: *Ulmus minor, U. glabra* (=*U. elliptica*), *Diospyros lotus, Fraxinus excelsior, Tilia begoniifolia, Pterocarya fraxinicolia, Pyrus caucasica, Malus orientalis, Acer campestre* and some species of *Salix*, etc. (Figures 123,124,125). On limestones appears *Laurus nobilis*, in the eastern part of Colchis occurs *Zelkova carpinifolia*, sometimes together with deciduous building form of trees, poor stands of a yew tree grow frequently (*Taxus baccata*) (Grossheim 1948).

As some botanists (Maleev 1940, 1941, Grossheim 1948) underlined, lianas are not a characteristic vegetation form for the virgin forests of Colchis and occur mainly in forest's margins. The fact that in Colchic forests they are met often, is the secondary occurrence. this is after destruction of the mentioned forests in vast areas by means of human agricultural activities from prehistoric times. West Georgia's lowlands and foothills are covered nowadays with exotic (for native environments) plant cultures, which are manifested by the following introduced plants: tea, citrus species, tung tree, maize, eucalyptus species, palm species, cryptomeria (*Cryptomeria japonica*), etc.

Ketschoveli (1960) emphasized that in the first half of 20th century an intensive introduction in Colchis of many subtropical plants made a profound reconstruction of primary biocoenosis. The latter one remained in a very limited area, for example, in Lake Paleastomi environs. Fragments of these communities are now restricted to very abandoned arable soils, thanks to abundant precipitation, triggering formations of secondary forests. Hence, communities with alder and hornbeams, in most situations are the derivatives of primary Colchian forests. Just these formations are so rich with secondary vines.

The following vines may be mentioned: *Hedera colchica, H. helix, Dioscorea caucasica* (Colchic endemics), *Diospyros communis, Smilax excelsa, Humulus lupulus, Clematis vitalba, C. viticella, Periploca graeca* and *Vitis vinifera silvestris*. Adventive lianas, such are *Pueraria stricta* and *Lonicera japonica*, display a successful competition capacity concerning the local, aboriginal lianas (Figure 122).

To the commonest representative of evergreen undergrowth belongs *Rhododendron ponticum*, which grows as a tall shrub or small tree. It participates in all types of Colchic forests and is spread in mountains higher than communities with canopy forests. The next characteristic species is *Prunus laurocerasus*, which makes the lower stratum rather than previous species and forms very dense groves. One of the most peculiar arboreal plants of Colchis is its endemic box-tree – *Buxus colchica*. Highly valued for timber, as in other regions of the Western Caucasus, it underwent intensive cutting considerably in XIX century.

Figure 122. Kudzu (*Pueraria stricta* (=*P. hirsuta*) (Fabaceae), climbing vine, another noxious weed, invasive species, a native of East and South-East Asia (2011).

The other evergreen shrubs are represented by *Daphne pontica*, *Ruscus colchicus, R. ponticus, Ilex colchica* and *Osmanthus* (*Phillyrea*) *decorus*. The latter one in the South Caucasus remained only in Adjaria.

It should be pointed out that in contrast to the deciduous component of understory, well developed and lush evergreen shrubs obviously prevailed in the Colchis region. In well preserved Colchis underwood, a dense shrub thicket causes a strong shading of ground, owing to which the herbage layer is poor in species.

The most characteristic feature of the local ground layer is a wide-spread occurrence of ferns, which, on the contrary, can rarely be met in forest's grasses. Herb vegetation of Colchis is very poor of endemic species (Grossheim 1948; Gulisashvili *et al.* 1975).

The relict components of Colchic forests; tree stratum of canopy are excluded at an altitude of between 600-800 m a.s.l., but poor stands of these communities occur on limestone ground. In such habitats frequently sticking out are the calcifuge plants – sweet chestnut, Pontian rhododendron (Gulisashvili *et al.* 1975).

In Colchic mixed forests a characteristic peculiarity is considered to be a wide distribution of epiphytes. Although, flowering plants are confined only to facultative epiphytes, it should be noted some ferns are occurring as typical obligative epiphytes (e.g., *Polypodium serratum*). There are a lot of epiphyllous lichens (e.g., *Usnea barbata*) (Grossheim 1948).

From the cenotical and floristical point of view, the lowlands, low-mountain foothills and lower mountain zone's forests up to an altitude of 500-600 m of Adjara are the richest in Colchis. The maritime area between Kobuleti – Chakvi and west slopes of the Adjaro-Imeretian ridge are covered by more typical Colchis forests with *Fagus orientalis, Castanea sativa, Quercus hartwissiana, Q. dschorochensis* and *Carpinus betulus* dominancy. Just in the mentioned territory of Adjara and adjacent to NE Turkey (till Malet – Ordu Gorge) are preserved rich forest's relict vegetation communities (Kaya & Raynal 2001). According to Manjavidze (1982), the natural forest vegetation of the mentioned zone is rather well preserved in the upper part of foothills and Kobuleti Sanctuary of the Kobuleti Lowland.

Figure 123. Colchis Lowland's forest with participation in canopy layer *Pterocarya fraxinicolia*. Guria province, W. Georgia (2016).

Figure 124. The fruit of wingnut. Guria province (2016).

Figure 125. Caucasian persimon (*Diospyros lotus*) grows in a second layer of Colchic forests. Samegrelo province (1987).

In the lower-mountains area of the Colchis, among large mammals species should be mentioned: jackal (*Canis aureus*), wild boar (*Sus scrofa*), roe deer (*Capreolus capreolus*), fox (*Vulpes Vulpes*), wild cat (*Felis silvestris*), and European otter (*Lutra lutra*). Among bird fauna the Colchic pheasant (*Phasianus colchicus*) attracts considerable attention.

According to Pollard (1977) cited in Sokos & Birtsas (2014), the Colchic black-necked pheasant (*Ph. colchicus colchicus*) was known to ancient Hellene scholars, for example, Aristophanes and Aristotle.

As well as the Argonauts, Colchis Medea and golden fleece, the Colchic pheasant was largely established in Greek mythology largely (Akymushkin 1973).

The scientific name *Ph. colchicus* is derived from the river Phasis (today Rioni River in Georgia) and the region Colchis (an ancient name for what is now a part of western Georgia) (Gegechkori 2007, 2013; Jelen 2009 cited in Braasch *et al.* 2011; Sakos & Birtsas 2014) (Figure 126). The first introduction of the Colchis pheasant into Central Europe dates back to between 500 and 800 AD. It was nominated as subspecies (*Ph.c.colchicus*) (Braasch *et al.* 2011).

Figure 126. Colchic pheasant (*Phasianus colchicus*), pure Colchic phenotype. It is a bird, which lacks a white neck ring, the distinctive anatomical feature for some other species. Riparian and lowland forests of Samegrelo province, W. Georgia (2009).

There are four subspecies of pheasants in the Caucasus: North caucasian pheasant (*Ph.c. septentrionalis*), Talysh pheasant (*Ph.c. talischensis*), Persian pheasant (*Ph. c. persicus*), and the aforementioned Caucasus pheasant (*Ph.c. colchicus*). Djanashvili (1963) considers for East Georgia an independent subspecies – Georgian pheasant (*Ph.c. lorenzi*). Populations of all four subspecies are seriously in decline. The reasons for the sharp decline have been boosted by heavy hunting pressure, loss of suitable habitat (deforestation of riparian and bordering forests and the use of pesticides) (Djanashvili 1963; Braasch *et al.* 2011 and references therein) (Figure 126).

The humid climate and hygro-thermophilous broad-leaved forests support the number of relict and endemic species of herpetofana: salamander (*Mertensiella caucasica*), Parsley frog (*Pelodytes caucasicus*), Colchic toad (*Bufo verrucossissimus*). During climate oscilation of the Pleistocene until today survived relict and endemic insect species: Schamyl's ghost moth (*Phasus schamyl*), butterfly (*Allancastria* (*Thais*) *caucasica* and others (Gegechkori 2000).

2.6.1.2.1.3. The Peculiarities of Talysh (Hyrcanian) Lower Mountain Broad-Leaved Forests

The Talysh region is a part of the Republic of Azerbaijan in the south-eastern part of Transcaucasia. It is located between 38^0 and 39^0N and 44^0 and 46^0E. It consists of two geographical units - the Lenkoran Lowland (Lankaran) and the Talysh Mountains. Azerbaijanian Talysh has a total of about 100,000 ha of the Hyrcanian forests (Schmidt & Uppenbrink 2009) vs 1.85 million ha of Iranian part of the Hyrcanian forests (Anon 2008 in Sagheb-Talebi *et al.* 2014). The Hyrcanian forests stretch in a large area as a narrow forest belt along the southern Caspian seashore from the Talysh Mountains and the Lenkoran Lowland (Azerbaijan) over the northern slopes of the Alborz Mountains and Caspian Lowland through Northern Iran (Gilan, Mazandaran and Golestan Provinces) (Browicz 1989; Sagheb-Talebi *et al.,* 2014; Knapp 2005).

Hyrcan is less humid than Colchis with about 1,400 mm of annual precipitation in the Talysh, and up to 2,000 mm in western Hyrcan adjacent to Azerbaijan's Talysh.

Forested slopes of the Talysh Mts and coastal lowland are characterized by a warm-temperate and humid climate. The annual precipitations, as mentioned, reach about 1.400 mm per year and peak in February and October (Safarov 1962, 1979).

The most important soils of the Talysh Mts are yellow, yellow-brown and mountainous brown - forest soils (MENR 2004 in Schmidt & Uppenbrink 2009).

The relict forests of Talysh are situated on the Caspian coastal lowland, foothills and lower mountain zone. These communities, in general, occupy an area from 50 up to 400-6 00m a.s.l., but some representatives through rivers gorges reach to 1,000-1,200 m a.s.l. (Gulisashvili *et al.* 1975).

Figure 127. Alexandrian laurel (*Danae racemosa*, Asparagaceae) along with other species of evergreen shrublands (*Ilex hyrcana, Ruscus hyrcanus*). Hyrcanian forests in southeastern Azerbaijan (1974).

According to Grossheim (1926, 1948), the Talysh forests enriched with relict vegetations, resemble other species-rich regions of Colchis forests, However they differ not only in quite different composition of woody species but also from a phytocoenological point of view. Botanists in unison emphasize the absence of dense evergreen understory within the local forests. Sparse evergreen shrubs (*Ruscus hyrcanus, Danae racemosa, Prunus laurocerasus*) sporadically occur mainly as the intrazonal formations (Grossheim 1926, Safarov 1962, 1979) (Figure 127).

The woody stands of Talysh form two types of forest formations: represented by mixture of some species (polydominant communities), and communities with predominance of one species (monodominant formations). Safarov (1962) and Gulisashvili *et al.* (1975) outline the main type of Talysh relict forests of mentioned zone.

According to Grossheim (1926), the primary type of the lowland forest is composed of Persian ironwood (*Parrotia persica*) (Figure 128).

Within forests, covering foothills and lower mountain belt (50 – 600m a.s.l.) an edificatory position is occupied by native oak - *Quercus castaneifolia* and Persian ironwood - *Parrotia persica*. These communities, typical of Hyrcanian forest, are represented either by monodominant or mixed phytocoenoses. The most striking feature of local forests, except ironwood, is growth including tropical-subtropical genera *Gleditsia* (*G. caspica*) and *Albizzia* (*A. julibrissin*).

It must be stated that the two belts just mentioned are populated by typical Hyrcanian relicts and sometimes relicts and endemics of some rare, for the Caucasus, species of animals (species of insects genera: *Brahmaea, Danaus, Isophya*; species of mammals –Indian crested porcupine (*Hystrix indica*), and until recent times – the Turanian tiger (*Panthera tigris virgata*).

Among extremely rare entomofauna of tropical-subtropical origin of the Hyrcanian refugium should be mentioned *Brahmaea christophi*. Based on the interdisciplinary study another humid-tolerant species *Brahmaea ledereri* does not inhabit prehumid habitats of Colchis. Therefore, the current orthodox view regarding the vicariant events between two species of moths as though they are indigenous of Colchis (*B. ledereri*) and Talysh (*B. christophi*) doesn't actually exist (Gegechkori & Didmanidze 2015).

Figure 128. Upper layer of paradigmatic Persian ironwood (*Parrotia persica*) (Hammamelidaceae) in Caspian forest. Rarest in the Caucasus tree species. Lenkoran, Talysh Mts, Azerbaijan (1974).

2.6.1.3. Mountain, Temperate Hardwood. Summer-Green Deciduous Forests

2.6.1.3.1. General Characteristics

Four types of deciduous forests are widespread over the Caucasus. Two types have the canopy dominated by only one species: the monodominant various species of oaks (*Quercus* spp.) and monodominant beech (*Fagus orientalis*) forests. The other two types are characterized by two canopy-building species: the different oaks-hornbeam (*Quercus petrae iberica/Q. robur imeretina - Carpinus betulus*) and chestnut-hornbeam (*Castanea sativa - Carpinus betulus*) forests.

2.6.1.3.2. Beech Forests

Beech (*Fagus* L.) is a genus of deciduous trees in the family Fagaceae. According to Fang & Lechowicz (2006 and references therein) the genus includes ten primary species and two minor segregates (Willis 1966 cited in Fang & Lechowicz 2006) broadly distributed in the Northern Hemisphere including the most famous and largest refugial centers: East Asia, Europe and West Asia, and North America. Six species are found in East Asia (four in China and two in Japan), two in Europe and West Asia, and one in North America. Two additional beech species *F. multinervis* from Korea and *F. mexicana* from Mexico, as new species are recognized in recent times (Fang & Lechowicz 2006 and references therein). Some of them (*F. multinervis* and *F. hayatae* from China) have a very limited range. European *F. sylvatica* is the most widely distributed of all beeches. Unlike beeches in other regions, Chinese beeches rarely form pure stands (Fang & Lechowicz 2006).

Beech trees are relict tree species which after the last deglaciations successfully restored its former range (Dolukhanov 2010), but the northerly distribution of beech in some regions of the Northern Hemisphere is characterized by either the lack (e.g., the island of Newfoundland, Canada) or has not reached its potential limit, (e.g., the British Isles) 'possibly due to insufficient time since deglaciation to expand its range' (Fang & Lechowicz 2006).

Beech forests of *F. orientalis* constitute the main vegetation type of the mid-altitudinal reaches of the Caucasus. Outside the Caucasus, the distribution area of *F. orientalis* comprises mountainous Crimea, the Pontian Mountains in Turkey and the South Caspian part of Iran, Alborz Mountains (Gulisashvili *et al.* 1975; Fang & Lechowicz 2016).

Oriental beech is broadly known as a canopy-building tree usually forming pure stands. It is one of the most important elements of forests in the temperate broad-leaved forest biome and 'represents an outstanding example of the recolonization and development of terrestrial ecosystems and communities after the last ice age, a process which is still ongoing' (Mosadegh, 2000; Marvie Mohadjer 2007 cited by Zoghi *et al.* 2011; Dolukhanov 2010).

In the Caucasus *F. orientalis* forms a monodominant forest zone. The species is shade-resistant and moisture-loving (1000-1500 mm of precipitation per annum, relative humidity > 70% L'vov 1964) and occurs on moderate to gentle slopes with more or less well-developed acid brown mountainous-forest soils. When they dominate the habitat, dense beech forests prevent sun beams reaching the ground and may hamper the establishment of other arboreal species. Only shade-enduring species can co-exist with beech to form beech-fir, beech-spruce-fir and beech-chestnut-hornbeam communities (Gulisashvili *et al.* 1975; Dolukhanov 1980, 2010).

Whereas the lower boundary is determined by air humidity, the upper limit of the altitudinal distribution of *F. orientalis* depends on temperature and on the amount of winter snow that protects the young beech trees against frost (Dolukhanov 1980). The lower limit of beech forests thus ranges from sea level (Colchis) to 1,000 m and under the most arid conditions even to 1,200-1,300 m. The upper limit varies in altitude between 1,400 m and 2,400 m a.s.l. Thus, in the moist Colchis *F. orientalis* occurs from sea level to 2,200-2,400 m. In the much drier East Georgia, the lower limit lies around 350-400 m (Alazani Basin), whereas in South Georgia (Trialeti Mts) beech forests only appear from 800-900 m upwards (Dolukhanov 2010).

Beech forests are found almost everywhere in the Caucasus and are only lacking in regions with a continental climate, i.e., Inner Dagestan (Tumajanov 1940) and South-East and South Transcaucasia. For the same reason beech forests are absent in the gorges of the Kubari and Malka Rivers on the western part of north macroslope of the Greater Caucasus (Gulisashvili *et al.* 1975).

In Western Georgia, beech forests are distributed on the north slopes of the Adjaro-Imeretian Ridge (Lesser Caucasus). Solitary old age (300-400 years) specimen trees are located directly on the shore of the Black Sea (Urushadze *et al.* 2010). In some districts with a dry climate on the south slopes of that range and on the slopes of the Trialeti Range (Lesser Caucasus) beech trees are found rarely and only occur as mixed stands with other tree species in the spruce zone and in subalpine crook forests (Ketskhoveli 1960; Dolukhanov 2010).

The wood formations of eastern Transcaucasia and eastern Ciscaucasia consist of the following arboreal species: *Quercus petraea, Q. p. iberica, Q. macranthera, Acer platanoides, A. pseudoplatanus, A. laetum, A. campestre, Tilia begoniifolia, T. cordata, Taxus baccata, Pinus kochiana, Betula pubescens litwinowii* and *B. pendula* (= *B. verrucosa*). In the undergrowth *Sambucus nigra, Corylus avellana, Euonymus latifolius, E. europaeus, Lonicera caucasica* and *Rhododendron luteum* are the most conspicuous species (Dolukhanov 2010).

In the north-western part of the Greater Caucasus and western Transcaucasia, the typical forest type consists of beech with fir (*Abies nordmanniana*), spruce (*Picea orientalis*), ash (*Populus tremula*) and maple (*Acer trautvetteri*). Within the underwood, characteristic shrubs like *Prunus* laurocerasus*, Rhododendron ponticum, Ilex aquifolium* and *Vaccinium arctostaphylos* are found. In western (Alazan-Agrichai district) and south-eastern Azerbaijan (Talysh) beech typically occurs with *Acer velutinum* and *Diospyros lotus* (Gulisashvili *et al.* 1975).

The herb layer of the beech forests is less differentiated. Common species include *Festuca montana, Galium odoratum* (= *Asperula odorata*), *Sanicula europaea, Rubus caucasicus, Oxalis acetosella, Poa nemoralis. Geranium robertianum, Galium rotundifolium* and various ferns (Gulisashvili *et al.*1975; Nakhutsrishvili 1999, 2013).

2.6.1.3.2.1. Classification of Beech Forests

The classification of beech forest communities is based primarily on studies of the botanists and foresters of the former Soviet Union' (Sokolov 1936; Sosnin 1939; Povarnitsin 1940; Tumadjanov 1940; Prilipko 1945, 1954; Grudzinskaia 1953; L′vov 1964, etc.), that were summarized by Gulisashvili *et al.* (1975) and Dolukhanov (2010). Dolukhanov (1989, 2010). They differentiate two classes of beech woods: beech woods without and beech woods with woody undergrowth.

2.6.1.3.2.1.1. Bare-Floor Beech Forests

Bare-floor beech forests (Fageta nuda) lacking undergrowth are widely distributed in the Caucasus especially at altitudes of 1,000-1,200 m a.s.l. The forest litter forms a thick layer that sometimes reaches 5-8 cm. They are characterized by an admixture of hornbeam (*Carpinus betulus*) and elm (*Ulmus glabra* (= *U. elliptica*). A herb layer is only rarely present. Forest rejuvenation is very weak, except where gaps in the canopy provide better light conditions (Gulisashvili *et al.* 1975; Dolukhanov 2010) (Figure 129).

Figure 129. Interior of a bare-floor Oriental beech forest, here showing its autumn colors. WLC, Guria province, W. Georgia (1975).

2.6.1.3.2.1.2. Beech Forest with Ferns

Beech forest with ferns (Fageta filicosa) are also widely distributed over the Caucasus, especially on the slopes of the Greater Caucasus. They develop under moist conditions and are the indicator of a humid biotope. Apart from beech, the tree layers consist of maple, lime-tree, elm and hornbeam. Woody undergrowth is almost absent. The herb layer is dominated by large ferns, especially *Dryopteris filix-mas*, *D. pseudo-filix-mas, Athyrium filix-femina*, and *Matteuccia struthiopteris* (the biggest size fern is distributed in South Caucasus) that almost completely cover the ground. Natural forest regeneration is suppressed by the thick fern cover. Some other species of ferns are common, but do not create a massive cover (Dolukhanov 2010) (Figure 130).

2.6.1.3.2.1.3. Beech Forest with *Galium odoratum* (=*Asperula odorata*)

Beech forest with *Galium odoratum* (=*Asperula odorata*) (Fageta asperulosa) is situated on slopes of medium steepness and northern exposure in relatively moist habitats, on well drained brown forest soils. Their altitudinal distribution ranges from 1,000 to 1,400 m a.s.l. Next to the dominant beech, hornbeam, maple (*Acer laetum*) and lime-tree occur in the tree layer. Woody underground is absent except for small, isolated stands of *Ilex colchica*. The sparse herb layer consists of – among other things - *Dryopteris filix-mas* and *Cephalanthera longifolia*. Natural forest regeneration is satisfactory. Dolukhanov (2010) differentiated it from European communities and called it Fegeta asperulosa caucasica.

Figure 130. Pristine Oriental beech forest with ferns in ground layer in the Lagodekhi NP. Native Protected Areas consist of emblematic mesic Caucasian flora and fauna. It was originally protected in 1912, under the Russian Empire as the first nature reserve in South Caucasus (1977).

2.6.1.3.2.1.4. Beech Forest with *Rubus hirtus*

Beech forest with *Rubus hirtus* (Fageta rubosa) is common in the north-western part of the Caucasus in Transcaucasia on slopes between 1,000 and 1,600 m. a.s.l. (less abundant in the Lesser Caucasus). They constitute the most common moist beechwood communities in the region, especially on the southern slopes of the Greater Caucasus. Next to the dominant species *Fraxinus, Ulmus* and *Tilia* participate in forming the canopy. The undergrowth and herb layer are poorly developed and include, next to the dominant *Rubus hirtus*, blackberry (*Vaccinium arctostaphylos*), *Symphytum grandiflorum* and *Matteucia struthiopteris*. The same moist beech forests are rather abundant with *Trachystemon orientalis* (a Colchian relict species) and *Pachyphragma macrophyllum* (an ancient, endemic monotypic Colchian genus) (Dolukahnov 2010; Nakhutsrishvili 1999, 2013).

2.6.1.3.2.1.5. Beech Forest with European Yew

Beech forest with European yew (*Taxus baccata*) and within this community with Russian ivy (*Hedera pastuchovii*) and understory (Fageta taxceto-ilicitosa-hederosa) is widespread on the Kakhetian Caucasus (Batsara and Pankisi Gorges), Aragvi Gorge of Pshavi (near Kanatia). Associated species are: *T. baccata, Quercus p. iberica, Carpinus betulus, Fraxinus excelsior*, etc. (Akhalkatsi & Tarkhnishvili 2012).

2.6.1.3.2.1.6. Beech Forest with Pontian (Pontic) Rhododendron (*Rhododendron ponticum*) Undergrowth

Beech forest with Pontian (Pontic) rhododendron (*Rhododendron ponticum*) undergrowth (Fagetum rhododendron) with *Rh. ponticum* occur on shady slopes of the moist regions of

Georgia and are very common in western Transcaucasia, especially in Colchis, but rare in eastern Transcaucasia (can be found in some gorges of East Georgia: Baniskhevi, Kvabliani and Nadzvistskali). In West Georgia the most favorable conditions are found at altitudes of 1,000 to 1,350 m, but the lower and upper limits of distribution range from sea level to 2,200 m a.s.l. Subdominant species in the canopy include lime-tree, chestnut, hornbeam and sometimes, fir. The undergrowth is well-developed and includes *Rh. ponticum* and *Vaccinium arctostaphylos*. The herb layer is poorly developed (Dolukhanov 2010).

2.6.1.3.2.1.7. Beech Forest with Ungern's Rhododendron (*Rhododendron ungernii*)

Some researchers distinguish a beech forest with *Rh. ungernii*, a relict and endemic of Colchis. *Rh. ungernii* grows under moist climatic conditions with a very high amount of precipitation amounting to 3000 mm. Usually these forests extend from 1,200 up to 2,000 m a.s.l. (Gulisashvili *et al.* 1975; Dolukhanov 2010; Nakhutsrishvili 1999, 2013; Goginashvili *et al.* 2018) (Figure 131).

Figure 131. Beech forests with extremely unique relict and endemic Ungern's rhododendron (*Rhododendron ungernii*), Adjara, W. Georgia, 1,600 m a.s.l. (2008).

2.6.1.3.2.1.8. Beech Forest with Cherry-Laurel (*Prunus laurocerasus* (=*Laurocerasus officinalis*)

Beech forest with cherry-laurel (*Prunus laurocerasus*) (Fagetum laurocerasosum) is common in western Transcaucasia, occurring also in the precipitation-rich north-western part of Ciscaucasia and cover of steep slopes of every exposition. The beech stand is accompanied by an admixture of hornbeam, lime-tree and mountain maple (*Acer platanoides*). The undergrowth is well-developed next to *Prunus laurocerasus, Rhododendron luteum, Ilex colchica* and *Vaccinium arctostaphylos*. The overground layer is poor (ferns, blackberry).

Some other beech woods are sometimes distinguished on the basis of their arboreal undergrowth, including Fageta ilicitosa (with *Ilex colchica*), Fageta muscosa (with *Ruscus colchicus*) and Fageta azalea (with *Rhododendron luteum*) (Gulisashvili *et al.* 1975, Dolukhanov 2010).

According to Akhalkatsi and Tarkhnishvili (2012), different from Pontic rhododendron, cherry-laurel grows well on limestone and well-illuminated slopes. Besides Colchis, it is common in the form of small populations far from the main area. For example, in East Georgia it is widespread in the Alazani basin and the River Ilto gorge (Kakheti). Existence of such a widely disunited area is related to ornithochoria, since birds feed on its fruit and disseminated seeds over large distances (Dolukhanov 2010).

The latter two communities sometimes are united within beech forests with Colchis understory (Fageta fruticosa colchica) (Akhalkatsi & Tarkhnishvili 2012).

2.6.1.3.2.1.9. Beech Forest with Viburnum (*Viburnum orientale*)

Beech forest with viburnum (*Viburnum orientale*) understory – Fageta viburnosa is characterized by a small range and, as a rule, holds the most moistened location. It can be common outside Colchis in the upper reaches of the rivers Aragvi and Alazani. This community extended from approximately 900 to 1,900 meters. This species of *Viburnum* (fam. Adoxaceae, previously – Carifoliaceae) is a Colchic relict and its close relative species *V. acerifolium* grows in the eastern part of the USA (Akhalkatsi & Tarkhnishvili 2012).

2.6.1.3.2.1.10. Beech Forest with the Caucasian Blueberry (*Vaccinium arctostaphylos*)

It is most widely distributed in Colchis. In East Transcaucasia it is common in the Kakheti (Lagodekhi) region and extends to the Zakatala Reserve (Azerbaijan), a small population is represented in the Caspian Sea coastal area – Talysh National Park (Azerbaijan). Besides blueberry, the understory consists of *Hedera colchica, H. pastuchovii, Rubus* spp., *Vaccinium myrtillus*, species of ferns and grasses (Dolukhanov 2010; Nakhutsrishvili 1999, 2013).

Worthy of attention are the following issues regarding the evolutional and cytological researches of species of beeches, including *F. orientalis*: Sochava (1949); Denk *et al.* (2002); Gailing & Wuehlisch (2004); McAllister (1993).

2.6.1.3.3. Oak Forests

There are seventeen species of oak in the Caucasus, of which fourteen are endemic and nine - forest building. Many oak forest communities can be distinguished and their undergrowth and herb layer contain more species than any other type of broadleaf forest in the region. The most widespread species are Georgian oak and high mountain oak (Maleev 1935; Menitski 1981; Dolukhanov 1966, 2010).

2.6.1.3.3.1. Forests of Georgian Oak *(Quercus petrae iberica)*

Forests of Georgian oak (*Q. p. iberica*) are distributed over almost all regions of the South Caucasus and even partly enter the North Caucasus (Dagestan, Chechnya and Ingushetia). The species is closely related to the widespread European *Q. petraea* (Dolukhanov 2010). *Q. p. iberica* are forests concentrated in the low and mid-mountain zone (500-600–1,000-1,500m a.s.l), whereas its lower and upper limits extend from sea level to 1,800 m (Svaneti, West Georgia). According to Vachnadze & Makhatadze (1989), the Georgian oak is very sensitive to saline, moist and badly drained soil and therefore avoids alluvial plains. Being drought-resistant and requiring much light, populations of oak are - on shady and moist slopes - completely replaced by beech, spruce and chestnut forests (Gulisashvili *et al.*1975). Tree species co-occurring with Georgian oak include *Carpinus betulus, C. orientalis, Zelkova carpinifolia, Ostrya carpinifolia, Acer laetum* and *Sorbus torminalis*. Depending on climate and orography, Dolukhanov (1992) differentiates three main categories of Gorgian oak forests: oak monodominant, oak-oriental hornbeam (*C. orientalis*), and oak-European hornbeam (*C. betulus*) communities.

The same author (Dolukhanov 1992, 2010) within the three main formations above, due to ecological demands of the same oak, specifies the following sub-formations: meso-thermophyte oak forest and xero-thermophyte oak forest. The first sub-group is represented by *Carpinus betulus – Quercus p. iberica* forests, which occupy a transition zone between oak forests and these of the Oriental beech. The second layer consists of shade-enduring *C. orientalis*. Xero-thermophyte oak forest populates the low-mountain (500-1,100 m) zone of Transcaucasia, while in humid districts of West Georgia extends from 350 up to 700-800m. They occur exclusively on southern slopes, where the above-mentioned oak is accompanied by Oriental hornbeam (*C. orientalis*) and grows in dry and sometimes limestone soil. Following Dolukhanov (2010), monodominant oak forests are controlled by ecological conditions, therefore can be outlined hemixerophytic (developing under the driest conditions), and xero-mesophytic (with developed woody undergrowth, including *Rhododendron luteum* and *Rh.. ponticum*). The former sub-group with sparse understory (species of *Cotinus, Spiraea, Pyracantha,* etc.), since prehistoric time has been strongly altered due to the impact of human activities.

As a result of human exploitation (unsustainable forest logging and overgrazing of domestic animals), Georgian oak forests initially turn into hornbeam forests, with increased pressure change to shibliak thorn vegetation, in which *Paliurus spina-christi* occupies the leading position, and finally become herbaceous communities dominated by *Bothriochloa ischaemum*. When grazing is forbidden, the shibliak vegetation gradually converts again into forest (Gulisashvili *et al.* 1975) (Figure 132).

Because of the open canopy and the heterogeneity of the physico-geographical conditions within the distribution area of *Q. p. iberica*, Georgian oak forests are floristically very rich (Dolukhanov 2010).

Figure 132. Mature forests of the Georgian oak (*Q. petrae iberica* (=*Q. iberica*). Algeti River Gorge, E. Georgia, 830 m a.s.l. (1990).

2.6.1.3.3.2. Georgian Oak Forests with European Hop-Hornbeam (*Ostrya carpinifolia*)

Among the communities with participant of Georgian oak, mention should be made of *O. carpinifolia*. The wood is very heavy and hard. The name *Ostrya* is derived from the Greek word 'ostrua', 'bone-like', referring to this very hard wood (*en.wikipedia.org/ wiki/Ostrya_carpinifolia*). It is Tertiary relict. It shows an extensive distribution from southern France to Bulgaria, Anatolia and Transcaucasia (Davis 1982; Çolak & Rotherham 2006). In the Caucasus it occurs in the Black Sea coastal area south of the river Tuapse (northern Colchis) and Ciscaucasia (gorges of rivers Chegem, Cherek and Baksan). In Transcaucasia it is found in Georgia, especially in the limestone and rocky ground regions, e.g., in its southern (Samtskhe, Borjomi gorge) and western regions (Abkhazeti, Svaneti, Racha-Lechkhumi, Samegrelo, Imereti, Guria, Adjara) (Gulisashvili *et al.* 1975; Tutin *et al.* 1964). The phytocoenoses that occur below 900m a.s.l. are mostly related to the sub Mediteranean type, typically forming stands in associations with Mediterranean evergreen oaks, hornbeams and maples. Communities dominated by the Fraxino orni–ostryetum carpinifoliae are the commonest in the Mediterranean countries. The communities that occur in more mesophitic conditions, mainly above 900m, include species typical for the mesophilic deciduous hornbeam – beech forests. In any case it is a light-demanding woody species (Popović *et al.* 1997; Griffiths *et al.* 2004) (Figure 133).

In South Georgia (Samtskhe) adjacent to northern Anatolia where the sub-Mediterranean influence is lesser, Georgian oak and hop-hornbeam forest (Quercetum-ostryosa) with the participation of *Carpinus betulus* can be found. It is seldom found in mixtures with light- and dark-coniferous forests. In the lower border of distribution of *O. carpinifolia*, when the hop-hornbeam and Georgian oak forests are cut down, vegetation of shibliak mixed with the oak is represented (Ketskhoveli 1960; Khintibidze 1990).

Figure 133. TBDF enriched with the European hop-hornbeam (*Ostria carpinifolia*). Meskheti region, Atskuri environs, 900 m a.s.l. (2013).

2.6.1.3.3.3. Forests with Rocky Oak (*Quercus petraea*)

Forests with rocky oak (*Quercus petraea*) are found on the slopes of the Greater Caucasus in the North Caucasus, where the species forms its own natural zone (Elagin 1953). In the western part of this region the upper limit of this zone occurs at approximately 700 m a.s.l., whereas in the eastern part it reaches 1,100-1,300 m. Some *Q. petraea* forests are very similar to those of *Q. p. iberica*. The following types of *Q. petraea* forests have been distinguished: oak forests with blackberry, oak forests with hornbeam, oak forests with hawthorn (*Crataegus lagenaria*) and *Carex divulsa*, oak forests with wild grasses, oak forests with ferns (*Pteridium aquilinum*) and oak forests with *Rhododendron luteum* (Grudzinskaia 1953).

2.6.1.3.3.4. Forests with Boissier Oak (*Quercus infectoria veneris* (= *Q. araxina*)

Forests with Boissier oak (*Quercus infectoria veneris* (= *Q. araxina*) are found in the south-eastern part of Transcaucasia (Armenia and Azerbaijan), where Boissier oak is found in both forest with closed canopy and in arid open woodland on slopes directed to the Aras River (Magakian 1941, sensu *Q. araxina*). Sometimes the forests penetrate into more or less mesic gorges of the Aras River Basin. *Q. i. veneris* is one of the most drought-resistant arboreal species of the Caucasus. At the lower limit of their distribution, Boissier oak forests border with deserts and create arid open woodland with *Pistacia atlantica*, *Celtis glabrata* and *Punica granatum* as a transition zone. The upper distributional limit of Boissier oak forests can be found at 200-1,000 m a.s.l, adjacent to Georgian oak forests. The following associations are being distinguished: oak forests with *Bothriochloa ischaemun* and *Onobrichis petraea*, oak forests with *B. ischaemum* and *Carex buschiorum* and oak forests with grasses and various herbs (Makhatadze 1957 sensu *Q. araxina*) (Figure 134).

Figure 134. Boissier oak. Meadle flows of the Aras River Gorge. Armenia, environs of Megri town (1973).

2.6.1.3.3.5. Forests with Chorokhian Oak (*Quercus dschorochensis*)

Forests with Chorokhian oak (*Quercus dschorochensis*) occur only in the valley of the Chorokhi (Coruh) River in Turkey and its right tributary the Adjaristskali River in the south-western part of Adjara. *Q. dschorochensis* is very closely related to Georgian oak and rocky oak. *Q. dschorochensis* forests extend to 800m a.s.l., preferring sunny and dry rocky mountain slopes of south exposure with a thin skeletal soil. Today only remnants of these forests have survived (Gulisashvili *et al.* 1975; Dolukhanov 2010).

A few types of Tchorokhian oak forests have been described: oak forests with grasses (with a dominance of *Bothriochloa ischaemum*), oak forests with Greek strawberry (*Arbutus andrachne*) (Dmitrieva 1990; Goginashvili *et al.* 2018) and *Cistus salviifolius* (representatives of the Mediterranean flora) that resemble arid open woodland, and oak forests with an undergrowth of evergreen *Rhododendron ponticum, Prunus laurocerasus* and *Ilex colchica* on more shady slopes (Dolukhanov 2010).

2.6.1.3.3.6. Forests with Hartwissian Oak (*Quercus hartwissiana*)

This species of oak frequently occurs as a typical component of mixed hardwood forests of Colchis. It is a characteristic canopy species of oligo- and polydominant forest communities of West Transcaucasia, therefore never forms monodominant stands. *Q. hartwissiana* grows on humus-rich and moist soils (Ketskhoveli 1960; Gulisashvili *et al.* 1975; Nakhutsrishvili 2013).

2.6.1.3.3.7. Forests with Caucasian (Persian) Oak (*Quercus macranthera*)

Forests with Caucasian (Persian) oak *(Quercus macranthera)* are found at an altitude from 1,350-1,600 up to almost the upper limit of distribution of the species (2,650m), mostly in Armenia and Azerbaijan. *Q. macranthera* is a light-loving, very frost- and drought-resistant species, which consequently grows in mid- and high-mountain conditions with a dry climate and much solar radiation. *Q. macranthera* forests grow on mountain forest brown soils. Close to the timberline they occur on mountain forest-meadow brown soils and meadow-steppe soil (Dolukhanov 2010).

In south-eastern Transcaucasia *Q. macranthera* forms on slopes of all exposures at a discrete zone above the belt of *Q. petrae iberica* up to the timberline (2,400-2,650 m a.s.l.). Here *Q. macranthera* (also called the 'highland oak') forms large tracts of monodominant forests, but also frequently co-occurs with hornbeam (Makhatadze 1957). In a western direction the belt of *Q. macranthera* becomes gradually narrower to be replaced on the north slopes by beech, then by spruce and fir until it is finally eliminated.

On the Lesser Caucasus (Trialeti Range) *Q. macranthera* does not form a zone, but is met as more or less big massifs on slopes of southern exposure.

A similar situation occurs on the Greater Caucasus Range. Rich stands of *Q. macranthera* can be found on its eastern part, particularly in the areas with a clear continental climate. In a western direction the belt gradually wedges out. In West Georgia this species of oak only occurs in the Greater Caucasus as small groups or single trees. The extreme west border of its distribution is the Enguri River Basin (Zemo (Upper) Svaneti) between 1,800 and 2,250 m a.s.l, where, according to Sosnovski (1915 cited in Dolukhanov 2010), Caucasian oak forests were formerly far more widely distributed than at present.

In Ciscaucasia *Q. macranthera* can be found in the subalpine zone in Dagestan and westward as far as the Elbruz summit, where its upper margin of distribution reaches 2,600-2,700 meters. The optimal height of distribution varies between 1,450-1,600 meters (Stepanov 1952). Next to the Caucasus *Q. macranthera* forests also occur in the Caucasus Ecoregion, in north-eastern Turkey (Davis (Ed.) 1965-1988), and northern Iran (Alborz Mts) (Khalegi 1998; Akhani *et al.* 2010; Kialashaki & Shaban 2011).

2.6.1.3.3.8. Forests with Grayish Oak (*Quercus robur pedunculiflora = Q. pedunculiflora*)

Forests with grayish oak (*Quercus robur pedunculiflora* (= *Q. pedunculiflora*) is closely related to the European species *Q. robur*. The grayish oak as a species is still debated, i.e., is a species controversial in taxonomy. Some botanists consider the drought-tolerant *Q. r. pedunculiflora* c. Koch a race of *Q. robur* L., 'the characteristics of which are results of an adaptation to drought climate and probable introgression of other species' (Carella 2013). But 'in the consolidated Index of Flora Europaea (Halliday & Beadle 1983) it is considered as a species' (Carella 2013). The same opinion about validity as a species of the mentioned oak is shared by other authors (Borza 1947; Tutin *et al.* 1993; Coldea *et al.* 2010 cited in Carella 2013) (Figure 135).

Q. pedunculiflora is considered a keystone species for the forest communities in the semi-dry environments (e.g., wood-steppe zone), because of the survival of many other living organisms depending on it (Donita *et al.* cited in Curtu *et al.* 2011).

In Eastern Transcaucasia forests composed of grayish oak occupy river valleys. In the arboreal components of these forests the following species should be mentioned: *Q. r. pedunculiflora* (which frequently occurs as an arboreal component of oligodominant forests), *Carpinus betulus, Ulmus carpinifolia, Acer campestre, A. velutinum, Tilia begoniifolia* and *Pterocarya fraxinicolia* (under moist conditions). Among lianas should be mentioned *Smilax excelsa, Periploca graeca, Clematis vitalba, Vitis vinifera sylvestris* and *Hedera helix*. Today only few remnants of these, once luxuriant forests can be found (Dolukhanov 1992 cited in Nakhutsrishvili 1999).

Figure 135. Solitary greyish oak in Alazani River Valley. Kakheti province, E. Georgia (2010).

Box 10. South Caucasus (Georgia) as a Region of the First Hominids (Genus *Homo*) Distribution from Africa into Eurasia

The ancient history of human activity in the Caucasus is fairly mysterious. It is one of the most impressive archaeological and palaeontological regions in the world. Especially, it should be noted, regarding the Pleistocene site of the ancient town of Dmanisi, southern Georgia. In the middle ages it was a strategically and economically important city, which was located on a caravan route leading to Asia (Kopaliani 1996). Here, the rivers Mashavera and Pinezauri cut canyon-like gorges in basaltic lava, resulting in the formation of a basaltic plateau. The latter was located at about 900m a.s.l. and surrounded by deciduous forest consisting of European hornbeam and Oriental beech (Gabunia *et al.* 2000b; Vekua & Lordkipanidze 2010). The fossil assemblage of the Dmanisi site is placed at an important junction, both geographically and temporally, for an understanding of the evolution of early *Homo* and the emergence of *H. erectus* (Wood 2011). This site is unique for the Lower Pleistocene hominin record in that it provides rich fossil materials derived from a tightly constrained locality. It consists of striking information, for discussions of hominin variability, dispersal, and ecology at the Pliocene/Pleistocene boundary (Arsdale & Lordkipanidze 2012 and references therein).

From 1936 this site has yielded rich archaeological and fossil records. The first partial skeleton (hominin mandible) of ancient *Homo* was discovered in 1991 (24.09). A team of archaeologists and paleoanthropologists, included Georgians and frequently foreign scientists, has successfully excavated more fossils. Between the 1990s and 2000s a paleoanthropological collection of Dmanisi consists of five skulls, twelve isolated teeth and about 50 other parts of human skeletons belonging to men and women of various ages (Vekua & Lordkipanidze 2010).

The assemblage includes one of the largest Pleistocene's *Homo* mandibles, one of the smallest Lower Pleistocene's mandible, a nearly complete sub-adult, and a completely toothless specimen. This archaic *Homo* exhibits strong sexual dimorphism with males being significantly larger than females but she was relatively short, about 1.5m height (Vekua & Lordkipanidze 2010).

These fossil remains date back around 1.8 million years. Of the archaic features of each discovered fossil materials the Dmanisi hominid manifested evident similarity with African contemporances (Vekua & Lordkipanidze 2010).

So, the Dmanisi hominid occupies an intermediate place between *H. habilis* and *H. erectus*, one of the most primitive populations of hominids, in particular *erectus/ergaster* stock. However, other precise investigations of size difference and surprising small brain ($600cm^3$) cases suggestive of an early form led scholars to propose a separate species, *H. georgicus* ('Dmanisi Man'), which they posited as a descendent of African *H. habilis* and ancestor to Asian *H. erectus* (Lordkipanidze *et al.* 2005, 2007). But this classification, however, was not supported, and the fossil is now classified as a divergent subspecies of *H. erectus* (Rightmire *et al.* 2006; Rightmire & Lordkipanidze 2011). The latter was either a species of *H. georgicus* and definitely a subspecies descendant of *H. habilis*. According to another evolutionary assumption, Dmanisian fossils represent a missing link in the evolution of *H. erectus/H. ergaster* and, finally, *H.e. georgicus* is a result of gene flow which occurred between African and Eurasian populations (Figure 136).

Therefore, *Homo* from Dmanisi belongs indisputably to the oldest remains of hominids found outside Africa. It can be argued that this population is closely related to *H. habilis* as known from the most outstanding sites of Africa (from Oldavai Gorge in Tanzania, Koobi Fora in northern Kenya, and Hadar in Ethiopia). The most striking importance of these discoveries was confirmation of suggestion of some scholars (of the first half of the twentieth century) rejecting the 'Out-of Africa' thesis. The latter anthropologists supposed that migration of the initial hominids from Africa to Eurasia took place a maximum of one million years ago. But the sensational finding in Dmanisi has shattered the ground of the firmly established point of view on the initial settlement of human beings.

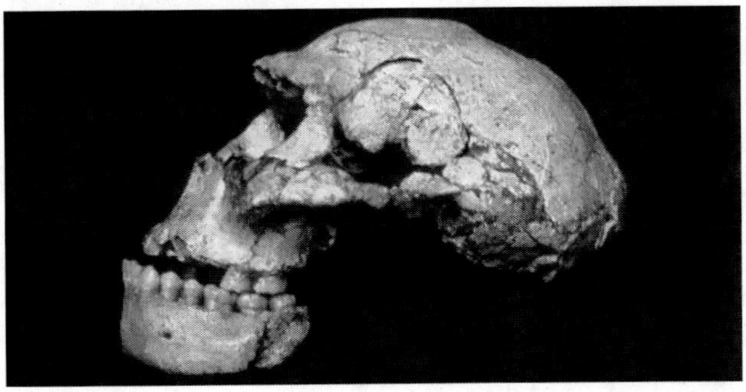

Figure 136. Cranium of *Homo georgicus* discovered in Dmanisi (E. Georgia) and deposited in the National Museum of Georgia (1992).

Palaeontological discoveries made by Georgian scientists, after some skepticism, were largely recognized by the international society of anthropologists, forcing scholars to rethink completely the scenario of evolution and dispersal patterns of *Homo*.

As it seemed after global climate changes the ancient hominid, *H. erectus/ergaster* – being fully bipedal ('Standing Man') spread out from Africa between one to two million years ago and moved rapidly across the globe adapting to diverse environmental conditions. The earliest hominids, highly social and big-game hunters may have followed their hunting prey, and as a sequence moved in different direction from Eurasia (Gamble 1994).

There have been many proposed explanations of the dispersion of *H. e. georgicus* (Agusti & Lordkipanidze 2011). It could be suggested that one successive wave of migration, to members of which was familiar band-societies and hunting in coordinate groups, occurred in the Middle East through the Levantine Corridor and settled in South Caucasus (Gabunia *et al.* 1995, 2000a; Gegechkori 2004; Gabunia & Vekua 1995; Vekua & Lordkipanidze 2010; Rightmire & Lordkipanidze 2011).

According to geologist Red Ferring and his colleagues (Ferring *et al.* 2011) African hominid reached the South Caucasus approximately 1.85 million years ago, during about 80,000 years while this hominid occupied the South Caucasus and was well established within this site. Well adapted to new environmental conditions this population evolved into *H. erectus*. On the other hand, from the point of view of Harvard University's anthropologist Philip Rightmire and his colleagues 'with the new Dmanisi dates, it certainly looks as though the African origin of *H. erectus* must be recognized' (Rightmire *et al.* 2006).

This striking discovery proves that by 1.8 million years ago, some earliest *H. erectus* were able to move from a high thermal savanna biome of Africa to the seasonally cold temperate zones of Eurasia.

The Early Pleistocene colonization of temperate Eurasia by *H. erectus* was not only a significant biogeographic event but also a major evolutionary threshold (Ferring *et al.* 2011). The ability of such a long migration, to live in a hunter-gatherer society and changes of climate zones suggests that the *H. erectus/ergaster* lineage of hominids had the combination of some traits by which they are differentiated from other mammals. As well as bipedism should be mentioned toolmaking, but it may not have been capable of producing sounds comparable to modern human speech unlike of that *H. ergaster* who likely communicated in a proto-language (Ruhlen 1994). *H. e. georgicus* resemblance to *H. ergaster*, was probably the first hominid care for infirm or weak companions (Lordkipanidze *et al.* 2005).

Such a long migration of the early hominids and change of climate zonality 'evoke in the genus *Homo* a change in foraging behavior and diet to include higher quality foods' (Pontzer *et al.* 2011). Today many anthropologists emphasize that the expansion of hominins into Eurasia had triggered a change in diet from herbivore foraging behavior into carnivore-based diet 'that the addition of meat as a significant portion of the diet in early *Homo* increased energy availability while simultaneously increasing selection pressure for improved ranging and cognitive abilities related to cooperative hunting and scavenging and stone tool manufacture' (Pontzer *et al.* 2011 and references therein).

Dmanisi is also a unique site with early stone artifacts, essentially the same as those from the Olduwan culture associated with *H. habilis*. The collection of primitive stone tools exceeds 2000 items (Vekua & Lordkipanidze 2010).

The current palaeofloristic spectrum of the hominid site of Dmanisi shows a considerable variety of vegetation cover and possibly some altitudinal zonality. An area which was occupied by the ancient hominid's settlement was represented by an arid mixed mosaic environment of arid open woodland (species of *Celtis, Ephedra, Artemisia* and some genera of Amaranthaceae), which perhaps resembles a modern savanna, riparian forests (consist of species of *Alnus, Salix,* and *Platanus*) along the river valleys and forested mountains (*Abies, Pinus, Ulmus*, etc.), where some altitudinal zonality was expressed (Klopotovskaja *et al.* 1989, Kvavadze 1997). According to reconstruction of the Dmanisi paleoenvironments, the Dmanisi faunal assemblage contains representatives of different habitats and consequently species with diverse ecological demands.

Implements and animal bones were found alongside the ancient human remains and along the river valleys an arid, open woodland ecosystem predominates. This suggestion is confirmed by the presence of *Struthio dmanisensis, Gazella borbonica, Equus stenonis, Mammuthus primigenius taribanensis* (in former literature as *Archidiscodon meridionalis*) (Bukhsianidze 2015) and others. Altitudinal zonality of the nearby hominid's settlement to forested mountainous areas is confirmed by the richness of the Cervid (*Cervus perieri)* fauna, where they represent more than 80 percent of Artiodactyla fauna. The species of the same autoecology could have lived in different belts of the forest biome, for example, *Sorex* sp., possibly wild cattle (*Bison* (*Dmanisibos*) *georgicus*). Carnivores were represented by *Canis etruscus, Ursus etruscus, Panthera gombaszoegensis* and some others (Vekua & Lordkipianidze 2010 and references therein).

Finally, Dmanisi is the only site in the world where a complete population of early hominid was discovered.

The listed outstanding discoveries soon became the subject of speculation as though 'Georgia is the motherland of the first Europeans. . . , it is not French, English or German, neither Spanish nor Italian. The first European is Georgian' (*Le Monde,* 19/IV. 2000). 'Georgian discoveries put an end to Spanish-Italian debates – who the first European was' (*Le Figaro, 14/VIII. 2000*) (see Vekua & Lordkipanidze 2010).

But from our point of view such type of speculation could be groundless because of *H. habilis/erectus* stock are unlikely to have been direct ancestors of modern humans – *Homo sapiens sapiens*. One of the undoubted confirmations of the above is: among existed hominids, as it seems the most successful, distributed within the largest range (Afro-Eurasia), *Homo erectus*, one million years later after its origin has become extinct having no living relatives (without direct offsprings). On the other hand, it is impossible that without 'missing links' *Homo ergaster* has remained a modern human race, including Europeans. Today it is broadly recognized that our own species appears to have evolved in Africa within the last 300,000 years, and then migrated to Europe and Asia. Phenotypical differences among human's populations in skin color and other anatomical features reflect adaptation to different environments (Gegechkori 2004, pp. 42-44).

2.6.1.3.4. Forests with Sweet Chestnut (*Castanea sativa*)

Sweet chestnut (also known as European or Spanish chestnut) is an outstanding representative of Arcto-Tertiary Geoflora (Wulff 1944). Results from pollen studies indicate that this remarkable tree managed to survive the Quaternary Ice Age and its peak – the Last Glacial Maximum.

According to Krebs *et al.* (2004) there are six regions (refugia), where sweet chestnut survived: (1) areas to the south of the Black Sea, (2) southern and middle Italy, (3) north-eastern Italy, (4) northern parts of the Iberian Peninsula, (5) southern Greece, and (6) north-western part of Syria. To quote Ketenoglu *et al.* (2010) 'the genus *Castanea* has been thought to have originated from Asia, Tertiary Period. Later, two ways of migration have occurred: its eastward migration gave rise to the American chestnut – *C. dentata*, while westward migration resulted in the European chestnut – *C. sativa*' (Zohary & Hopt 1988 cited in Ketenoglu *et al.* 2010) (Figure 137).

Sweet chestnut is a species of the genus *Castanea* (fam. Fagaceae). The Latin species name *sativa* means 'to sow', i.e., 'cultivated by humans'. Today it is widely cultivated and naturalized in Europe and some localities in temperate Asia (Sales & Hedge 1996; Dolukhanov 2010).

According to Villani *et al.* (1994), the evolution and spread of sweet chestnut in Europe has occurred in: (1) eastern Turkey, the supposed center of origin of the species; (2) western

Turkey, the area in which human domestication started; (3) Italy, where domesticated chestnut was first introduced to the rest of Europe by the Romans; and (4) France, representing the latest phase of expansion, close to the northern limit of the taxon.

Figure 137. Pure grows of sweet chestnut. Akhmeta municipality environs, Kakheti province, E. Georgia (1982).

Genetically very close to sweet chestnut is *Castanopsis* with over 100 species represented in subtropical-tropical SE Asia with only two species occur in western part of North America (Sales & Hedge 1996).

C. sativa is a valuable tannin-rich timber, which has many uses. Numerous culinary and medicinal uses, mainly for its prickly fruits that contain the nuts, go back to Roman times. Popular as a landscaping tree in the 18th and 19th centuries, sweet chestnuts were planted in parklands and large estates (Haltofova & Jankovsky 2003).

C. sativa is a deciduous, long-lived (up to 500 years old) tree up to 35m tall, with a trunk up to about 2m in diameter. The bark often has a net-like pattern with deep grooves running up the trunk. Toothed leaves are 16 to 28cm long, 5 to 9cm wide and 10 to 20cm long catkins hold the flowers of both sexes. *C.sativa* has been cultivated for over two thousand years for its edible starchy nuts ('food for poor people') (Abashidze 1962). Chestnut is nutritionally very similar to wheat, except for a lack of the protein gluten, a binding agent. Chestnut is an extremely durable timber; it has been traditionally used to make furniture, barrels, vineyard-stakes, fencing posts, rails, gates, coffin boards and roof beams (Abashidze 1962).

In western Georgia (Colchis, Kolkheti) with high levels of moisture, usually are built the traditional (home owned) one–and generally two-storeyed wooden houses with a richly decorated timber-balcony and a conjunction of poles with the roof corners. A type of oak and especially sweet chestnut as an incomparable timber are common building raw materials throughout the area. The high natural durability of chestnut and the tendency of iron fastenings to corrode in moist conditions makes chestnut the builders choice as a consequence of the ability of this hardwood to resist decay in high moisture area and its outdoor use without the need for preservative treatment (Ketskhoveli 1982).

As a rule, many of them have lasted every bit as well as others built in stone or brick in eastern Transcaucasia. Some of them have reached a level of development, which can hardly be improved by modern technology, and this alone has ensured their survival well into the 21th country, where the majority of Americans and some Europeans still live just in wooden houses (Bramwell (Ed.) 1976; *en.wikipedia.org./wiki/house*).

Like the box tree, in prehistoric time the natural range of sweet chestnut noticeably prevailed on the one today. Due to anthropogenic factors, first of all logging for timber, the chestnut's forest has always (until today) been subject to destruction (Figure 139). Unsanitary conditions became the main reason due to the threat of mass fungal and viral pathogens and has resulted in severe losses of chestnut forest communities (Ketskhoveli 1960, 1982, Dolukhanov 2010).

Sweet chestnut can be attacked by some diseases and pests on the bark, leaves and fruits. The most serious threats for chestnut communities today is microflora, e.g., Chestnut blight (*Cryphonectria parasitica*) (formerly *Endothia parasitica*), supposed to cause bark cancer (Dolukhanov 2010). The fungus is native to China also occurs on chestnut trees in Japan and northern India. In Europe the chestnut blight was observed in Italy in 1938 and then it spread rapidly throughout Europe (Hreidarsson 2011 and references therein).

Two major fungal diseases *Melanconis modonia* and *Phytophthora cambivora* are causing ink disease and significant problem in many countries of Europe, Asia Minor and the western Caucasus (Lyubenova *et al.* 2008 and references therein, Dolukhanov 2010) (Figure 138).

It is noteworthy that chestnut ink through the fungal decay includes very large areas in the western Caucasus. Sweet chestnut trees could be weakened by drought and then became a victim of the fungi giving losses totaling hundreds of trees. Diseases include stem lesions, leaf loss, crown dieback and a large percent (33.7) of empty, immature fruits (Abashidze 1952).

Figure 138. Sweet chestnut forests. Many specimens in the canopy have been killed and damaged by saprotrophic fungi. Racha-Lechkhumi province, environs of the Nakerala Mountain pass (about 1,100m a.s.l.) (2010).

C. sativa has entered into Red Data Books of Azerbaijan, Georgia and the Russian Federation.

Thanks to its vital capacity now in certain abiotic conditions of the Caucasus the sweet chestnut participates in the formation of broad-leaved forests′ climax vegetation communities. Therefore, this relict species of hardwood is of particular interest (Bebia 1989).

Forests composed of chestnut with an admixture of other deciduous trees (mainly with beech or hornbeam) have a more limited distribution in the Caucasus than oak and beech forests. Pure stands of the chestnut forest are rare. *C. sativa* is a warmth-and moisture-loving species. These requirements limit the distribution of *C. sativa* to the south-western part of the region (Gulisashvili 1950).

In the North Caucasus chestnut is rare, but occurs as a component of oak forests and in its western part, on the slopes of the Greater Caucasus, as small populations (Maleev 1941, Grudzinskaia 1953). Chestnut forests are widely distributed in Western Transcaucasia, in Kakheti (East Georgia) and in the Zakatala district (Azerbaijan). In East Transcaucasia, it originates in the natural belt of *C. sativa* forests; covering both the southern slopes of the Greater Caucasus Range and on the other hand in the Adjaro-Imereti Range the northern slopes of the Lesser Caucasus (within the borders of Adjara, Guria and Imereti in Georgia). They also occur on the Surami Range, the 'natural bridge' between the Greater and Lesser Caucasus (Gulisashvili *et al.* 1975).

Chestnut forests are mainly restricted to shady slopes. They extend from 500 up to 1,000 m a.s.l, though small groups or single trees may also occur above and under these altitudes. The absolute upper limit of *C. sativa* is at 1,400-1,500 m, in West Georgia, whereas the lower limit of forests is about 100 m (Dolukhanov 2010). This range is determined by the specific autoecology of *C. sativa*. The upper limit is not determined by winter frosts, but by insufficient warmth during the vegetation period. The latitudinal distribution of chestnut in the Caucasus depends completely on mean annual precipitation, with 600-800 mm as the minimum quantity. Mean annual temperature of the distribution zone of *C. sativa* ranges from 9.8° to 12.2° (Gulisashvili 1950). Therefore, the chestnut tree is a typical representative requiring a moderate type of climate with sufficient precipitation and rather warm winters (average January temperature between -0.4° and +3.2°) (Grudzinskaya 1953; Gulisashvili *et al.* 1975).

Figure 139. Timber logging (actually poaching), using hard equipment. W. Georgia, Imereti province (2013).

Chestnut forests primarily occur on heavily leached brown or yellow-brown soils. They avoid calcareous soils, although chestnut can be found on a limestone substrate in several localities in the western part of Transcaucasia (Gulisashvili *et al.* 1975).

Most chestnut forests in the Caucasus have no arboreal undergrowth and a poorly developed herb layer (with *Hedera colchica, Paris incomplecta, Primula vulgaris sibthorpii, Viola silvestris*) (Gulisashvili *et al.* 1975; Dolukhanov 2010). In the canopy, chestnut is frequently accompanied by Caucasian beech (*Fagus orientalis*) and particularly by hornbeam (*Carpinus betulus*). Such forests are well-represented on steep slopes in deep moist and shady gorges (Gulisashvili *et al.* 1975).

According to Gulisashvili *et al.* (1975) and Dolukhanov (2010), *C. sativa* forests lead to the formation of various associations.

2.6.1.3.4.1. Chestnut Forest with Cherry Laurel (*Prunus laurocerasus*)

Chestnut forest with cherry laurel (*Prunus laurocerasus*) in the undergrowth also have a characteristic undergrowth of *Rhododendron ponticum* which forms dense thickets of up to 3-4 m high.

2.6.1.3.4.2. Chestnut Forest with *Rhododendron luteum*

Chestnut forests with *Rhododendron luteum* occupy drier habitats than the previously mentioned types. In the canopy chestnut mixes with hornbeam and Georgian oak (*Quercus p. iberica*). The underwood mainly consists of *Rh. luteum,* but rarely also *Crataegus monogyna* and *Vaccinium arctostaphylos* occur. The poor herb layer consists of *Pteridium tauricum* and *Rubus caucasicus*.

Rh. luteum is widely distributed in the Caucasus. Depending on the soil type the shrub participates in an understory of various forest communities (deciduous, coniferous). As mentioned, it is a single deciduous species of its genus in the region. It is a shrub growing to 3m tall, rarely 4m. It doesn't form a discontinuous shrubland. Yellow flowers are strongly perfumed, however, despite the sweet perfume, the nectar of the flowers is toxic, causing the poisoning of people eating the honey produced by honey bees.

On the rocky habitats of the lower flow of the Adjarustskali River gorge (Adjara) rarest to the Colchis region, is the Mediterranean proper origin shrub, *Osmanthus decorus* occurs (Gulisashvili *et al.* 1975; Dolukhanov 2010).

2.6.1.3.4.3. Chestnut Forest with *Vaccinium arctostaphylos*

Chestnut forest with *Vaccinium arctostaphylos* is widely represented between altitudes of 800 and 1,200 m a.s.l. In this type *Castanea sativa* forms pure stands and is only rarely accompanied by oak, beech and maple. The main component of the undergrowth, the Caucasian blackberry (*V. arctostaphylos*), forms a rather dense thicket. This high shrub (1.5-2.0 m) forms a mixed shrub layer with *Rhododendron ponticum* and *Rh. luteum*. The herb layer is poorly expressed and includes *Rubus ibericus* and *Festuca montana* (Gulisashvili *et al.* 1975; Dolukhanov 2010).

2.6.1.3.4.4. Chestnut Forest with *Festuca montana*

Chestnut forest with *Festuca montana* has *Rh. luteum* and *Corylus avellana* in the shrub layer, whereas the well-developed herb strata are represented by - among others - *Festuca montana, Aruncus vulgaris* and *Athyrium filix-femina* (Gulisashvili *et al.* 1975; Dolukhanov 2010).

2.6.1.3.4.5. Chestnut Forest with Fern Cover

Castanea sativa–Dryopteris filix-mas communities occupy rocky slopes. Associated species are: *Fagus orientalis, Carpinus betulus, Alnus barbata, Ulmus carpinifolia, Acer laetum* (Gulisashvili *et al.* 1975; Dolukhanov 2010).

2.6.1.3.4.6. Chestnut Forest with Herbal Cover

Castanea sativa with the following representatives of ground cover: *Galium odoratum, Paris incompleta, Sanicula europaea, Festuca drymeja*. From vines *Hedera helix* is typical in this formation (Gulisashvili *et al.* 1975; Dolukhanov 2010).

2.6.1.3.5. Hornbeam (*Carpinus betulus*) Forests

C. betulus (= *C. caucasica*) is widely represented in the Caucasus, where the thermophilic species grows on fertile soils with good drainage and plays a dominant role in the second stratum of hornbeam-beech and hornbeam-oak forests (Gulisashvili *et al.* 1975).

Figure 140. Pure stand of *Carpinus betulus*. Caucasian deer and European roe-deer decorate the winter landscape - characteristic ungulates of the Lagodekhi NP, Kakheti, E. Georgia (1982).

Hornbeam forests are phytosociologically very diverse. According to Dolukhanov (2010). There are two main types of *C. betulus* forests - hornbeam forests with Colchian undergrowth and hornbeam forests without Colchian underwood. Those with Colchian undergrowth occur on foothills and in the low-mountain zone, where they are not outcompeted by beech woods and dark coniferous forests. Dolukhanov distinguishes four main associations: forests with *Rhododendron luteum*, forests with *Rh. ponticum*, forests with *Vaccinium arctostaphylos*, and forests with *Prunus laurocerasus*.

The hornbeam forests without Colchian underwood have a herb layer with *Poa angustifolia* or with *Trachystemon orientalis* (Figure 140).

2.6.1.3.6. Forest with Caucasian Elm (*Zelkova carpinifolia*)

Among estimated different species of trees of the Caucasus, Caucasian zelkova (elm) is especially worthy of attention. In botanical literature zelkova is frequently pointed out as a 'unique tree', 'remarkable', etc., but little seems to be known about its Latin name. It is noteworthy that 'zelkova' is native, i.e., comes from one of the Kartvelian (Georgian, the Caucasus) languages. For example, the Georgian name ძელქვა (*dzelkva*) is composed of two following words: ძელ (*dzel*) meaning 'bar', and ქვა (*kva*) meaning 'stone' ('rock') (Ketskhoveli 1980b; *https://en.wikipedia.org/wiki/Zelkova*).

Six extant species of elm - *Zelkova* Spach. (Ulmaceae) are recognized in Eurasia; one in the Transcaucasian region (Z. *carpinifolia*), two in southern Europe (Sicily - Z. *abelicea,* Crete - Z. *sicula*), and three in East Asia (China - Z. *schneideriana,* Z. *sinica*, China, Japan, Korean Peninsula - Z. *serrata*) (Denk & Grimm in Kozlowski *et al.* 2012; Kozlowski & Gratzfeld 2013; Maharramova 2015). Kozlowski and Gratzfeld (2013) comment: 'Based on the findings from chloroplast and nuclear DNA analyses, the genus was found to be divided into two main sections: a southwest Eurasian cluster with Z. *sicula,* Z. *abelicea* and Z. *carpinifolia*, and an East Asiatic group with Z. *serrata,* Z. *schneideriana* and Z. *sinica*'. These authors also pointed out that two main clades of Z. *carpinifolia* were identified in the Ponto-Caspian area: (1) eastern clade of the Hyrcanian origin (southern Caspian basin: Azerbaijan and northwestern Iran). This clade includes also disjunct population of eastern Georgia (Babaneuri in Kakheti region), (2) western clade – populations from Colchis (western Georgia and northeastern Turkey). Z. *carpinifolia* from Babaneuri is genetically very close to the Hyrcanian populations (Kozlovski & Gratzfeld 2013).

According to molecular analyses similarities between the two groups suggest close relationships during earlier geological periods. It is also reported from North America in the Early Cenozoic (Burnham 1986 in Kozlowski *et al.* 2012). From mid-Miocene it became fragmented regarding the increasing aridity in Central Asia. Consequently, the eastern Asiatic species has been isolated from their European and western Asiatic counterparts. Furthermore, quaternary climate deterioration began to break up and separate the European range of *Zelkova* (Wang *et al.* 2001). Botanists (Wang *et al.* 2001; Kozlowski & Gratzfeld 2013) indicate that 'fossil evidence confirms that there was once a much wider distribution of closely related taxa across the entire Northern Hemisphere'. To quote Kozlowski *et al.* (2012) 'This made the genus *Zelkova* an important and emblematic object for multidisciplinary research and conservation of plants'.

Z. *carpinifolia* is a relict of the Arcto-Tertiary flora. It is a tree of Euxino-Hyrcanian origin. Its disjunct, refugial distribution in Transcaucasia (Colchis, Karabakh, Talysh) indicates that in past geological times zelkova occupied a far larger area between the Black and Caspian Seas (Abashidze 1962; Nakhutsrishvili 1999, 2013).

Caucasian zelkova is distributed in the Caucasus (Georgia, Azerbaijan (Talysh, Karabakh[5]), Kachkar (Turkey) and the Alborz and Zagros Mountains (Iran) (Dolukhanov 2010).

Except Hyrcanian and Colchian forests, *Z. carpinifolia* through smaller, isolated sited stands expanded its range in northwest and East Georgia, in south Karabakh, in East Turkey and in northern and western Iran (Bobek 1951, 1968; Abashidze 1962; Gulisashvili *et al.* 1975; Dolukhanov 2010; Nakhutsrishvili *et al.* 2011; Gerber & Kozlovski 2014; Maharramova 2015). 'This disjointed pattern of distribution confirms more recent palaeobiological research that indicates, for example, that 6000 to 5500 years ago, forests with *Z. carpinifolia* in Transcaucasia were more widespread than they are today' (Gerber & Kozlovski 2014).

Figure 141. Relict (Arcto-Tertiary) forest stand composed of extensive groves of *Zelkova carpinifolia*, mixed with some other hardwood species. The stand is preserved in the Babaneuri part of the Batsara-Babaneuri Protected Areas; foothills of the GCR, E. Georgia (1976).

A monodominant forest of zelkova in the Caucasus is extremely rare. Such a forest is located in Babanuri Reserve (770ha), East Georgia (Abashidze 1947; Kikodze & Gokhelashvili 2007; Dolukhanov 2010) (Figure 141).

Z. hyrcana A. Grossh. et A. Jarm. was originally described as a different species in Hyrcan (Azerbaijan), but was later treated as a synonym of *Z. carpinifolia*, because it was shown to be

[5] The territory remains internationally recognized as part of Azerbaijan, which has not exercised power over most of the region since 1991.

an ecological or age form of the latter species (Sokolov 1951; Sosnowsky 1952 cited in Maharramova 2015).

Maharramova *et al.* (2015) observed no clear genetic differentiation of *Z. carpinifolia* shared between long-lasting, isolated Colchis and Hyrcania; in addition, they suggest that long-distance pollen dispersal (pollination occurs by the wind) is an important factor of gene flow among populations of the mentioned species of the zelkova tree. According to Kvavadze and Conor 2005 cited in Maharramova *et al.* 2015), pollen of *Z. carpinifolia* is characterized by long distance – up to 100 km – dispersal, but the findings of Maharamova *et al.* (2015) highlighted the possibility for zelkova's pollen dispersal over larger distances, 200 km or more. The authors suggest that before the climate oscillation, origin of topographical barriers and recent human-induced fragmentations of the relict species should be possible with more stepping-stones throughout the range of the populations of the relict *Z. carpinifolia*. So, such long-distance pollen dispersal, which is described in some papers (Petit & Hampe 2006; Heslewood *et al.* 2014 cited in Maharramova *et al.* 2015) can explain pollen-mediated gene flow among long-lived zelkova's populations within and between Colchis and hyrcan.

To quote Maharramova *et al.* (2015); 'The results suggest that spatially isolated populations are not isolated genetically as a result of gene flow…'.

In the lowland and lower parts of the forest belt of West Transcaucasia, considerable attention is attracted by the elm forest. Caucasian zelkova is a medium-size to large deciduous tree growing to 20-35 m tall, with a massive, strong and durable trunk of up to 2m in diameter. In climatically and edaphically favorable locations of the Talysh and Alborz (Iran) mountains, however, trees can reach up to 40m with a diameter of 3-4m (Safarov 1966; Kozlowski & Gratzfeld 2013). Huge individuals of it can be found quite often in Imereti (West Georgia) (Ketskhoveli 1960, 1982). The bark is greenish-grey in young trees and brownish-grey in older trees (Maharramova 2015). Caucasian zelkova is a long-lived tree, for example, 800 to 850 - year-old individuals have been reported from the Ajameti Nature Reserve (Western Georgia) and Talysh Mountains (Azerbaijan) (Safarov 1966; Ketskhoveli 1982; Kozlowski & Gratzfeld 2013).

Zelkova has a large, vase-shape multiple branched crown with a short broad trunk. It is grown as an ornamental tree. With its symmetrical canopy almost no other trees in the Caucasus will truly match the grace and elegance of the Caucasian zelkova. The girth of this majestic tree increases with age. The surface of the trunk of a mature tree begins to separate into 'independent' constructions' adding to this species of Ulmaceae an unusual, striking, grandeur appearance. Single ancient big trees inspire with awe, and their individuality and historical importance are recognized by the trees being given proper names (e.g., 'beladi' in Babaneuri), which means 'chief' (Abashidze 1962; Kacharava (Ed-in-chief.) 1982; Ketskhoveli 1982; Kikodze & Gokhelashvili (Eds.) 2007) (Figure 142).

Zelkova prefers a warm climate and good lighting habitats (Abashidze 1962; Akhalkatsi & Kimeridze 2012; Nakhutsrishvili 1999, 2013). It is confined to the foothills at altitudes of 430-500 m a.s.l., however, in the Talysh mountains it occurs at 1,200-1,500m or can even grow above 1,500m a.s.l. (Grossheim 1926; Safarov 1962, 1966; Gurbanov 2006). In Anatolia, as a mesoxerophytic tree, dwarf forms are found on stony soils (Kozlowski & Gratzfeld 2013; Denk *et al.* 2001).

Figure 142. Archetypical specimen of the Arcto-Tertiary tree – Caucasian elm. About a 2,000-year-old tree preserved in Lenkoran NP (Zuvand), 1,200m, Azerbaijan (1974).

In West Georgia (Colchis) zelkova forests frequently develops on limestone ground at an altitude between 200-700m. Here a zelkova forest is characterized by the admixture of oaks (*Quercus r. imeretina, Q. petrae iberica, Q. hartwissiana*) and hornbeams (*Carpinus betulus, C. orientalis*).

Two sub-types and more than 10 plant communities are described by botanists (Sharashidze 1967; Akhalkatsi & Kimeridze 2012).

The botanists and foresters specify zelkova as a species with admixture of other deciduous relict trees of the Caucasus (Grossheim 1926, 1948;; Gulisashvili *et al.* 1975; Gerber & Kozlowsky 2014).

Zelkova is timber-dense, light, stable and rot-resistant. Timber is used for various tools, but in general elm's timber is associated with building and construction materials (shipbuilding, various construction), also paper production, etc. Apart from hardwood timber, fine in texture, it is largely used in traditional furniture (Ketskhoveli 1982; Kozlowski & Grafzfeld 2013).

Over centuries the low- and mid-mountains forests of Transcaucasia have been under strong pressure from extraction of timber and firewood, from grazing and from browsing. Caucasian zelkova as a rare tree became threatened or endangered as a natural component. Today it is protected in some reserves and Natural Parks within Caucasus Ecoregion and adjacent areas (e.g., Babaneuri Nature Reserve (Georgia), the Hyrcan NP (Azerbaijan), Golestan NP (Iran).

It should be pointed out that many regions of Iran, which are covered with Caucasian elm, have faced the problems of fungal diseases (e.g., *Ophiostoma novo-ulmi*). It caused the host plants extinction. Until now the fungal pathogen has infested mainly species of *Ulmus*, another genus of Ulmaceae. *O. novo-ulmi* causes leaf wilting and a subsequent mortality event with high rate (Ahmadi *et al.* 2014).

Figure 143. The elm tree is a pagan symbol in Georgia. It symbolizes eternal life and immortality. This huge specimen of tree decorated with many colored stripes. Imereti province, W. Georgia, near Rioni River riparian forests (2012).

Z. carpinifolia is classified as 'Lower Risk/Near Threaten' in the IUCN Red List.

In addition to the ecological and socio-economic benefits zelkova has cultural, religious and spiritual values as well (Kozlowski & Gratzfeld 2013). It should be suggested that *Zelkova* is a pagan symbol in Georgia. It is identified with the concepts of incredible beauty and at the same time majesty, the soul's immortality. For this reason, elm trees are often planted in cemeteries (Figure 143).

In 2010, the Botanical Gardens of the University of Freibourg (BGF, Switzerland), in collaboration with the Natural History Museum Fribourg (NHMF, Switzerland) and the Botanical Gardens Conservation International (BGCI, United Kingdom), initiated an interdisciplinary project to undertake the scientific review and coordinated conservation actions on relict, rare and threatened woody species, with a main focus on the relict tree genus *Zelkova*. Today in this project, an international team participates consisting of 20 institutions from 10 countries worldwide (head of team Prof. Dr. G. Kozlowski).

2.6.1.3.7. Mesophilic Wild-Growing Pears (*Pyrus* L.) of the Caucasus

The genus *Pyrus* has fruit of great importance for the agriculture of countries of a moderate latitude (Silva *et al.* 2014). Asian pears are thought to have been cultivated in China about 3000 years ago (Lombard & Westwood 1987 cited in Bell & Itai 2011). European pears are thought to have been cultivated in Europe from about 1000 BC (Bell & Itai 2011).

The Caucasus, Asia Minor, Iran and Turkmenistan represent a center of pears diversity (Silva *et al.* 2014). The name pear is derived from Latin, *pere* or *pira* as wild type and *apios* as cultivated for wild ancestors of edible European pears (*P. communis*) (Vavilov 1931,1992).

P. caucasica is progenitor of many local pear varieties and cultivated pears (Brezhnev & Korovina 1981; Likhonos *et al.* (Eds.) 1983). In a recent classification system of the genus *Pyrus*, Caucasian *P. caucasica* and European *P. pyraster* are assigned to the rank of subspecies. However, Caucasian wild pear is represented as a separate species by many authors (Grossheim 1946; Ketskhoveli 1960; Brezhnev & Korovina 1981, etc.). Caucasian wild pear from the Caucasus region is believed to be ancestral to the cultivated pear, *P. communis* (Asanidze *et al.* 2011; Asanidze *et al.* 2014) and *P. caucasica* and *P. pyraster* are thought to be most likely ancestors of the domesticated European pear (Challice & Westwood 1973 cited in Volk *et al.* 2006). European *P. pyraster* originates from the areas west and south of the Black Sea (Balkan countries, Turkey and other European countries). It is a vicarious form for *P. caucasica* occupying its range in the Caucasus and the Crimea (Volk *et al.* 2006) (taxonomically both species of wild pears are accepted as subspecies by the cited authors).

The third European species *P. nivalis* is used to make wine and has been of great importance in Britain and France for over 400 years (Silva *et al.* 2014).

During this period, ancient Mediterranean fruits such as grape, olive, fig, and pomegranate have been domesticated. 'The earliest mention of growing pears in Europe was made by Homer in ancient Greece, a little less than three thousand years ago, who wrote that 'pears are a gift of God' (Layne & Quamme 1975 cited in Silva *et al.* 2014).

Many pear cultivars occur in Georgia from the pre-historic period (Javakhishvili 1930 cited in Asanidze *et al.* 2011) indicating the early domestication event of this cultivated fruit tree in this country (Asanidze *et al.* 2011).

As already mentioned, *Pyrus* has 11 species in Georgia (Kutateladze 1980; Gagnidze 2005). Among these species two (*P. caucasica, P. balansae*) are mesophilous (sect. Archos Kochne emend. Fed). If the former species is endemic to the Caucasus, *P. balansae* is restricted within some provinces of Georgia (Adjara, Lechkhumi, Samtskhe-Javakheti, Samegrelo), also in northern Turkey (where it grows in beech forests with Colchic understory), and on Greek islands (Chouliaras *et al.* 2003 cited in Asanidze *et al.* 2011, 2014) (Figures 144, 145).

Figure 144. Growths of *Pyrus balansae*. Lechkhumi, surroundings of Khvamli limestone massif, 1,750 m a.s.l. (2016).

In a recent work (Asanidze et al. 2014), the authors consider that the features for description of *P. balansae* as an independent species (Decaisne 1920s cited in Asanidze *et al.* 2014) are not sufficient to rank it as a distinct species. Both species have close relations with domesticated pear cultivars in Georgia (Khomizurashvili 1973 cited in Asanidze *et al.* 2011).

Figure 145. Fruits of *Pyrus caucasica*, Tianeti municipality, E. Georgia (2016).

Of recent species, *P. balansae* is allopatric with *P. boissieriana*, *P. hyrcana* and *P. mazanderanica* in Hyrcanian Province (Azerbaijan, Iran) (Zamani *et al.* 2008).

The wild pear, Caucasian mesophyte *P. caucasica* is one of the dominant tree species in the forests (broad-leaved, mixed coniferous) of the Caucasus, in mountainous and foothill areas (300-1890m) (Asanidze *et al.* 2011) with sufficient moisture. It is usually part of the stands in first- and second-story levels of deciduous and mixed forests. It is deciduous tree to 20-25m tall and 60 cm in diameter. There are a large number of fruits which are edible when ripe, melliferous, honey plant, fruits are also used to make vodka; the tree is a light demanding plant preferring mostly open habitats. Despite occurring in sufficient moisture, it is cold- and dry climate-tolerant species (Ketskhoveli 1960; Abashidze 1962).

2.6.1.3.8. Pattern of Altitudinal Distribution of the European Yew (*Taxus baccata*) in West Georgia

The Latin name 'Taxus' (i.e., 'toxic') is derived from the Greek 'taxis'. From the word taxus the Latin word 'taxicum' meaning 'poison' and the word 'textile', woven with the fibers of the bark (*www.european-trees.com/irish-uew.html*); *baccata* is Latin for *bearing red berries*. It was known to Theophrastus as an evergreen shade-tolerant species (*https://en.wikipedia.org/wiki/Taxus_baccata*).

It should be stressed that 'Taxaceae radiated from southwest China and evolved into several species in Europe by the Tertiary period' (Spjut 2000 cited in Dhar 2008).

The genus *Taxus* encompasses around seven closely related species scattered throughout the northern temperate regions.

The European (English) yew (*T. baccata* fam. Taxaceae) is, as an evergreen, non-resinous gymnosperm conifer tree species up to 20-28 m, trunks to 1.5m diameter, with a spherical or pyramidal canopy. Below a dense canopy, yew character is extremely shade-tolerant even at maturity (Dolukhanov 2010), but can withstand full exposure to the sun and dryness (Kikodze *et al.* 2007, Dhar 2008). The yew is less demanding from the soil. It grows on well-drained, weak alkaline calcareous soils (Dolukhanov 2010). It reproduces with seeds (yew is a dioceous species), but is characterized with better vegetative reproduction (Abashidze 1962; Ketskhoveli 1982).

The current geographic distribution of English yew is known in the northward up to 63^0N latitude in Scandinavia (Vidakovič 1991 in Thomas & Polwart 2003). The southern limit includes the Iberian Peninsula, also Madeira and the Atlas Mts (Algeria, 33^0N), the Pontic and Taurus Mts in Turkey, the Caucasus and Alborz Mts, in northwestern Iran. Today yew grows sparsely and does not correspond to the places it used to occupy millions of years ago (Charles 1982 in Kassioumis *et al.* 2004). Large populations of English yew still are found in northern Spain and Italy (Kassioumis *et al.* 2004), in Central Europe and the Caucasus (Bugala 1978 in Thomas & Polwart 2003). But the majority of natural yew woods, as mentioned, are fragmented and isolated (Quézal & Médial 2003 in Kassiomus *et al.* 2004).

According to Goginashvili & Tvauri (2016), English yew occurs in all mountains forests of the Caucasus and altitudinally reaches up to 2,000m a.s.l. The maximal distribution is recorded in the Himalayan Temperate forests (Kashmir Himalayas), where it occurs at 1,800-3,400m (Shaneen *et al.* 2015 and references therein).

T. baccata is a rare relict. According to fossil data, evolution of the genus *Taxus* occurred in the Northern Hemisphere since the Cenozoic Era.

Figure 146. Climax and pure stand of the European yew. Batsara part of the Batsara-Babaneuri Protected Areas. Akhmeta district, Kakheti province, E. Georgia (1989).

There are 130 known yew stands in the Western Caucasus (some of the most recorded data in the North Hemisphere), of which a Sochi yew stand in the southwestern Caucasus encompasses 301 ha, and a Khosta yew-box grows on 190 ha (Prydnya 2002), the Batsara Reserve stand in Kakheti (East Georgia) also is one of the largest (237 ha) (Gulisashvili 1937) (Figures 146, 147).

Figure 147. Soft, bright red, berry-like 'fruit' (aril) of the European yew, Batsara-Babaneuri PA (1973).

T. baccata occurs especially in beech forests and more rarely in oak forest zones, where *Fagus orientalis* is wedged out (Gulisashvili *et al.* 1975; Dolukhanov 2010; Goginashvili & Tvauri 2013). Over these vast territories, yew may form pure stands or forests with other species like beech, hornbeam and lime-tree in the canopy. The latter, mixed – Oriental beech forest with yew (Fageta-taxceto) – in East Georgia (Kakheti district) in the Alazani River's left tributary, as mentioned, occurs in the Batsara River Gorge. In this area are relatively pure stands of *T. baccata*, which have still remained preserved in the Batsara Reserve. Here it forms one of the most extensive and best virgin stands in the world (Prydnia 2002; Dolukhanov 1948, 2010).

To protect the area with the unique monospecies stands of yew in 1935 the Batsara Natural Reserve was founded (Dolukhanov 1948; Gigauri 2000).

The European yew in Transcaucasia occurs also in Armenia, the Tarsa-Tchai gorge (25 ha) and vil. Dilizhan environs (about 10 ha), in Azerbaijan (e.g., yew stand of the Talysh Mts) (Ketskhoveli 1982).

T. baccata in the Caucasus extends within the limits of forest distributions in high-mountains (2,200m) - in the treeline ecotone of the Greater Caucasus. Here, in the subalpine zone, it grows as solitary shrub (1-1.5m tall) (e.g., Okhachkue limestone outcrops (2,200m a.s.l.). This is a common habitat of the Caucasian rhododendron represented as an elfin stand (Gegechkori 2018) (Figure 148).

Khvamli (top of mountain is 2,002m a.s.l.) is an isolated area. It gives the impression that it is really the single subalpine site in the limestone massif of the Greater Range, where a gentle top-flat mini-plateau (1,800-1,850 m a.s.l.) is associated with the homogenous semi-prostrate patch of *T. baccata* with a small straight-trunk yew (about 5-6 meters in length). In a local, harsh environment the yew shrub inhibits self-regeneration. Observation revealed that yew as a shrub canopy occurred in patches within calcareous rocks in high density, without the support of a closed canopy of neighboring forests (ecologically selfshading effect). So, to discover as a solitary form (Okhachkue limestone massif, 2,200m a.s.l.), such rare plant populations of relict plant communities of yew (Khvamli) holds great scientific interest for living nature (vegetation cover) of the Greater Caucasus (Gegechkori 2018).

Figure 148. Solitary, low shrub of yew in the Greater Caucasus, 2,200 m a.s.l. (Samegrelo) (2015).

2.6.1.3.9. Forest with Colchic Box Tree (*Buxus colchica*)

Two refugial areas of the South Caucasus are represented by two species of box tree with a disjunct range between Colchis and Hyrcania (*B. hyrcana*), which surely is considered as a relict species (Grossheim 1936, 1948; Gulisashvili *et al.* 1975; Browicz 1989; Nakhutsrishvili 2013).

B. colchica is a native of West Georgia with its main occurrence in the limestone localities of the Greater and Lesser Caucasus in Abkhazeti, Samegrelo, Racha-Lechkhumi, Imereti, Guria, Adjara (W. Georgia) and partially extends into NE Turkey. In East Georgia boxwood stands are in the Aragvi River Gorge, Saguramo, the Bursa River bank, Story River Gorge and other places of Kakheti (Ketskhoveli 1960; Abashidze 1959).

There are different opinions on the primary origin of the boxwood in East Georgia. Some scientists consider that it is Tertiary relict, which is preserved in refugial sites (Troitsky 1928; Matikashvili 1953). The majority though suggest that the box presence is mainly attributable to a secondary wild form of cultivated (planted at the surroundings of churches) specimens (Tugushi 1972).

B. colchica grows in moist gorges from sea level up to 1,400 (1,600) m. On some mountain slopes it frequently forms pure stands of box groves. Under a deciduous forest canopy (hornbeam, beech, etc.) it may form the undergrowth, but in some upper reaches of the rivers of West Georgia as a middle tall (10-15m) tree it attains a role as canopy builder (make the first layer of montane forest). Colchian box, highly valued for its timber, was subject to intensive logging, especially in the 19th century. Therefore, some groves untouched by human hand are preserved only in some inaccessible riverheads in West Georgia.

Figure 149. The primeval stands of Colchic box tree, completely covered with mosses and lichens (*Usnea barbata*), GCR, Samegrelo, Abasha River tributary - source of the Toba River, W. Georgia. This growth almost inaccessible to visit (1986).

The flora of forests with box tree on limestone slopes is characterized by a wealth of local endemics, including *Ruscus ponticus, Hedera helix, Cyclamen abchasicum, Carex divulsa, Veronica peduncularis, V. persica,* etc. (Dolukhanov 2010) (Figure 149).

Numerous alien species – pests and diseases (plants, fungi, or animal species) have become successfully established in various habitats in the South Caucasus. They cause loss and disturbance not only to the environment, but also to the human economy. Some of them were introduced recently, from the 1990s.

Currently, among other deciduous and evergreen trees and shrubs boxwood suffers from several introduced pests. Recently Colchic box tree has declined or has been completely eliminated in some areas of Colchis by a combination of below characterized eradicating pest and disease, which independently act together.

The newly emergent disease of box tree, boxwood blight is caused by the invasive fungal pathogen – *Calonectria pseudonaviculata* (Crous *et al.* 2002). (syn. *Cylindrocladium buxicola, C. pseudonaviculata*) (Guo & Olsen 2015) (Figure 150).

The fungus severely infested plants may show sudden and complete defoliation (they kill the shrub and tree, therefore devastate boxwood in natural habitats or destroy ornamental horticulture and nursery (Matsiakh 2016; *https://www.cabi.org/isc/datasheet/17414*).

Figure 150. Saprotrophic fungi causing death of box trees in Western Transcaucasia, Mtirala NP, Adjara, W. Georgiaq.

Box tree moth (*Cydalima perspectalis*) (Lepidoptera, Crambidae), similar to boxwood blight, is an invasive pest introduced to Europe in the last decade. It causes serious damage (up to extinction of the plant) to its natural habitat and ornamental box (*Buxus* spp.) shrubs and trees, including relict species (*B. colchica*) represented in Colchis region (Matsiakh 2016) (Figure 151).

Figure 151. Box tree moth (photo T. Matsiakh, 2016).

Nowadays, there are no effective methods of protection of Colchic box from the above characteristic pest and disease (Patarkalishvili 2017).

Conclusion

- As a terrestrial biome, nemoral or temperate broadleaf deciduous forests (TBDF) are located in the mid-latitude areas, particularly in the Northern Hemisphere. It should be mentioned that northern and southern limits of the biome largely depend on soil types and microclimates. The three major areas of this thermophilic dendroflora are; eastern North America, East Asia and southern Europe. To other, globally distinctive regions also belong: the Himalayas, the Caucasus, and the Russian Far East.

- The hardwood arboreal plants in winter-deciduous forests are angiosperms. The key factor of these forests is the distinct seasonality of life processes, which is significantly expressed in the seasonal appearance and disappearance (in the dormant period) of the canopy.
- The essential abiotic factor is the average annual precipitation, which exceeds 1000 mm and falls during 4-6 or 6-8 months. Such moist forests are largely free of the threat of wildfire.
- Regions with an oceanic climate, experience high humidity (precipitation exceeds 1500 mm per year and falls during about 10 months). Aside from deciduous forests with mesophilic species, it is hypothesized that such climate peculiarities provide the growth of large coniferous evergreen forests of the Arcto-Tertiary Geoflora (ATG) origin (e.g., magnificent redwood forests in California, yew and yew-box forests in the Black Sea coastal area of the western Transcaucasia).
- In the temperate regions of the Northern Hemisphere representatives of the Fagaceae (species of deciduous *Fagus*, *Castanea* and *Quercus*) are the dominant plants of the TBDF biome. At the same time genus *Quercus* is the marker of distribution pattern of the summer green forests.
- Mature hardwood forests of southeast inland Europe include single – species – dominant or monodominant forest stands (e.g., species of *Quercus, Fagus*); forests of the coastal northwest of Europe comprise two- or three co-dominants (oligodominant forests); the central part of the Colchis Triangle (Adjara, Guria) with oceanic climate consists of five co-dominants species; finally, eastern Asia contains of 10-12 co-dominant species (polydominant forests). Similar species-rich and co-dominant nemoral summer green forests occur in eastern North America.
- Structurally deciduous forests can have four-five layers of plants: (1) top layer of large trees on canopy, (2) shorter layer of trees, (3) understory, (4) forest herbs (forbs), (5) mosses (that live in the shady forests), and lichens growing on tree trunks or infrequently on ground area; hardwood woodlands also contain several vine species. In addition, in early spring annuals and ephemeroid perennials appear.
- In the northern temperate regions, under the TBDF are developed nutrient-rich brown forest soils (podzols or alfisols). More heavily leached, distinctive red or yellow forest soils (ultizols) are found in countries with a warmer climate.
- Nutrient-rich fruits, foliage, a rich decaying litter, etc. provide opportunities for fruits, nuts, acorns and other frugivores, herbivores, carnivores and omnivores. The majority of birds, mammals and invertebrates shows adaptations to an arboreal life. Four seasons of TBDF, rhythm of plant production and insect activity determine vertebrates' behavior (hibernation, migration and so on).
- The southern counterpart of summer green forests in the Southern Hemisphere (southernmost South America) and New Zealand is represented by southern beech - *Nothofagus* forests. In these floral communities, thanks to an oceanic climate, leaves of one or few species fall as the cold season approaches.
- Winter-deciduous forests, supposedly descendent from evergreen subtropical forests, are represented in the Cenozoic Era, in high-latitude and high-altitude areas.

- In the Northern Hemisphere from the Mesolithic until today there are no natural TBDF ecosystems or vegetation coverage found, which remained unimpacted by human agricultural activity.
- The Caucasus TBDF biome, equally with the biomes of arid and semi-arid climates (significantly deserts and semi-deserts) has primary importance.
- Forests with formations of Oriental beech are the most common hardwood community in the Caucasus. At higher elevations (2,200-2,400 m a.s.l.) at the limit of forests, *F. orientalis* tends to turn into crummholtz. The majority of species of thick undergrowth is composed of evergreen broadleaf shrubs.
- In mid-elevations Oriental beech occurs in mixed stands with *Carpinus betulus, Castanea sativa* and needleleaf species – Oriental spruce (*Picea orientalis*) and Nordmann's fir (*Abies nordmanniana*) (hardwood-conifer forests).
- Despite a comparatively dry climate in eastern Transcaucasia, deep and humid gorges of the Greater Caucasus harbor species which tolerate high rainfall. The largest remnants of pure stands of beech are found in the Lagodekhi National Park (adjacent Zaqatala State Reserve) and Batsara Nature Reserve.
- Batsara reserve is incredible due to the presence of some key species of the ATG flora. For example, zelkova (*Zelkova carpinifolia*), sweet chestnut (*Castanea sativa*) and well-preserved yew tree (*Taxus baccata*) forests. As is broadly known, deciduous broadleaf forests of temperate areas of western Eurasia have suffered greatly by past climate oscillation during the Pleistocene glacial periods peaking during the Last Glacial Maximum. All three species mentioned managed to survive in patches of glacial refugia.
- It should be speculated that populations of mentioned three species occurring in patches of glacial refugia in East Georgia were probably affected by severe bottleneck, i.e., went through harsh biogeographical filter.
- Warm-temperate deciduous forests are found in two refugial areas in Colchis ('Colchis Triangle') and Talysh.
- On the territory of Colchis with a warm oceanic type of climate, below 600m a.s.l. are encountered two types of vegetation and associated animals: (1) Colchis wetlands and riparian communities and (2) mixed broad-leaved forests, sheltering thermophilic woody flora.
- The most typical vegetations of Colchic type are situated in the middle part of the Colchis Lowland, in the Odishi-Guria Lowland (Western Georgia). The present relief was shaped mainly during Plio-Pleistocene, when the whole, young, accumulated plain had emerged. In the framework of the Ramsar Convention, Kolkheti National Park ranks among the most important ecosystems of the North Hemisphere.
- Unique thermophilic dendroflora that faced the Last Glacial Maximum thanks to extraordinary relief of the Colchis Tringle, has managed to survive. Among others should be pointed out canopy building species consisting of yew (*Taxus baccata*) and boxtree (*Buxus colchica*) forests, which still keep the past climate (from the Oligocene time).
- Mixed broad-leaved (polydominant) forests of Colchis belong to the eastern part of the Euxinian province. Most of the Colchis region, which typically encompassed the south

- eastern part of the Black Sea coast between about 41^0 and 45^0 north latitudes, falls within the oceanic climate zone producing rainforest.

- The most outstanding biogeographic feature of Colchic is its rich and dense undergrowth, where evergreen shrubs predominate (with high levels of endemism, e.g., some species of *Rhododendron*). The same pattern is seen in countries of East Asia (with a far richer diversity of mixed deciduous trees).
- At mid-elevations deciduous forests are frequently dominated with conifers (Oriental spruce, Nordman's fir) and constitute broad- and needleleaf forests.
- In the Caucasus broadleaf forests are highly violated through human disturbance. However, in some areas, significantly in the southern slopes of the Greater Caucasus with limestone substrate in the upper reaches of the local rivers, there still remain virgin forests, which have avoided human impact through the centuries.
- Talysh is another refugium of the South Caucasus. The main notable fact of the Hyrcanian forests (within the borders of the Republic of Azerbaijan), polydominant forests, unlike the Colchis, which is distinguishable by their overstory. The canopy layer of native noble hardwood consists of typical Hyrcanian relicts and other relicts and endemics of the warm-temperate Arcto-Tertiary origin (e.g., keystone species, Persian ironwood (*Parrotia persica*). The Hyrcanian Floristic Province presented a single region of the Earth, where forest communities with parotia tree have been preserved. The uniqueness of flora and fauna of both refugia of global importance is still the subject of exhaustive systematic studies.

2.6.2. Coniferous (Boreal) Forests: Dark and Light Coniferous Forests

2.6.2.1. Dark Coniferous Forests with Oriental Spruce (Picea orientalis) and Caucasian Fir (Abies nordmanniana)

The dark coniferous forests include two dominant species - spruce (*P. orientalis*) and fir (*A. nordmanniana*). These conifers frequently co-occur with deciduous species, forming mixed spruce-fir-beech forests. The latter two coniferous species are frequently outcompeted by noble hardwood, with a significant presence of *Fagus orientalis* (Dolukhanov 2010).

Dark coniferous and dark coniferous-deciduous forests are distributed in the mountains of the western part of the Caucasus on crystalline schists and rocky soils (Gulisashvili, 1964). These forests are dominated by the Caucasian fir (*A. nordmanniana*) or by Oriental spruce (*P. orientalis*) that have a similar distribution (Gulisashvili *et al.* 1975). However, due to its wider abiotic niche *P. orientalis* is distributed more eastward than the fir, which is more sensitive to moisture, and is missing in the most continental parts (Makhatadze 1966). In East Georgia spruce penetrates into the upper reaches of the Didi (Greater) Liakhvi River and the Khevsuretian Aragvi Valley in the Greater Caucasus (Metreveli 1954). To the east it is common in separate groups and reaches the most extreme border of its distributed area in the Aragvi Gorge of the river Pshavi. It populates the northern slopes of the Trialeti Ridge in the Lesser Caucasus, the upper reaches of the Tedzami Basin, and the sources of the Algeti River (East Georgia). Spreading spontanously, the Oriental spruce almost reaches the eastern border of the distribution of fir on the Greater Caucasus and flows in the Didi Liakhvi Gorge, whereas the corresponding border on the Lesser Caucasus is in the Algeti Reserve, found in the upper part of the Vere River, west of Tbilisi (Dolukhanov 1964) (Figure 152).

Figure 152. Coniferous forests composed of *Picea orientalis*. Meskheti, S. Georgia, about 1,600m a.s.l. (1981).

In the North Caucasus small patches of spruce stands occur in the riverhead of the Kubani River and up to the Terek Basin (Gulisashvili *et al.* 1975). Its southern and south-western distribution borders lie outside the Caucasus in north-east Turkey (Turna 2004; Ucler *et al.* 2007).

Fir and fir-beech forests are widely distributed in the South Caucasus on the slopes of the Greater Caucasus from the Sochi-Adler district and western Abkhazeti to the Rioni River Basin. In East Georgia, in the Greater Caucasus, fir mixes with spruce, but the small populations are restricted to the upper part of the rivers gorges and do not reach the Didi Liakhvi River Basin. Extensive beech-fir and pure fir forests grow on the north slopes of the Adjaro-Imeretian Range in the Lesser Caucasus and on the north slopes of the Shavsheti Range, the watershed of which represents the state border between Georgia and Turkey. Similar to spruce, the southern and south-western distribution limit of fir lies in Turkey, in the Pontic Range. In South Georgia fir forests penetrate weakly, they can be found in the environs of the Goderdzi pass up to Abastumani resort.

Dark coniferous forests never form a continuous belt, though their occurrence is restricted to a definite altitudinal-climatic zone. This zone extends from 1,400-1,600 m up to 2,000m in the South Caucasus and from 1,100-1,200 m up to 1,850-1,950 m a.s.l. on the north slopes of the Greater Caucasus. They form the most common type of vegetation between 1,400 and 1,900 m (Gulisashvili *et al.* 1975; Dolukhannov 2010).

Under natural conditions the dark coniferous forests rarely border or mix with subalpine tall grass meadows. Usually between the dark coniferous forests and the subalpine meadows there is a belt of deciduous open woodland and crook-stem forest with birch, beech and high-mountain maple. The single examples where dark coniferous forests reach the timberline are attributable to human activity and the result of an anthropogenic lowering of the upper limit of forest distribution (Dolukhanov 2010).

Floristically the dark coniferous forests of the Caucasus are quite different from the typical dark taiga (Dolukhanov 1980). Some typical species of taiga, such as larch and boreal species of spruce and fir, can be found in West Europe, but are absent from the Caucasus. In some regions of the Caucasus, i.e., in Talysh (Azerbaijan) and Kakheti (East Georgia), the climatic conditions are quite favorable, but these regions lack dark coniferous communities. Gulisashvili (1940) pointed out that during the Pliocene the climate promoted maximum development and wide distribution of dark coniferous forests, also in the mentioned part of East Georgia, where today they are absent. Pollen analysis of Würm glacial deposits from Lake Sevan in Armenia (Delle 1962) confirms the wide distribution of *Picea orientalis* and clearly shows that *P. orientalis* grew in the eastern part of the South Caucasus not only in the Late Tertiary, but even in the recent past (Dolukhanov 2010).

The climate deterioration during some of the Pleistocene period, especially during the Last Glacial Maximum (LGM), promoted maximal development and wide distribution of dark coniferous forests. In Colchis the dark coniferous forests were partially replaced by broadleaf beech woods, which Kolakovski (1961) attributes to changes of the climate (warming) during the Late Quaternary. In the east Oriental spruce decreased in favor of invading beech or was replaced by light conifers, for example in Borjomi district (South Georgia), where due to cutting of spruce forests, communities dominated by *Pinus kochiana* started to appear (Dolukhanov 2010 and references therein).

2.6.2.1.1. The Classification of Dark Coniferous Forests: a Brief Review

Gulisashvili *et al.* (1975) differentiate the Oriental spruce formations on the basis of the composition of their herb and moss layers into spruce communities with mainly: bare floor (Piceeta sicca), spruce forests with *Festuca montana* and abundant mosses, spruce forests with a well-developed moss layer that covers 80-100% of the soil (Piceeta hylocomiosa), spruce-fir forests with ferns (e.g., *Dryopteris filix-mas*, *Matteuccia struthiopteris*).

Authors subdivide the fir formations into mainly: fir forests with bare-floor, fir forests with *Festuca montana*, fir forests with *Luzula sylvatica*, fir forests with *Rubus caucasicus*, fir forests with *D. filix-mas* and *Euphorbia macroceras*, fir forests with fern (*M. struthiopteris*), fir forests with subalpine forb herbs, fir forest with borage cover (Abieta trachystemosa), fir forests with the cover of various herbs (Abieta subalpina heteroherbosa), fir forests with thin herbal cover (Abieta rariherbosa); dark coniferous forests with Colchic understory (Piceeto-Abieta fruticosa colchica), dark coniferous forests with an undergrowth of *Rhododendron ponticum*, dark coniferous forests with cover of *Vaccinium arctostaphylos*, dark coniferous forests with *Ilex colchica*, and dark coniferous forests with *Prunus laurocerasus* in the undergrowth (Figure 153).

Figure 153. *Abies nordmanniana* dominating in forest stand with *Struthiopteris filicastrum* in ground layer (Abkhazeti, Audhara resort, W. Georgia, 1,800 m a.s.l. (1978).

2.6.2.1.1.1. Abiotic and Biotic Factors Affecting the Dying Back of Mixed Coniferous Forests in Borjomi Gorge (South Georgia)

The trigger mechanism of the dieback of mixed conifer forest in unique Borjomi Gorge in recent decades was a combination of climatic anomalies (summer drought) and different factors favorable for pest organisms.

Pine woodland, as it is broadly known, produces highly flammable litter (needles and cones). The semi-dry area of the southern Caucasus is characterized by enough fire frequency. We analyzed fire severity data from the Borjomi Gorge (southern Georgia) with regard to the pure stands of Oriental spruce and Caucasian pine forests, mixed (spruce-pine) dominated coniferous forests as well. This site has been severly burnt by Russian forces during the Russian-Georgian war (August 2008), when 950 ha of mixed forest of the Tori Mountains (the Lesser Caucasus) was dramatically altered – destroyed.

Additional to attracting and destroying military and civilian infrastructure, the Russian military aggression (via dropped incendiary bombs from military helicopters) caused ecological disaster. Hundred of hectares of unique forest have been purposely burned.

Later, in August 2017, after an unusual continuous drought a massive wildfire in the same place (Borjomi Gorge) lasted about one week. Complex strategies (including comprehensive supporting materials) were drawn up for handling the fire. International firefighting assistance should be pointed out came from Azerbaijan, Armenia, Turkey and Belarus in this action to extinguish the fire (partially, for the purpose of cross-border cooperation against the cross-boundary consequences of wildlife). The massive forest fire broke over 5 days. Estimated fire areas due to wildfire almost doubled. These two events had negative impacts on one of the most popular, international tourist destinations in Georgia and have resulted in severe environmental damage including to the unique native biodiversity. Here should be pointed out that after the Russo-Georgian war beginning successional processes became blocked. In the near future different successional pathways, in order to predicting global climate warming, are largely dependent on local relief, future climate regimes plus biotic factors. Government and non-government institutions opened bank accounts for donations to restore the forest burnt following the recent devastating man-induced destruction and wildfire (pers. observations) (Figure 154).

Among biotic factors threatening mixed coniferous forests should be pointed out the great spruce bark beetle (*Dendroctonus micans*). In 1956 it was accidentally introduced into southern Georgia (Borjomi Gorge) with unprocessed logs from Siberia (Kantschaveli & Supatashvili 1968, Mukhashavria 1984).

Figure 154. 'Hiroshima in woodland'. The dried forests have burned away, leaving felled logs and standing killed trees; Borjomi Gorge, S. Georgia (2009).

Over 10-15 years the beetle extended its range to almost all other regions of Georgia (except Abkhazia) and adjacent Turkey (Mukhashavria 1984). The pest infests Oriental spruce, and sometimes attacks Caucasian pine – both of which are canopy-builder trees and economically important conifers. *D. micans* damages host trees by tunneling into the bark of the living trees in order to lay eggs under the bark. Developing larvae feed on the inner woody layers. Such kinds of damage weaken, and in some cases kill, the infested tree. Unlike its natural range, in Georgia eruptive populations (i.e., when populations reach epidemic level) of this beetle during 60s-70s affected large areas and caused high levels of tree mortality resulted in timber losses. Drying of trees with increasing frequency has been noted in 1960, therefore *D. micans* became the major cause of tree mortality. (Mukhashavria 1984). Integrated control systems included the felling, removal and processing of infested trees (this method was found to be an important factor for prevention of spread of tree pest); chemical control (however, treatments were not successful enough); biological control (most successful way of control of the pest beetle). The natural enemy of *D. micans*, a specific predator, *Rhizophagus grandis* was introduced from Russia and Chechoslovakia (Kantschaveli & Supatashvili 1968). More than four million specimens of predator were breed in *in situ* conditions and then infested *D. micans* in natural conditions. During a short time 'Beetle vs Beetle' proved very efficient at controlling pest beetle populations (reduced them during 1970s by 70% to 90%). In addition woodpeckers prey on the larvae and pupae of the pest (Kobakhidze 1965, Tvaradze 1984). During approximately thirty years from when *D. micans* was introduced in Georgia, regarding mutual efforts of scholars and natural enemies it converted natural components of native forest ecosystems and enabled homeostasis mechanisms in local environments (Gegechkori 2018).

2.6.2.2. Light Coniferous Forests with Caucasian Mountain Pine (Pinus kochiana)

There are 126 recognized species of pines together with 35 unsolved species (*https://en.wikipedia.org/wiki.pine*).

The light coniferous forests of the mountain consist of Caucasian pine, *P. kochiana* (=*P. hammata, P. sosnowskyi*) (Tumadjanov 1980, Dolukhanov 1992, cited in Nakhutsrishvili 1999). Caucasian pine is the most widely distributed tree species in the Caucasus. In Europe it is replaced with Scotch pine (*P. sylvestris*) (Gulisashvih *et al.* 1975; Dolukhanov, 2010). It is an "aggressive" species that easily expands into new sites and a pioneer tree species that occupies naked stony ground (Medvedev 1915; Ketskhoveli 1960), Grossheim (1948) suggests that the ancestor form of the Caucasian pine invaded the Caucasus from its North Boreal center at the beginning of the Quaternary Glaciation and then gradually populated all suitable habitats.

The current distribution area of *P. kochiana* includes the Caucasus, the Crimea and northern Asia Minor. Forests comprising *P. kochiana* are typical of the mountains of the Greater Caucasus.

Their distribution in the Lesser Caucasus mainly involves South Georgia (Meskheti district, Borjomi-Bakuriani Gorge). The western distribution border of *P. kochiana* on both ranges is the Black Sea coastal area, in the east it reaches up to the Kusarchai River on the north slope of the Greater Caucasus and the Zakatala Reserve (Azerbaijan) and on the south slope of the same range. In the Lesser Caucasus the species occurs up to the Karabakh Range (Gek-kel Lake, Azerbaijan). In south Armenia *P. kochiana* is absent.

In both the Greater and Lesser Caucasus, Caucasian pine reaches the alpine zone. Pine forests extend from 700 up to about 2,400 m a.s.l. with the most favourable conditions being provided at an altitude of 1,000-2,200 m. At some sites, Caucasian pine has been recorded at 2,500-2,600 m (Tumajanov 1938).

The lower limit of mountainous pine forests varies depending on the climatic conditions (Gulisashvili *et al.* 1975). On the north slopes of the Greater Caucasus, particularly in its central part where glaciations were much more expressed, the lower distribution limit of *P. kochiana* reflects the lower border of maximum glaciation during the Ice Ages. At that time, glaciers descended to 800 m a.s.l. in the Kuban River Basin and to 700-650 m in the Baksan River Basin (Mushketov 1896; Dolukhanov 2010 and references therein). The lower limit of pine tree distribution is determined by high summer temperature and warm mild winters with very few frosty days. Caucasian pine is not only frost-resistant, it is an explicit cold-loving species. Hence, at lower altitudes, in the western part of the Greater Caucasus *P. kochiana* is replaced by the maritime pine - *P. brutia* var. *pitiysa* and further to the east by the former maritime pine

- *P. brutia eldarica* (Gulisashvili *et al.* 1975). In Dagestan, the North Caucasus, due to the driest conditions the lower distribution limit of Caucasian pine forests occurs between 1,200-1,400 m a.s.l., while in the humid districts of the western part of the South Caucasus the same limit is recorded at 250 m (Baghdadi environs, Rioni River Valley, western Georgia) (Gulisashvili 1964; Gulisashvili *et al.* 1975).

Among the mountain forests, pine tree forests are the most xerophytic and light-demanding. Therefore, they not only occur with a well-developed canopy, but also as open communities that resemble open woodlands, e.g., the pine grove of the Tetrobi Gorge in South Georgia (Ketskhoveli 1960). *P. kochiana* grows on both limestone and acid magmatic rocks (stones and rocky slopes). On these soils pine forests can outcompete both dark coniferous and broad-leaf forest communities (Dolukhanov 2010).

In the high mountain district of Tusheti (East Georgia, north macroscope of the Greater Caucasus), the climax pine tree forests occupy a limited area, but are very diverse because of the lithological peculiarities of the rocks and the chemical character of the soil (Tumajanov 1938; Dolukhanov 2010). In the Tusheti and Pirikiti Alazani River Gorges of the district, Caucasian pine forests, which are typical of stony and rocky slopes, also occur in habitats with a humid, well-developed soil that normally would be covered by spruce and beech (Nakhutsrishvili 1999, 2013; Dolukhanov 2010). Although presenting the widest natural distribution among pine species, the presence of a native population in Tusheti has long surprised botanists, ecologists and biogeographers. These occurrences are explained by the orographic and geographic isolation of the Tusheti Hollow. Such geographical isolation is allowing pine forests to survive in modern time as extraordinary complete remnants of the last Ice Age with many relicts in both flora and fauna (Gegechkori 1983, 1984a, 1990, 2007; Dolukhanov 2010). To quote the words of Dolukhanov (2010): 'it escaped being replaced by beech and spruce only because their seeds never got to this area' (Figures 155, 156).

Figure 155. Pure forest with *Pinus kochiana* dominancy, relict population of the last Ice Age, Würm (Tusheti region, north slope of the GCR, E. Georgia (1974).

Figure 156. Crossbill (*Loxia curvirostra*) – relict of the Last Glacial Maximum feeds on closed-cone pine. Pine tree (*Pinus kochiana*) forest, Tusheti region, E. Georgia (1976).

Apart from pure stands, pine forests grow together with birch (*Betula raddeana*) to form birch-pine forests. Sometimes a small grove of *P. kochiana* comes into contact with oak and beech forests (Mariamjvari pine grove in East Georgia), or with beech forests (Zakatala Reserve in Azerbaijan) (Dolukhanov 2010).

Doluhkanov (1992) in Nakhutsrishvili 1999) distinguishes four categories of *P. kochiana* forests.

2.6.2.2.1. Caucasian Mountain Pine Forests of Dry Biotopes

Caucasian mountain pine forests of dry biotopes can be found in the western part of the distribution area of *P. kochiana* where they occupy south-exposed slopes and grow on skeletal and poorly-developed soils. In their canopy, birch (*B. raddeana*) also occurs. The second stratum consists of *Quercus p. iberica* and *Q. macranthera*. Their herb layer consists of *Poa* and *Carex* species, in dry habitats also of xerophytic herbs. In East Georgia, pine forests with tragacanthic species of *Astragalus* (*A. caucasicus*) and junipers (*Juniperus oblonga, J. hemisphaerica* (=*J. depressa*) mainly occur at 1,900-2,000m a.s.l. Pine forests with *Chamaecytisus salviifolius*, as previously mentioned, are found in Adjara and pine forests with *Ch. caucasicus* in the Trialeti Mountains in the Lesser Caucasus (southern and south-eastern Georgia). In Abkhazeti the undergrowth of *Ch. caucasicus* is replaced by *Ch. hirsutissimus.*

Park-like landscapes of mountainous open woodland are mainly found on the Javakheti Upland Plateau, where mountain steppes and mountain meadows are scattered with *P. kochiana*. These forest communities extend between 1,700-2,400m a.s.l. (Ketskhoveli 1960) (Figure 157).

Figure 157. Open stand of Caucasian pine with low herbaceous layer in Meskhet-Javakheti, Tetrobi-Chobareti site, E. Georgia (1974).

According to Ketskhoveli (1960), the pine forest of limestone on Tetrobi-Chobareti Range, which extends from 1,800m up to 2,000m a.s.l. is considered to be quite different in comparison with other formations of pine forests. Firstly, physiognomically the Tetrobi pine formation, as mentioned above, is close to arid open forest. Secondly, earlier the local population of pine restricted of limestone ground in Javakheti was considered as endemic *P. kochiana*, whereas a far more widely distributed population of this conifer was called as 'Sosnowski pine' – *P. sosnowskyi*. At the present time both populations are considered as one species – *P. kochiana* or subspecies – *P. sosnowskyi kochiana* (Dolukhanov 2010). Thirdly, the Tetrobi pine forest is enriched with Anatolian Upland's (Asia Minor) xerophytes, therefore, the pine forest of Tetrobi with its composition is considered as a peculiar refugium, where pine tree's stand is mixed with shrubs and grasses of mountain steppes. In this community 48 species of vascular plants contribute to the formation of a local steppes' community. From the first stratum of arboreal vegetation pine is accompanied by *Quercus macranthera* and *Sorbus caucasigena.* Besides the mentioned species with the sparse distribution are characterized *Betula pendula, Malus orientalis,* and on the south slope occurs *Rhododendron caucasicum*, a striking feature of this site of southern Georgia. It should be pointed out that the commonest species of shrub layer is represented by juniper (*Juniperus hemisphaerica*). The ground layer is made of xerophytic species some of which are endemics of limestone of Tetrobi plateau (e.g., *Medicago dzhawakhetica*) (Khintibidze 1990). One of the most striking feature of the open pine forest of the Tetrobi-Chobareti site is the local ant (*Formica rufa*). They build large nests, some of which exceed 1-1,5 m (Figure 158). Among herpetofauna should be emphasized the long-nosed viper (*Vipera ammodytes transcaucasiana*) (Figure 159). The range of both species is restricted to South Georgia.

At higher altitudes (about 1,800-2,000m a.s.l.) between the Azaveti and Tetrobi sites shrubby communities of rare willow (*Salix armeno-rossica*) occur. Stands of willow in this treeless area are characterized by distinct microtopography. Its spread in the mountainous area of Georgia aside from Javakheti Plateau is found only in Kartly (East Georgia), and abroad in the adjacent territory of Turkey (Khintibidze 1990; Shetekauri & Chelidze 2016). Interesting that on this rare species of willow was discovered a new, willow-feeding species of psylloids (Hemiptera), *Cacopsylla iphigenia** (Gegechkori) (Gegechkori 1974).

Figure 158. The most striking feature of the open pine forest of the Tetrobi- Chobareti site is the large number of ant (*Formica rufa*) nests, some of them are as tall as 1-1.5m (1974).

Figure 159. Transcaucasian long-nosed viper (*Vipera ammodytes transcaucasiana*) (2014).

2.6.2.2.2. Caucasian Mountain Pine Forests of Moderately Moist Biotopes

Caucasian mountain pine forests of moderately moist biotopes occur mainly on the north-exposed slopes with deep, moist soils, where pine is associated with spruce as well by *Acer platanoides*. The herb layer (with species of *Briza, Poa*, etc.) is also well developed (Gulisashvili *et al.* 1975; Dolukhanov 2010).

2.6.2.2.3. Caucasian Mountain Pine Forests of Swampy Habitats

Caucasian mountain pine forests of swampy habitats occur under moist conditions. The most moist phytocenosis is represented by pine forests with *Sanicula europaea*. In these forests pine frequently mixes with spruce, birch and aspen (*Populus tremula*). Arboreal components in the undergrowth are *Lonicera caucasica* and *Chamaecytisus caucasicus*, whereas the herb layer, next to *S. europaea*, includes *Primula vulgaris sibthorpii* and *Festuca montana*.

Scattered trees of the Caucasian pine are restricted to the peat marshes in Kobuleti (Adjara). The appearance of this sparse forest as well as the boreal floristic composition consisting of *Calluna vulgaris, Vaccinium arctostaphylos, Drosera rotundifolia*, etc., is very similar to those of Eurasia boreal biome. Some authors consider it as a refugium, which hasn't changed as a result of the Ice Age (Akhalkatsi & Tarkhnishvili 2012).

2.6.2.2.4. Caucasian Pine Forest with *Vaccinium myrtillus*, *V. uliginosum* and *V. vitis-idaea*

Caucasian pine forest with *Vaccinium myrtillus*, *V. uliginosum* and *V. vitis-idaea* are found on the north slopes of the Greater Caucasus. Forests with *Rhododendron caucasicum* or with *Rh. luteum* are situated mainly on the Greater Caucasus and extend in altitude between 1,700-2,100 m a.s.l. There also occur pine forests with a well-developed moss cover of *Hylocomium splendens*, with a herb layer, consisting of species like *Poa*, *Geranium* and *Festuca* (Dolukhanov 1989, 2010) (Figure 160).

Figure 160. Solitary tree of the Caucasian pine represented up to the treeline ecotone. Zemo Svaneti province, W. Georgia, 2,150m (1984).

Mountainous Dagestan is characterized by pine forests with *Pteridium aquilinum*. Next to *Pinus kochiana* the tree layer consists of *Betula pubescens litwinowii* and *Sorbus caucasigena*, whereas *Juniperus oblonga* and *Lonicera caucasica* form the undergrowth. The ground layer is covered by moss (Dolukhanov 2010).

2.6.2.2.5. Caucasian Mountain Pine Forests Found on Stony Screes and Rock Streams

Under dry climatic conditions Caucasian pine rock-tree forests are distributed on the slopes of rocky southern exposition or limestone layers. Two main forest communities, which reached the subalpine belt, involve: (1) rock pine forest with scattered ground cover (Pineta rariherbosa) occurs on clear rocky slopes with a thin soil layer. A typical stand, frequently of dwarf shrub-like pines, is confined in Khevi province (East Georgia) with the canyon-like gorges of the Terek River; (2) rock pine forests with xerophyte herbal and semi-shrub cover (Pineta xeroherbosa) is represented under more continental climatic conditions, found in southern Georgia. Here they reach 1,500 m elevation and often are found in contact with spruce forests on northern slopes (Khintibidze 1990; Akhalkatsi & Tarkhnishvili 2012).

Although Caucasian pine covers a vast area in the mountains, it is sensitive to human impact. In the regions with a dry climate pine communities after cutting the forests are replaced by mountain xerophytes (e.g., *Astragalus caucasus*) in the mid-mountain zone and by *Nardus glabriculmis* in the high mountains (Ketskhoveli 1960).

2.6.2.3. Forest Fauna

Forests with rich plant and other food sources are the most important biome for a lot of animal species in the Caucasus.

Broad-leaved and coniferous forests harbour many mesophilic focal (i.e., keystone, indicator, umbrella, flagship, charismatic, and vulnerable animal species) including six priority species: Persian (Caucasian) leopard, brown bear, Caucasian red deer, bezoar goat, European bison and Caucasian salamander (Ecoregional Conservation Plan for the Caucasus (ECPC 2006, 2012).

Among the large carnivores the leopard (*Panthera pardus ciscaucasica*) is the most endangered species. In the 1950s the species still occupied forest habitats next to the high-mountain and (sub-) arid landscapes, where it still lives. It occurred in the mountain forests of both the Greater and the Lesser Caucasus. In South Ossetia and in the Shida (Inner) Kartli region of East Georgia (Zedazeni, near Mtskheta town), two specimens were killed both in deciduous (Oriental beech, Georgian oak) forests according to the collections of the Georgian National Museum. Today the flagship species is almost extinct in such habitats in the Caucasus (ECPC 2006).

The brown bear (*Ursus arctos*) is a keystone species and top predator in the food chain in most habitats and especially in the forests of the region. The total number does not exceed 3,300-3,500 individuals (Chestin *et al.* 1992). The taxonomy of the Caucasian bears is one the most complicated ones (Lordkipanidze (2003) and references therein). Specialists recognized from one to four subspecies of brown bear in the Caucasus Ecoregion with complicated interrelations and overlapping ranges. However recent results of genetic studies show that all populations in the Caucasus belong to the Eurasian brown bear (*Ursus arctos arctos* L.) (Murtskhvaladze *et al.* 2010) (Figure 161).

The common otter (*Lutra lutra*) and European mink (*Mustela lutreola*) are associated with riparian forest ecosystems.

Caucasian populations of the European wild cat (*Felis silvestris caucasica*) and pine marten (*Martes martes*) are relatively abundant in forests and their populations are important for conserving the species worldwide (ECPC 2006). The typical carnivore of the forest is the lynx (*Lynx lynx*).

Figure 161. About one year old, this brown bear cub (*Ursus arctos*), the largest land mammal of the Caucasus.Batsara Gorge, Akhmeta district, E. Georgia (1984).

Several species of ungulates have to be mentioned. The Caucasian subspecies of red deer (*Cervus elaphus maral*) is represented in Georgia by 240, in Azerbaijan by less than 600, and in Russia by several hundred individuals. Turkey and Iran also shelter small populations of this species. In almost every country deer populations have decreased; in Russia red deer is legally hunted, while in the South Caucasus the species is legally protected, but poached.

There are just over 70 European bison (*Bison bonasus*) – the largest herbivore in Europe, including hybrids of different subspecies, remaining in two nature reserves in the Russian Caucasus. The European bison, as mentioned, was reintroduced to the North Caucasus after its extinction in the wild in the coniferous forests of Abkhazeti in 1927, but is once again under threat (ECPC 2006, 2012).

The highly endangered bezoar or wild goat (*Capra aegagrus*) is found in the pine-birch forests of the Greater Caucasus Range and in the southern part of the Lesser Caucasus Mountain Chain and Southern Highland, which carry mainly mountain-xerophyte vegetation.

Today, several highly isolated populations of bezoar goat exist in different regions. Around 1,000 animals live in Dagestan and about 3,000 in Armenia and adjacent Azerbaijan in the middle reaches of the Aras River. Some 300 individuals live in Georgia, approximately 2,500 goats in North-East Turkey, and probably twice that number in Northern Iran (ECPC 2006, Abbasian *et al.* 2004).

Conspicuous among the small mammals are several endemic rodents and insectivores, such as Robert's snow vole (*Chionomys roberti*), Caucasian mouse (*Apodemus ponticus*), Caucasian mole (*Talpa caucasica*), Transcaucasian water shrew (*Neomys teres* (=*N. schelkovnikovi*), small mole (*Talpa levantis*), Caucasian shrew (*Sorex caucasicus*), Radde's shrew (*S. raddei*) and Caucasian pigmy shrew (*S. volnuchini*) (Bukhnikashvili 2004). Several species of bats (*Myotis, Barbastella, Rhinolophus*) are recognized as Vulnerable and Endangered (IUCN

2002). A keystone animal typical for these types of forests is the Persian squirrel (*Sciurus anomalus*) (*www.senckenberg.de/.../10_vertebate_zoology_64-1_amr_...*).

The bird fauna of the Caucasus forests is also rich, although the majority of species is widespread in European forests and the number of endemic or globally threatened species is relatively low (Gulisashvili 1964; Zhordania 1999).

Reptiles associated with the forest landscapes include the endemic Caucasian adder (*Vipera* (*Pelias*) *kaznakovi*) (Figures 162), nose-horned viper (*Vipera ammodytes transcaucasiana*) and endemic rock lizards of the genus *Darevskia* (Akhalkatsi & Tarkhnishvili 2012). Amphibians include the Caucasian salamander (*Mertensiella caucasica*), which has evolved independently for 15 million years (Tarkhnishvili & Gokhelashvili 1999; Tarkhnishvili *et al.* 2008) (Figures 163), bended newt (*Ommatotriton ophryticus*), the Caucasian mud-diver (*Pelodytes caucasicus*), and the Caucasian hoptoad (*Bufo verrucosissimus*). Many endemic invertebrates (e.g., *Carabus caucasicus* (until the recent time it was considered as *Carabus* (*Procerus*) *scabrosus*, *Helix buchi* and *Spanioneura caucasica*). A high diversity of soil invertebrates is also exclusively dependent on forest ecosystems (ECPC 2006, 2012, Gegechkori & Djibladze 1976).

Figure 162. Kaznakov's viper (*Vipera* (*Pelias*) *kaznakovi*), an endemic species to the Caucasus (has penetrated into NE Turkey) [EN]. Abkhazeti, W. Georgia, GCR, dark coniferous forests, about 1,000m a.s.l. (1982).

Figure 163. Caucasian salamander (*Mertensiella caucasica*) is one of the most archetypical species (endemic and relict of the Late Miocene time) [VU]. Borjomi district, S. Georgia, 1,350 m a.s.l. (1987).

Mountain forests are frequently used by pure orophile species that use forests below the timberline as feeding and wintering habitats.

The forests of Colchis and Talysh are largely isolated from other tracts of forests in Europe, Siberia and Middle Asia and therefore contain most of the narrow endemic species of the Caucasus (Gegechkori 2011, 2012a, 2012b).

2.6.2.3.1. Hybrid European Bison (*Bison bonasus* × *B. bonasus caucasicus*)

The Caucasian bison (wisent) was a subspecies of the European bison (*Bison bonasus*) dwelling in the forests of the Greater Caucasus Mountains (Vereshchagin 1947).

The European bison is unique and the largest herbivore in Europe (sometimes genus *Bison* H. Smith is treated as a synonym of *Bos* (Linneaus) (Groves 1981). It is a single species of wild Bovinae of Europe, which survived till the present day. It is a very large species from the group of Bovini (Belousova 2000). Three subspecies are recognized: the Lowland bison (*B.b. bonasus*), Carpathian bison (*B.b. hungarorum*), and the Caucasian bison (*B.b. caucasicus*) (Pucek *et al.* 2002).

In the beginning of the 20th century, in the territory of the Caucasus, there were two subspecies of European bison, Lowland bison (*B.b. bonasus*) and Caucasian bison (*B.b. caucasicus*) (Figure 164).

The Caucasian wisent in many ways differs from the lowland European wisent by its size, a slender constitution, a flat frontal skull surface, more slender legs, the hooves are relatively narrow, its main slightly developed, pelage is lighter, it has beautiful short and curly fur on the back, front body and on the head and a brush at the tail; horns are thinner. Caucasian bison living seasonally in harsh highland's environment represents a climax in the capacity of a such large ruminant to adapt to alpine areas (Vereshchagin 1947; Heptner *et al.* 1961).

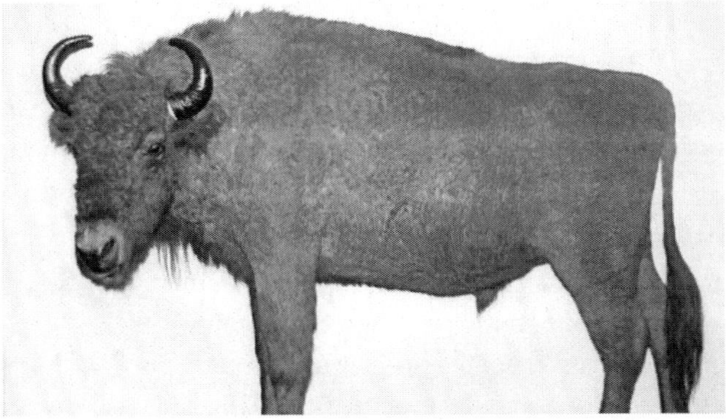

Figure 164. Caucasian bison (wisent), an extincted specimen as stuffed specimen can today be seen at the Georgian National Museum (1975).

In the 17th century, the Caucasian wisent still populated a large area of the Western Caucasus. The range of the Caucasian bison (occurred in the mountains of North Caucasus) from the Pshish River in the west to the Terek River in the East, and from the Enguri River in the south to Kuban River in the north (Radde 1839; Vereshchagin 1947; Zablotskaya 2004).

After that, human settlement in the mountains intensified and mass poaching took place.

Local poaching continued and in 1927, the last three wild Caucasian bison were killed by shepherds on Mt. Alous (North Caucasus) (Bashkirov 1939).

In 1940 five hybrids between European-North American bison (wisent-American bison hybrids) were introduced to the Caucasus State Nature Reserve and in 1959 into the Nalchik Forestry Game Management Unit (Kabardino-Balkaria). These animals mated between themselves up to 1950. The European-American bison hybrids occurring in the Caucasian mountains also contain genes from one Caucasian bison bull. Later some pure-blood wisents of the lowland-Caucasian breeding line of the Caucasian-Bielowiezh bison line (*B. bonasus bonasus X B. bonasus caucasicus*) were released in the reserve to form a single mixed herd. In this way, the American bison admixture in the gene pool of the local population was reduced to several percent (Zablotskaya 2004). In the 1960s the latter herd reached 189 animals and became a free-ranging, completely independent Caucasian population in the natural ecosystems of the Greater Caucasus. Thirty years of natural selection finished with the formation of the Caucasian mountain wisent (Heptner *et al.* 1961). In 2000 this modern breeding line was named the highland bison, i.e., a new subspecies of the European bison (*Bison bonasus montanus nova*), although questionable (with no scientific basis) subspecies (Rautain *et al.* 2000). The latter one was included in the Red Data Book of the Republic Adygea (the Russian Federation).

Therefore, the genetic purity of the Caucasian mountain subspecies has been lost and the core of the present herd is made of hybrids – Bielowieza (Poland/Belarus) - Caucasus bison and North American-European bison.

Currently, about 600 bison live in the Caucasian Biosphere Reserve.

2.6.3. Non-Wood Forest Products in the Caucasus

Forest (also wood or woods) ecosystems of the Planet Earth have come to be seen as the most important component of the biosphere (The forest encyclopedia 1985). Tree forests cover approximately 9.4% of the Earth's surface, or 30% of the total land area. Environmental services of the temperate, boreal, and particularly tropical forests, include the regulation of water flow and its quality, stabilization of soils by means to avoid the risk of flash floods, soil erosion and landslides. Forests contribute to climate relevant cycles, help to mitigate greenhouse gas emissions by absorbing and storing carbon dioxide and also to biodiversity related processes (Costanza *et al.* 1997; Pearce *et al.* 1996).

Forests are among the most successful of all natural communities, as is evidenced by their survival since the earliest eras of organic development (Ponting 1991).

For thousands of years native peoples of the different regions of the world have been coexisting with the natural environment around them. Interconnections and interaction of human beings and forests have required traditional forest - related deep knowledge from time immemorial in Eurasia (Bocharnikov *et al.* 2012). Our early ancestors' conscious dependency on forests and their goods led them to possess a deep awareness of their environment (Nasi *et al.* 2002).

Forests have important cultural values for people, especially for forest dependent cultures. Through the long history of religions and mythology wood and trees have played an important role, and have been given deep and sacred meanings throughout the ages.

For early human societies, trees have been viewed as having souls and spirits. The most ancient cross-cultural symbolic representation of the universe's contraction is the world tree. For different peoples, paganism in temperate climate countries appeared to have involved cultic practice in sacred groves, especially those of oak (Jonson 1976; Taylor 1979; Altman 1994; Elashvili 2008).

According to Nasi *et al.* (2002), 'The belief that forests and trees are homes of the gods can be found in nearly every culture. Just this point of view has led to both respect and reverence for sacred forests or trees, which were often protected from cutting or dismemberment'. The yew, box tree and other trees are associated with Christianity, both of them are largely value in southern Caucasus, mostly in Armenia and Georgia. But in recent decades the forest biome of the world faces the dual threats of deforestation and stress from climate change, and this negative trend includes the forests of Amazonia, one of the Earth's greatest biological treasures and a major component of the Earth system (Malhi *et al.* 2008).

The importance of the mentioned and other problems is supported by Kyoto Protocol (UN Framework Convention on Climate Change, Kyoto 1997), which was ratified by more than 140 nations of the world. This recent activity by the UN is determined by greenhouse effect caused by two main events, changes in forest cover on a global scale and increases of energy-intensive activities by means of accelerating carbon dioxide emissions from developing countries, since 2000 (Raupach *et al.* 2007).

However, in many regions of our planet, according to timber overharvesting, the traditional timber consumption is changed by non-wood production. For example, Siberian cedar (*Pinus sibirica*) is a natural pride and wealth of Russia. In recent time high demand for cedar nut products results in higher prices paid for trees. One hectare of cedar stand gives five tones of nuts, the value of which in Russia amounted to seven thousand roubles, while the timber from the same area is costing two thousand roubles (value in roubles in the year 2000). The annual income from cranberry (*Vaccinium oxycoccus*) harvests in one hectare of acidic tundra should be 5-7 times higher than the income from 80-100 years old forests, which are growing on the same site. According to the specialist's data, the annual market price of utilized production for wild berries totaled more than 10 billion US dollars and the value of annual utilized natural production of mushrooms – 5 billion US dollars (*Forests of Russia 2002*).

According to Burmistrov (1995), at present forests cover more than 55% of the territory of the former USSR. Wild fruit, small fruit and nut-bearing plants are distributed over the area of 57 million hectares and can annually yield 3 or 4 million tons of valuable food and vitaminic raw matter. However, only the same 10-12% of potential forest bioresources are now utilized. For berries this quota is 4-5%.

So, if the total economic value of forests was really taken into account, then people and what is more important, the developing state governments like the former Soviet Union's countries would recognize their importance and better protect and manage forest biomes.

Out of 374,000 of which about 308,312 are vascular plants (Christenhusz & Byng 2016) existing on the Earth, only about 3,000 species have been used as agricultural crops and only 150 species have been cultivated on a commercial basis. However, developing new crops by domestication of currently wild edible species, offers considerable potential (Heywood 1999).

Despite agricultural societies' primary reliance on crop plants, the tradition of eating wild plants has not completely disappeared. More than 300 million people throughout the modern world depend on wild resources including wild palatable and edible plants from forests (Pimental *et al.* 1997).

Various kinds of edible wild plants are crucial for some rural people, especially in periods of food crisis. Many edible wild plants are nutritionally rich; therefore they are expected to reduce the serious problems of nutritional requirements of malnutrition sufferers, especially vitamins and micronutrients (FAO 2004).

Forests are also vital to the quasi-public goods – income and social life, some of which are essential for their daily lives, while others contribute to the longer-term economic stability. The most important and used goods is the timber, which provides construction materials and firewood. The world's annual industrial round wood production is estimated at 1.52 billion m^3. In 1999 some 1.75 billion m^3 of wood was extracted for fuelwood and conversion to charcoal (FAO 2001).

Other goods, which may be as important as wood for humankind are known as non-timber forest products (NTFPs) or non-wood forest product (NWFPs). NTFPs usually are considered as the object of collated use of all forests, which does not necessitate harvesting trees. Non-timber forest products occupy in many cases a prime place in rural people's livelihoods for products sold (cash income) and for auto-consumption, in some cases constitute the main source of income for refugees and other landless or unemployed people. These products including food (edible mushrooms, berries, nuts), medicinal purposes, decorative plants, and grazing for domestic animals (silvipasture), honey and hay productions and other biotic products, harvesting of game animals or sport fish (CAREC 2006; Wikipedia 2013; REC Caucasus 2013). Wild harvested plant food includes: roots and underground parts; shoots and leafy greens; berries and other fleshy fruits; grains, nuts and seeds; mushrooms, lichens, algae and other non-animal origin organisms. Use of any of these species requires special skill regarding harvesting, preparation, cooking and other forms of processing (Turner *et al.* 2011).

Forests hold a wide range of recreational opportunities such as hiking, bird watching and camping, eco-touring, and many other activities.

Many people have a deep appreciation of natural forests. That is apparent in the art, literature, nature photography and film making. Some sites attract large numbers of visitors. According to Kreiger (2001) and Barnhill (1999) cited in Nasy *et al.* (2002), in 1996 recreational activities accounted for a value of US 1 billion in five national forests in the southern Rocky Mountains.

In addition, NTFPs harvesting promotes better physical health of people engaged in this field of occupation. A study of residents in Siberia found that people, who regularly pick berries and mushrooms have less sick leave, and their labor productivity is 5.2% higher than average (Weigard *et al.* 1995).

However, if we take into consideration the opinion of Nasy *et al.* (2012), 'a very high diversity is not necessarily linked to more useful or valuable forest goods and services. The existing 'Vavilov' centers of annual crop genetic diversity, for example, are mainly in areas with low forest diversity'.

As mentioned, people have gathered valuable non-wood wild things elsewhere since remote times. However, only in the last decades have non-timber forest by-products gained more attention in world trade (Mousavi 2012). It should be stressed that they are not legally regulated, hence, woodland's rare, endemic, relict and other endangered species of plants remained unprotected by legislation.

Forest conservation makes sense from the viewpoint of minimizing high environmental risk under considerable consumption of its wide range of goods and services. Therefore, a key aspect of any campaign for the conservation and sustainable use of biological resources (sustainable yield) is enhancing community awareness and participation. Population pressure and non-wood product gathering activities are causing a rapid depletion of forests. As a result, countless plant species are facing a considerable danger of extinction (Lubchenco 1991).

Most forest plant species of each forest biome's layers, depend on animal pollination for reproduction. The well-being of these pollinators depends on the existence of a wide spectrum of habitats needed for the successful completion of their life modes (Nabhan & Buchman 1997 in Nasy *et al.* 2012).

The Caucasus is one of the most biologically rich regions on Earth and is home to a large number of forest formations of rich ecological composition.

This most important biome for biodiversity conditions in the Caucasus, the forests, are covering nearly one–fifth of the region (Zazanashvili *et al.* (Eds.) 2011).

Forests play a significant cultural, social, environmental and economic role in the Caucasus. For thousands of years native people of the region have been coexisting with the natural world around them. As well as the food gathered and stored, medicinal plants were largely used when required. Every inch of the forested landscape featured useful materials, requiring deep knowledge of the environment (Grossheim 1952a,b).

Wild-growing fruit and berries are a source of vitamins, proteins, carbohydrates, organic acids, mineral and other materials necessary for the human well-being (Grossheim 1952a,b, Fidarova 2006).

Following the demise of the Soviet Union, combined with social, economic, political and cultural factors, which threaten the health and well-being of a myriad of smallholders with weak property, the local populations were occupied by over exploration of indigenous biological resources. From this viewpoint the most important factor is forest decline. In the Caucasus human–induced causes include forest degradation or near–complete deforestation in some regions, permanent conversion of forest to cropland and pasture, overgrazing, unsustainable logging practices, over extraction of fuelwood, overharvesting of non-timber forest resources, including some herbs, shrubs and trees listed in the local countries' red books, harmful practice of trees by browsing cattle, overfishing and overhunting, recreational activities, national booms of bungalow construction in recreation zone covered by forests and so on.

All the mentioned negative relations to forests goods cause higher forest ecosystems loss.

The best way to avoid the forest degradation or more, deforestation proper, is to enlarge the net of National Parks and strict reserves, the further development of forest-based tourism through international high- class ecotourism.

The decline of forests is caused also by the logging of evergreen conifer trees (Christmas trees) and cutting branches of box tree in pre-Christian feast days. In recent time natural conifers are replaced by artificial Christmas tree and unique box tree – by artificial plantation of the mentioned tree (Grossheim 1952b, Molchanov *et al.* 1989).

Currently the level of study of the non-wood products in the Caucasus is insufficient. Statistics are very dispersed, especially at the small-scale farmers level, and concerning the most marginalized members of the rural population consumption of the raw material. They are not aggregated at the national level. After the disintegration of the USSR, some marketing studies of NTFPs resources in the Caucasus were carried out by local research institutes and universities together with international environmental organizations (Kikava *et al.* 1997).

All the mentioned and other fruits and berries, as in other countries, are used by Caucasian peoples in two main preparation types: unprocessed and processed.

Based on a lot of information about the plants as food purposes of different regions of forest biome of the Caucasus, some of them were the most utilized plant species. These plants have been traditionally used as food, medicine and other ethnobotanical purposes.

Many of the widely used species, as in other countries of the world, are categorized as weeds-species that grow and reproduce readily in disturbed or cleared land, and are common around human settlements and agriculture areas (Turner *et al.* 2011).

The world's genetic diversity of wild fruit and small-fruit plants comprises 5,320 species representing 55 genera of 22 families. Out of this number only 818 species are utilized all over the world. One of the world's richest centers of diversity of wild fruit species is in the Caucasus: over 260 species of 37 genera occur in Caucasian forests (Burmistov 1995).

Plant and animal domestication have been called '. . . the most important development in the past 13,000 years of human history' (Diamond 2002 cited in Beridze *et al.* 2011).

The southern Caucasus, together with the Fertile Crescent, is one of the areas from which many plant species originated. Archaeobotanical material dating back to the Early Neolithic, about 7,000-6,000 millennium BC confirms that Transcaucasia was an independent and early site of numerous, useful crop product domestications. This refers in particular to wheat, barley, rye and grapevine. Armenia, Azerbaijan and Georgia are important areas for domestication of some wild cereals and grapevine (Batello & Azzu (Eds.) 2010).

N.I. Vavilov (1926) determined 18 centers of crop origin and diversity. Among them was the fourth center, which included the South Caucasus, Asia Minor, Iran and Turkmenistan. The main crops domesticated in this center (including Georgia) are wheat, rye, oats, seed and forage legumes, herbs, fruits and grapes for winemaking. According to Vavilov (1926), the origin of ancient crop varieties and landraces in South Caucasus coincides with the period of their primary domestication.

Vavilov regarded the Transcaucasian area as the major formation base of species and varieties for a number of wild fruit and plants (Burmistov 1995). Here the forests are formed by a chestnut (*Castaea sativa*) and a walnut (*Juglans regia*). Pomme fruits - an apple (*Malus orientalis*) and pear (*Pyrus caucasica*), except the Caucasus, have their primary center of origin in Asia Minor and Central Asia. In the second stratum of forests should be mentioned persimmon (*Diospirus lotus*).

The underwood here is represented by some shrubs. *Corylus avellana* or European hazelnut has its primary center in the Caucasus, where a number of other *Corylus* species are also found. In deciduous forests of the Caucasus fig (*Ficus colchica*) is widespread, cornel (*Cornus mas*), wild alycha (*Prunus divaricata*), plum (*Prunus spinosa*), and other representatives of the genus *Prunus* (*P. spinosa*), medlar (*Mespilus germanica*), species of hawthorn (*Crataegus* spp.), wild species of cherry (*Prunus (C.) avium, P. (C) mahaleb*), some subspecies of quince (*Cydonia oblonga*) and pomegranate (*Punica granatum*) are spread in southern Daghestan, Azerbaijan and eastern Georgia.

Several species of blackberry (*Rubus* spp.) are spread on the Caucasus in wild conditions, and the European raspberry (*Rubus idaeus*) in culture too. In underforests of mountain regions grow species of currant (*Ribes* spp.), some of them has been cultivated in the South Caucasus for a long time. In wild environments of the Caucasus, growth and use by the population include: blueberry (*Vaccinium myrtillus*), gooseberries (*Grossularia* spp.) and others. Bladdernut (*Staphylea pinnata* flowers used for marinade), hazelnut (*Corylus avellana*), almond (*Amygdalus communis*) and walnut (*Junglas regia*). In the river valleys and on the banks of the Caucasian rivers there is extensive sea-buckthorn (*Hyppophae rhamnoides*) and so on (Grossheim 1952a, Brezhnev & Korovina 1981, Burmistov 1995).

One of the most typical liana (vines) in the deciduous forest of the Caucasus is *Vitis vinifera sylvestris* (Grossheim 1948), the wild ancestor of the cultured vine.

Among domesticated plants, vine growing gives one of the most ancient and widely cultivated and economically important fruit crops in the world (Mattia *et al.* 2008; McGovern 2003).

The distinction between wild and cultivated forms of *Vitis vinifera* L. is mainly based on different morphological traits – plant sex. Wild grapevines are dioecious, while cultivated forms are mostly hermaphrodite plants (Zohary 1995, This *et al.* 2006).

Archaeological and historical evidences suggest that South Caucasus (Armenia, Azerbaijan and Georgia) together with Eastern Anatolia and some adjacent countries of the Fertile Crescent have been considered as the birth place (dated ~6,000 BC) for viticulture and winemaking. Other archaeological evidence of early winemaking was found in the northern Zagros Mountains dated to about 5,400-5,000 BC and in the countries of the Fertile Crescent (Levant) dated to 4,000-3,200 BC (Zohary & Hopf 2000, cited in Akhalkatsi *et al.* 2012).

At the same time, recent genetic studies have shown that secondary domestication events had occurred in the Mediterranean basin (Garcia & Revilla 2013 and references therein).

Today vines are found: (1) in natural habitats (wild lianas), (2) cultivated forms, (3) hybrids, i.e., cultivars mixture with wild vines, and (4) rootstocks escaped from vineyards (Garcia & Revilla 2013 and references therein).

Until the middle of the 19th century grape vine cultivars were abundant in their indigenous range in Europe and Caucasus. Later a sharp decrease of cultivars occurred for two reasons: firstly, when accidentally introduced from North American a pest, an aphid, *Phylloxera vastatrix* and mildew epidemics; secondly, in the 20th century the growing dominance of commercial wine varieties (new cultivars have higher yields). Both reasons dramatically reduced overall genetic variability in cultivated grapes. Currently, almost all Georgian grape varieties are grafted on rootstocks of American grapevines (*Vitis* spp.), which are resistant to *Phylloxera* (Beridze *et al.* 2011; Akhalkatsi *et al.* 2012; Pipia *et al.* 2012; Garcia & Revilla, 2013).

In 2003 only 248 of the 524 autochthonous Georgian varieties remain. This extreme situation made it necessary to undertake urgent steps for *ex situ* conservation of autochthonous grapevines via establishing living collections in Georgia. During the period of existence of the USSR about 1,030 cultivars were grown in this country. But after the fall of the Soviet Union, instead of state ownership, in the early 1990s a market-oriented system of vineries was formed in Georgia. In recent years several living collections have opened in this country (Akhalkatsi *et al.* 2012).

Apart from a broad living collection (about 1,900 accessions), Georgian cultivars are represented in small living grapevine collections as well. For example, the G. Eliava National Museum in Martvili district, Samegrelo province. It was founded in 1972 and contained 24 old Colchic unique grapevine varieties (Eliava 1992 in Akhalkatsi *et al.* 2012).

Currently the same severe problems of genetic erosion exist in many endemic and ancient cereals of South Caucasus (for instance, the tetraploid and hexaploid wheat species *Triticum timopheev*, *T. zhukowskyi* from Georgia) (Akhalkatsi *et al.* 2012).

The first authors who indicated the Caucasus as a center of grapevine origin were Hehn (1870) and De Candolle (1883) (Akhalkatsi *et al.* 2012).

To quote Vavilov (1931): 'All existing data indicate that the Caucasus is the main heart of origin of the wild and cultivated grapevine. A great number of various autochthonous varieties in Georgia, which have striking diversity of color and shape of berries and seed, indicates about the concentration of the processes of form origin here'.

Recent archaeological research from the South Caucasus indicates that winemaking dates back to the early 6th millennium BC in Transcaucasia (McGovern 2003; Barndard *et al.* 2011 cited in Pipia *et al.* 2012).

The grapevine was among the first fruits to be cultivated in Georgia (Javakhishvili 1987).

Georgia, has an abundant 540 endemic varieties of grapes out of 2,000 known in the world, which in itself makes a big statement about the importance of wine in Georgian culture (Ketskhoveli 1960). The oldest pips of cultivated vine (*Vitis vinifera vinifera*) were found in southeast Georgia in the vicinity of Shulaveri village (Ramishvili 1988 cited in Akhalkatsi *et al.* 2012) and they belonged to the period of 6,000BC.

The forests of the Caucasus contain rich medicinal arboreal plants and herbs, which by local population from pagan times are used for treatment of all kinds of illnesses. In this region a lot of species of medicinal plants are recorded (Zemlinski 1958; Miller *et al.* 2005).

2.6.3.1. The Impact of Genetic Erosion on the Forest's Rare Flora

Forests provide a range of wood and non-wood products as well as social and environmental services. Meeting these various, often competing, demands and striking a balance between consumption is the most crucial problem from an environmentalist's viewpoint (*www.worldagroforestry.org/.../15_Domestication...World Agroforestry centre*).

According to the severe anthropogenic press non-wood forest products in the Caucasus including crop wild relatives (wild fruits and berries, also medicinal environmental and ornamental plants) are increasingly reducing (Kikava *et al.* 1997).

Deforestation depletes biodiversity by destroying habitat, degradation and fragmentation ('edge' effects) of forests. Animals dependent upon the trees and other vegetation of the forest also disappear. Historically the forest lost, i.e., deforestation activities placed in foothills and riparian areas, but recently loss of forest areas encompasses all natural belts in the Greater and Lesser Caucasus Ranges except some riverheads, where modern techniques cannot be used (Ketskhoveli 1960; Gigauri 2000; Gegechkori 2008).

Additionally, human impact disturbances are caused due to road, pipeline and dam construction works (Akhalkatsi et al. 2012). To prevent harm and less to environment *ex-situ* conservation (germplasm collections) is the chief mode for preservation of genetic resources. Today several gene banks and living collections occur in Georgia (total number of germplasm accession is 6,286 in Georgian gene banks today) (Akhalkatsi *et al.* 2012).

Biodiversity is an important term and it is broadly accepted today (e.g., Cardinale et al. 2012). To protected biological diversity of the Caucasus, including forest ecosystems, the primary aim is creation of protected areas. This is *in situ* conservation. It includes any levels of biological diversity – species, genetic diversity within species and of habitats and ecosystems.

Protected areas are one of the most effective tools for slowing the rate of biodiversity loss and many species continue to survive only because of the protection provided by national parks, nature reserves and other protected areas (Dudley & Stolton (2015) and references therein).

But, on the other hand, strict reserves, national parks and other nominally protected areas are not immune regarding plants and especially wildlife to these threats (Vilcove 1986 cited in

Caroll *et al.* 2004). The main reason that causes such habitation in the level of contrast, which exists between national park ('island') is a protected unit of environment and the landscape matrix ('mainland'). Usually, most protected areas are a resemblance to 'land-bridge islands within a sea of hostile habitats' because 'human-associated development transforms the surrounding landscape' (Caroll *et al.* 2004 and references therein). Large vertebrate, i.e., keystone species may survive in a large landscape (Daak 1995 in Caroll *et al.* 2004). Therefore, discrete habitat islands (i.e., smaller and isolated parks) embedded within a matrix transformed by human landscape includes high level risk of large vertebrate extinction or loss of more species than those of less-fragmented big protected areas (landscapes) (Newmark 1987, 1996 cited in Caroll *et al.* 2004). To avoid such risks, protected areas except buffer zones around these sites should have a 'critical matrix habitat', i.e., corridors, which provide for large animals 'to move from simple patch-matrix models to models that incorporate landscape structure' (Wiens 2001 in Caroll *et al.* 2004).

The Caucasus, as previously underlined, is an internationally recognized hotspot and one of WWF's Global Ecoregions. Since the beginning of the 20th century the primary aim in this region was creation of protected areas. Although within the understanding of the context of the modern view of conservation biology and the concept of biological diversity the Caucasus Biodiversity council, a regional body, has been established. This council is called upon to promote within the Caucasus Ecoregion transboundary conservation (TBC) net. However, at present in this Ecoregion no transboundary protected areas (TBPA) have been established. A number of sites for future TBC have been identified. The proposed sites include collaborations between Georgia and Armenia (Samtskhe-Javakheti National Park and Lake Arpi National park), Georgia and Turkey (Machakhela National Park and Jamil's Biosphere Reserve) and Georgia and Azerbaijan (Lagodekhi Nature Reserve and Zakatala State Reserve (www.iucn.org/.../International Union for Conservation of Nature and N...).

Until today there were already-protected areas, which included all natural heritage of the Caucasus. But nowadays steps to developing an innovative solution to a problem to create transboundary protected areas with natural borders of various landscape units, can bring both local and global benefits providing *in situ* conservation of the region's wildlife and biodiversity in general.

Recognized *in situ* conservation as a primary approach to conserve biodiversity, in the background of a large anthropogenic threat to the environment, botanical gardens and zoos is the most conventional methods of *ex situ* conservation.

Among others Tbilisi National Botanical Garden was officially established in 1845, it occupies an area of 161 ha; Batumi Botanical Garden opened in 1912, consisting of an area of 107.7ha, it has one of the richest collections of flora comprising 2,037 taxa representing every continent of the world; Sochi, Arboretum Botanical Garden (Dendranium) officially was founded in 1922, covers 49 ha area and consists of 1.500 trees and shrubs; Sokhumi Botanical Garden opened in 1938 (30.000 ha and 5.000 taxa of plants, correspondingly).

Conclusion

- In the Caucasus, coniferous forests ('mountain taiga') occur as natural forest biome in the belt between the mountainous broadleaf hardwood forests and timberline ecotone ('alpine tundra'), although, infrequently, as a patch, except maritime pines, Caucasian pine is found in lowland areas as well.

- Coniferous forests form the most common stands between 1,400-1,900 m a.s.l., where they envelope a very broad spatial range in the Greater and Lesser Caucasus ranges. They have zonal distribution and constitute a climax type of natural community.
- Noteworthy, the coniferous forests of the region by their structure and composition in general are quite different from the typical dark or light coniferous taiga, found primarily in the Circumboreal Region around the North Pole.
- The mid-elevation belt of the Greater Caucasus and Lesser Caucasus harbors three species of indigenous conifers: Oriental spruce (*Picea orientalis*), Caucasian fir (*Abies nordmanniana*) and Caucasian pine (*Pinus kochiana*), as well as the deciduous Oriental beech (*Fagus orientalis*), which is found either as canopy builder independent belt or as mixed with dark conifer species.
- Notable, in forest structure of native conifers, in overstory and mid-story, there are essentially no co-dominant species, apart from mixed-coniferous-broadleaf (beech) stands. It should be pointed out that historically and ecologically, as dark conifers, two species *A. nordmanniana* and *P. orientalis* are co-dominating.
- Ecologically, three species of conifers are primarily contrasted. Caucasian fir and Oriental spruce are obviously shade-tolerant species, whilst the Caucasian pine is a light-demanding (shade-intolerant) species of tree.
- In the high moisture area, in the western part of the mountainous Caucasus the mesophytic or hygromesophytic *Picea-Abies* forests are the most common type of local environment and are generally found in areas where annual rainfall exceeds 1,000mm. On the contrary, in the eastern part of the same region with fewer precipitations, the semi-xerophytic Caucasian pine stands become superior (sometimes mixed with hardwood noble species (e.g., *Quercus macranthera*).
- According to the local, harsh continental climate, coniferous forests are lacking in the Armeno-Javakhetian Plateau. Nevertheless, in the Holocene, this highland, in accordance with fossil materials harbored forests composed of Caucasian pine and Caucasian (Persian) oak (*Quercus macranthera*), moreover, Oriental beech (*Fagus orientalis*), as a small natural relict stands as living evidence of recent geological time, remains in highland area of the plateau.
- Absence of large conifers (*Picea, Abies, Pinus*) in the Talysh region (south-eastern Transcaucasia) should be explained by the fact that this region (in a broad sense, Hyrcanian refugium) avoids the Quaternary multiple glaciation events and, therefore, expansion of organisms associated with boreal forests.
- It should be pointed out that all three species of coniferous forests of the Caucasus reach the vertical limit of tree distribution, i.e., the uppermost outposts in the Greater and Lesser Caucasus. Anatomical and biological adaptations to harsh climate conditions in timberline and tree line ecotones are displayed in different way in each species of Caucasian conifers.
- As a biome, coniferous forests of the Caucasus are a complex formation which occurred in the recent (Plio-Pleistocene) geological time. In the Holocene time, according to human pressure, they generally lost their primeval ecosystem features.

- In the Caucasus the largest share of the Europe-Siberian (Boreal) biota, i.e., the highest diversity of living organisms of today's Circumboreal Region of the Earth's polar area was found in the Pleistocene, significantly in the Last Glacial Maximum (LGM).
- Studies of botanists and foresters have shown that ecologically, among conifers, Caucasian pine is a much more "plastic" softwood tree.
- Pine trees exhibit higher levels of a behavioral plasticity (ranging from anatomy, physiology, phenology, breeding system, so on). *P. kochiana* is highly adaptable to various types of soils and climate shifts (nature hazards, including wildlife) compared to the other mentioned shade-tolerant two coniferous species. Therefore, Caucasian pine is able to produce different ecotypes such as woodlands with close canopy (most distributed in the Caucasus), highland's open woodland, petrophyte of rocky and stony slopes, patches ('islands') found in wetlands, growth of crook-straight-and flag stem trees in subalpine-alpine ecotones.
- Large mammals of the coniferous forests include *Panthera pardus ciscaucasia* and *Bison bison bonasus*, recently thought to be extinct and then reintroduced as Wicent-American bison hybrids.
- The temperate coniferous forests have experienced a dramatic increase in tourist destinations, recreational activities and are subjected to heavy recreation pressure.
- Today large-scale deforestations are threatening even in previously inaccessible sites of the Caucasus, significantly in the limestone massifs of the Greater Caucasus. Some private foreign companies are occupied by construction of logging mud roads using of heavy equipment, resulting in notable damage to local, until recently, pristine biodiversity.
- Caucasian pine seedlings create a valuable resource in reforestation of degraded forest and will play a key role in reestablishing a forest cover through natural succession in areas including man-made damage (clear-cut, overgrazing) and natural disasters caused by abiotic (avalanches, landslides, mudflows, wildfires, etc.) and biotic (pest insects, fungi) factors. In such areas, on the other hand, the Caucasian pine serves as native fast-growing pioneer species for afforestation sites, i.e., performs a necessary role to restore former ecological sustainability (example how to use native species versus an introduced one). For example, black locust (*Robinia pseudoacacia*).
- The Caucasus is a mountainous region. Hence, the steepest slopes of mountains are under constant complex threat from various nature hazards. Hence, undoubtedly, coniferous forests of the Caucasus in the mid-elevation belt of the native mountains occupy an important ecological and hydrological role.
- Protected areas' network for the coniferous forests, especially in calcareous sites, if we take into consideration the high endemism of plant and animal species of the Greater Caucasus, is inadequate.

2.7. The High Mountain Biome (Orobiome)

"Below me the silver-caped Caucasus lies….
A stream at my feet rushes foaming and roaring.
I watch a lone eagle, o'er the peaks calmly soaring,

Drift near as the motionless circles the skies.
Here rivers are born that tear mountains asunder
And landslides begin with a crash as of thunder…"

Aleksandr Pushkin 'The Caucasus' (1829).
(In 'Three centuries of Russian poetry'. Moscow, '*Progress Publ.*' 1980).

2.7.1. General Characteristics

The world's mountains encompass some of the spectacular landscapes that are often an inexhaustible source of beauty, breathtaking views of the world nature heritage.

The paintings, for example, Himalayas by Nicolas Roerich, and majestic examples of the Mountains' photography of the many art-photographers underline the aesthetic appeal of the property (Gvozdetski & Golubchikov 1987; Gegechkori 1997b).

Some mountains, especially limestone formations like the Dolomites superbly exposed geology, provide marine and terrestrial life of the Paleozoic, Mesozoic and Cenosoic Eras. These are the remains after the greatest dramatic extinction events recorded in the history of life on the Earth. This combination of geomorphological and geographical values creates a property of global scientific significance (*www.aapg.org›EXPLORER›archieve›January 2012*).

Mountains provide the raw materials for the soil and the source of much of the water that sustain life and the biodiversity of the planet. They are often called nature's 'water tower'. They also store water in various ways, including the formation of snow and ice, which is later realized as melt-off during warmer periods. All the major rivers in the world (Rio Grande, Nile, etc.) have their headwaters in mountains. 80% of global water resources are located in mountains. Mountains fulfil the freshwater needs of more than half of humanity as well as provide the natural resources needed as a source of energy, or income for growing food, e.g., needed to meet the daily needs of over a fifth of the world's population (Messerly & Ives (Eds.) 1997).

Water is becoming more and more sparse in many areas and regions of the planet. An alarming situation about freshwater resources is emphasized in the latest data from the World Water Council's Report on sustaining water (1996): 'In 1950, only 12 countries with 20 million people faced water shortage; by 1990 it was 26 countries with 300 million people; by 2050 it is predicted to be as many as 65 countries with 7 billion people, or about 60 percent of the world's population, mainly in the developing countries. Hence, mountains play a crucial role in the supply of freshwater to humankind, in both mountains and lowlands' (Liniger *et al.* (Ed.) 1998).

The main sources of water in the Caucasus generate the river flow originating from the glaciers and other 'water towers' of higher altitudes of the Caucasus. This water supply largely covers annual needs for drinking, for industrial and urban use and, most importantly, for irrigation of the local population of Ciscaucasia and Transcaucasia. But seasonally, in winter and summer months, rivers stream flow is minimal in the lowland rivers of the eastern Caucasus. Therefore, in semi-arid and arid zones of this region water abstractions for irrigation and other demands of population have dramatically increased (Ioganson *et al.* 1966, in Gerasimov 1966).

Despite this the alpine catchments of the Caucasus and especially north-eastern part (Russia) and southern slopes (Georgia) of the Greater Caucasus Range are characterized by much higher and constant mean annual discharge per unit area than other mountain systems of the mentioned region. It should be underlined that Georgia has excessive resources of perennial rivers' supplies of freshwater. In the case of careful management of certain resources, without damage to the unique biodiversity, this country is able to satisfy the basic human needs. The growing world's population's demands about 15 billion liters of the clean and reliable water (physiological standard for freshwater for humans per person is 2-2,5 liter) (Giorgadze & Kikalishvili 2013).

Mountains are found on every continent, they extend as latitudinal or vertical (altitudinal) gradients. They cover, depending which definition is applied, between 22-27% (a quarter) of the world's land surface (Messerli & Ives (Eds.) 1997). A new definition of mountains by the Global Mountain Biodiversity Assessment is using ruggedness as a simple and pragmatic proxy for steepness (slope angle to the horizontal), therefore excluding all plateaus, arriving at a 15.5% calculation for global mountain areas. According to this definition 12.3% of the terrestrial surface belongs to the global mountain area outside of Antarctica (Kőrner 2002).

Mountains offer vertical environmental gradients for life otherwise are only seen over several thousand kilometers of latitudinal distance (Gvozdetsky & Golubchikov 1987). The steep gradient, stretching from the lowland to the nival zone, represents compressed latitudinal zones and consequently determines a rich biodiversity in mountain biome (Nagy *et al.* (Eds.) 2003).

Mountains are stratified into thermal belts, because increasing elevation causes decreasing mean temperature. The land below the natural climatic limit of trees (the tree line) is called the 'montane' belt, the altitude from the climatic alpine, tree line and the nival zones (treeless alpine meadow and herbaceous vegetation) is called the 'alpine'; transition belt. Between tree line and treeless alpine belt, with the dominancy of tall herbaceous (forb) vegetation, is called the 'subalpine'; the high mountain part, where snow stays all year round, is called the 'nival' belt. The tree line is also called 'tree line ecotone'. It is in reality not a sharp line, but a zone of vegetation transition between two communities – contiguous subalpine forest and open alpine tundra. However, treeless vegetation with the same floristic communities also occurs widely in subalpine areas of mountains. Trees are excluded from these areas because of some abiotic and anthropogenic factors (Gagnidze 1966, 1974; Kőrner 2003, 2012; Nagy & Grabherr 2009).

Treelines are determined by climatic conditions, but have been modified and lowered by as much as several metres in some areas because of the anthropological impact. In upper mountain belts, the trees thin out and are crooked and twisted into krummholz (German *krumm*, 'crooked' and *holz* 'wood' or *knieholz* (knee timber). The tree lines of the world vary along geographic scales from boreal to tropical biomes, participate in formation of altitudinal zonation depending to elevation above sea level and local climate peculiarities (Kőrner 1998).

In the Caucasus common trees showing krummholz formation include birches, Oriental beech, Caucasian pine, Oriental spruce, etc. The tree line in the Caucasus varies from about 1,800-1,900m up to 2,400-2,500m a.s.l. (Central Greater Caucasus) (Dolukhanov 1978; Nakhutsrishvili 2003; Nakhutsrishvili *et al.* 2006).

Mountains vary vertically and exposure to the sun and prevailing wind direction produces a range of different habitats the composition of which may vary dramatically with minor variations in altitude.

Extremely diverse topography, a wide variety of ecosystems and mosaic habitats, complex interactions between abiotic and biotic factors, lead to striking diversification of mountains' biota (Kǒrner & Spehn (Eds.) 2002).

Insularity of highlands ('sky islands') that are isolated by unfavorable surrounding lowlands are recognized as a geographical pattern of adaptive evolution of new species. Such tend towards speciation of localized populations of plants and animals. These kinds of mountains serve as a refugia for boreal species which have survived since the last ice age (e.g., *Vaccinium vitis-idea, V. myrtyllus, Loxia curvirostra* and so on in the highlands of the Caucasus) (Gegechkori, 2007).

Mountains support about one quarter of the world's terrestrial biological diversity, and half of the 34 world's biodiversity hotspots are found in mountain areas. Of the 20 plant species that supply 80% of the world's food, six species (maize, potato, barley, sorghum, tomatoes and apples) originated in mountains. Large portions of domestic animals (sheep, yak, llama and alpaca) originated in mountain regions (Schild 2008). Much of the important genetic and agrobiodiversity found in the mountains is not found elsewhere (Price 1995; Messerli & Ives (Eds.) 1997).

With regard to alpine zone, the scale of its expanse represented above a treeline ecotone largely depends to the continuous mountain landscape. Of all land area of the globe (about 151 million km^2), alpine life zone covers around 3% or 4.5 million km^2, which is one-third less than the area of the vegetational arctic life zone in the Northern Hemisphere (largely tundra) (Körner 2003). The alpine zone occupies around 3% of Europe's land surface (Nagy *et al.* (Eds.) 2003).

Alpine vegetation as 'alpine tundra' represents the only biogeographic unit on land. Depending on altitude, it can be found at all latitudes (Körner 2003).

Around 10,000 alpine species are confined to this alpine life zone and comprise about 4% of the global flowering plant species (out of a total terrestrial flowering plant species of 295,383) (Christenhusz & Byng 2016). The greatest diversity of vascular plant species occurs in tropical and subtropical mountains (Price 1995; Woodward 2003).

One of the largest expanses of the alpine zone in Europe is found in the Caucasus, especially in the Greater Caucasus Range (Grossheim 1948; Gulisashvili *et al.* 1975; Nakhutsrishvili 2003, 2013).

Except for this, more than 600 floristically defined associations have been described, for example, for the Alps. Biosphere reserves should be designated to protect this natural heritage of biodiversity (Messerli & Ives (Eds.) 1997). Except nature conservation the same reserves can serve another interest, namely those of economic development of the rural population or something in between. At present, 84 of Europe's 158 Biosphere Reserves are located within mountain ranges (Borowski & Munteanu 2011). There are three Biosphere Reserves in the Caucasus Ecoregion: Caucasian (Kavkazsky) State Nature Biosphere Reserve in the extreme western end of the Greater Caucasus Range (Russia, over 275,000 ha), Camili Biosphere Reserve of the Turkey's Karchal Mountains (Turkish Caucasus, total area 27,152 ha), and Arasbaran Biosphere Reserve of the Iranian Highlands (total area 72,960 ha). All of them have alpine belts, especially the Caucasian Biosphere Reserve that has not experienced significant human impact, and highland's pasture only been grazed by wild animals (Williams *et al.* (Eds.) 2006).

Over thousands of years the mountain areas have been affected by loss of biological diversity as a result of human activities, largely modified in land uses. Cultivated fields can be found at altitudes of up to 4,000 meters in the tropical Andes and in the Alps up to 1,800 meters (Grabherr & Messerli 2011), and in the Caucasus up to 2,000 meters (till 2,200m, Zemo (Upper) Svaneti, West Georgia). However, some mountains of the world, where higher zones have remained in a natural, sometimes even pristine state according to characterized harsh environments. Here should be mentioned Tibet and the Himalayas, Pamir, Tien-Shan, etc. Unlike these regions in the Caucasus and especially in Georgia and the extreme western part of the Great Caucasus (Russia), are supported a number of extremely rare natural communities in almost pristine condition. The most unique examples of the western Caucasus environment are its almost virgin high mountain forests and timberline ecosystems spread over approximately 800-1000 thousand hectares. Almost virgin forest ecosystems and biodiversity primarily occur in remote high-altitude areas and inaccessible gorges on protected areas (strict reserves and National Parks) and are usually represented on sites situated along the riverheads, on very steep slopes of the subalpine zone. This rich, natural heritage occurs in the upper reaches of the following rivers: Bzipi, Kodori, Enguri, Tskhenistskali, Rioni in western Georgia, and Alazani and Liakhvi in eastern Georgia. These upper reaches of the Greater Caucasus' mainly southern slopes include sites which are unique or uncommon not only in the Caucasus, but are not necessarily common in Europe. The sites that have the same ecological condition are believed to have existed prior to prehistoric settlement. About 12% of the original vegetation is considered pristine (National Biodiversity Action Plan, 2005). According to narrow and inaccessible gorges they are buffered enough to support and permanently protect the natural community consisting of deciduous and coniferous forests and typical upland biota. Even the surrounding landscapes of protected areas are only moderately disturbed by anthropogenic press. Biodiversity in virgin forests is characterized by a great variety and also by the interaction and mutual influence of both the components of local biota–plants and animals. As a model of the mentioned protected areas should be mentioned the mountainous part of the Caucasian Biosphere Reserve in the western Caucasus (southern Russia), extending from the Black Sea to Mount Elbrus (Zabelina *et al.* 1998; Gigauri 2000; Nakhutsrishvili 2003; Gegechkori 2008; Nakhutsrishvili *et al.* 2009).

It must be stated that mountains offer a stronghold and refuge against enemies. In medieval times the Caucasus was overrun by Arabs, Mongols, Persians, and Turks as they tried to control this strategically important area. The people of the Caucasus have a reputation of being fiercely independent, and some highland regions were never conquered (Wertsch 2013).

Armenia and Georgia have been Christian since the fourth century. With the fall of Constantinople in 1453, Georgia as the most mountainous country in the South Caucasus, became the only Christian stronghold in a region of Muslim kingdoms (Culture of Georgia – countries and their cultures: *www.everyculture.com›Ge-it*).

Due to constant military invasions, the royal treasury and the treasury of religious centers – manuscripts and priceless samples of Christian art were largely hidden in highland peoples' churches and pagan temples temporarily, West Georgia (e.g., Khvamli Mountains, Lechkhumi district) (Maruashvili 1981).

High mountain ecosystems are among the most complex and fragile in the world. Except for natural hazards (seismic activities, avalanches, floods), etc., the highland ecosystems are highly vulnerable to anthropogenic press. In this sense the Convention on Biological Diversity (CBD) of the United Nations requires the support of development cooperation (Messerli & Ives (Eds.) 1997).

Since the 1970s, mountain ecosystems have received increasing recognition in several research programmes. On the basis of a resolution of the UN General Assembly (1998), the International Year of the Mountains (IYM) was celebrated by the world's environmental society (Rio de Janeiro 1992, 2012; Johannesburg 2002; Kuala Lumpur 2004). The International Mountain Society (IMS) was founded in 1980. The International Center for Integrated Mountain Development (ICIMOD) was founded in 1983.

According to an unprecedented level of collaboration among governmental, non-governmental organizations and research institutions, a comprehensive report (Messerli & Ives (Eds.) 1997), and a companion policy document (Bisaz *et al.* 1997) were presented to a special session of the General Assembly, Rio+5, in 1997.

As previously mentioned, major centers of plant and wildlife species diversity are in tropical and subtropical mountains. Secondary centers of biodiversity occur in the Mediterranean mountains, the Alps, the Caucasus and Southeast Asia (Messerli & Ives (Eds.) 1997).

In terms of geology, climate, vegetation, wildlife, and human interference, each mountain environment is somewhat unique. The Caucasus Ecoregion is one of the most outstanding regions in the world, which is well known for its diverse landscapes and aesthetic, cultural heritage, biological and hydrological values (Krever *et al.* (Eds.) 2001), Williams *et al.* (Eds.) 2006).

Fascinating alpine environments of the Greater and Lesser Caucasus with their aesthetic resources: open landscapes, dominated by shrub thickets, heaths and grasslands, with their orophile wildlife, fresh water supplies, mineral springs. The cultural heritage multiplies these natural peculiarities, such is the production of ecologically clean products by local farmers; historical monuments: fortress and churches, a multitude of archeological sites. All these benefits are an important potential economic factor as a mountain forest, timberline, tree line and alpine destination of various kinds: eco-, scientific, ethno-, gastro-, and agrotourism (Krever *et al.* (Eds.) 2001).

A huge infrastructure was created by the Soviet Government in the last century and millions of citizens of the Soviet Union spent their holidays in the Caucasus, which by its beautiful beaches of two seas and the mountains was converted into one of the attractive resort areas and one of the main centers for different kind of tourism and alpinism in the USSR. The breakup of this state, the emergence of violent conflicts, the sharp decrease in living standards resulted in the collapse of the once flourishing tourist infrastructure (*www.osce.org/baku/27857*); O'Loughlin *et al.* 2007, Henze 2013).

In the past decade the North and South Caucasus, the Black Sea coast area, especially Adjara (Batumi city) have again become a major recreational zone. Despite this, Gudauri skiing center (Khevi district, East Georgia) rock and ice climbing in Svaneti (West Georgia) and Karachaevo-Cherkessia (North Caucasus, Russia) have mushroomed foreign tourists as well, only small groups of tourists from abroad have come to the Caucasus countries at present. Poorly planned and implemented tours and poor awareness of visitors with local people's tradition and culture can create negative perceptions of tourists in local communities and lead to an inhospitable atmosphere (Henze 2013; *en.wikipedia.org/wiki/tourism_in_Russia*).

However, these activities can provide foreign nature lovers, and adventurous-minded people, still less residents of the region with a way to experience the beauty, uniqueness and sometimes thrilling sites of high mountain ecosystems. But if not properly managed, they can threaten the fragile state of the alpine environments, cause significant damage to the sensitive sites such as breading or nesting areas of local animal life, to the habitats of endemic and other rare vegetations.

This problem is especially topical and urgent for many regions of the Caucasus, where the institution of National Parks is comparatively new and, therefore, it is sometimes problematic to ensure adventure activities or tour groups by professionals follow designated routes, Also they need to avoid overwalking or camping of sensitive environment sites, to be aware of avoiding any cultural sites of significance, not to deteriorate the virgin landscapes with litter or domestic animals waste. Mainly, on the base of 'An ecoregional conservation plan for the Caucasus' (2006), to rise up the environmental awareness and work in partnership with the local authorities, tourism board and community groups in the development of sustainable tourism policies and strategies.

Finally, as long as the region is generally received as dangerous due to the potential for armed conflict and terrorism, the tourist trade will remain weak. Therefore, in general, the Caucasus Region is not competitive in the global tourism market (Henze 2013).

The term 'high mountain' was used by Troll (1935) cited in Müller 1986) to describe plant communities above the closed forest limit.

In the Caucasus all altitudinal belts (zones) characteristic of the moderate climate regions of Eurasia are represented: subalpine, alpine, subnival and nival. The height distribution of these belts in various parts of the Caucasus depends on their geographic positions and abiotic factors, and ranges from 1,800-1,900 up to 3,500-3,700 m a.s.l. 'One of the most pronounced peculiarities of high mountains is the diversity of habitats within even a very small area' (Nakhutsrishvili 2013).

High mountain biome covers 17% of the Ecoregion. About 1,500 vascular plant species are found in the Greater Caucasus high mountains and around one third of these are endemics. This is alongside, approximately 80% of the plant species in rock and scree communities on Colchis limestone ridges in the Greater Caucasus are endemic (ECPC 2012).

One group (the majority) of the Caucasian botanists (e.g., Grossheim) suggests that the high mountain flora of the region consists mainly of the Boreal (Boreo-Alpine) and Arcto (tundra) - Alpine elements, which occurred in the Caucasus Isthmus during the Pleistocene's climate oscillation. Another group of scientists considers that the most common representatives of the local subalpine-alpine flora are of native autochthonous origin and have a tertiary nucleus (e.g., Kharadze 1960 in Nakhutsrishvili 2013).

2.7.2. Subalpine Vegetation

2.7.2.1. General Characteristics

The subalpine belt is defined as the areas above the climatic limit of dense forests up to the lower limit of the alpine grassland (Grossheim 1948; Shiffers 1953).

Some botanists consider the subalpine belt and even dark coniferous forests to be part of the alpine zone, (Schröter 1926; Landolt 1969).

The subalpine belt is the transition zone between full canopy forests and the alpine open zone. This natural belt unites both arboreal (shrub-woods, crook-stem, straight-stem and park-like forests) and tall and low herbaceous communities (Gulisashvili *et al.* 1975).

In the treeless areas of Javakheti Plateau, mountain steppes compete with alpine meadows. Except the mentioned region, highland steppes within the subalpine belt having characteristic formations in the Caucasus are absent (Gulisashvili 1964).

The subalpine belt is characterized by specific climatic conditions: high air humidity, low temperatures, short vegetation period, small daily temperate variations, rich soils and (compared to the alpine belt) no high solar radiation. A thick layer of snow protects the low-growing evergreen vegetation from destruction by frost, but prevents the development of trees (Makhatadze & Urushadze 1972).

The subalpine belt possesses a considerably high floristic and phytocoenotical diversity; it is very rich in endemic species and even genera of plants and animals (Kharadze 1974; Gagnidze 1966, 1974; Gegechkori 2007).

According to Caucasian botanists (Bush 1935; Makhatadze & Urushadze 1972; Ketskhoveli 1960; Gulisashvili *et al.* 1975; Dolukhanov 1966; Nakhutsrishvili 1999, 2013) subalpine zone includes the following types of vegetation:

1. Subalpine open woodland (park-like forests);
2. Subalpine crook-stem forests;
3. Elfin woodlands;
4. Subalpine tall herbaceous (forbs) vegetation;
5. Subalpine meadows (including high- mountain meadow steppe).

2.7.2.2. Subalpine Woody Vegetation

The upper limit of distribution of the arboreal component of the subalpine belt depends on the climatic dryness (aridity) and the massiveness of the mountain systems. In mountains in a continental climate, for example, the eastern part of the northern slope of the Greater Caucasus Range (Tusheti, Dagestan), the upper borders of the subalpine belt are situated higher (2,400-2,500 m a.s.l.), but in mountains under moist climatic conditions, such as in western parts, the Black Sea coastal Caucasus where the mountain slopes are directed to the Black Sea, the upper forest limit is lower and rarely exceeds altitudes higher than 2,000-2,150 m (Dolukhanov 1956; Gulisashvili 1964; Makhatadze & Urushadze 1972).

2.7.2.2.1. Subalpine Open Woodland (Park-Like Forests)

2.7.2.2.1.1. Subalpine Forests with High-Mountain (Caucasian) Maple (*Acer trautvetteri*)[6]

Subalpine forests with *A. trautvetteri* are found usually in the western part of the Caucasus both on the north slopes of the Greater Caucasus and in Transcaucasia. In the eastern part of the region this species is replaced by *A. hyrcanum*, a species that requires a drier climate. In the western part of the South Caucasus *A. trautvetteri* is found in the subalpine belt, excluding the western part of the Adjaro-Imeretian Ridge. *A. trautvetteri* forms impressive open woodlands, extending from 1,800 up to 2,200m a.s.l. It is present as pure stands, but is also associated with beech and rarely with birch and sometimes forms mixed communities with Oriental fir. Up to

---

[6] In recent times *Acer trautvetteri* Medw. is treated as a subspecies of *A. heldreichii* (*A. heldreichii* subsp. *trautvetteri* (Medw.) A.E. Murray).

approximately 2,000m it grows as a straight-trunk tree; at higher altitudes it turns into crook-stem, semi-creeping and dwarf shrubs. When the trunk makes contact with the ground, the species propagates by gemmation. Around 2,100-2,200m this species of maple was replaced by *Betula pubescens litwinowii* (Makhatadze & Urushadze 1972; Gulisashvili *et al.* 1975) (Figure 165).

Figure 165. A multi-trunked Trautweter's highland maple. Timberline ecotone of Abkhazeti. Audkhara surroundings, W. Georgia, 1,800 m a.s.l. (1986).

Park-like open woodland consists of *A. trautvetteri* mixed with *F. orientalis* frequently occupying an area from 1,700 up to 2,050 m. The constant shade-enduring plants of these communities frequently occur as subalpine, tall, herbaceous, and thickets of large ferns (*Athyrium distentifolium, Dryopteris oreades, Oreopteris limbosperma*, etc.) grow in the herb stratum as well (Dolukhanov 2010).

2.7.2.2.1.2. Subalpine Forests with Persian Oak (*Quercus macranthera*)

Subalpine forests with *Q. macranthera* occur under dry climatic conditions, where Caucasian (Persian) oak occupies altitudes between 2,100 and 2,350 m a.s.l., including the subalpine belt. Until recently the Caucasian oak groves were widely distributed in the eastern and southern parts of Transcaucasia, but due to human impact, these forests have been widely destroyed.

Today the distribution area of *Q. macranthera* is significantly narrowed and replaced by xeric mountain vegetations (Gulisashvili *et al.* 1975; Dolukhanov 2010) (Figure 166).

Figure 166. Park-like forest composed of *Quercus macranthera*. Trialeti Ridge, Lesser Caucasus, Algeti River Gorge, E. Georgia, 1,750 m a.s.l. (1968).

2.7.2.2.2. Subalpine Crook-Stem (Krummholz) Forests

With the increase of altitude towards the upper limit of dense forests on steep slopes with badly developed soils, shrub like, creeping and semi-creeping forms of beech, birch and several other deciduous species develop. The reason for the formation of crooked or twisted trunks (krummholz) in highland ecotone's is 6-7-meter-thick snow, which permanently presses against the ground flexible trunks of highland's ecotone beeches, birches and pines (Dolukhanov 2010).

At a young age the trees are pressed to the ground under the heavy snow mass and the lower parts of their trunks get rooted. All stages and forms of metamorphosis from straight-trunk vegetation demonstrate the remarkable adaptive ability of the vegetation under severe climatic and orographic highland conditions (Dolukhanov 2010).

Quercus macranthera, *Picea orientalis* and *Acer trautwetteri*, which mainly lack the ability to turn into shrubs and to propagate by germination, get gradually scattered with increasing altitude and decreasing air temperature and lose their role as the dominant forest species and are replaced by highland meadows (Gulisashvili *et al.* 1975).

The vegetation near the timberline is very sensitive to human impact. Arboreal high-mountain vegetations have become very rare and in some localities have been completely eliminated under anthropogenic influence, including firewood harvesting by shepherds, timbering and overgrazing. As a result, the near timberline vegetations have almost everywhere retreated from their natural upper limit of distribution. On the other hand, the opposite picture can be observed in some reserves, where due to global temperature rise the subalpine forests expand into the subalpine meadows. Within protected areas and on inaccessible sites the natural borders of high-mountain forests are still preserved (Ketskhoveli 1960; Nakhutsrishvili 1999; 2013). Botanists distinguish different types of arboreal highland.

2.7.2.2.2.1. Concerning the Timberline and Alpine Tree Line (Krummholz) Ecotone Definitions and Terminology

Climatically determined changes in vegetation cover ('natural laboratories') are most visible at different types of ecotones, i.e., between adjacent vegetation types of two ecosystems (sensu Risser 1995). 'Ecotones, the narrow transition zones between extensive ecological systems, may serve as sensitive indicators of climate change, because they harbor species that are often at the limit of their physical and competitive tolerance' (Wasson *et al.* 2013). Therefore, ecotones are more sensitive to climate changes (even to short-term climatic fluctuations).

Among others should be mentioned forest-alpine tundra transition, which is one of the most visible intermediate ecotones.

Researchers have long studied what they call 'timberline' and 'tree line'. According to Troll (1973), it began in the Swiss Alps during the 16th and 17th centuries. For a transition zone extending from contiguous closed subalpine forest to the uppermost open alpine tundra is used the term 'alpine tree line ecotone (ATE) or tree line' (Körner 1998; Holtmeier 2003; Zeng 2010).

Körner & Paulsen (2004) define another term 'timberline' (or 'forestline') as the upper limit of closed canopy forest. The tree line as the extreme limit of trees and tall shrubs should not be confused with a lower timberline with its forest-forming tree species (Jorgensen 2009).

Concerning to another definition an alpine tree line ecotone consists of the lower (timberline) and upper (tree line) parts, where *timberline* is found in the upper limit (edge) of continuous forest. It involves tall, erect timber sized trees, but *treeline* encompasses the highest patches of forest at higher elevation of mountains, where trees may be stunted or not (Wardle 1974; Körner 1998; Daubenmire 1954 cited in Richardson & Friedland 2009). Some authors consider an area between the timberline and tree line as an independent transitional ecotone. Depending on highlands topography, steepness of the latter ecotone may vary from tens to hundreds of meters in width (e.g., Körner 1998).

In the alpine tree line ecotone trees are able to survive from local areas that are too harsh for them to live. Tree life-form of plants is limited to grow in height by main ecological drivers – inhospitable climate - low temperature stressed conditions and alpine land use (especially grazing). Among few, key abiotic factors, low temperatures are ubiquitous, affected forests' edges in both directions: altitudinally (alpine) and latitudinally (arctic), i.e., globally common (Richardson & Friedland 2009).

However, currently many aspects regarding timberline and tree line (alpine) ecotone, especially at regional levels, still remain unclear. On a local scale, regional climate, local topography, i.e., topo climatic variables produce complex stressful conditions for both – timberline and tree line biotas. Here should be mentioned the following abiotic factors: sunshine radiation, strong winds, daily variation in local wind speed, thin layer of soils, avalanches, landslides, etc., combined with other ecological factors. For example, biotic factors: impact of excessive grazing (domestic – livestock, wildlife, i.e., foraging by red deer, roe deer and other wild ungulate species); anthropogenic factor: for centuries (from 5,000-4,000 BC) and human activities (alpine farming, forest pasturing) can cause not only degradation events of tree lines, but humans are found to be a major driver of current tree line shifts.

Observations indicate that the average growing season - worldwide average temperature - along the tree line is 6.7^{0}C (Körner & Pausen 2004).

Tree line elevation ranges from sea level in high latitudes areas (across the Polar Circle), up to 4,700 m in South Asia (Tibet) and 5,000m in South America (Andes) (Troll 1973). According to Humboldt's Law (1817), 'in terms of temperature, latitude (lower elevation) compensates for altitude', and elevations with tree lines decrease from the south as one moves from equator to Arctic (Salisbury & Ross 1992 cited in Richardson & Friedland 2009).

Climate conditions strongly determined the function and distribution of plant species around the world (Lenoir *et al.* 2008). On the contrary, plant communities are thought to be the most sensitive indicator of climate ('to be in quasi-equilibrium with climate') (Walther *et al.* 2002).

Accordingly, as tree species in alpine ecotone are usually scattered, the height gradient of trees is limited: about 2m (Wardle 1974), or lower than 3m (Körner 2012). Increasing negative abiotic factors largely impacted on tree physiognomy, inducing various transformations. On more exposed sites of highlands strong and constant winds produce flag (banner) trees. In other cases, native scattered trees are characterized by dwarfing (stunting), twisting of stems, sub-prostrate, prostrate and dense low branching (clonal growth) tree mats. As it stressed, these stunted, wind-blown trees (natural bonsai) in rugged environments of tree line zone are known as krummholz or elfinwood (Zwinger & Willard 1996).

2.7.2.2.2.1.1. Colchian (Western Georgia) Subalpine (Crook-Stem) Forests

Colchian (Western Georgia) subalpine (crook-stem) forests are characterized by the presence of *Fagus orientalis*, *Betula medwedewii*, *Quercus pontica*, *Acer trautvetteri*, *Picea orientalis* and *Abies nordmanniana*. In the shrub stratum *Rhododendron ponticum*, *Rh.. caucasicum*, *Prunus laurocerasus*, *Sorbus caucasigena* and *Corylus colchica* occur. The slopes directed to the Black Sea receive a large annual precipitation (2,000-4,500 mm) and have a mild snowy winter. Such climate peculiarities determined the survival of many representatives of Tertiary flora. *A. trautvetteri, P. orientalis* and *A. nordmanniana* frequently remain with a straight-trunk even in tree line ecotone.

Subalpine forests are divided into three categories: Colchian, Boreal and Hyrcanian-Caucasian (Gulisashvili *et al.* 1975).

2.7.2.2.2.1.1.1. Subalpine Beech (*Fagus orientalis*) Forests

The upper limit of straight-trunk beech (*F. orientalis*) reaches about 2,000-2,100, rarely 2,300 m a.s.l. At altitudes higher than 2,200-2,300 m, beech forests are replaced by dwarf semi-creeping and creeping formations (Dolukhanov 2010) (Figure 167).

Vast areas of crook-stem forests of beech and elfin woodlands occur mainly in southern Transcaucasia and in the southern part of the north slopes of the Greater Caucasus Range. In the moist Colchian climate crook-stem beech forests appear at 1,800-2,000 m, gradually turn into ground-creeping forms to change above 2,300-2,400 m into subalpine meadows or elfin woodlands. On steep rocky slopes above the tree line, crook-stem birch forests occur. Almost the same formations are found in Teberda Reserve on the north slopes of the Greater Caucasus. In the eastern part of the South Caucasus crook-stem beech forests rarely occur (Dolukhanov *et al.* 1945; Tumajanov 1960; Makhatadze & Urushadze 1972).

Unlike high-mountain with the increase in altitude birch formations with a canopy layer, subalpine *F. orientalis* stands rarely form open woodlands and they form dwarf, crook-stem trees, no shrubs, although two to four trunks of the same size may sprout from one trunk (Dolukhanov 2010).

Figure 167. Subalpine timberline ecotone covered by Oriental beech in late spring. The beech's 'clever' adaptation to the harsh environment is displayed in crook-stem anatomy of the plant. Goderdzi pass neighborhood, Adjara, W. Georgia, 2,025 m a.s.l. (1978).

Several communities of crook-stem beech forests are known: *Fagus orientalis* forests with *Vaccinium arctostaphylos*, *F. orientalis* forests with *Rhododendron caucasicum*, *F. orientalis* forests with *Galium odoratum* and *F. orientalis* forests with tall herbaceous vegetation (the latter divide into various subgroups), (Ketskhoveli 1960; Gulisashvili *et al.* 1975).

In areas with less contrasting conditions, crook-stem beech forests are in contact with *Sorbus-Betula* forests as well as with thickets of Colchian endemics as *Rhamnus imeretina*, *Corylus colchica*, *Sorbus subfusca* and *Salix caprea* (Dolukhanov 2010).

2.7.2.2.2.1.2. Boreal Subalpine (Crook-Stem) Forests

Boreal subalpine (crook-stem) forests are characterized by woody plants of Boreal origin. They are situated on the northern slopes of the Greater Caucasus and in the south-western part of Transcaucasia, south of the famous Borjomi resort. During the glaciations the cryophyte flora migrated into this area from the north Boreal Center and partially from Asia Minor. After deglaciation some of them remained in the Caucasus highland as relict species of Quaternary glaciations. Boreal crook-stem forests are composed of birches (*Betula raddeana, B. pubescens litwinowii*), pine (*Pinus kochiana*) and mountain ash (*Sorbus caucasigena*). The shrub understory includes species of *Ribes, Rubus* and *Salix*, with a frequent participation of bilberries (*Vaccinium* spp.).

2.7.2.2.2.1.2.1. Subalpine Birch (*Betula* spp.) Forests

Betula L. is a genus of approximately 60 species, subspecies or varieties with a wide distribution in the Northern Hemisphere (Wang *et al.* 2016). Recent representatives of the genus *Betula* section *Lentae* are closer to the ancestors of the genus described from the Early Tertiary Period (Ashburner & McAllister 2013).

In the Caucasus five species of birch are distributed (Gulisashvili et al. 1975; Gagnidze 2005). This genus is divided into two sections: (1) *Costatae* Regel and (2) *Albae* Regel. The first section includes three endemic and Tertiary relicts to the Caucasus species: *B. medwedewii,*

B. megrelica, and *B. raddeana*. The second section encompasses two endemic but Boreal relationships: *B. pendula* and *B. pubescens litwinowii.*

The birches in the Caucasus reach the timberline. The lower limit of distribution on the north slopes of the Greater Caucasus and in eastern Transcaucasia is 800-900 m a.s.l. In western Transcaucasia with its moist climate the birches come down lower.

They are commonly associated with *Pinus kochiana, Sorbus caucasigena, S. boissieri, Salix caprea, S. kazbegensis, Vaccinium vitis-idea, V. myrtillus, V. uliginosum, Daphne glomerata, D. mezereum, Populus tremula*, and *Acer trautwetteri*; from the herbs *Anemone fasciculata, Festuca drimeja, Cicerbita racemosa,* etc. (Gulisashvili *et al.*1975; Nakhutsrishvili *et al.* 2006).

According to Gulisashvili *et al.* (1975), one of the most characteristic species of the birch forest communities is Caucasian pine (*P. kochiana*). Where they mix, depending on climatic conditions and relief dissection, they frequently compete resulting in the prevalence of one taxon over the other and *vice versa. Betula pendula* and *B. pubescens litwinowii* are far more resistant to warmth and coldness than Caucasian pine. In lower latitudes (from 1,000 to 1,500 m a.s.l) pine outcompetes birches. In high altitudes, up to 1,700-1,800 m, the ecological conditions are different. There the climate favors the growth of both species of birches, and the pine groves are eliminated. Only on the slopes of almost all exposures in the highland, pure birch forests dominate in the arboreal vegetations. Only on slopes with south exposure, being the driest and warmest, near the upper forest limit, pine forests prevail.

Tumajanov (1937) distinguished birch forests with *Calamagrostis arundinacea* in the herb layer (mainly in the North Caucasus), birch forests with *Rhododendron luteum* (frequently occurring in the South Caucasus on the south slopes of the Greater Caucasus) and birch forests with *Rh. caucasicum* and *Vaccinium myrtillus*, chiefly on the north slopes of the Greater Caucasus (Ochiauri 1965). Questions of the origin of birches in the Caucasus were carried out by Sochava (1949).

For Litvinov's birch (*Betula pubescens litwinowii*) Dolukhanov (2010) has proposed the term 'white-birch forests'. These forests are constituted as the most typical communities of the subalpine belt. The distribution area of this endemic to the Caucasus species varies in altitudes between 2,000-2,400, rarely 2,600m a.s.l. Inclination of slopes doesn't exceed 10-25^0 that determines stable cover of snow during winter. In specific habitats (moraines, rocky slopes) it comes down to lower altitude. As a light-demanding species it occurs under relatively continental climatic conditions. In East Georgia it inhabits exclusively dry mountains (Makhatadze & Urushadze 1972; Dolukhanov 2010). On limestone substrates (rocky outcrops, steep limestone hills of western Georgia, it clings to thin soil and almost bare rock, similar to Caucasian pine in other regions of the Caucasus, where no other climax species of trees will grow; it is found in relatively moist habitats (pers. observations) (Figure 168).

Crook-stem *B. pubescens litwinowii* forests are rarely monodominant. Most frequently they are associated with *Sorbus caucasigena* and *S. boissieri*. In the lower part of the subalpine belt Litwinow's birch is found together with the high-mountain maple *Acer trautvetteri*; in Colchis with *Sorbus subfusca, Rhamnus imeretina, Salix caprea* and *Ribes biebersteinii*. The herb layer is well developed and consists of wild grasses and herbs of herb meadows and tall herbaceous vegetations, species of *Festuca, Lilium, Aconitum, Geranium, Aquilegia, Valeriana, Oxalis* (Dolukhanov 2010).

According to Dolukhanov (2010), it is worth mentioning that subalpine crook-stem white-birch communities with *Rh. caucasicum* are characterized by the presence of representatives of the flora of taiga. It should be mentioned the following species peculiar for this type of phytocoenoses are: *Huperzia selago, Listera (Neottia) cordata, Pyrola minor, Vaccinium vitis-idae, V. myrtillus, Oxalis acetosella*, etc.

Figure 168. *Betula pubescens litwinowii*, upper limit of tree species distribution. Zemo (Upper) Svaneti, Ugviri pass environs. General view from 2,200m a.s.l. (1988).

Ecologically, the distribution area of *B. pubescens litwinowii* and silver birch (*B. pendula*), native to Europe and parts of Asia, approximately coincided with that of the Caucasian pine (*Pinus kochiana*), growing in habitats with continental climate (cold-tolerant trees or shrubs).

One of the most interesting birches is Medvedev's birch, *Betula medwedewii* (=*B. medwediewii*). It is a very distinct relict of the Tertiary and confined only to the south-western part of Transcaucasia, and adjacent north-eastern Turkey. Frequently they are situated in straight-trunk and crook-trunk beech forests, but usually found on subalpine mixed forests at elevations between 600-2,400 m a.s.l. This species of birch is rare across its range, very small populations are fragmented and unstable (Shaw *et al.* 2014) (Figure 169).

Figure 169. Elfin stand of incredible relict birch (*Betula medwedewii*) in upper limit of distribution. Kintrishi Reserve, Adjara, W. Georgia, 2,100m, near to alpine Lake Tbikeli (2,217 m a.s.l.) (2009).

Another endemic and Tertiary relict birch species is the Megrelian birch (*Betula megrelica*). It is a close living relative to former species. Resident of moist, limestone slopes (inland cliffs and rocks) of the mountains. This species of birch is known from very restricted gorges of Samegrelo – Mount Mingaria and Mount Jvari (West Georgia) (Ashburner & McAllister 2013, pers. observation).

The species grows in altitudes mainly between 1,800-2,000 m a.s.l., forming small groves (Figure 170).

Figure 170. Specimen of unique relict and endemic Megrelian birch. Limestone ground area of the southern slope of the GCR, with a lot of specifically alpine herb plants, which are found nowhere else. Mingaria limestone massif, Samegrelo, W. Georgia, 1,850 m a.s.l. (1975).

B. megrelica faces threats from stochastic events as well as forest clear cutting. The companies harvesting timber on forest lands have made plenty of dirt roads damaging the ecological integrity of private and even state forests. Paradoxically, but the rarest plant species (e.g., *B. megrelica*) must be protected from scholars who illegally (similar poachers) collected *in vivo* birch specimens (!) for laboratory tests (pers. eyewitness).

Another silver birch, *Betula raddeana* (= former *B. verrucosa*) is relict and endemic to the Caucasus Ecoregion (Azerbaijan, Georgia, northwestern Iran (Alborz), Russia).

Medium sized trees (about 10 m in height) with whitish-pink peeling bark on the trunk. It grows in small populations on the north slopes of the Greater Caucasus, especially in the Terek River (Khevi region, Kazbegi site, East Georgia) and Sulak River gorges (the Russian Federation) and in the Tusheti Hollow (Georgia). On the south slopes of the Greater Caucasus this species covers the upper reaches of the Didi Liakhvi, Duruji and Aragvi rivers. Typical orophyte appears at 1,400-2,800m a.s.l. (Ketskhoveli 1960; Abashidze 1962).

It makes either monodominant stands, or co-occurs in highland forests with *B. pendula* and *B. pubescesns litwinowii*. For Dagestan also some small groves of *B. raddeana* within oak and beech forests have been reported (Abashidze 1962; L'vov 1964; Ochiauri 1965; Galuŝhko 1978) (Figure 171).

It is a vivid witness of the former continuous distribution of low canopy birch in the high mountains, but nowadays in Georgia, it infrequently occurs as the holy shrine forests, which are protected by the local people for spiritual reasons (Dolukhanov 1978,1980).

These forest 'islands' have remained untouched through the ages and give a good picture of the composition of former subalpine birch forests. Troitsky (1933) dedicated a special work to some 'sacred groves' in the upper reaches of the Aragvi River (Figure 172).

Two relict species, *Betula medwedewii* and *B. megrelica* of section *Costatae* are considered as Near Threatened (NT) for former species and Endangered (EN) for latter species and are included in the IUCN Red List (Beech *et al.* 2015); *B. raddeana* has another status – Least Concern (LC). All three species are included in the Red List of all countries of the Caucasus, where they are represented.

Figure 171. Stand of Radde's birch enclosed timberline area of the Khevi region, Kazbegi district, Mkinvartsveri ('Glacier Peak') summit (5,047m a.s.l.) environs, northern slopes of the GCR, 2,100m a.s.l. (1982).

Figure 172. Shrine decorated with rings and red deer's horns. In background: an 'island' of Holly Forest. Gudamakari region, vil. Chokhi environs, E. Georgia, 1,660 m (1985).

2.7.2.2.2.1.2.2. Subalpine Forest with Caucasian Pine (*Pinus kochiana*)

Subalpine forests with *P. kochiana* are situated in some localities with a continental climate. In West Georgia they occur in deep gorges well preserved from the Black Sea air masses (e.g., in the upper reaches of the r. Inguri, Svaneti). They cover a narrow strip to the upper limit of the distribution of pine-birch forests. Pine trees frequently develop crook-stem trunks under the influence of fierce winds. Rarely pines are represented as semi-creeping shrubs that are partially pressed to the ground. Some botanists (e.g., Medvedev 1919) consider this pine as an independent form: *Pinus montana caucasica.*

Within the subalpine zone of the Caucasus, the following types of pine forests can be distinguished: pine forests with subalpine meadow vegetation; dry pine forests with *Poa nemoralis*, *Agrostis alba* and other grasses within juniper shrubs in the forest margins that are pressed to the ground; pine forests with subalpine steppe vegetations, occurring only in South Georgia, with *Festuca sulcata* and *Stipa tirsa* as dominant grasses in the herb layer; pine forests with *Vaccinium vitis-idaea*; pine forests with *V. myrtillus* (Gulisashvili *et al,* 1975; Dolukhanov 2010).

Dolukhanov (2010) comments: 'The Tushetian pine forest is not only a representative of boreal flora, but a specific florogenetic complex distributed throughout the taiga, circumboreal biome'. Some genera and species are inseparable participants of Hylocomiosa types of pine forests, including the following communities: Pineta myrtillosa, Pineta vacciniosa, Pineta oxalidosa. They populate steep slopes of northern exposure and beyond Tusheti are not found in the Caucasus (Tumadjanov 1938 cited in Dolukhanov 2010) (Figures 173, 174).

Figure 173. *Vaccinium vitis-idaea*, relict of a borealization event of the Caucasus. Tusheti Hollow, W. Georgia, (1991).

Figure 174. *Vaccinium myrtillus*, relict of last glaciation, Tusheti Hollow, W. Georgia (1991).

Box 11. Northern Georgia – an Interesting Source of Unique Pictograms

'Mountain societies are not the sort that tend to originate novelties, to agitate for change. On the contrary, they are often relics. Until well into the 20th century an obscure Caucasian mountain tribe called the Khevsurs still wore armor like crusader knights' (Milne & Milne 1976).

Mountainous regions of northern Georgia in general and Khevsureti in particular remained until nowadays completely archaic ritual sites which originated in the pre-Christian world and were performed absolutely unconsciously. They are astonishing in their variety.

Khevsurs and adjacent Tushetian and Pshavian peoples are known for their outstanding oral literature, significantly poetry, folksong, history, stories, who kept ancient mythology, spiritual life and other ethnographic features ('Ballads recorded from the mouths of Khevsur peasants take their place in poetry anthologies beside the masterpieces of the Georgian classical and romantic periods…') (Tuite 1994).

The pearl of folk poetry is the heroic poem 'Verse of the leopard and the young man' – a folk ballad of hunting demonstrating the character and the creed of northern mountainous Georgian people. Pshav-Khevsuretian folk high school poetry has been a fertile source of literary masterpieces (lyrical and dramatic poems) for the outstanding Georgian poet Vazha-Pshavela (from Pshavi).

The main regions, which host pictograms in northern Georgia (the eastern Caucasus) are represented by Khevsureti, Pshavi, Tusheti, and Khevi. Khevsureti is located on the southern and northern slopes of the Greater Caucasus Range, Pshavi region occupies southern slopes, Tusheti – northern, and Khevi – mainly northern slopes of the range. Settled in fragile ecosystems, i.e., high-risk environments (avalanches, landslides, harsh climatic and environmental conditions, fragile soils and vegetation cover, remoteness make mountain areas vulnerable and native people disadvantaged. Despite these harsh, unpredictable conditions which contributed to the marginalization of the mountain, these indigenous peoples, despite ethnic minorities, kept impressive testimonies to a sophisticated mountain culture, including enigmatic pictograms.

Fragmentary rock art with abstract signs in Khevsureti, which have a resemblance to pictograms dating back thousand years, allow scholars to make random glances into the vague dawn of human cultures of people of the ancient Near East, the home of early civilization.

For the first time data about stone carving of Khevsureti and Tusheti were registered by famous Georgian ethnographers and philologists. The 2010s of the twenty-first-century were marked with a new discovery of rock carving.

An amateur ethnographer (but an artist by his primary occupation), Gia Gigauri undertook major field expeditions in northern Georgia. Through online corporation – 'Magthi' funded project he had yielded cryptographic materials of great importance, fixed and drew a great number of unique stone images made, it seems, by the Early Medieval Period's settlement. Based equally on his own field expeditions and written sources as an artist and researcher Gigauri has successfully published a monograph entitled 'Cryptographic signs from mountainous area of eastern Georgia' (2010). This book is the first attempt to describe the mentioned cultural remains in Georgia (photographs, illustrations, graphic drawings). This scientific work includes the following number of cryptic figures – 'stick messages': Khevi region (Thruso River gorge) – 25 stones (with 78 figures), Pshavi region – 4 stones (6 figures), Khevsureti – 60 stones (204 figures), Tusheti region – 74 stones (197 figures). At present, in total, more than 480 precise drawings have offered an excellent case of the study of symbolic and sacred meanings.

Gigauri thoroughly described the location of the pictograms – sacred carvings, which are made by ancient native peoples on the stony walls of dwelling houses, fortresses, sacred and burial sites (gravestones), etc. The motifs of pictograms made by artist are generally divided into two: (1) animal and (2) symbols and geometric figures, though other motifs have been identified in this part of Georgia. For example, pictograms with ritual-magic scenes and anthromorphic figures (images of men – hunters, warriors armed with spears). Numerous geometric figures include cross-shaped inscriptions, circles and semi-circles and a diagonal line, pictures inscribed in circles and quadratic shapes (sacrifice scenes), pictures of the sun and stars. Researchers believe that the function of these figures is quite important including among others zodiac symbols (Khazaradze 2007, cited in Gigauri 2010).

All of these wall engravings depicting humans, animals and subjects of astronomical cognition are an exceptional testimony to a way of life that has almost disappeared, graphic representations of activities connected with the spiritual life of residents of the mountains of eastern Caucasus.

All these facts give evidence that ancient myths and legends that exist in the modern-day Khevsureti and Tusheti have preserved the echoes about the relations of rock art with the cult of pagan Gods. This phenomenon still remains especially in mountainous northern Georgia (and with adjacent peoples of Chechen-Ingusheti settled northern slopes of the Greater Caucasus, Russian Federation). Frequently the religious importance of paganism is embodied within the stone cult (stony pyramidic shape shrines) or mysterious engraving in the stones.

Regarding current evidences, the mentioned motifs have a rich background in term of different groups of residents in this part of Georgia from prehistoric periods up to the present time.

It is important to note that the motifs of pictograms found on some stones of northern Georgia, are of the same style found as geometrized images in southern Georgia, in Samtskhe-Javakheti region.

As it is mentioned, the richest stone carvings (images) were registered in Khevsureti. Predominantly these stones are built in fortresses, house-fortresses stones and pagan shrines. On the surface of some stones Gigauri counted about 20 sign-figures. The stones with numerous mystical figures, Khevsurs called ‘colorful stones’ (Gigauri 2010).

Among Christian symbols, some of the most famous are the cross and the fish but both of them are descendants of paganism (see ‘Origin of the ‘Christian’ Fish Symbol’).

Among symbols and geometric figures, most leading motifs in Khevsureti occur as the cross – a remarkable shape usually consisting of just two intersecting lines – with four points. In Georgia, similar to innumerable religions, crosses of various kinds as a religious symbol were the favorite motifs in the art and architecture of the pre-, early- and later periods of Christianity.

The Cross - as Emblem of Christianity has also evolved from its Pagan roots. It has associations with hatred and despair love, grief and heroism.... (*www.seiyaku.com/customs/**pagan-symbols**.html*).

It should be mentioned about the early Christian symbol, the fish (ancient Greek ichthys). It resembles the profile of a fish. Pagans (Greeks, Romans, and many others) used the fish symbol before Christians. ‘In certain non-Christian beliefs, the fish also has been identified with reincarnation and the life force’ (https://en.wikipediaoorg/wiki/Ichthys).

The next figure represents a double helix, which as ethnographers believed is the unique symbol of the sun.

Another stone image of Khevsureti which attracts relations with the Earth’s environments and cosmic space is the hand. Hands as a symbolic inscription on the stones are of two kinds: (1) as independent motif and (2) it is included in other geometric figures (Figure 175).

Figure 175. Symbolic inscriptions from Khevsureti (N. Georgia) (Photo by Gigauri, 2008).

From 'hand', in Georgian *Kheli are* derived numerous and important words. For example, *khelmtsife* – king, *xelisuphali* – ruler, *xelmarjve* – skilful, *kheltkmnili* – made by human hands, *khelapkrobili* - with hands raised (prayer), *khelbaraqiani* - with golden (lucky) hands, *khelgaushleli* - stingy, *gakheleba* - making mad, *xelmotsaruli* – empty-handed, etc.

The 'colorful' and other stones built in various Pshav-Khevsureti-Tushetian fortresses and other buildings are scanty remains of former stones with enigmatic signs and mystical figures. Open-air monuments of northern Georgia are exposed to high winds and will gradually deteriorate through deflation and human (natives, tourists) activity. According to Gigauri, cradled in the folds of the mainly north slopes of the Greater Caucasus, especially Khevsuretians and Tushetians have relied on their geographic isolation to protect their ancient culture. But because of terrible disease–bubonic plague, forced migration during the Soviet Union's time from mountains in lowland area of eastern Georgia, the mountain dwellers culture today is virtually destroyed. Only populated are a few house-fortresses, kept as pure remains, the sacred places (sacred woods of special religious importance), and pagan shrines. Despite these disastrous events the natives struggle to keep the fragile traditions of their 'sacred mountain' during continuing urban development in recent ages. Today there is a national will in native populations to embrace strongly in the field of their cultural heritage protection including the stone images language of the symbols.

Famous ethnographer S. Makalatia in his book 'Khevsureti' (1935,1984, cited in Gigauri 2010) considered that many carvings and printings were done by the native shamans of previous centuries, and many of these sacred signs are manifestations of their attempts to commune with the supernatural phenomenon, i.e., transcendental power.

Khevsureti still survived the institution of paganism and, therefore, the chief priest of a pagan shrine (*Khevisberi* in Georgian, i.e., the 'old man of the tribe'). Chief priest's – prayers initial words begin: 'Glory to God', Glory to the Sun and its Satellites Angels'…

Traditional clothing, carpets and bags of indigenous people of northern Georgia woven with warp and weft of different color are adorned with geometric ornaments, a pastry (filled with flour and sugars), wooden chests on legs (to store all food) are adorned with astral signs, at the same time they remained all-Georgian's ornaments (G. Bochoridze, 'Tusheti' unpublished work, cited in Gigauri 2010).

Similar rock carvings with contour images and straight lines, signs of zodiac and astrology are common with the Kura-Aras culture (N. Shanshashvili 'Kura-Aras signs-symbolics', 2000, cited in Gigauri 2010) and are common also with Etruscan, Sumerian, Egyptian civilizations (Z. Kiknadze, 'The system of Georgian mythologic narrations' 1985 cited in Gigauri 2010).

'Rock arts are most important and most mysterious archaeological historical remains and it is believed that it can be the 'common human language' and 'the live archive of history' (Mirzaei et al. 2016), 'Rock arts are the earliest records known to us of human forms of communication and date back to prehistoric times' (Kiknadze 1985 cited in Gigauri 2010). Taking into consideration the value of northern Georgia's stone images – stick messages - as vital importance to understanding the culture that existed in prehistoric and historic world, it is a significant challenge of modern time to achieve sustainable preservation of mentioned art, hence the further cataloguing and scientific survey should be continued by Gigauri and his like-minded.

2.7.2.2.3. Subalpine Elfin Woodlands

Very special formations of the Caucasian highland are the elfin woodlands that are very diverse with respect to their flora, their biocoenoses and their origin. Several widely distributed types consist of high-mountain, low shrub thickets.

2.7.2.2.3.1. High Mountain Heaths Dominated by Dekiani (*Rhododendron caucasicum*) Thickets

The word *dekiani* is derived from Georgian 'deka', name of shrub found in the highland area, *Rhododendron caucasicum*. From this word comes another word, 'dekiani', i.e., plant communities, which consist of Caucasian alpide *Rh. caucasicum* – dominated heath (Figure 176).

Figure 176. Alpine belt with elfin growth of the Caucasian rhododendron. North Caucasus, Teberda River source area. In background: subnival belt, 3,100m a.s.l., Teberdinsky Reserve, S. Russia (1974).

Caucasian rhododendron is distributed in the subalpine and alpine belts from 1,800 up to 3,000 (3,100) m elevation throughout the highlands of the Caucasus (The Greater Caucasus, The Lesser Caucasus, and far more infrequently in Javakheti upland). Beyond the region it occurs in the Pontic Mountains (i.e., Asia Minor, north-eastern Turkey). Among four evergreen relict species, the Caucasian rhododendron is more widespread in the Caucasus isthmus, although its whole range is narrower than that of *Rh. ponticum* (Ivanov 2002).

As undergrowth it grows in mesophyllous subalpine forests, on the tree line (krummholz community) or is forming in open subalpine and alpine areas of pure scrub frequently on north-facing slopes and is sometimes found along the wide bottoms of glacial valleys. This smaller canopy of shrub is represented as a dense, almost impenetrable thicket (Gulisashvili *et al.* 1975). Thickets of *Rh. caucasicum* inhabit peat-humus soils, rarely on limestones. In the highlands of the Caucasus evergreen *Rh. caucasicum* is keystone, the most typical 'landscape make shrub', significantly in its western part (Ivanov 2002).

Currently, flora of dekiani of the highlands of the Caucasus estimated 277 vascular plants from which 245 species belong to angiosperms. 111 species (41.2%) of dekiani thickets are endemics of the Caucasian flora. In this climax of orophyte plant communities the Caucasian rhododendron is a dominant species, where the subcanopy is composed of the smaller species: dwarf ericoids and prostrate juniper species. Caucasian rhododendron appears to create its own phytocoenosis, microclimates and organic soil-peat-forming plant species (Ivanov 2002 and references therein).

When the Caucasian rhododendron predominates in high-mountain heat communities, this shrub grows as large (1-1.5 m height) individuals with many leafy branches, which make dense scrubland impassable (persn. observation).

2.7.2.2.3.2. Elfin Woodland of Junipers

Junipers (*Juniperus pigmaea, J. hemisphaerica, J, depressa, J. sabina*) are found south-east of the distribution area of *Rh. caucasicum*, i.e., in more arid continental climates. The typical habitats for junipers woodland are screes and rocks, where the shrubs are often accompanied by highland xerophytes (Ketskhoveli 1960; Gulisashvili *et al.* 1975; Imkhanitskaya 1990) (Figure 177).

Figure 177. Stand of *Juniperus depressa*. Okhachkue limestone massif, Samegrelo, W. Georgia, 2,100m (2015).

In Colchis, the fragmentary *Juniperus* communities rarely intermingle with *Rhododendron* thickets.

2.7.2.2.3.2.1. The Peculiarities of the Hyrcano-Caucasian Subalpine Forests

These high mountain forests differ considerably from the Colchian and Boreal subalpine (cruuk-stem) forests being found in the regions distinguished by a very continental climate and thin snow cover. The dominant straight-trunk tree species are highland oak (*Quercus macranthera*) and Hyrcanian maple (*Acer hyrcanum*), which in some places are accompanied by *Pyrus syriaca* and *P. zangezura*. The shrubs layer mainly consists of *Sorbus persica*, *S. armeniaca*, *S. caucasigena* and *Juniperus pigmaea*. Thorny cushion shrubs, representatives of tragacanthic communities: *Astragalus* (subg. *Tragacantha*) and *Acantholimon* are characteristic of open woodland composed of *Quercus macranthera* that occurs on south-exposed stony and dry slopes. At higher altitude these open forests change to mountain steppes and cushion shrub communities.

Subalpine meadows are absent here. Mountain dry steppes are replaced by grasslands over 3,000 m a.s.l.

2.7.2.3. Highland Herb Vegetation

2.7.2.3.1. Subalpine Tall Herbaceous (Forb) Vegetation

The term 'subalpine tall herbaceous' was first proposed by the soil scientist Zakharov (1914). Caucasian vegetation of tall forbs is a unique high-mountain community in the moderate climate of the Northern Hemisphere (Gagnidze 1976). Unlike many other mountain systems (Alps, Pamir, Himalayas, Rocky Mountains), the formation is well represented in the subalpine belt of the Caucasus (Nakhutsrishvili 1999, 2013).

The association consists of dense stands mainly of broadleaf herbs of 1.5-2 m (sometimes 3-4 m) high. A turf layer, so characteristic of grassland, is absent: the ground cover consists of dead leaves and remnants of plants like forests litter (Ketskhoveli 1960; Gagnidze 1966). Subalpine vegetation due to climatic and soil conditions extends from 2,350-2,500m to 2,450-2,500m a.s.l. (central part of the Greater Caucasus) (Nakhutsrishvili 1999, 2013).

The flora of subalpine tall-herb communities may be dominated by species or more frequently may consist of several species with any one species reaching clear dominance (polydominant tall-herb vegetation) (Figure 178). Altogether 69 species form the Caucasian tall-herb vegetation, of which 52 are endemic to the Caucasus and the other 17 occur in Europe (e.g., *Pyrethrum macrophyllum, Telekia speciosa, Campanula latifolia*) (Gagnidze 1966, 1974, 1976).

Most tall herbs are not grazed by cattle, because they are toxic (poisonous), i.e., unpalatable (*Veratrum lobelianum, Delphinium flexuosum, Aconitum orientale* and species of *Senecio*), therefore they are preserved in intensively grazed places (Gulisashvili *et al.* 1975).

Tall herb vegetation is characteristic of the northern slopes, with some depression in the relief, deep rich soils with enough moisture, high air humidity and high solar radiation. In Ciscaucasia it can be found from the Black Sea to the Caspian Sea on the north slopes of the Greater Caucasus (Onipchenko 2002a).

Figure 178. Subalpine forb (hayfield) in Zemo Svaneti. Greater Caucasus characterized with the largest continuous alpine area among other mountain systems in the Caucasus. Vil. Adishi environs, 2,250 m a.s.l. In local subalpine-alpine transition ecotone pristine wildflower displays striking bloom of flora. Left side: dense grow of *Betula pubescens litwinowii* (1985).

Figure 179. Subalpine forb vegetation. Pambak Ridge with a view of the Ararat Mts in Turkey. Great Ararat summit, 5,165 m, Lesser Ararat summit, 3,925 m (1974).

On the highlands of the Lesser Caucasus the vegetation is distributed towards the east in the northern part till the Zangezur Range (Goris town). In the south it is found at the bottom of deep gorges (Magakian 1941, Gulisashvili *et al.* 1975). The vegetation is typical of open subalpine woodland and develops into dominant vegetation after deforestation (Ketskhoveli 1960) (Figure 179).

Figure 180. *Centaurea* (= *Grossheimia*) *macrocephala* (Asteraceae) native to the Caucasus. Tskhratskaro environs, Borjomi district, S. Georgia, 1,827 m (2018).

In the West Caucasus the following typical species of highland forbs include: *Heracleum mantegazzianum*, *Senecio platyphylloides* (sometimes treated as *Caucasalia pontica*), *Aconitum orientale*, *Delphinium flexuosum*, *Cnidiocarpa alata*, *Inula magnifica* (Gagnidze 1974; Gulisashvili *et al.* 1975).

In the East Caucasus the following typical species can be mentioned: *Heracleum sosnowskyi*, *Aconitum orientale*, *Ligusticum alatum*, *Symphytum asperum*, *Galega orientalis*, *S. platyphylloides*, *D. flexuosum*, *Centaurea* (*Grossheimia*) *macrocephala*, *Cephalaria gigantea*, *Campanula latifolia*, *Telekia. speciosa* (Gagnidze 1974, Gulisashvili *et al.* 1975) (Figures 180).

Dolukhanov *et al.* 1964 (in Nakhutsrishvili 1999) mentions three associations with the following characteristic species:

1. *Heracleum mantegazzianum*, *H. grossheimii*, *H. sosnowskyi*;
2. *Senecio platyphyllus* (sometimes treated as *Caucasalia macrophylla*) *Inula grandiflora*;
3. *Rumex alpinus*, *Senecio othonae*.

2.7.2.3.2. Subalpine Meadows

'Subalpine meadows' is a term introduced by the Caucasian botanist Bush (1898). They occur above subalpine forests and consist of more or less mesic herbaceous associations with relatively high plants (1m high and more). The very dense stands have several levels of herbs (Dolukhanov *et al.* 1942; Ketskhoveli 1960).

Subalpine meadows develop on the territories with optimal humidity that do not exhibit drought or high moisture. Therefore, they are lacking in the south-eastern part of Transcaucasia, because of lack of precipitations and low air humidity. Instead of forests or over highland forests their position is occupied by meadow steppes. In the extreme south-east of the region steppe vegetation and the driest type of mountain xerophytes (especially - *Astragalus* subgenus *Tragacantha*) is dominant. The altitudinal zonation of south-east Transcaucasia is similar to the vertical zonation in the mountains of Central (Middle) Asia, North Iraq and North Iran (Takhtajan 1941).

Typical of meadow vegetation are turf formation and the accumulation of organic materials in soils (Magakian 1941).

The subalpine meadows of the Caucasus can be divided into primary subalpine meadows and secondary meadows formed after forest cutting and overgrazing. With respect to their flora and biocoenoses both types are very similar (Ketskhoveli 1960).

Subalpine meadows can further be divided into groups according to the ratio of grasses and dicotyledons. Grass meadow communities are distinguished by the dominance of *Festuca versicolor.* This s found in an area on the south slopes of the eastern part of the Greater Caucasus and in the south of the Lesser Caucasus, where they occur on skeletal cobble soils in the driest environments. Typical grass species include *Zerna variegata, Calamagrostis arundinacea, Festuca djimilensis, F. varia, Agrostis alba, Trisetum pratense* and *Hordeum violaceum* (Ketskhoveli 1960).

Forbs meadows are very diverse but rare, and occupy the moister localities. The main components are *Ranunculus caucasicus, Pulsatilla violacea, Veratrum lobelianum, Anemone fasciculata, Betonica grandiflora, Alchemilla oxysepala, Geranium ibericum, G. gymnocaulon, G. platypel, G. platypelatum, Inula glandulosa, Iris grandiflora* and *I. orientalis*. In very wet conditions *Trollius patulus, Myosotis alpestris, Silene ruprechtii, Gentiana gelida* and *Poa alpina* are the predominant species.

Associations with a predominance of Fabaceae (e.g., *Trifolium* spp.) are characteristic of the south slopes of the western Trialeti Range (the Lesser Caucasus) (Ketskhoveli 1960).

Ketskhoveli (1960) and Nakhutsrushvili (1999) distinguish about 20 types of subalpine meadows.

2.7.3. Alpine Vegetation (Alpine Turfs, Grasslands and Alpine Carpets with Admixture of Dwarf Caucasian Rhododendron and Bilberry Heaths)

Alpine vegetation includes the alpine meadows and prostrate shrubs and occupies slopes of all expositions. The lower limit is the timberline, its upper limit the eternal snow. The extent of alpine vegetation varies with location, i.e., with the altitudinal change of atmospheric factors and the differences in the relief (Nagy *et al.* (Eds.) 2003).

Alpine meadows occupy vast areas in the Greater Caucasus. In the Lesser Caucasus they do not form continuous formations. In areas with a continental climate (East Georgia) alpine vegetations occupy higher altitudes, while in the regions with a prevailing oceanic climate (West Caucasus, particularly Adjara) alpine vegetations occur at lower elevations. The alpine belt extends from 2,300-2,700 up to 2,900-3,700 m a.s.l. (Gulisashvili *et al.* 1975).

Trees are absent because of the short vegetation period (2-3 months) and because of insufficient warmth for wood formation and ripening. The native communities are characterized by short grass meadows, carpet-like alpine meadows, alternating with *Rhododendron caucasicum* thickets and rock-scree vegetation.

The alpine belt is characterized by special abiotic conditions and it must be realized that no uniform climatic conditions prevail in the high mountain. Important factors include the intensive light (with approximately double intensity of short wave and ultraviolet radiation from that of the lowlands) and the lower atmospheric pressure, which increases evaporation and decreases the water-carrying capacity of the air. These factors make the environment relatively dry: plants intensify their transpiration rate and many species are xeromorphic (Nakhutsrishvili 1999, 2013).

Other important factors include the large differences between surface soil and air temperatures. Overnight the surface soil temperature drops below zero, forcing the plants to recover rapidly as soon as the temperature gets above freezing point. During the day time the soil surface temperature increases quickly and plants are stimulated to grow where the air temperature is higher, i.e., in the relatively thin warm layer just above the ground. This may be one of the causes of the prostrate habit of alpine plants.

As the growing season is short and plants must devise ways of growing strategies that allow them to develop quickly for example starting to grow while still under snow cover and growing and developing very fast (Dolukhanov 1980).

These different climatic and relief conditions determine the mosaic-like distribution pattern of vegetation cover and the same abiotic factors cause the coexistence of completely different vegetation types (Ketskhoveli 1960; Gulisashvili 1964).

Because of low temperatures and the long period of vegetation dormancy, soil formation is slow and physical processes are more important than chemico-biological factors. The former often result in rock-scree ground formation, preventing the formation of mature soil profiles. Unlike the subalpine belt, turf formation in the alpine belt takes place slowly. The differences between the two belts are expressed in the herb layer: a pure alpine coenosis usually consists of one or two strata, while that of the subalpine belt includes three or four layers (Ketskhoveli 1960).

Following the viewpoint of Grossheim (1946) and Gulisashvili *et al.* (1975) our biogeographical studies (Gegechkori 1984a, b,1985,1997,2008) give no reason to distinguish the subnival vegetation as an independent belt, and we consider the latter to be an impoverished type of alpine vegetation.

Gulisashvili *et al.* (1975) divide the alpine vegetation of the Caucasus into four groups:

1) Alpine turfs, where grasses and sedges are dominant;
2) Alpine carpets, where grasses and sedges are subdominant and the main turf forming is performed by the herb meadow;
3) Species-poor vegetation, where psychrophilic plants of cold, raw soils predominate;
4) Rock and scree vegetation.

2.7.3.1. Alpine Turfs and Grasslands

Alpine turfs and grasslands are widely distributed at high altitudes throughout the Caucasus. The grasses usually form dense and short turfs. Alpine meadows mainly occur in mountain wetlands and they contain local endemics (Bedoshvili 1985 cited in Nakhutsrishvili 2013). The following types can be distinguished: tussock sedge meadows, forbs meadows and tussock grass meadows. The typical component is *Festuca airoides,* which is often accompanied by *Carex meinshauseniana* in the western Caucasus or by *C. tristis* in the eastern Caucasus. In addition, the following species can be mentioned: *Zerna variegata, Koeleria caucasica* and *Lotus caucasicus*. In the southeastern Transcaucasia the species of *Carex* usually dominate over those of the genus *Festuca*. Overgrazing leads to degradation of alpine turfs (Kimeridze 1975).

2.7.3.2. Alpine Carpets

Alpine carpets occur in the upper part of the alpine belt in places with permanent moisture from melting snow (near the constant snow), such as moraines and cirques, where soil formation is hampered by a lot of factors (Fedorov 1942; Ketskhoveli 1960; Narynyan 1962). Therefore, alpine carpets form an eumesophytic phytocoenosis. The term 'alpine carpets' (or 'carpet-like alpine meadows') (Grebenshikov 1957; Bush 1897, 1898, 1905 cited in Grossheim 1948; Gulisashvili *et al.* 1975) coincides with 'Dicotylen-Tepiche' of Ellenberg (1996) (Nakhutsrishvili 1999, 2013) (Figure 181).

Carpet-like alpine meadows differ from true meadows by their floristic composition and ecological peculiarities. They consist of small 3-4 cm high herbs; hence some authors consider these types of meadows as nanoherbata. The turf is made by forbs, not by sedge and grasses and typically has a two-layer structure (Mikeladze 1960).

Figure 181. The most spectacular displays occur in mid-summer when blossoming alpine anemone (*Pulsatilla aurea*) and Caucasian rhododendron carpet cover the steep slopes of the highland. Arkhiz part of Teberda Reserve, Ciscaucasia, S. Russia, 2,350m (1984).

The main mass of the turf is concentrated in the upper 5-6 cm. The deeper layer consists of separate deep roots (20-25cm). Alpine carpet meadows are distinguished by large and brightly colored flowers and inflorescences (Ketskhoveli 1960).

Alpine carpets are species-rich communities, with up to 200 species. The most widely distributed meadows are formations with the leading species *Taraxacum stevenii*, *Alchemilla caucasica*, *Potentilla crantzii* and nanoherbata with dominance of *Campanula tridentata, C. tridens* and *C. latifolia* in South-East Transcaucasia. In the Lesser Caucasus the associations predominantly consist of *Taraxacum stevenii, P. crantzii, P. raddeana* and *Thymus kotschyanus* (Gulisashvili *et al.* 1975).

On the slopes and rocky habitats, alpine carpets form less dense stands and represent transition to rocky-scree, more xeric formations. These species-poor carpets are characteristic of some localities of south Armenia (Takhtajan 1941) and Dagestan (Gulisashvili *et al.* 1975). The most typical species is *Sibbaldia parviflora*. Carpet meadows with the endemic *Ranunculus helenae* occur on limestone in Colchis. These communities developed around snow spots and at the edges of glaciers and include the following species: *Ranunculus oreophilus, R. baidarae, Primula algida, Gentiana pyrenaica* (=*G. djimilensis*), *G. angulosa*, *G. nivalis* and *Cerastium cerastoides*. Cryophytes growing at the banks of brooks include *Primula auriculata, Poa alpina, Pedicularis crassirostris, P. nordmanniana*. Carpet meadows are characterized by patchy distribution (Kolakovski 1961; Sokhadze 1982; Akatov 1999; Adzinba 2000; Gagnidze & Davitadze 2000).

Besides the climax carpet-like alpine meadows, secondary carpets, enriched by meadow herbs, occupy a vast area and are characterized by *Alchemilla* spp., *Sibbaldia parviflora* and *P. crantzii*. These secondary carpets are formed by overgrazing. In natural conditions wild endemic goats, turs (*Capra cylindricornis, C. caucasica*) graze on the alpine meadows of the Caucasus (Grossheim 1948; Ketskhoveli 1960; Gulisashvili *et al.* 1975).

2.7.3.3. Heathland or Species-Poor Vegetation

Heathland or species-poor vegetation is characterized by *Nardus stricta* that by its xeric nature is adapted to the high cold, physiological dry conditions (psycrophytes), belonging not to typical meadows, but to heathland, i.e., species-poor vegetation. Frequently this heathland develops as a result of degradation, often by human impact. Other secondary heathland meadows are comprised of *Kobresia persica*, *K. capillifolia* and *K. schoenoides* with other species. Dolukhanov (1980) linked *Kobresia* and similar cryophytes to mountain tundra vegetation. Associations with dominance of *Sibbaldia parviflora and S. semiglabra* are heathland meadows that may originate from grazing degradation (Shiffers 1953).

The heathlands commonly have a raw, nutrient-poor mineral soil over which mosses are well developed. Lichen heath formations develop on ridges with low or no snow cover, where soil freezes deeply.

Shrub heath vegetation is characterized by the presence of some mono- or oligodominant dwarf shrub formations with *Vaccinium vitis-idaea, V. myrtillus, Empetrum caucasicum*, species of willow (*Salix* spp.) and juniper (*Juniperus sibirica*).

Rhododendron caucasicum is at the highest altitudes, restricted to cirques and valleys, where high humidity and profound snow cover protect the species from frost (Kolakovski 1961).

The vegetation of highland limestone in Colchis includes a subshrub community formed of *Dryas caucasica*. Its typical habitat is cold stony slopes of northern exposure (Sokhadze 1982).

2.7.3.4. Rock and Scree Vegetation

Rock and scree vegetation are an azonal phenomenon in alpine and subalpine belts that is closely associated with alpine climate, in particular with the rather high precipitations and high air moisture. This rock and scree vegetation of the belt greatly differs from the xerophytic vegetation of rocks at lower altitudes. The flora is very diverse and depends on the physical and chemical composition of the bedrock (limestone, acid rocks), the character of its weathering and other physical conditions such as the color (light or dark) and the consistency (hard or hygroscopic).

Figure 182. *Sempervirum pumilum* – succulent plant of highmountains. Pirikita Khevsureti, vil. Shatili environs, E. Georgia (1972).

The severe climatic and edaphic conditions are triggered for the local vegetation to develop various adaptations. Many of them have prostrate, cushion forms or possess various kinds of hairiness or succulent form (Gagnidze & Shetekauri 1988).

The habitats of rock and scree of the central part of the Great Caucasus, the Kazbegi region are distinguished by quite high species richness among which is found a high number of stenotopic (mono- and oligotype) endemics (Ivanishvili 1988 cited in Nakhutsrishvili 2013).

Characteristic species of rock and scree vegetation, including crush rock ridge, landslides and moistened cliffs, consist of some cushion plants - *Draba bryoides, D. mollissima, D. siliquosa, D. incompta, Saxifraga laevis, S. exarata,* and the succulent species - *Sempervirum pumilum, Sedum stevenianum, S. tenellum;* herbs - *Campanula petrophila, C. ciliata, Dianthus* spp., *Alchemilla sericea;* grasses - *Phleum alpinum* and *Poa alpina* and species of sedge represented by *Carex oreophila, Juncus atratus*. Another ecological type - chasmophytes are peculiar for coenetic variability. This community is rather rich in endemic species. Examples are *Saxifraga cartilaginea, S. flagellaris, S. sibirica, Viola biflora, V. caucasica, Campanula hypopolia, Primula nivalis* and others (Kimeridze 1966, 1975; Ketskhoveli 1960; Gulisashvili *et al.* 1975; Gagnidze & Shetekauri 1988; Shetekauri *et al.* 2006) (Figure 182).

2.7.4. Subnival Vegetation

This covers the area above the climatic limit of alpine vegetation, i.e., in the zone of melting snow and the side of the glacier and free rocks coming out of snow. The most striking features are the presence of plants without the ability to form communities and the absence of mature soil. Therefore, almost all the plants grow on rocky and talus soil. In addition, the snow beds do not melt everywhere (especially between large stones and moraines) in the subnival belt during summer, but where it melts, the ground may become free from snow for a very limited time (Nakhutsrishvili 1999, 2003; Akhalkatsi & Tarkhnishvili 2012).

As a rule, in the subnival belt severe climatic conditions determine the growth conditions. Individual plants are often scattered over considerable distances (which limits competition), the growth period is very short and plants must often start their growth while still under snow cover. Some species of *Ranunculus* can develop ripe seeds 17 days after flowering (Nakhutsrishvili 1999; Akhalkatsi & Tarkhnishvili 2012).

Subnival vegetation is found at the altitude from 2,900-3,000 up to 3,500-3,700m a.s.l. In the Greater Caucasus the lower limit of the subnival belt varies in altitude between 2,900 and 3,100m, increasing from west to east. The upper limit is 4,000m on Mount Aragats (the Lesser Caucasus, Armenia). Due to the low position of the eternal snow line (about 2,800m), in the western part of the Greater Caucasus, the boundaries of the subnival belt are fragmentary (Nakhutsrishvili 1999). (Figure 183)

The role of temperature, as a limiting factor, is more important than in the alpine belt. Excessive solar radiation raises the temperature of the soil surface and of the boundary air layer, and daily temperature fluctuations also increase. The role of wind (its direction) increases considerably (Korner 2003; Larcher *et al.* 2010 cited in Nakhutsrishvili 2013).

Despite the harsh climate, over 300 species occur here with 109 of them exclusively growing in the subnival belt (Kharadze 1965). 'In high-mountains some species of plants (chionophobes) are well adapted to low temperatures and others (chionphiles) have low frost resistance' (Larcher 1977, 1987; Sakai & Larcher 1987 cited in Nakhutsrishvili 2013). Plants are adapted to a great variety of ecological niches even within small territories. Besides the climate, the edaphic composition has a great impact on the flora (Larcher 1980).

The location of the upper distribution boundary of flowering plants depends on the level of climate continentality and on the height of the mountain. A local endemic species *Draba araratica* grows on the summit of Mount Aragats (Lesser Caucasus, Armenia) at 4,095m and only nine species of higher plants reach above 3,800m there (Voskanjan 1976). In Kazbegi (Central Caucasus) *Cerastium kazbek* appears at 3,950 m (Nakhutsrishvili 1999, 2013). In the entire Caucasus 94 species grow over 3,200m (Prima 1974; Nakhutsrishvili 1999, 2013).

The typical subnival species found at 3600-3800m on the Greater Caucasus Range are *Draba supranivalis, Saxifraga moschata, S. flagellaris, S. sibirica, Tephroseris* (=*Senecio*) *karjaginii, Veronica minuta, V. telephifolia, Cerastium pseudokazbek* and, *Colpodium versicolor. Carum caucasicum* sometimes creates a pure growth. At the side of the stream can be found *Primula auriculata, P. algida, Ranunculus oreophilus, Gentiana nivalis, G. angulosa, Cerastium cerastoides, Poa alpina,* etc. (Grossheim 1948; Nakhutsrishvili 1999, 2013).

In the alpine and subalpine zones, sometimes in the places of melting snow, some occur on the banks of the stream and also by the rivers that start at the glacier. The highland's typical vegetation is manifested by the communities of *Cardamine uliginosa, Caltha polypetala, Alchemilla tredecimbola, Primula auriculata, P. algida, P. ruprechtii, Scilla rosenii, Pulsatilla violacea, Carex oreophila, C. verna, Fritilaria colina* and others (Kharadze 1944; Ochiauri 1965; Nakhutsrishvili 2013) (Figure 184).

Figure 183. Typical subnival belt. Kazbegi district, GCR, 3,500-3,700m (1974).

Figure 184. *Fritilaria colina* growing in subnival belt. Kazbegi district, GCR, 2,700m (1974).

A large number of endemic species (including narrow endemic species) and even endemic genera occur in the subnival belt. Some of them are monotypic genera: *Pseudobetckea* (Valerianaceae), *Symphyoloma* (Apiaceae), *Pseudovesicaria* (Brassicaceae); from the oligotypic genera: *Coluteocarpus, Didymophysa, Eunomia* (Brassicaceae), *Vavilovia* (Fabaceae) should also be mentioned (Kharadze 1965, Adzinba 2000, Gagnidze & Davitadze 2000, Nakhutsrishvili 1999, 2013). A high percentage of Caucasian endemics is connected with mergel slates (Dolukhanov 1969). On the limestone rocks of the Western Caucasus a number of local endemics can be found (Sokhadze 1982). Also, many species of mosses and lichens occur under harsh lithologic conditions (Kolakovski, 1961).

2.7.5. The High Mountain Fauna (Ultra-Orophiles)

Subalpine and especially alpine open landscapes are completely dominated by herbaceous vegetation. Tree and shrubs are generally limited to the southern margins of the subalpine belt. Similar to the vegetation cover of the orobiome, the fauna of the highlands is rather rich and characterized by a high ratio of endemism. Eight priority species use high mountain habitats: Persian (Caucasian) leopard, bezoar goat, West Caucasian tur, East Caucasian tur, Gmelin's mouflon, Caucasian chamois, Caucasian black grouse and golden eagle. High mountains are a habitat for some small mammals: Nehring's mole-vole (*Nannospalax nehringi*) (VU), voles (*Microtus arvalis* (LC), *M. nazarovi, M. daghestanica*), endemic birch mice (*Sicicta caucasica* (VU), *S. kazbegica* (VU), and snow voles (*Chionomys gud*) (LC), *Ch. nivalis* (LC). The small mammals of the Caucasus include both newly evolved species and relict species, many of which are primitive mammals (e.g., species of *Mesocricetus* and *Sicista*). The unusual long-clawed mole-vole (*Prometheomys schaposhnikovi*) (VU) is typical of the high altitude of the Greater and Lesser Caucasus. It is the only representative of an endemic genus in the rodent fauna, Muridae.

Some other mammals visit the orobiome temporarily for foraging or making dens or burrows. Among such animals are: badger (*Meles meles*), weasel (*Mustela nivalis caucasica*), otter (*Lutra lutra*), brown bear (*Ursus arctos*), wolf (*Canis lupus*), fox (*Vulpes vulpes*), and lynx (*Lynx lynx*). There is very scant information concerning the leopard (*Panthera pardus ciscaucasica* (CR) (Figure 185), inhabiting the highlands. Some locations of orobiome, sometimes as high as 3,100m a.s.l. are visited by red deer (*Cervus elaphus*). The same biome includes key habitats for large scavengers such as bearded vulture (*Gypaetus barbatus*) (VU), black vulture (*Aegypius monachus*) (EN), and griffon vulture (*Gyps fulvus*) (VU). To exclusive rare small birds belong rock nuthatch (*Sitta tephrnota*), white-winged snow finch (*Montifringilla nivalis alpicola*) and, of course, dozens of other songbirds. The commonest birds are *Corvus monedula, C. corax, Pyrrhocorax pyrrhocorax*; raptors are represented by *Buteo buteo, B. rufinus rufinus* (VU and, *Aquila chrysaetus* (VU), whereas through gorges and deep rocky ravine the alpine zone is reached by the rock (Chukar) partridge (*Alectoris chukar*), falcon (*Falco peregrinus*) and hawk (*Accipiter gentilis*) (Djanashvili 1963, Gulisashvili 1964 and references therein, Abuladze 2013).

Crushed rock ridge and other similar microhabitats are shelters for the green toad (*Bufo viridis*). High mountain meadows are home to three species of adders: *Vipera* (*P.*) *darevskii*, an extremly rare species with fewer than 250 mature individuals, distributed in the Javakheti range along the border of Armenia and Georgia (CR), *V. renardi renardi* (South Caucasus, EN) and *V. dinniki* (Caucasus proper, VU); several (5-6) endemic lizards of the genus *Darevskia*. The most common reptiles and amphibians are *Lacerta agilis, Anguis fragilis, Coronella austriaca, Natrix natrix, Rana macrocnemis*, also, *Triturus cristatus* can be found (Tarkhnishvili & Gokhelashvili 1999) (Figure 186).

Figure 185. Leopard's skull and tail found in Pshavi region, environs of vil. Matura, 1,850m (E. Georgia), GCR. It was identified as the victim of an avalanche in 1970. These remains of the beast are the last accident when, by scholars (Dr. Arn. Gegechkori), was pictured in Pshavi (07.1970) and then enriched collections for the Georgian National Museum.

To rare and spectacular insects belong Apollo butterflies (*Parnassius* spp.), Capricorn beetle (*Rosalia alpina*) (Akhalkatsi & Tarkhnishvili 2012 and references therein), *Carabus caucasicus colchicus* which is usually restricted to broadleaf and mixed forests, but rare specimens, it seems, reach up nearby to the top of karst (limestone) mountains of North Colchis (pers. comm.).

In the subalpine zone of the Lesser Caucasus, after the findings of other biologists (Mumladze & Tarkhnishvili 2008 and references therein), the *Helix buchi* was detected by me in Meskheti Mountain Range, Kintrishi Protected Areas (Adjara), in subalpine tall herb (forb) vegetation above the tree line ecotone about 1,900m a.s.l. (collected by Gegechkori 2009, det. Mumladze). Usually this species of snail is known as dwelling exclusively in mountainous broadleaf forests of the Western Lesser Caucasus (Mumladze *et al.* 2008).

The local ecosystems provide shelter to the following endemic species of psyllids (Hemiptera and Psylloidea). Species restricted to the Greater Caucasus including new species of psylloids have been described by the author. All of them are endemics for the Greater Caucasus: *Aphalara nigrimaculosa**, *Craspedolepta caucasica**, *C. georgica**, *C. promethea**, *C. iltoica**, *Cyamophila megrelica**, *Cacopsylla montanica**, *C. loginovae**, *Bactericera liberta** (described by Loginova 1969), *B. chevsurethica**, and *B. annae**. Many of them have close, vicarious species in Europe and Siberia (Gegechkori 1984a; Gegechkori & Loginova 1990).

High mountain habitats are affected by overgrazing, which impacts on the plant species diversity and reduces the food base of mountain ungulates. Poaching puts significant pressure on large mammals and endemic birds in high mountain regions (ECPC 2012). About 12% of the high mountain habitats are afforded some sort of protection. The most important parts of the Ecoregion for conservation of high mountain habitats are the Greater Caucasus Mts and parts of the Lesser Caucasus Mts and the Javakheti Upland. The long-term aim should be to grant protection equivalent to IUCN categories I-IV to at least an additional 8% of the high mountain habitats (especially limestone habitats) in the Ecoregion (ECPC 2012).

Figure 186. Meadow viper (*Vipera renardi renardi*) encompasses from Tusheti Hollow, N. Georgia (1990).

Box 12. Shota Rustaveli, 'The Knight in the Panther's Skin' – To the Sources of the Renaissance

In one of the most outstanding literary monuments, the Georgian national epic poem 'The Knight in the Panther's Skin', the greatest Georgian poet Shota Rustaveli (XII-XIII) describes the Medieval European humanistic ideals: chivalry, feeling of love, friendship, courtly love, courage and bravery, philanthropy and generosity. One of the main heroes of the poem, Tarieli, is dressed in a panther skin. This extraordinary dress reminds the knight of his beloved-one, Nestan, who by her beauty and character resembles the wild cat - leopard. Rustaveli is considered to be one of the standard-bearers of the Early Renaissance. The poem was translated into many languages, including Persian, Hebrew, Russian, English, German, French, Italian, Spanish, Japanese, Chinese, Hindi and Arabic. Translation by Lyn Coffin (2015) is a recent version of a beautiful poetic translation to English of the epic poem.

Today the bust of Georgia's prominent poet is located in one of the best-known landscape gardens of Villa Borghese, Rome (Italy), among the world's greatest writers: Homer, Ovid, Ferdows (Firdausi), Goethe, etc. The bust was erected in 9 April, 2010. This fact once again serves as a proof that the iconic Georgian medieval writer occupies an honorable place in the world's immortal literature (Gegechkori 2009a,b, 2010a) (Figure 187).

Figure 187. The monument to Shota Rustaveli in Villa Borghese, Rome.

2.7.5.1. Some Umbrella Species of the High Mountain Zone of the Greater Caucasus

The Greater Caucasus Range is the exclusive home to two endemic species of wild (stone) goats (turs): *Capra caucasica* and *C. cylindricornis* with a small hybridization zone (Kopaliani & Gurielidze 2009). This situation is typical for many other ungulates. It is now impossible to restore parts of the historical range, due to intensive anthropogenic pressure (Magomedov & Yarovenko 2009).

2.7.5.1.1. The West Caucasian or Kuban Goat (tur) (*Capra caucasica*)

The West Caucasian or Kuban goat (tur) (*Capra caucasica*) is unique to the western section of the Greater Caucasus in Georgia and south-western Russia. It has one of the smallest living ranges (about 250 km^2 and 450 sq. km^2) of all large ungulates (Radde 1899, Satunin 1920). The tur is a typical orophyle animal that lives at elevations ranging from 2,000 to 4,200m a.s.l. Typical habitats are the subalpine and especially alpine meadows adjacent to glaciers and rocky slopes (Figures 188, 189).

In winter the animal migrates to lower elevations, to subalpine open woodlands and slightly lower altitudes. The body length of adult males is between 120 to 165 cm, with shoulder height between 78 and 109 cm and weight between 65 and 100 kg. Both males and females have horns. Those of the males are three times larger than those of the females and are about 75cm long. Their pelage is rusty grey to rusty chestnut (Morozov 2009). In November, when the mating season begins, the two sexes unite in herds that consist of over 500 animals. The female produces only one offspring and rarely two (Heptner *et al.* 1989; Weinberg *et al.* 1997; Gegechkori 2007). In the late 1980s the population was estimated to be around 12,000 animals, while in 2004 it was estimated to be 5,000 to 6,000 (Large Herbivore Foundation 2010). The species has Endangered (IUCN Red List) (Weinberg 2008), status within countries of the Caucasus – Vulnerable.

Figure 188. *Capra caucasica*, the best known animals of the alpine cold climate area. Zemo Svaneti, Dolra River Gorge, W. Georgia, 2,700m (1983).

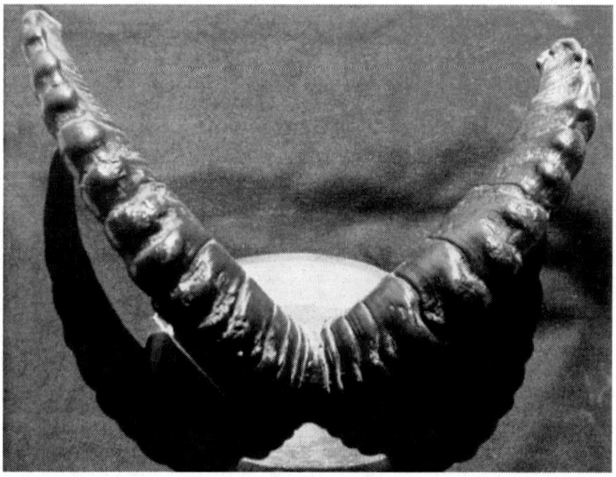

Figure 189. West Caucasian goat's horns from Zemo Svaneti (collected by G. Radde, 1864; from collections of the Georgian National Museum (1985).

2.7.5.1.2. The East Caucasian Goat (Tur) (*Carpa cylindricornis)*

The East Caucasian goat (tur) (*Carpa cylindricornis*) is endemic to the eastern part of the Greater Caucasus Range, where it inhabits a relatively small area approximately 500 km^2 at elevations from 2,000 to 4,000 m and is distributed in Azerbaijan, Georgia and Russia (Radde 1899; Satunin 1920). Its summer habitats are alpine and subalpine meadows, steep rocky slopes and other areas that are in close proximity to year-round mountain snow and glaciers. The species migrates from higher elevations to lower elevations (timberline and mountain forest zone) in winter. In spring herds consisting of 100 or more turs return to the higher mountain grazing pastures. *C. cylindricornis* is the largest ungulate among the Caucasian wild goats and sheep (one species). It has a massive, somewhat elongated body, a comparatively small head, a short neck and short, thick legs. Males have a body length of 190 cm, shoulder height of 104 cm, and weight of 140 kg. The horns of the male are three times larger (70-90 cm in length) than those of females (20-22 cm). The coat of the males varies seasonally, from chestnut-brown in winter to light rust-brown color in summer (Morozov 2009).

One can distinguish the two different species of turs by their horn structure and form. Horns of the east Caucasian tur are more crooked, complex and are elegantly helix-shaped with some variation within the range (Morozov 2009; Weinberg *et al.* 1997). The horns of the West Caucasian tur are far straighter and directed upright. In the central part of the Greater Caucasus the two species sympatrize and intermate (Figure 190).

Turs are gregarious animals. During summer the males flock roam separate in the most isolated and highest mountain ranges (Figure 191).

In the mid- 20th century, during particularly snowy winters, herds of tur were even observed in the foothills of 500 m. There are some essentially forest populations that include adult males (Guliev *et al.* 2009).

Active daily migration of wild goat has several causes, among which the activity of predators (*Ursus arctos, Lynx lynx, Canis lupus, Panthera pardus* and *Aquila chrysaetos*) and even sheepdogs is of decisive importance. Wolves hunt turs throughout the year, causing the most significant damage to their populations. Predators hunting turs especially during the breeding period and females with yearlings are the most vulnerable to predators (Kudatkin 1982; Filonov & Kaletskaya 1985; Magomedov & Yarovenko 2009). The breeding season lasts from the second half of November until January. The latest data (2004) suggest the total current population was estimated at 25,000 animals, but this number may be optimistic.

In pagan time, turs were widespread in the Caucasus not only in the Greater Caucasus (Dinnik 1882), but also in the Lesser Caucasus, which was noted by fossil bones found in human settlements of the Paleolithic period in Georgia (Kokhodze 1991).

Figure 190. Herd of *Capra cylindricornis*, large mammalians, which live exclusively in an alpine environments Lagodekhi Reserve, Kakheti province, E. Georgia, 2,900m (1964).

Figure 191. The horns of *Capra cylindricornis* hunted in 1848 in the mountains of Kakheti, E. Georgia, GCR; from collections of the Georgian National Museum (1985).

Figure 192. Aged man from Kvemo (Lower) Svaneti, W. Georgia, GCR, dressed in wild goats short fur jacket (1976).

Box 13. The Face of a Wild Goat in Ancient Georgian Highland Pottery

In the highlands of Tusheti, archaeological excavations have discovered numerous specimens of pottery adorned with the heads and horns of the tur ('zoo morphological pottery') and stemming from various ancient periods. A dish adorned with a head and horns apparently had some spiritual function in local cult since the Early Bronze Age. In particular, the unique ancient pottery with tur figures from Shenako village has drawn great interest from Georgian archaeologists. On the figures of wild goat sacred signs are depicted, in particular circular signs and swastikas grooved on the forehead of the animal.

The round sign on the tur forehead could be an astralizing symbol connected with the cult of the moon, the host of the beast. The ancient, totemistic belief in the godhead of the beast is the personification of man.

According to the Georgian ethnographical and archaeological materials, in some mountain regions of Georgia, swastikas are a symbol of fire. This is confirmed by oral testimony of Tushetian people that the expression of swastikas in daily or spiritual life remains mainly connected with fire (Gegechkori 2007).

The swastikas on the pottery of Shenako village thus conceptualize life, an analogue of heavenly flames (Dolaberidze 1984) (Figures 193a, b, c).

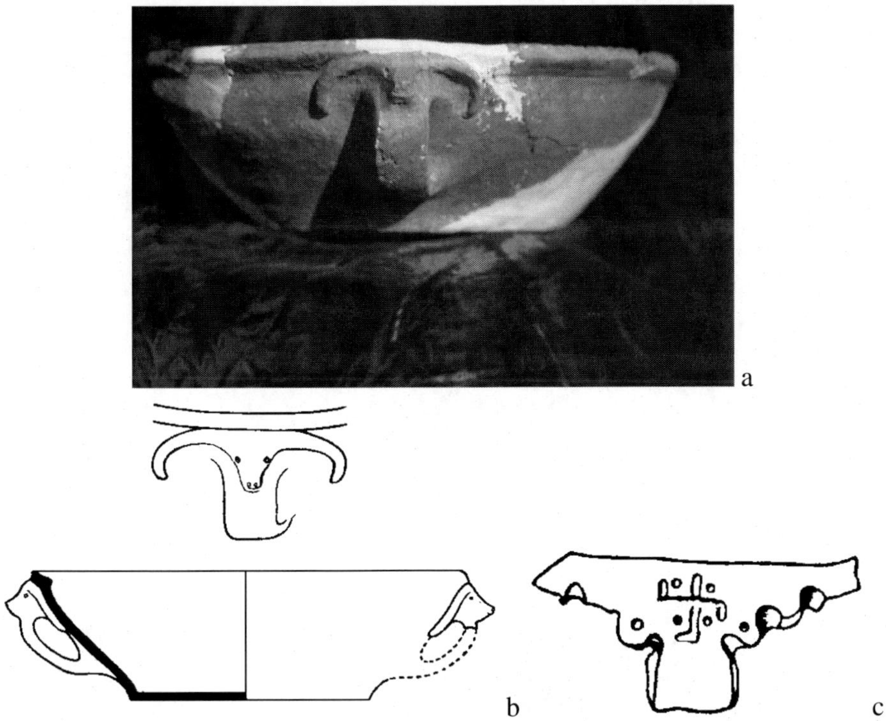

Figures 193 a,b,c. Pottery adorned with male wild goat's head (193a); graphical depiction of the specific signs of the tur's head. Decorations include either circular sign (imitation the Sun or Moon) (193b), or swastikas (193c). Excavation finding in Tusheti, vil. Shenako by archaeologists of the Georgian National Museum (1976) (photo by Dr. Gegechkori, 1978).

The turs are the pride of the Caucasus highlands. The relationship of native people to these herbivores is ambivalent: on the one hand, they worship these incomparable representatives of wildlife, on the other hand, the tur populations have sharply decreased under human impact by hunting and poaching (Figure 192).

Religious belief is an important driver of the latter. In the mountain regions of Georgia shrines are decorated with tur horns (sometimes with deer horns). The sacrificed horns are regularly replaced by freshly hunted horns, especially bigger ones. To hunt a bigger animal and sacrifice it to the shrine is at first glance quite romantic, but this obsession causes a decrease in the population of this endemic and impressive species, leading to a negative selection (Gegechkori 2007). Drinking vessels made of tur's horns are popular in Georgia. Hence, poaching and then grazing are the main threats to tur in the Caucasus (Weinberg *et al.* 1997).

C. cylindricornis is listed as vulnerable (VU) on the IUCN Red List (IUCN 2013).

2.7.5.1.3. The Wild (bezoar) Goat (*Capra aegagrus*)

The Wild (bezoar) goat (*Capra aegagrus*) can be found with various subspecies on islands of the Aegean Sea, in Asia Minor, Iran, Turkmenistan, western Afghanistan and north-western India. The bezoar goat is rather common in the Greater and Lesser Caucasus (Vereshchagin 1959), where it is (except Asia Minor, southwest Asia, and Iran} represented by the common subspecies *Capra aegagrus aegagrus* (Sokolov 1959).

The whole population has between 3,500-4,000 animals with 1,500 in the Greater Caucasus and the rest in the Lesser Caucasus. In the adjacent land to South Caucasus Iran's territory, according to statistics from the year 1991, 2,500 individuals were then living in Golestan National Park and 4,000 individuals in the Central Alborz area (Abbasian *et al.* 2004 and references therein).

Figure 194. Bezoar goat, and other rare stuffed animals attracts visitors at the Georgian National Museum (1972).

The species inhabits rocky outcrops and cliffs between 900 and 3,000 m a.s.l. that adjoin mountain steppe meadows, where it feeds on herbs, grasses (Poaceae are the main diet), twigs and other plant material (Djanashvili 1963; Abbasian *et al.* 2004).

The length of the male is about 150 cm, the shoulder height 80-95 cm, the weight rarely exceeds 40 kg. Males are larger than females and have long black horns that are curved in a saber-shape and much flattened on the sides. The male horns may extend to more than 100 cm in length, the female horns are shorter, about 40 cm. The males are also distinguished for having a long beard. In summer the fur of the back and sides has a chestnut-brown and in winter grayish-brown color. Adult males and females spend most of the year separately. In autumn the flock starts to aggregate, initially between 2 and 7 and increasing up to more than 30 goats. One or two lambs appear in April-May (Gegechkori 2007; Akhmedov *et al.* 2009) (Figure 194).

The IUCN Red List has categorized the species as Vulnerable because of the large human impact. It has also been included in the Red Data books of the USSR, the Russian Federation,

Dagestan, Armenia, Azerbaijan and Georgia (el. version – CR) (Gegechkori 2007; Akhmedov *et al.* 2009; Khorozyan *et al.* 2009; Talibov *et al.* 2009; ECPC 2012).

2.7.5.1.4. The Caucasian Black Grouse (*Tetrao* (=*Lyrurus*) *mlokosiewiczi*)

The Caucasian Black Grouse (*Tetrao* (=*Lyrurus*) *mlokosiewiczi*) is a large bird of the grouse family (Tetraonidae) and closely related to the Eurasian black grouse (*T. tetrix*). The cock is larger than the hen (50-55cm in length compared to 37-42 cm) and very distinct, with all-black plumage, red eyebrows and a long, deeply forked tail. The female is grey with dark stripes and has a cackling call (Djanashvili 1963; Zhordania *et al.*1999).

The Caucasian black grouse is a sedentary species, breeding in the Caucasian mountains, northeast Turkey and extreme northeast of Iran. In the Great Caucasus Range it occurs continuously but on the Lesser Caucasus only very sporadically. It populates areas above the timberline, open slopes with elfin *Rhododendron caucasicum* or other scrubs. This bird has group display in early spring, the hen lays up to 10 eggs in a ground scrape (Djanashvili 1963; Zhordania *et al.* 1999).

Figure 195. The Caucasian Black grouse in Lagodekhi Reserve, GCR (photo: www.Lagodekhipa.ge).

This is perhaps the least-known of all grouse species in the world, and classified as vulnerable (VU) by the IUCN. Recent research has shown that the population is declining according to habitat loss and fragmentation due to unsustainable land use and poaching. This species is listed to some extent and the species is consequently listed as Near Threatened (NT). The Red Data Book of Georgia (el. version), Armenia, Azerbaijan, Russia, Turkey and Iran have listed the species as Vulnerable. Hunting and habitat destruction have greatly reduced the population in recent years (Zhordania *et al.* 1999; Gokhelashvili *et al.* 2003a, b; ECPC 2012) (Figure 195).

2.7.5.1.5. The Caucasian Snowcock (*Tetraogallus caucasicus*)

The Caucasian snowcock (*Tetraogallus caucasicus*) belongs to the pheasant family (Phasianidae). It is endemic to the Caucasus mountains, particularly to the western Caucasus, where it breeds at altitudes from 2,000 to 4,000 m on bare stony mountains. Its food consists of seeds and vegetable matter. The bird is 50 – 60 cm long and weighs from 1.6 to 2.4 kg. Its plumage looks grey from a distance, but is in fact patterned with grey, brown, white and black. It has a white throat and a white patch on the side of the neck. Male and female plumages are similar (Djanashvili 1963; Zhordania *et al.* 1999).

It forms small flocks when not breeding. Before mating (in early spring) the males begin courtship with spectacular display. The nest is an unlined bare scrape in the ground under the shelter of some rocks and stones. Typically, 5-8 eggs are laid and the female incubates the eggs alone. The species is classified as Vulnerable in the Georgian Red Data Book (Djanashvili 1963; Zhordania *et al.* 1999). (Figure 196)

Figure 196. The Caucasian snowcock with chicks. Lagodekhi Protected Areas, 2,150 (1966).

In the highest mountain systems of Eurasia, from the Caucasus to the Himalayas, four more species of snowcocks are distributed: the Caspian snowcock (*Tetraogallus caspius* (VU), which populates mountains of Asia Minor, Iran, the Lesser Caucasus and Kopet-Dagh (Middle Asia); the Himalayan snowcock (*T.himalayensis*) which occurs in the ranges of Central Asia from the Himalayas to China (Nan Sham Mountains), Afghanistan, the Saur-Tarbagatai Mountain range (China). Thirdly, the Altai snowrock (*T. altaicus*), inhabiting the mountains of the Altai, the Sayan (Siberia) and the Hangai (Central Mongolia), and finally the Tibetian snowcock (*T. tibetanus*) from eastern Pamir and Ladakh to Tibet and the Central Hymalayas (*Great Soviet Encyclopedia*, vol.26. p.1793).

Mountains as 'island in the sky' provide one of the best examples of adaptive radiation of living organisms as a result of orogeny. Snowcocks are a young group of species that diversified from one common founder stock after the separation of the mountain systems of the Palaearctic. Their allopatric speciation is closely linked to Late Pliocene-Quaternary mountain building, the recent display of the Cretaceous-Cenozoic Alpine-Himalayan Orogeny.

It should be pointed out that the genetic founder species of snowcock diverged into two stocks: (1) Mongolian-Tibetian with *T. tibetanus* and *T. altaicus* and (2) Caucasian-Himalayan with *T. hymalayensis*, *T. caspius* and *T. caucasicus* (Kozlova 1952; Kuzmina 1974 cited in Zhukov 2004).

2.7.5.1.6. Some Charismatic Species of Insects of the Highlands of the Caucasus

2.7.5.1.6.1. The Apollo or Mountain Apollo (*Parnassius apollo*)

Insects are represented by a rich orophile group from various families. The most beautiful of the order Lepidoptera is *P. apollo*, one of the most amazing inhabitants of the Caucasus. Its wingspan is 64-86mm., its ground color is milk-white. On the forewings there are several black patches, and on the hindwings two big, red eyes with black fringe. These species feeds on alpine flowers, resulting caterpillars feed on species of *Sedum* and *Sempervivum* (Figure 197).

Figure 197. *Parnassius apollo*. Tusheti Hollow, E. Georgia, 2,200m (1989).

The majority of the ultra-oreophile cold-resistant relict *Parnassius* species populate the mountains of Central Asia. The Europe-Siberian-Caucasian *P. apollo* is widely distributed in the Caucasus. Typical of high altitudes range is from 1,800-2,000 meters up to 2,300 meters (Nekrutenko 1990; Didmanidze 2005; Tshikolovets & Nekrutenko 2012). Its population is sharply diminished due to extensive logging of open forests, overgrazing of alpine pastures and butterfly poaching for illegal trade. It is included in the Red Data Books of many countries; IUCN Red List Category – VU (Didmanidze 2005).

2.7.5.1.6.2. Nordmann's (Caucasian) Apollo (*Parnassius nordmanni*)

Nordmann's (Caucasian) Apollo (*Parnassius nordmanni*) is a rare relict species that is endemic to the Caucasus and one of the most beautiful butterflies of this region. The wingspan is 64-70 mm. The ground color is milk-white with transparent ranges. In the central cell of the forewings there are two large black patches, on the hindwings two big, black-fringed red-orange eyes. The butterfly is a resident of both the Greater Caucasus and the Lesser Caucasus, penetrating into the north-eastern part of Turkey. The larva feeds on *Corydalis* sp.

From the alpine belt, where it feeds on flowers, its range stretches into the snow line (4,000m a.s.l.) (Nekrutenko 1990; Didmanidze 2005; Tshikolovets & Nekrutenko 2012). It is included in the Red Data Book of the former USSR. IUCN Red List Category EN (Didmanidze 2005).

2.7.6. Upland Rangeland of the Caucasus (Traditional Pastoralism) and Effects of Grazing Practices on Local Biodiversity (Social-Economic Issues)

Striking wilderness and the purest environment, frequently far from the civilized world, give mountains an aura of mystery and sanctity. Over centuries, indigenous people were inspired to open themselves to the spirit's transformation, because mountains have an unusual power of awakening a sense of the sacred (Bernbaum 1998).

The Greater Caucasus is one of the spiritual sites of multi-ethnic Caucasus represented by churches of orthodox Christians and at the same time sacred pagan buildings (shrines).

Along the strikingly beautiful landscapes these churches, shrines, towers and other fortified buildings develop an aesthetic approach of nature which inspires countless artists and other creative persons.

There are some places in the world, where the natural beauty of high mountains is perfectly integrated with historic buildings. There are unusual impressive testimonies to a sophisticated mountain culture, including medieval architectural achievements.

One such place is Zemo (Upper) Svaneti, which is the highest inhabited area in the Caucasus. It is situated on the central part on the southern slopes of the Central Greater Caucasus. The Svans, the indigenous population of Svanetia, are an ethnic subgroup of the Georgians. The famous Svanetian towers, strictly distinctive cultural phenomena, were created mainly in the 9th-12th centuries. Architectural monuments of Upper Svaneti are included in a list of UNESCO World Heritage Lists.

Yes, 'Since the beginning of human history, mountains have had an extraordinary power to evoke the sacred in our lives' (Bernbaum 1998), however mountain people are among the world's poorest and most disadvantaged (FAO 2011).

Some of the Eurasian temperate mountain regions have been settled since prehistoric times (Alps, Pyrenees, the Caucasus). The best evidence for this is the ice mummy found in the Tyrolean Alps (Grabherr & Messerli 2011). Despite this, compared with the lowland, alpine vegetation and wildlife have been less modified by human activity. Even Europe's mountains are still populated by the wilderness long lost in the lowlands (Nagy *et al.* (Eds.) 2003).

Except the traditional agriculture and comparatively new activities such as tourism, mountains have one most important source of uses, such is another traditional form of human activity – transhumance system, which has been practiced with summer pastures on highland and winter pasturing in lowland steppes and more arid landscapes.

The specific regional climatic, topographical and ecological conditions are largely determined by which kind of grazing-based animal husbandry would be developed in various regions of all over the world according to domestic animals biological and behavioral characteristics. For example, cattle require higher quality pasture and more water than sheep or goats, and are not well suited to drought conditions (Bendrey 2011, *http://www.pastoralismjournal.com/content/1/1/8*). Historically the need for adaptation of livestock systems led to pastoralism, either seasonal movement of herders between fixed summer and winter pastures (pattern of vertical transhumance), or horizontal transhumance – migration of people and livestock frequently over long distances (Ferret 2009).

Rangelands are the primary source of the human's meat and other supplies of domestic animals. Of these the mountain grasslands are of particular interest due to their relatively recent origin, evolutionary trends, and particularly close coevolution with grazing ungulates and changes, coupled with more recent human pastoralism activities (Rawat &Uniyal 1993). Current evidence of zooarchaeological studies indicates domestication of sheep, goat and cattle occurred in separate centers of the Fertile Crescent in the Near East between the 9th and 8th millennium BC (Zeder 2008; Vigne 2011).

Mountains, especially their high-altitude zone, where an area by expansion of pastoral people and other kind of human settlements dates back a long time with large-scale human impacts started in Neolithic times (c.3,000-1,100 BC). Through agro-sylvo-pastoralism and seasonal transhumance mountains people have modified parts of their natural habitats into cultural landscapes (Emanuelsson 2009).

Grazing by domesticated animals is one of the most important human impacts on ecosystems throughout the world, with approximately 47% of the world's land surface currently used by livestock, and 80% of that considered degraded (National Research Council 1994).

As the mountain region of the Caucasus has been populated since ancient times and the highland grasslands have been used for thousands of years, grazing or overgrazing may be regarded as one of the major diversity threats (Nakhutsrishvili et al. 2006). On the other hand, all summits, for example, in the Central Caucasus represent deforested areas since the Late Holocene. Mountains forests and timberline's arboreal plants, where thinning and disappearing generally more by human's impact than by worsening of the climate conditions (Djanelidze & Margalitadze 1977).

As mentioned, the highlands of the Caucasus region are well known for their diverse landscapes, aesthetic, cultural, biological, and hydrological values, which are largely used for recreational opportunities like hunting, hiking and camping. But historically, the primary use of rangeland except mining, hydroelectric power, etc. has been to provide forage for livestock and wildlife. While tourism contributes today much to Europe's Alps (e.g., Swiss Alps), seasonal migration or nomadic pastoralism to high pastures is still practiced in the Caucasus and they still exist today in familiar scenes over the grassy landscapes (Erschbamer *et al.* 2010).

The Caucasus region has witnessed a series of changes in its geomorphology, climate, flora and fauna since the origin of the Great Caucasus Range and then the other mountain systems during Cretaceous-Pleistocene Periods. These changes, in parallel with more recent human activities, have given rise to present day vegetation cover, which ranges from lowland to the subnival belt and several secondary formations of Javakheti Plateau (Grossheim 1948; Gulisashvili *et al.* 1975). From the rangelands and pastoralism point of view, the natural and semi-natural grasslands are of particular interest due to their relatively recent geological origin and close co-evolution with grazing ungulates, especially domestic one. The latter group of animals is a main biotic factor to reduce the biodiversity of the natural grassland represented as a priceless refugium for the wild grazing and browsing ungulates including the keystone species of local biota (endemic Caucasian turs).

The bones of domestic cattle dating back to the 5th millennium B.C. were found in archeological excavations in Georgia. The domesticated form of aurochs (*Bos primigenius*) have served the Georgian people for more than 7,000 years (Burchak-Abramovich 1955; Batello *et al.* 2010).

According to both legends (Greek mythology about the Argonauts and the Golden Fleece) the measurement of wool remains, might refer to the old method of washing gold from streams, which was well attested in the region of Colchis (modern-day part of West Georgia), on the easternmost shore of the Euxine (Black) Sea (Hirbo *et al.* 2006). Sheep fleeces, sometimes stretched over a wood frame, would be submerged in the stream, and gold flecks borne down from upstream placer deposits would collect in them. The fleeces would be hung in trees to dry before the gold was shaken or combed out. The story is of great antiquity and was current in the time of Homer.

A sluice box with submerged sheep fleeces is still used in placer mining in riverhead sites of the Inguri River and its tributaries in Zemo (Upper) Svaneti (West Georgia) (private information of Svans).

According to Uerpmann (1996), the Caucasus should be part of the area, where the domestication of small ruminant breeds – sheep and goats took place. This is supported by the fact that the rock carvings of the tribal society of the Ghegham Mountain Range (Armenia) was discovered and studied in 1969-1972. It dates back to the 5th-1st millennium B.C. (*http:/www.iatp.am/resource/artcult/rockar. . ./Rama/book-e.htm*).

Bezoar goat and then Armenian mountain sheep were the first animals to be domesticated in Armenia. This domestication is likely to date back to the 7th and 6th millennia B.C. This is supported by the archaeological findings (Martirossian 1982) and by the fact that endemic breeds of sheep were recognized as early as in the 9th century B.C. (Batello & Azzu (Eds.) (2010).

Petroglyphs in Gobustan (Azerbaijan) confirm the early (about 3,000 B.C.) development of crop cultivation and animal husbandry including goat's carving in the southern Caucasus (Batello & Azzu (Eds.) 2010).

Finally, Armenian mouflon together with bezoar goat are still found in the southern part of the country (Gegechkori 2007).

2.7.6.1. Human Migration Patterns in Rural Communities in the Highlands of the Caucasus and Their Negative Effect on Pastureland and the Natural Environment

Pastoralism is the basis of the economy in the Caucasus as a mountainous region (Agladze & Zotov 1987). The obvious difference in proportions of livestock husbandry between various countries of this region may be explained by their largely contrasting climates. As mentioned in previous chapters, the western part of the Caucasus and especially Georgia benefits from westerly airstreams originating over the Black Sea and the orographic effect of the Caucasus Mountains, which can receive more than 2,500 mm of rainfall each year. Therefore, the western Caucasus is preferred for cattle raising based on limited movements between the Greater and Lesser Caucasus alpine pastureland in summer and cattle shelters in lowland settlements area during winter. To the east of Mount Elbrus, mainly in the eastern Caucasus the mountain ranges form a rain-shadow over much of this area, which can receive as little as 400 mm annually. In the semi-deserts and desert lowlands of the southern and northern Caucasus annual precipitation decreases to ~ 250 mm. Hence, in this area of the Caucasus sheep husbandry dominates. This way of pastoralism based on transhumance that lasts much longer, is used with most of the herds of Georgia, Azerbaijan, Dagestan (North Caucasus), and partially Armenia.

This kind of transhumance occupies a time from October to May to the arid landscapes of the lowlands and then back to the high mountains (O'Loughlin *et al.* 2007).

During the Soviet Era, all administrative units of the Caucasus (South Caucasus and 'The Russian South') were one of the most prosperous Soviet Republics and Autonomic Republics. Soviet planners were not in error, when they developed a rational economic policy and use of the indigenous knowledge, which led to rational nature and land use developing cereal grains and sunflowers in the north, while promoting specialized southern agriculture (wine, tea, citrus fruits and cotton) in the republics of the South Caucasus (O'Loughlin *et al.* 2007).

A quite different situation faced pastoralists of the Caucasus following the collapse of the Soviet Union.

Traditional agriculture in the countries of the Caucasus in the past was sustainable and did not seriously threaten the biodiversity in this region. The main impact was generalized in the Soviet era. In 1979 agriculture in the USSR was organized into about 28,000 collective farms (average size of $64km^2$), and 18,000 state farms (average size of $194km^2$) (Heywl *et al.* 1981). This caused the establishment of new branches of agriculture such as citrus and tea plantations. With these new by-products it was mainly Georgia which supplied Russia and other Soviet republics. Demands in wool and cheese caused an increase in sheep and cattle herds up to several million heads on the relatively small territory (compared with Central Asia) of the region.

While still shepherded almost in the traditional way, the herds became so large that they threatened many high mountain pastures and caused overgrazing. This problem reduced after the disintegration of the Soviet Union in the early 1990s, when drastic political and socio-economic changes occurred in the former USSR. Over recent years, as the economy came to a standstill, individual farms have replaced collective farming and subsistence agriculture and livestock breeding have become common (O'Loughlin *et al.* 2007).

The Caucasus and the neighboring (in an eastern direction) five countries of Central (Middle) Asia, historically has been the bridge between Europe and Asia Minor with China via the Silk Road. The region is home to a large diversity of livestock and plant species and is, very likely, the area where some species were domesticated (Gromova 1935).

Following the demise of the Soviet regime in the early 1990s, this brought changes that severely impacted on the agricultural sector of the whole region. The most important involving the privatization and liberalization of formerly state-owned livestock. The collectives provided transport for nomadic moves between seasonal pastures, supplementary feed for the harsh winter and spring, provided veterinary and specialized animal breeding services. The downfall of the Soviet Union led to a dramatic reassertion of the importance of pastoralism, which attracted within fragmentation of the former large collective or state-owned farms into numerous, unproductive smallholdings that formed the basic social unit in this region (Suleimonov *et al.* 2006). The change that occurred during 10 years after the collapse of the USSR was crucial. The declines in sheep population were Kazakhstan (76%), Kyrgistan (74%), Georgia (63%), Tajikistan (41%), Azerbaijan (41%), and Armenia (30%). Less pronounced population changes (0% and 11%) were recorded in Turkmenistan and Uzbekistan, where the application of reforms was more conservative (Suleimanov *et al.* 2006). Within the North Caucasus the small cattle numbers almost disappeared in North Ossetia, decreased to a lesser degree or increased in other regions, where wool is a traditional source of income (Gracheva & Belonovskaya 2010).

In response to this unprecedented decline, ICARDA recognized the urgent need to assess the genetic diversity of the region's livestock and its suitability for the new demands of the emerging economic and production systems (Iniguez & Mueller (Eds.) 2008).

Multiregional and multiethnic Caucasus unique settlements – social groups are associated with the different transhuman lifestyle, and frequently identified as remnant of an older ethnic culture. Over the centuries, pastoralists have developed their own ecological behavior, skill treated sparingly of local (highland) vulnerable environment, in other words, they are the custodians of traditional landscapes, They have developed local livestock breeds that are well adapted to the local climate and steep, sometimes dry, mountain slopes, these breeds are specialized and feed on exceptionally native grasslands. Usually this is local poor-quality pastures. In this way, local livestock breeds and native flora species have coevolved. For example, the Khevsureti region (East Georgia) with its highly mountainous nature, where winter's temperatures are extremely low (falling between -15 and -20^{0}C), snow cover is thin and lasts about seven months, feeding resources are scarce and very poor quality, only the Khevsuretian cattle breed can survive. The cattle breed is very small with an average weight of cows 185-200kg, bulls – 340kg, an average withers height of 98-100cm. Georgian mountain cattle – Khevsuretian, Phshavian and Ossetian breeds remained on the southern, partially on the northern (Khevsureti) slopes of the Great Caucasus Range and mountainous Adjara (Mitichishvili *et al.* 2013).

During about 20 post-soviet years the high mountain land use in the Caucasus changed profoundly. Following attempts of forced collectivization in the 1920s, in the early 1990s, the indigenous knowledge called only for family survival. Farmers were encouraged to turn back to the traditional independent households and to join voluntarily into private partnerships, i.e., herd composition was being restructured along the commercial lines. Privatization in some countries of the South Caucasus, for example in western Georgia, began as attempts to increase the head of cattle and land in private ownership. This action was unsuccessful because of a lack of forage and markets, but its consequences are very significant (Gracheva & Belonovskaya 2010).

The South Caucasus countries achieved their independence, after which they began reformation of the mentioned socio-economic system and transition from centrally planned to a market driven economy. The North Caucasus, – a region of traditional mountain livestock farming - remained within the border of the Russian Federation, however, the difficulties of the transition period were aggravated by the separatist conflict, which remains unsolved until today. During USSR times, traffic between Russia and the Transcaucasia was made possible by circumventing the Caucasus mountainous chain at its two coastal ends. The collectives regularly used trailers to transport livestock to the far destination sites. Today this livestock mobility between seasonal grasslands, due to new state borders is being significantly reduced or becoming separate. The same should be said with respect to the age-old traditions of cultural interactions between mountains and plains. The main reason is the questionable intra-montane roads within rough terrain of the mountainous chain. As regards tunnels through the Great Caucasus Range: Roky Tunnel – the only road joining the North and South Caucasus, was unused after the 1990s (but renewed in 2008 during the Georgian-South Ossetian War supported by the Russian forces). In the 1970s, a tunnel was started between Ingushetia (the North Caucasus) and Georgia, but now lies abandoned. The other routes between Georgia and Russia include the Georgian Military Road and the Gantiadi-Adler crossing in Abkhazeti but they lost their pasture rotation importance because of their permanent conflict situations (Battello *et al.* 2010; Gracheva & Belonovskaya 2010).

During the difficult political and economic times following independence in the 1990s, the interdependency associated with economic relations including animal husbandry died out (O'Loughlin *et al.* 2007; Gracheva & Belonovskaya 2010).

Currently, over 80% of the grasslands in Georgia and more than 86% of natural pastures and hay fields in Armenia are degraded, some 50% of the total pastures and rangeland in Azerbaijan are subject to erosion processes (Battello et al. 2010). 50-80% of alpine ranges are abandoned in different regions of the North Caucasus ethnic (non-Russian) republic (Gracheva & Belonovskaya 2010).

There are two main forms of transhumance in the Caucasus: nomadic pastoralism and semi-nomadic pastoralism.

Traditional or fixed transhumance (nomadic pastoralism) is the important branch of the economy of many indigenous ethnic groups of the Caucasus with their unique pastoral cultures.

Transhumance involves relocation of livestock: cows, sheep, goats, horses. The mountain period of shepherds starts in May and ends in early or mid-October. In winter time herds of the eastern Caucasus use mainly ranges of steppes and semi-deserts and deserts in the lowland area.

As mentioned, the southern Caucasus was one of the prehistoric centers of sheep breeding. In the ancient Urartu state, fine-woolled and coarse-haired sheep breeds were known. Sheep and goats are inextricably linked to the spiritual life and culture of the countries of southern Caucasus (Battello *et al.* 2010). Today, there are about 25 local sheep breeds in the South Caucasus. The following should be mentioned Tushin, Imeretian, Gala, Shirvan, Balbas, Masekh, Bozakh, and others. All these sheep breeds originated from the Caucasian fat-tailed sheep and are grouped as merino breeds (Hirbo *et al.* 2006 and references therein).

The genetic diversity of Caucasian sheep, as Hirbo *et al.* (2006) emphasized, can be attributed to four major events: the silk road trade; the Mongol invasion of the 13th century and disperse trends of their livestock; import of German sheep to the Russian Empire by the Czar to provide wool to clothe his army in the 19th century and finally the contribution of transhumance of southern Caucasian people since pagan time.

With other ethnic groups of the eastern Caucasus the practice of the mentioned system of transhumance in Tusheti (East Georgia) has dated back to the Early Medieval Period or even earlier (Roschupkin 1958). This remote area is one of the most pristine high-mountain regions (average height is 2,000m a.s.l.) in the Caucasus (Gegechkori 1983). Located on the northern slopes of the Greater Caucasus Mountains, the people are ethnic Georgians called Tushetian (Tush). Traditionally, Tushs are sheep herders. For example, in the 18th-19th centuries the number of sheep reached 250,000 (Roschupkin 1958). Tushetian Gouda cheese and high-quality wool were famous and were exported to Europe and Russia. The local shepherds spend the summer months in the highland areas of Tusheti, but lived in the lowland villages of Kakheti region (East Georgia) in winter time.

This inaccessible mountainous region is the place of origin of various domestic breeds of animals, such as the Tushetian horse and sheep and the Georgian or Caucasian sheep-dog. Sheep-breeding also promoted the development of a special school of folk craftsmanship. Using wool and natural dyes, Tushetian women produce a variety of beautiful items, including socks, and especially carpets, etc. (Makalatia, 1933).

Tushetian shepherds migrated annually with their flocks between the mountains of Tusheti and their winter pastures, as they have for centuries. During the Soviet Union, the rural population settled on the northern slopes of the Great Caucasus; like some other shepherds of the eastern Caucasus region, they used the Nogai drylands (Dagestan) (Rcheulishvili 1957). The distance travelled annually was between 450 and 500km. After the 1990s the minority of Tushs continued living a tough transhumant life. They take three days to cover the transhumance route. They have reduced livestock mobility and with other pastoralists of eastern Georgia use as a winter pasture, steppes and semi-deserts of this part of the country including the buffer zone of wooded grasslands of the famous Vashlovani National Park. Because farms in the lowland are too small to support a larger herd in a comparatively small area, the winter pasture adjacent to Azerbaijan became a source of overgrazing, habitat degradation and increased human-wildlife conflict. As underlined above, all of this contributed to a dramatic decline in large predators, the best shelter of which is now Vashlovani National Park (Gegechkori 1982).

Thanks to the Black Sea coastal wet marine climate, herds of Megruli red cattle use alpine mountain pastures. In winter they pasture mainly in Colchis bogs without supplementary food (Mitichishvili *et al.* 2013).

Following the collapse of the Soviet Union, some Megrelians migrated to Russia. A private small herd of about 20 animals were allowed to run wild in one of the numerous tiny islands of Colchis wetlands. Some years later the farming attempt had failed and the cattle were abandoned. The cattle avoided being exterminated during the comparative mild winter of Colchis. They persist as a ‘wild’ population until today with little human interference in the Colchis National Park’s area. The island is predator-free. The Megruli red breed of cattle here offers a rare opportunity to study their natural behavior.

In a semi-nomadic pastoralism system, domestic animals are grazed on a rangeland around permanent settlements until summer, and thereafter the herds are driven not far from a settlement located at comparatively higher altitudes sites until autumn. A semi-nomadic animal raising system was commonly employed in the upland of Southern Georgia and adjacent Armenia (Javakheti Volcanic Upland). As already mentioned, Javakheti mountain steppes are in most cases semi-natural ecosystems which were created by man through logging of local high mountain forests.

The Baku-Tbilisi-Ceyhan oil pipeline, the South Caucasus natural gas pipeline, and the Kars-Tbilisi-Baku railway (under construction) pass through the region. This system of construction may disturb local ecology – land and marine ecosystems.

Another apparent change of pastoralism is the dynamics of the horse population, which has increased with tourism development in the Khevi area. Kazbegi district is famous for its unique alpine meadows dotted with Caucasian rhododendron, junipers and grassland, mineral springs. The most prominent geographic feature of the area is Mount Kazbek, locally known as Mkinvartsveri (i.e., ‘ice-capped’ in English), a dormant 5,047 - meter high volcano. It is a part of the projected Khevi-Aragvi Biosphere Reserve. Among the important cultural sites of Khevi is the Gergeti Holy Trinity Church, a domed church with a bell-tower. It represents an outstanding monument of Georgian architecture of the 14th century. The church is situated on the left bank of the river Tergi, at an elevation of 2,170m a.s.l., under Mount Kazbegi, in front of the Glacier. Therefore, it is the monument built on the highest spot in Europe. Its isolated location on top of a steep mountain surrounded by the vastness of a striking beauty of the Great Caucasus highland’s nature, the steep slopes of the glacier valley, lakes and waterfalls, make it one of the most spectacular places.

Since ancient times, Khevi has been of great strategic and military importance due chiefly to its immediate neighborhood to the Darial Pass, which connects the North Caucasus with the South Caucasus. Here is represented a part of the Georgian Military Road, which runs between Tbilisi (capital of Georgia) and Vladikavkaz (Russia) and follows the traditional route used by invaders and traders throughout the ages. The road stretches some 208 kilometers through the Terek valley, crosses the Rocky Ridge in the Darialy Gorge, passes Mount Kazbek, then in Khevi area leads to the Jvari Pass (2,379m). Known since antiquity (it was mentioned by Strabo and by Pliny), the Georgian Military Road in its present form was begun by the Russian military in 1799 and completed in 1863 (Gvasalia 1987; *en.wikipedia.org/wiki/Khevi*).

According to its landscapes and historical importance, the Georgian Military Highway always had romantic connections in both the Russian and Georgian imagination. The high mountainous part of the highway inspired world-famous poets and writers – Pushkin, Lermontov, Tolstoy, Gorky – to feature it in their writings. Inspired by the contemporary liberal movements in Europe, among Georgian intellectuals called "tergdaleulebi" (i.e. those, who have drunk the water of the Terek River) due to a clash of imperial Russian-Tsarist and feudal Georgian cultures on the migration from educational institutions from Georgia to St. Petersburg and back during the second half of the 19th century. They were founders of the Georgian National Identity. Ilia Chavchavadze celebrated the savage freedom of the region in his patriotic manifesto, 'Letters to a Traveler' (1861-1871). In 1987 he was canonized as saint by the Georgian Orthodox Church and Georgians revere him as *Pater patriae* of Georgia (Reisner 2009; *en.wikipedia.org./wiki/ilia_chavchavadze*).

Biodiversity decreases world-wide in an unprecedented rate. Traditionally managed mountain grasslands in the Alps are still species-rich ecosystems that developed during centuries of livestock grazing (Spiegelberger *et al.* 2006).

Biodiversity of grassland is important not only as a tool to protect plant and animal communities, but also in sustaining their agriculture productivity. Extensively managed pastures and meadows are of crucial importance for grassland biodiversity across Europe. Unfortunately, biodiversity of such biocoenoses is currently threatened either by traditional extensive use or by abandonment (Spiegelberger 2006; Metera *et al.* 2010 and refences therein). This trend is noticeable since 'The man in the ice', and later, after retreating ice above 3,000m a.s.l. (Kőrner 2000).

According to Callaway *et al.* (2000), during the Soviet Union, large portions of the Caucasus Mountains experienced grazing intensities that were much higher than those previously maintained by the native people. In the Caucasus from the 1950s through the 1980s sheep herds numbering in the millions were driven on seasonal migrations from winter to summer pastures. These migrations caused severe overgrazing and erosion, and large-scale invasions of weedy plants (Gulisashvili *et al.* 1975).

After the breakup of the USSR the major expansion of grazing lands has mainly occurred in the highland area (Gracheva & Belanovskaya 2010). On the other hand, sheep grazing in the winter ranges of the eastern Caucasus has nearly tripled in the past decade (WWF-Caucasus 2008).

Mountain pastoral lands are about 5 million ha in the North Caucasus. They occupy 30-40% of the total area in the eastern Caucasus and over 25% in central and western regions (Gracheva & Belanovskaya 2010). Azerbaijan has the most pastureland within the Southern region, totaling 3.8 million ha (Battello *et al.* 2010).

According to Gracheva & Belanovskaya (2010), among the main reasons for deterioration of the high mountain rangeland's area in the Caucasus should be mentioned: breakdown of seasonal pastures rotation according to locally specific patterns of customary land tenure, selective high mountain expensive grazing (for example, in 2002 in Azerbaijan 43% of winter pastures and 29% of summer pastures were used by sheep with high grazing densities (Batello *et al.* 2010), of the nearest pastures and hayfields, neglecting traditional regulations of nature production.

High mountain pastureland risks have triggered a chain of environmental disasters. The most visible and significant degradation processes include soil surface disturbance. Continuous grazing creates channels or paths on hill slopes, which remove huge quantities of soil during rains. Along with the increase of individual farming, and largely used high mountain areas as pasture grounds for livestock, it is leading to moderate or severe soil erosion and maintenance of hydrologic cycles and ecosystem goods, like tourism and recreation, sometimes evoking avalanches and mud-slides (Figures 198, 199).

The current uncontrolled grazing causes soil erosion. In this situation anthropogenic impact is usually much faster than soil formation. It means irreversible loss of fine soil matter (Gracheva & Belanovskaya 2010). A third of pastureland in the Caucasus is subject to erosion (Kurtubadze 2008).

Intensive grazing often has a dramatic effect on plant community composition and structure. The latter usually has become worse.

Intensive grazing of grasslands has a selective character, when the most palatable species are eaten first in preference to less palatable ones. When less palatable plants dominate the grazed area, this is the last stage of vegetation degradation and, as a rule leaving the formation of unproductive, unhealthy grasslands (Valentine 1990). These plants have been described as 'invaders' (Callaway *et al.* 2000 and references therein).

Figure 198. Soil surface disturbance. The Tianeti region, environs of Kartli Ridge, part of the Greater Caucasus, E. Georgia, 1,800m (1974).

Figure 199. The overgrazing timberline is frequently driven by avalanche processes. Abkhazeti, Audhara site, W. Georgia, 2,150m a.s.l. (1982).

Furthermore, as animals move around they trample the soil with their hooves, exerting pressure and resulting in soil compaction, which leads to reduction of soil infiltration rates causing higher surface water runoff. These soils are then vulnerable to the wind and water erosion, which removes nutritionally rich upper layers of the soils.

Highly degraded lands are often nutritionally marginal, barren and usually abandoned by herders. Overgrazing may cause the development of other unpalatable species, weed invasion and a decrease in forage availability (Spiegelberger *et al.* 2006; Parolo *et al.* 2011).

Overgrazing is a serious problem for most of the mountainous areas of the Caucasus (Nakhutsrishvili *et al.* 2009).

Plants can use two main strategies to defend against damage caused by ruminant herbivores, these are spines (thorns) and toxins. The thorns on the stem of some plants, serve as a mechanical defense against herbivory. Chemical composition of toxins can act as repellents or toxins to herbivores, or reduce plant digestibility. Some plants immature fruit have a bitter flavor, etc. (Grossheim 1948; Fritz & Simms (Eds.) 1992).

High unpalatable weed densities, in particular, of weeds toxic to cattle, reduce the fodder quality and therefore the economic value of a grassland (Spiegelberger *et al.* 2006).

Grossheim, concerning the flora of the Caucasus (1948), recognized three groups of synanthropic vegetation: (1) the segatal plant communities of cultivated habitats, (2) the ruderal vegetation formed by plants growing mainly in disturbed lands, and (3) the pastoral communities' group of pasture weeds (*paskual* vegetation).

According to Grossheim (1948) and Gulisashvili *et al.* (1975), the main natural pastures and hay meadows are represented in subalpine and alpine zones of the Caucasus. Among the plants (weeds) poisonous to livestock of high mountain areas should be stressed the following species: false hellebores – *Veratrum lobelianum*), aconite (*Aconitum nasutum*) and larkspur

(*Delphinium* spp.). Very thorny grooved stem and branches are represented usually in many species of Plume thistles (*Cirsium* spp.).

Figure 200. Pasture infested by *Veratrum lobelianum*. Kazbegi district environs, E. Georgia, GCR, 2,100m (2012).

Veratrum consists of 45 species in the family Melanthiaceae. All parts of false hellabores are poisonous to livestock. The toxins are only produced during active growth. Then, after fall in early autumn, the defense chemicals are either retranslocated below ground or break down. Once *Veratrum* losses its toxicity the livestock will eat it (Zamtaradze 1985). Subalpine meadows of *V. lobelianum* are abundant in western and partially Central Greater Caucasus, as well as in the Lesser Caucasus. It is a pasture weed, usually rejected by animals. It is a largely invisible unpalatable weed, which reduces the biomass production of desirable grasses and legumes (Grossheim 1948; Nakhutsrishvili 2013) (Figure 200).

Aconytum nasutum is a genus of over 250 species (fam. Ranunculaceae). All parts of this herbaceous plant species are poisonous and were used in the past for killing wolves. All parts of young plants and seeds of *Delphinium* (fam. Ranunculaceae) are poisonous, often fatal.

Cirsium (fam. Asteraceae) is a thistle, and is defended by the sharp spines on the slightly hairy leaves and stems. Species of this genus are noxious weeds of high mountain pastures, commonly formed into thickets (Grossheim 1948).

In overgrazed pastureland all species of the mentioned genera are characterized by an abundant population, therefore, the forage productivity of invaded rangeland for livestock invariably decreases (Grossheim 1948; Zamtaradze 1985).

To selective palatable species belongs matgrass (*Nardus strict*, fam. Poaceae), and tussock grass. It is native to Eastern Europe, the Caucasus Mountains and Siberia. Matgrass has spread in West Europe's countries, and recently (late 1960s or early 1970s) naturalized in North America. This weedy grass is grazed only by sheep and goats (Perkins 1968).

It is worth mentioning that from Callaway *et al.* (2000) viewpoint one of the most important positive effects that one unpalatable plant can have another in protection from herbivory. By associating with weed and a toxic neighbor, a tasty species may avoid being eaten and increase in size and have a reproductive output under unpalatable plants. From this point of view two dominant unpalatable invaders, *Cirsium obvalatum* and *V. lobelianum* were examined in subalpine meadow plant communities of the central part of the Greater Caucasus within the borders of Georgia. Certain research of the mentioned authors indicates that unpalatable plants, secondary dominant species of unhealthy rangelands, have positive effects to preserve the diversity of many other subalpine meadow species heavily grazed areas. By providing refugia in which, according to positive interactions ('plants helping plants'), palatable species can recolonize overgrazed pastures by seed.

Figure 201. Pasture severely infested with weeds (mainly *Rumex alpinus*). Askhi limestone massif, Samegrelo province, W. Georgia, GCR, 1,850 m (2015).

Many unpalatable weeds on mountain grassland, including all the above-mentioned genera, are tall (among them should be stressed ruderal species as well, e.g., *Rumex alpinus, Urtica dioica,* etc.). At the same time some of those species reach high densities undergoing a process of gradual abandonment and therefore play a role of extensification weeds (Spiegelberger *et al.* 2006) (Figure 201).

Pastoralism is a most important business to pastoralist societies of high mountains. The native population is limited to self-sufficiency in most basic food products.

This fact that residents of a high mountain area frequently face political, social and economic marginalization and lack access to such basic services as health and education, most mountain households pursue several activities to generate income. As a rule, migration has always played a prominent role in mountain livelihoods, particularly for young men.

The author of the present work during the last 50 years is eyewitness to completely or near completely abandoned rural sites with a lot of ruined houses.

In Conclusion:

- Rural abandonment is a general problem that has to face, for example, Europe (Rois-Diaz *et. al.* 2006), but especially the majority of countries of the former USSR.
- Globalization represents the greatest threat to the modern rural areas due to excessive internal/external urban areas migration. Globalization appears to be responsible for the disappearance of local historical and cultural features shaped for centuries or millennia.
- Rural abandonment by ethnic-majorities due to the breakup of the former Soviet Union and responding mainly as the deteriorating socio-economic situation continues

especialy by young people, and those villages which are still inhabited are mainly populated by elderly people.

- These events of intensive depopulation had the largest effects in the mountain areas of the Caucasus countries and regions.
- Similar to arid lowland, high mountain pasture land areas, after founding a new economic landscape, are under severe pressure from two contrasting trends: increasing intensity of livestock grazing on the pastures to 2-4 times more than their carrying capacity in the past decade and secondly abandonment of overgrazed meadows with secondary plant communities in others.
- To avoid a further destruction of unique highland's ecosystems of the Caucasus through both reasons – increasing intensity of livestock grazing and abandonment of overgrazed meadows. In detail, the primary environmental goal from the local and international environmentalists is that local authorities should be focused on elaborating certain rural development policies and pasture management, which should be adjusted to the local conditions to implement the appropriate to the Caucasus region's strategy of biodiversity conservation. For instance, it should be mentioned the following project of rural development: 'Rural development in the Racha-Lechkhumi Region' (2009-2011), supported by the Switzerland Agency for Development and Cooperation'.

2.7.6.2. Transhumance Beekeeping of the Caucasus, an Additional Ecological Role for the Biodiversity of Highland Environments (Social-Economic Issues)

Pollinators are an essential group of animals for the benefit of agro-and biocoenosis. The most recognized biotic pollination is entomophily, i.e., pollination by insects (Proctor et al. 2003). Among insects as pollen vectors should be stressed wild and domestic species of honey bees, *Apis* (Hymenoptera: Apidae: Apini), which are plainly adapted to pollination (Wilson 2004). Plants and their pollinators began evolving over 100 million years (Cretaceous Period) with the origin of the flowering plants (angiosperms). Co-evolution between entomophilous plants – flowers and pollinator insects have determined a complete domination of flowering plants and some of their pollinator partners (Cenozoic era – last 65 million years) (Leppik 1957; Brackenbury 1995).

At least 80% of our world's crop species require pollination to set the seeds. For human nutrition, the benefits of pollination include not just abundance of fruit, nuts and seeds, but also their variety and quality (Free 1970; Battello *et al.* 2010). Honey bees – obligatory flower feeders – are very important insect pollinators. As pollination is one ecosystem service, therefore, pollinating insects also play a critical role in maintaining of wild ecosystems and natural plant communities within them and ensuring production of seeds in most flowering plants, including highland's ecosystems (Kevan 1999).

Beekeeping is one of the oldest agricultural occupations. It is known that honeybees have existed for 50 million years. The first *Apis* bees appear in the fossil record at the Eocene-Oligocene (65-25 Mya) boundary, in European deposits (Wilson 2004). Collecting honey from wild bee colonies or honey-tree hunting is one of the most ancient human activities.

There are nine species of honeybees, although only one – *Apis mellifera* is commercially viable (Free 1970). It was domesticated for honey and wax in medieval Europe (Michener 2000). Our ancestors started collecting wild honey, which is usually found in hollow trees.

Honey hunting - plundering wild nests of honeybees to obtain crops of honey and bee wax is still practiced by poor aboriginal societies in some regions of Africa, Asia, Australia and South America. The colonies of honeybees, depending on species, may be nesting in the wild in tree cavities, in trees, or rocks, etc. Honey hunting may be seen as part of the lives of the world's remaining hunter-gatherer societies. These forms of honey collecting entailed the destruction of the entire colony, when the honey was harvested. Some of the earliest evidence of gathering honey from wild colonies is from rock paintings. As evidenced by cave paintings – honey seeker – dating at least 8,000 years ago - is known near Valencia, Spain (Crane 1983). It is generally accepted, that beekeeping emerged in ancient Egypt and Mesopotamia as early as 3,500 years ago (Buchmann & Nathan 1996). Honey - bee products have been preferably consumed for health protection throughout human history (Dadant 1984).

The oldest remains of honey have been found in southern Georgia. Archaeologists have found honey remains on the inner surface of clay vessels of an unearthed ancient tomb, dating back to some 4,700-5,500 years ago (Kvavadze *et al.* 2006).

As mentioned, not far from Kutaisi, the main city of western Georgia, is located the Sataplia State Nature Reserve. In this site occur famous caverns of Mount Sataplia, where billions of wild bees once swarmed (the name comes from the Georgian word for honey, *tapli*). As this place has long been famed for its bees, Georgians have been visiting Sataplia for *tapli* for centuries. The trade in honey and wax was of great economic importance to the Kutaisi region, which is confirmed by the largely known very ancient trade route via river Phasis (Rioni) in Colchis from the remote past.

Even today western Georgians make an excellent honey vodka, *taplis araki*, famous for its honey flavor and extremely well-tested liquid (www.vavilontravel.com/en/national-park-sataplia).

The European scientific study of bee colonies began in the 18th century. In the 18th and 19th centuries a major revolution in beekeeping occurred which allowed the bees themselves to be preserved whilst taking the harvest at the same time through the invention and perfection of the movable comb (Ruttner 1988).

Genus *Apis* with four species – *Apis florea, A. dorsata, A. cerana,* and *A. mellifera* – were originally found only in the old world. There is only one fossil species documented from the New World, *A. nearctica*, known from a single Miocene deposit (14 million years old) specimen from Nevada, USA (Engel *et al.* 2009). Only *A. cerana* and *A. mellifera* are suitable in apiculture, with modern movable comb hives.

Today apiculture is a widely known activity of honeybees worldwide except the polar areas (Dadant 1984). Honeybees have adapted to a wide range of ecological conditions on the earth showing a broad spectrum in physiological and behavioral attributes. Except crossbred, there are purebred honeybee races in the world.

Among two dozen pure geographic races of the *A. mellifera*, four races are considered for beekeeping in a cool temperate climate, namely, Italian or Ligurian (*A.m. ligustica*), Carniolian (*A.m. carnica*), the nearest relative of the Italian, Caucasian (*A.m.caucasica*) closely resembles Carniolian, and the Dark European honey bee (*A.m.mellifera*). The yellow Italian honey bee is the most widely distributed of all honey bees (Gould & Gould 1995; Wilson 2004).

Known worldwide, grey mountain bees of the Caucasus (Georgia) (*A.m. caucasica* Gorbachev) are an autochthonous race of honey - bee to the mountain range and southern valleys of the Caucasus (Georgia, part of Azerbaijan) and of the Black Sea coast in Anatolia (Turkey) (Kara *et al.* 2012) (Figure 202).

Figure 202. The elite Megrelian race of the Georgian bee. Chkhorotsku district, Samegrelo, W. Georgia (1986).

The climate within the range of the Georgian bee varies from humid, almost subtropical on the coast to cool temperate in the mountains. Local strains of bees reflect the different climates; the bees from the mountains being larger and darker than those from the lowland region. It is well adapted to the higher altitude's sites and long and snowy winters (Kara *et al.* 2012).

In the eastern part of the Caucasus, towards the Caspian Sea and Armenia, the grey bee is largely replaced by yellow Iranian bee, amr (*Apis mellifera meda*).

Georgia is the birthplace ('central homeland') of the Caucasian bee (Fert 2001; Corso 2013). The pure populations of this race occur in mountainous sites of Georgia: river Khobistskali gorge (Samegrelo), Abkhazeti, Guria, Imereti, and Kartli (Corso 2013; Tsitlidze *et al.* 2013); the north-eastern part of Turkey also has completely pure strains of the mentioned race (Kara *et al.* 2012).

The history of Georgian beekeeping, one of the ancient branches of agriculture, is intimately linked with the history of its people. Traces of that activity are to be found in the remotest written works. In prehistoric Georgia, there existed a system of high-status apiculture, as can be concluded from the previously mentioned finds or honey remains. Later, in the 4th BC, advanced honey industry is discussed in the historical work 'Anabasis' written by Xenophon. The ancient Greeks, to ensure the preservation of game on their way back from Colchis, would immerse the meat in honey-filled amphoras bought locally. The famous medical text 'The Karabadini – a book on Doctoring' by Zaza Phanaskerteli a Georgian philosopher, medical author and physician of the 15th century was the reference book in Georgian. It recommended using the produce of the hive to cure the toughest pathologies. According to some ethnographic and historical sources the Georgian people regard the bee as sacred, close to God, they see God's creation of bees and all that they do as a miracle. Bees were idealized by many Christian church fathers including Armenians and Georgians (Fert 2013; Tsitlidze *et al.* 2013).

As a race, the Caucasian race is the gentlest, i.e., peace-loving natural bee in the world. The distinctive external features of this bee are a long proboscis (7.2-7.3mm), which is the longest among the true honey-bees on the Earth, allowing them to exploit sources other races bypass. The grey-bee performs better in the flows with very deep corollas such as acacia and red lucerne together with their ability to fly in cooler and wetter weather than other bees, allows Caucasians to take advantage of marginal forage, and make them suitable in the areas with long, drawn out nectar flows. These bees generally go into winter with a smaller cluster than warmer climate bees and conserve winter stores. Due to their capacity for work Caucasian bees are far superior to Italian bees, which was proved more than once by subjecting them to a special test. In 1923-1931 the comparative test by means of honey yield production, which was held in Wyoming State (USA) finished profitable for the Georgian bee. It gathers a very respectable crop of honey, on average it collected 36.5kg. more honey than Italian bees. Another reason why this race of bee is so popular lies in its other useful economic and behavior characteristics, which include as well as honey, high wax productivity, fertility of queen bees, a weak swarming instinct, active defense of nests. The production rate of the Caucasian bee is the highest among bee species.

Disadvantages associated with this race include an increased susceptibility to *Nosema* disease (this leads to heavy winter losses) and their excessive use of propolis, increased robbing tendency.

The Caucasian race of bees (grey mountain bee) for its many good qualities is one of the best bees on the market today. Particularly docile and remarkably well adapted to long and harsh winters, it is exported to over 40 countries and has been widely used worldwide for hybridization (Fert 2001; Kara 2012 and references therein).

Intensive studies have been carried out on this bee race in Russia. Interest in this species was aroused by the Russian researcher A. Boutlerov, who published an essay about it in 1877, emphasizing its pacific attitude. The scientific description of *Apis mellifera caucasica* was made by Gorbachev (1916). The starter materials of Megrelian, Abkhazian, Imeretian, and Kartalineian populations were brought from Georgia. Later on, the base of the Caucasian (Georgian) bee race, the Apiculture Research Institute and Krasnopoliansk nursery (north-western part of the Caucasus) were founded during the Soviet era (Vinogradov 1969).

It should be noted that, in 2012, commercial beekeeping in the United States lost between 40 to 50% of their hives, the worst year for bee-colony collapse since 2005. A lower bee count reduces the supply of fruit, vegetables, nuts, and beans dependent on pollination, which, consequently, increases prices.

As grey mountain Caucasian bees are more resistant either to the mites (*Varroa* spp.) or pesticides that could be causing the deaths of American bees, entomologist Walter S. Shepperd (Washington State University) has started taking bee semen from Georgia (especially the Megrelian variety) to create more variety in American bee populations (Corso 2013).

The Georgian honey bee is a Grand Medal winner in 1961 at the International Horticultural and Apicultural Exhibition in Erfurt (GDR). Two bee families collected 97.3 kg of honey leaving all other bees far behind. Georgian bees received a Grand Gold Medal in 1965 in Bucharest at the International exhibition sponsored by the World Congress on Apiculture.

Finally, they were awarded their third gold medal in 1971 at the 23rd International Congress on Apiculture in Moscow (Andguladze 1971; Tsitlidze *et al.* 2013; *beevenomlab.com/Georgian – bee*).

Figure 203. Example of transhumans bee keeping. Highmountain meadow, Kazbegi environs, N. Georgia, 1,950m a.s.l. (2014).

In the Caucasus, especially in Armenia and Georgia, traditional beekeeping is based on a vertical migratory following temperature variation, in which apiaries are moved from the lowland during winter to a higher altitude during the summer months, alternating between orchards and agricultural fields and high mountain's forests and meadows. Understanding this fact that each local ecotype of the Caucasian bees has been well adapted to its natural ecosystem, beekeepers extend the range of forage in much the same way as is done in transhumance. Experienced beekeepers are able to separate the type of honey by the area of pollination, and thereby yielding a product with species characteristics. In lowlands and foothills area the nectar-rich acacia – black locust (*Robinia pseudoacacia*) tree is one of the bee's favourite trees. Other most important honey plants are species of *Castanea, Rhododendron, Ribes, Tilia, Taraxacum*, and *Trifolium* (Kara *et al.* 2012) (Figure 203).

In the Caucasus with the modern wooden boxes and posts, an ancient tradition of apiary still remains. In terms of this tradition it should be pointed out that in Adjara (Western Georgia), especially some gorges of the Adjara-Imeretian range of the Lesser Caucasus, for example, Machakhela, Mtirala, Kintrishi, etc. Here the history and cultural past are well correlated with its luxuriant broadleaved forest and beautiful environment in general. For this remote rural area, it is characteristic developed farming and then apiculture as well. In the mentioned gorges of Adjara resembling many other remote, settled regions of the world, the indigenous people have devised different ways to exploit bees for their honey and other products. This range of bee practices can be categorized into four working definitions: honey hunting, beekeeping in natural woody vessels, beekeeping of a semi-wild nesting colony and modern type beekeeping.

Beekeeping in a hollow tree – a natural vessel was used till the domestication of honeybees. This way of honey yielding was called 'futi'. The next stage of making of the traditional hives was called 'geja' (or 'rogo'). This is a common name of the vessel familiar for husbandry (e.g., for cattle's drinking water). After domestication of honeybees from the long trunk it was easily made into a primitive hive. Fert (2001) considered 'that the tragic situation in beekeeping caused disappearance of the traditional hives called 'geja'.

Figure 204. 'Geja' hive from Adjara, Shuakhevi district, W. Georgia (1977).

These hives are mounted high in trees at a height of 15-20m (mainly in beech, *Fagus orientalis*). In addition, sheet metal is wrapped around the tree trunks to prevent bears for climbing to the hives, bear populations of which in Adjara are so dense that they can cause significant damage (Figure 204).

Russian apiarist T.D. Palagin (1988) suggests that unlike the common, standard beekeeping 'apiculture by the rules of a hollow trees' is getting closer to wild conditions.

All villages and Camili Basin of Artvin province (Turkey) were declared as an isolated area and the Georgian bee race was taken under conservation and development in the year 2008 within the context of Domestic Animals Gene Conservation Scheme (Kara *et al.* 2012).

As previously mentioned, in 2007 the Machakhela National Park was founded as part of the Western Lesser Caucasus corridor of Georgia. Apart from protection of the high endemism ration of this territory, currently the main focus is to support the local population of *Apis mellifera caucasica* and development of horizon beekeeping activities via transboundary cooperation with Kamili National Park in Turkey in the same upper stream of the Machakhela River.

So, people represented in one of the most humid area of Georgia and Turkey, may use their natural resource wealth in both–traditional and modern ways.

With the extraordinary way of apiculture, local customs and traditions have become associated with semi-wild beekeeping and this is an interesting field for ethnologists to study. Indeed, in this remote area rural tourism which was founded in very recent times should be based on viewing the best traditional semi-wild honey hunting spectacles. Harvesting of a bee colony nesting in a tree would be described almost as honey hunting.

Once prized, Georgia's unique species of bees have been largely forgotten. However, mentioned together with the crisis with honey bee populations in the United States and other countries, this could unexpectedly offer the Georgian race of bees a chance to return to past glory. American entomologists began to import genetic material from the Georgian bees (Corso 2013).

The Caucasus is rich in other types of related bees. For example, bumble bees, which are found mainly in northern temperate regions. Many species of the bee genus *Bombus* (Apidae) are restricted to the high mountains of the Caucasus at timberline and beyond (Battello *et al.* 2010). Butterflies and flies often replace bees at high altitudes on temperate mountains (Buchmann & Nabhan 1996).

- After disintegration of the FSU, the urban population frequently more than doubled. On the contrary, to avoid crucial problem to be out-migrated abroad, in some regions of the former independent Republics, the rural population increased. To some rural populations with lower incomes (compared to urban areas), the most urgent problem become to create competitive markets (regional, local level).
- After privatization of lands, various branches of agriculture, depending on climate and relief (soil) conditions, became fundamental. Among others, transboundary apiculture began to be an important component of development of the rural husbandry. Therefore, significantly through individual farming, developing their skill, apiarists managed produce honey and a number of other hive products rapidly.
- Among insects, which are usually pollination generalists (pollen bees and bumble bees butterflies and moths, flies and beetles) are of key antofile insects including honey bees. So, further development of transhumance beekeeping of the Caucasus is an additional vital pathway to the functioning highland's mainly non-woody plants biodiversity.

Conclusion

- The Caucasus is mainly a mountainous region. It is a central part of the Alpine-Himalayan orogenic belt. About 15% of local terrain is above the tree line. This is its main geographical feature.
- This region is well known for its main mountain system - the Greater Caucasus Range (GCR). This range in a southward direction is joined with other mountain system, the Lesser Caucasus via the Surami (Likhi) range. The western portion of the Lesser Caucasus converges with north - eastern Turkey (Anatolia). To the south of the Lesser Caucasus rises the Armeno-Javakhetian plateau, a small spur of the Armenian, Iranian and Turkish highland plateaus. In the south-eastern part, the Talysh Mountains extend, which are sharply delimited geographically and geologically from other highlands of the Caucasus (today the Aras River separates the Lesser Caucasus from the Talysh Mountains and, at the same time, from the other highlands of the region). Part of the Talysh mountains occurs in northwestern Iran.
- At high altitudes of mountains or 'sky islands' autochthonous living organisms, whether they are plants (essentially cryophytes) or cold tolerant animals, are predominant. Therefore, formation of an elevational gradient toward the permanent snow line may have provided the primary driver of striking biodiversity and endemism found on 'sky islands' in the Greater and Lesser Caucasus surrounded by Colchis and the Kura-Aras lowland's biota.

- The GCR, unlike the other highlands of the region, is a native of the Caucasus. It extends for about 1,200 km (between the Taman Peninsula, westwards to the Absheron Peninsula, eastward). It is a watershed between Eastern Europe and Western Asia.
- The tallest mountain of the Greater Caucasus (the tallest point of the region) is the Elbrus (5,642 m a.s.l.). Among other tall mountains of the Caucasus Isthmus reaching over 5,000m should be mentioned Dykh-Tau (5,205m), Shkhara (5,201m), Koshtan-Tau (5,151 m), Kazbek (5,047m).
- In the Caucasus, predominantly, within the borders of the GCR, all altitudinal belts (zones) common to the temperate climate zone of the North Hemisphere – subalpine, alpine, subnival and nival – are represented.
- High mountain biomes include all characteristic ecosystems and diverse habitats, which occur in temperate mountain systems. Significantly, such a complex spectrum of habitats is found in the Greater Caucasus. The timberline and tree line ecotones and all treeless high mountain belts are worthy of mention.
- In the Caucasus typical subalpine open woodlands (park-like forests) as a timberline ecotone may consist mostly of mesophitic red-bud subalpine maple (*Acer trautvetteri*) and hemi-xerophitic Persian (Caucasian) oak (*Quercus macranthera*). Both species may be single or in groups. Among needleleaf species the mountainous Caucasus hosts extensive stands of the Caucasian pine (*Pinus kochiana*), both closed canopy and open park-like stands at higher elevations.
- Subalpine crook-stem (krummholz) forests occur in the transition, tree line ecotone area, where local abiotic conditions dramatically influence tree growth, converting erect trees into krummholz shrubs. Therefore, krummholz ecotone lacks true tree life form. Shrubland in this ecotone is remarkable for its peculiar adaptations to local harsh environments that frequently are exposed to extreme climatic factors. Among other particularly important factors is low temperatures during the vegetation period, extreme frosts, heavy snow duration, high speed winds, radiation intensity, etc. Among biotic-driven factors, it's worthy to mention reproductive failure and competition with herbaceous vegetation.
- Among crook-stem forests, Caucasian beech (*Fagus orientalis*) is a very common species. At its upper limit straight-trunk trees are largely replaced by crook-stem tall shrubs and then, at elevations higher than 2,000m, gradually turn into ground-creeping forms. Four species of birches (*Betula*) of the Caucasus are typical expressions of the highland's crook-stem arboreal communities. *Betula medwedewii*, similar to *Quercus pontica* within beech forests, grows as a small crook-stem tree, although on the border of subalpine and alpine belts (Adjara, the Lesser Caucasus) it appears as a dwarf elfin population.
- Very characteristic formations at higher elevations of the Caucasus are the elfin (stunted) stands, predominantly composed of the Caucasian rhododendron (*Rhododendron caucasicum*) – dekiani. In the upper limits of distribution of arboreal shrubs, at 2,200-2,300m, this (sub-) endemic and relict species of genus *Rhododendron* forms impenetrable, exclusively primary formation.
- *Rh. caucasicum* is a typical, keystone orophilic (subalpine and alpine) plant species. In this climax dwarf plant communities *Rh. caucasicum* is the dominant shrub, where a subcanopy is composed of other, smaller ericoid species (*Vaccinium* spp.) and

ground-creeping junipers (*Juniperus* spp.). The Caucasian rhododendron is significant by its extraordinary features: in highland it is capable of creating its own phytocoenosis (245 species, 41,3% of which are endemics of the Caucasus). Scrubland composed of the Caucasian rhododendron is able to create its own microclimate and along with the other ericaceous and other shrubs, semi-shrubs and mosses in the GCR produce organic matter, forming peat.

- Above the tree line 69 species of polydominant tall-herb vegetation of which 52 are endemic to the Caucasus prefer habitats with high air humidity and high solar radiation. They are found in the subalpine belt of the Greater and Lesser Caucasus ranges (in the GCR elevation of the subalpine belt reaches its peak and ranges from 2,350 to upper limit 2,500m a.s.l.). Here should be mentioned species of the following genera: *Heracleum, Senecio, Aconitum, Delphinium, Inula*, etc. Unpalatable communities composed of *Veratrum*, also *Delphinium, Aconitum* and *Senecio* are dominating in intensively grazed places and are frequently represented as secondary primary vegetations.
- Among mountain systems of the Caucasus in elevations of 2,300-2,700 up to 2,900-3,700 m a.s.l., specific flora and vegetation are confined as alpine meadows. Here should be mentioned short grass, carpet-like meadows (nanoherbata), cushions, mats, and dwarfed low woody shrubs (e.g., Caucasian rhododendron, species of *Vaccinium*, etc.) and rock-scree vegetations. In addition, mentioned of the fact that different habitats of alpine tundra form continuous formations most notably in the GCR.
- Alpine carpets are polydominant, species-rich communities (with up to 200 herb species). Among leading species should be mentioned species of *Taraxacum, Primula, Gentiana, Ranunculus, Alchemilla, Campanula*, etc.
- Rock and scree vegetations are developing to the severe climatic and edaphic conditions. Characteristic plants include species of *Draba, Saxifraga, Sedum, Phleum, Dianthus* and others.
- In the subnival belt, in altitudes between 2,900-3,000 and 3,600-3,800 m in the GCR, habitats are represented by melting snow and immature highland soils. Here, in patches with plant-chionophobes (high frost resistance), they are adapted to micro-ecological niches. The typical genera of plants include: *Draba, Saxifraga, Primula, Ranunculus, Cerastium, Poa*, etc. The subnival belt is enriched with endemic and strictly endemic plant species and even genera (*Pseudobetckea, Pseudovesicaria*, *Symphyoloma* etc.). Here should be stated that according to our own observations the subnival biota should be considered as an impoverished type of alpine vegetation.
- Similar to the vegetation cover of the subalpine and alpine belts, the fauna of highland is frequently represented by stenobiont species, restricted with high altitude area and by a high ratio of endemism. To umbrella species belong wild stone goats of the GCR (*Capra caucasica, C. cylindricornis*), massive game birds Caucasian snowcock (*Tetraogallus caucasicus*), Caucasian grouse (*Tetrao mlokosiewiczi*) subendemics of the Caucasus, species of small mammals, vipers, many invertebrate animals.
- Upland rangeland of the Caucasus is a grazing area for wildlife and traditional transhumans pastoralism, therefore, highland has been settled since prehistoric times.
- In the Northern Hemisphere during the Pleistocene multi-glaciations, boreal biomes – taiga and tundra with characteristic living organisms are broadly distributed in lower

latitudes. Among other mountain systems, they extended within the Caucasus. During interglacial times, as glaciers retreated, they survived in the highlands of the Caucasus – alpine and subalpine zones. It should be mentioned there are some remnants of the hypercriophilic species of plants and animals (kwasiarctic group). Some species under the press of harsh abiotic factors were formed as nanophytes – dwarf plant forms (e.g., species of the genera *Vaccinium* and *Empetrum* can be found in high altitude belts of the Caucasus). Regarding the insects, among others, *Cacopsylla sibirica* was found in the highlands (north slopes of the GCR) of the region. Usually glacial relicts are forming an extraordinary arcto-alpine and boreo-alpine disjunction area. Examples of the arcto-alpine disjunction range were discovered for some plants and animals (above mentioned nanophites, insect), typical boreo-alpine disjunctive area is manifested by birds, *Loxia curvirostra* and *Pyrrula pyrrula.*

CONCLUSION

- The Caucasus, as a natural bridge between Europe and Asia, has an extraordinary crossroad geopolitical, cultural and biodiversity position.
- The Caucasus as a part of the Caucasian-Arabian belt belongs to the central portion of the huge Late Cenozoic Alpine-Himalayan orogenic belt. In consequence, its complex geomorphological structure - high mountain ranges, volcanic highlands and intermountain depressions, is a result of the collision of the Eurasian and the African-Arabian plates.
- The Caucasus region in the Late Proterozoic, Paleozoic, Mesozoic and the Early Cenozoic is located in the now transgressed Tethys Ocean (Prototethys, Paleotethys, Tethys, Paratethys), and its Eurasian and Gondwanian margins. Within the ocean-continent convergence zone, there existed a system of island chains. Among them was represented a tiny island – the future Great Caucasus Range – backbone mountain system of the region.
- In the Sarmatian Epoch (10-7 mya ago), according to plate (Afro-Arabian and Eurasian) movement the Paratethys regressed and the Caucasus Island (known from the Oligocene as Jephethis) joined with the part of southern mainland (nowadays Asia Interior and Asia Minor), when it was still a part of Gondwana and turned into its peninsula. Later (mid-Pliocene, about 3 mya), with regard to the same tectonic events, the Caucasus Peninsula located northward via the Stavropol elevation and has converged with the northern Euro-Siberian (former Laurasian) mainland and became an isthmus, that is the Caucasus Isthmus as we know this region at present (between the Black and the Caspian Seas).
- From the biogeographical point of view modern biota of the Caucasus consists of two major groups – allochthonous and autochthonous. The former one is represented by the migrant taxa of flora and fauna since the Paleogene-Neogene boundary; the latter one envelopes the local endemic forms of plants and animals. The allochthonous group is prevailing over the autochthonous in all major taxa of flora and fauna of the Caucasus.
- The allochthonous group includes four biogeographical units, from which two have a primary importance. The first major one encompasses temperate climate-tolerant (nemoral) and cold climate-tolerant (boreal) groups of living organisms. In past geological ages they entered the Caucasus from the northern, Boreal center. Firstly (the Oligocene-Miocene) as nemoral, later (from mid-Pliocene) as boreal proper

vegetations and associated animals. Depending on the climate fluctuations, a cold climate-tolerant complex of organisms in the Holocene, are still expanding ranges from their ice age refugia (e.g. *Castanea sativa*). The second biogeographical group involves species (genera) of the hot climate-tolerant, Irano-Turanian centre of origin. Their wave of expansion took place from adjacent to the Caucasus arid center, primarily from the Turan Hollow (south-eastern part of the Caspian Sea) and earlier (the mid-Miocene), it should be speculated, from Asia Interior (more precisely, nowadays Iran's territory).

- Xerophilic species prevail over humid climate-loving forms in the eastern Caucasus and on the contrary, mesophilic species obviously dominating in the western Caucasus (in spite of these trends, in mountainous areas these ecological differences gradually become negligible).
- The representatives of the third biogeographical group – the steppe dwellers – perform a more or less important role in the Caucasus biota. Finally, the last, Mediterranean proper group, today exists as a small population – relicts - which found their shelters in some micro-refugial areas (patches) around the coastal areas of the Black Sea.
- In accordance with the endemic (autochthonic) taxa, they first and foremost found protection in the highlands of the Greater Caucasus.
- Nowadays in the Caucasus relict and relict-endemic genera and species of flora and fauna of quite different origin are present. The following remnants of past geological epochs should be mentioned (chiefly, the Neogene): (1) moist tropical-subtropical (e.g. *Hymenophyllum tunbrigense*), (2) Mediterranean subtropical (maquis) (e.g. *Arbutus andrachne*), (3) initial temperate broadleaf deciduous forests (Arcto-Tertiary Geoflora) (e.g. *Quercus pontica*), (4) relicts of the Pleistocene multiple glaciations (e.g. *Vaccinium vitis-ideae*), and (5) relicts of deserts of the Irano-Turanian origin (e.g. *Nitraria schoberi*).
- Being restricted within refugial areas, these striking representatives of natural heritage of the Caucasus, especially relict and relict-endemics, are the key species to understand the past biotas of the Caucasus.
- The Caucasus with its highly topographic and climatic diversity and habitat heterogeneity, as it was stressed above, is famous as one of the world's richest biodiversity's in the temperate climate Eurasia, making it one of the 34 Hotspots of the Earth.

References

Abashidze, J. L. 1947. Zelkova in Georgia [Dzelkva v Gruzii]. *Soobch. AN Gr.SSR*, 7(1): 29-32 (in Russian).

Abashidze, J. L. 1952. On the fruiting of sweet chestnut [K plodonosheniyu kashtana posevnogo]. *Proceed. Forest. Inst. Ac. Sci.*, Tbilisi, 4:55-77 (in Russian).

Abashidze, J. L. 1959. *Dendrology [Dendrologia]. Vol.1.* Sas. Sam. Inst. Gamomc. [Publ. House of Agric. Inst.], Tbilisi, 348pp. (in Georgian).

Abashidze, J. 1962. *Dendrology on the basis of plant geography [Dendrologiya mtsenareta geografiis safudzvlebit].* Part II. *Pterocarya fraxinicolia* (Lam.). (pp34-35). Georgian Agricult. Inst. Publ., Tbilisi, 42pp. (in Georgian).

Abbasian, H., Kiabi, B. H., Kavousi, K. 2004. Food habits of wild goat, *Capra aegagrus aegagrus*, in the khorramdasht area, Kelardasht, Iran. *Zoology in the Middle East,* 33:119-124.

Abbott, L. B., Bettmann, G. T., Sterling, T.M. 2006. African rue (*Peganum harmala*) seedling response to herbicides applied under water-deficit–stress. *Weed Sci*, 54:198-204.

Abdieva, R. T. 2005. The mosaics of vegetation of solontchak deserts in Gobustan [Mozaichnost' rastitel'nosti solonchakovykh pustyn' v Gobustane]. *Nat. Acad. Sci. Azerb. Inst. Bot.*, Baku: 1-32 (in Russian).

Abramova, T. A. 1972. *The reconstruction of paleogeographic conditions of Quarternary's transgression and regression of the Caspian Sea (according to palaeobotanical research) [Rekonstruktciya paleogeograficheskikh uslovii Chetvertichnoi transgresii i regresii Kaspiiskogo moriya (na osnove paleobotanicheskikh issledovanii]* (*PhD thesis*) *geogr. sci.* Moscow (1-24pp.) (in Russian).

Abuladze, A. 2013. *Birds of prey of Georgia. Materials towards fauna of Georgia [Khischnye pticy Gruzii. Materialy k faune Gruzii]. issue 6.* Inst. of Zoology, Ilia State University. Lasha Khvichia Publ. Tbilisi, 118p.

Abutalibov, M. 1976. *Vegetation of Azerbaijan [Rastitel'nost' Azerbaijana].* Tshygi, 256p. Baku (in Russian).

Adamia, Sh., Akhvlediani, K. T., Kilasonia, V. M., Nairn, A. E. M., Papava, D., Patton, D. K. 1992. Geology of the Republic of Georgia: a review: *Inst. Geology Review*, 34,5: 447-476.

Adamia, Sh., Alania, V., Chabukiani, A., Chichua, G., Enukidze, O. Sadradze, N. 2010. Evolution of the Late Cenozoic basins of Georgia (SW Caucasus): a review. In: Sosson, M., Kaymakci, N., Stephenson, R. A., Bergerat, F., Starostenko, V. (Eds.). *Sedimentary*

Basin Tectonics from the Black Sea and Caucasus to the Arabian Platform. Geol. Soc., London, Spec. Publ. 340, 239–259.

Adamia, Sh., Alania, V., Chabukiani, A., Kutelia, Z., Sadradze, N. 2011a. Great Caucasus (Kavkasioni): a long-lived North-Tethyan Back-Ark basin. *Turkish Journal of Earth Sciences*, 20:611-628.

Adamia, Sh. A., Chabukiani, A., Kakabadze, M., Kotetishvili, E., Lordkipanidze, M., Tabidze, D. 1991. Geological background of the Caucasus and adjacent area. Geological events of the Cretaceous-Paleogene boundary. In: *Proceed. of Intern. Sympos.*, Tbilisi, 12: 18-5.

Adamia, Sh. A., Chkhotua, T., Kekelia, M., Lordkipanidze, M. B., Shavishvili, I., Zakariadze, G. 1981. Tectonics of the Caucasus and adjoining regions– implications for the evolution of the Tethys Ocean. *J. Struct. Geology.* 3 (4):437-447.

Adamia, Sh., Zakariadze, G., Chkhotua, T., et al. 2011b. Geology of the Caucasus: a review. *Turkish Journal of Earth Sciences*, 20, 489-544.

Adamovich, L. 1899. The Mediterranean elements of the Serbian flora [Die Mediterranen elemente der serbischen Flora]. *Engl. Bot. Jahrb., Bd.* 27, Leipzig, 351-389pp.

Adamovich, L. 1905. The development of the Balkan flora since the Tertiary period [Die Entwicklung der Balkanflora seit der Tertiärzeit]. *Engl. Bot. Jahrb., Bd.* 36, Beibl 81.

Adamovich, L. 1909. The vegetation conditions of the Balkan countries [Die Vegetationsverhältnisse der Balkanländer]. *Vegetation der Erde*. Leipzig, 1909, Bd. 11.

Adams, K. 2013. Genomics. Genomic clues to the ancestral flowering plant. *Science,* 342, 6165:1456-1457.

Adams, R.P. 2014. Junipers of the world: the genus *Juniperus. 4th edition* Trafford Publ. North America & international, 413pp.

Adem, C., Arancli, S., Zeydanli, U. 2007. Camili Biosphere Reserve, Turkey. *GEM-CON-BIO Case Study Report*, 64pp.

Adveev, B., Niemi, N.A. 2011. Rapid Pliocene exhumation of the central Greater Caucasus constrained by low-temperature thermochronomentry. *Tectonics*, 30, TC 2009, doi:10.1029/2010TC00280.

Adzinba, Z.I. 2000. Calcareous endemism of Colchic flora. In: *Mater. Sci. Session dedicated to 90th iubil. Date A.A. Kolakovski*. Sukhum:28-38.

Afanasjeva, I. N., Vasetskii, S. G., Vasina-Popova, E. T. et al. 1979. English-Russian biological dictionary. *Russian Language Publishers*, Moscow, 732pp.

Afkhazava, I. S. 1969. Lakes – the beauty of our nature [Tbebi-Čveni bunebis silamaze]. *Sabchota Saqartvelo*, 167pp. (in Georgian).

Agadjanov, S. D. 1962. The solonchak deserts of the coastal sands of Azerbaijan [Solontchakovye pustiny pribrezhnykh peskov Azerbaijhana]. In: *Sci. Notes Az. State Univ. Ser. Biolog.* 2: 16-221, Baku (in Russian).

Agadjanov, S. D. 1967. Flora and vegetation of the sea coastal sands of Azerbaijan and its importance for fixing and development of sands. *PhD thesis biol. sci.*, Baku: 1-32 (in Russian).

Agaguliev, I. (Ed.). 2000. Flora and vegetation of south-east Shirvan [Flora i rastitel'nost' jugo-vostochnogo Shirvana]. *Baku State Univ.*, Baku, 147pp. (in Russian).

Aghababyan, K., Khanamiryan, G., Ghukasyan, E., Badalyan, N. 2015. Abundance of narrow-clawed crayfishes (*Astacus leptodactylus* Esch., 1823) and its trends in Lake Sevan, Armenia. *J. Aquacult. Engin. Fisher.* Res., 1,3:116-124.

Aghasyan, L. A. 2014. Contemporary status and perspectives of the Herpetofauna of 'Goravan Sands' Sanctuary. *Proc. Int. Conf. 'Biol. Divers. and Conserv. Problems of the fauna of the Caucasus -2'*. 16-19pp. Sept. 23-26, 2014. Yerevan, Armenia (in Russian).

Agladze, G. D., Zotov A. A., 1987. *Mountainous rangelands and haylands of the Caucasus [Gornye pastbyscha i senokosnye ugodiya Kavkaza].* Soviet Georgia Publ. Tbilisi, 198pp. (in Russian).

Agusti, J., Lordkiphanidze, D. 2011. How 'African' was the early human dispersal out of Africa? *Quat. Sci. Revi.*, 30,11-12:1338-1342.

Ahmadi, A., Kavosi, M. R., Soltanloo, H. 2014. *Zelkova carpinifolia* reservoir from Hyrcanian forests, Northern Iran, a new sacrifice of *Ophiostoma novo-ulmi*. *Biodiversitas*, 15,1:48-52.

Akatov, V. V. 1999. The insular effect as a factor of high-mountain phytocoenosis formation in the Western Caucasus [Ostrovnoi effect kak factor formirovaniya vysokogornykh fitocenozov Zapadnogo Kavkaza]. *MGTI*, Maikop: 33-42 (in Russian).

Akatov, V. V., Akatova, T. B., Bebin, A. R. et al. 2009. Natural complexes of the Imeretian Lowland: bioecological diversity; zoological importance and recommendation on the protection [Prirodnye kompleksy Imeretinskoi nizmennosti: biologicheskoe raznoobrazije, sozologicheskaiya znachimost', rekomendacii po sokhroneniyu]. Kopi-Print Publ. Sochi-Maikop, 93pp. (in Russian).

Akhalkatsi, M. 2015. *Forest habitat restoration in Georgia, Caucasus Ecoregion,* Orchis Publ. Tbilisi, 101pp.

Akhalkatsi, M., Kimeridze, M. 2012. Implementation of the classification system of forest habitats in accordance with the 'Natura2 000' standarts in the Georgian Legislation. *In: 'Legal Aspects of European Forest Sustainable Development'. Proceed. of the 12th Intern. Symp.,* Cypros (Šulek et al. - Eds.).

Akhalkatsi, M., Tarkhnishvili, D. 2012. Habitats of Georgia. *Natura*, 2000.118pp.

Akhani, H. 2006. Biodiversity of halophytic and sabkha ecosystems of Iran:71-88pp. In: *Ajmal Khan, M, Boer, B., Kust, G. S., Banth, H.-J. (Eds.), Sabakha ecosystems.* vol. II: West and Central Asia. Springer, 259pp.

Akhani, H., Djamali, M., Ghorbanalizadeh, A., Ramezani, E. 2010. Plant biodiversity of Hyrcanian relict forests, N.Iran: an overview of the flora, vegetation, palaeocology and conservation. *Pak. J. Bot. Special Issue*, 42:231-258.

Akhani, H., Ghorbanli, M. 1993. A contribution to the halophytic vegetation and flora of Iran. In: Lieth, H. & Al Masoom, A. (Eds.). *Towards the rational use of high salinity tolerant plants*, 1: 35 - 44. Kluwer Academic Publishers. Dordrecht.

Akhmedov, E. G., Yarovenko, Y. A., Nasrullaev, N. I., Babaev, E. A., Akhmedov, S. G. 2009. Conservation of the Bezoar goat in the Eastern Caucasus. 25-31pp. In: Zazanashvili, N. & Mallon (Eds.). *Status and Protection of Globally Threatened Species in the Caucasus.* Tbilisi: CEPF, WWF, Contour Ltd.

Akhmedova, S. Z. 2004. Biodiversity of the ecosystem of the vegetation cover of Djeiranchel and Adjinauri regions of Azerbaijan [Bioraznoobrazie ekosistem rastitel'nogo pokrova Djeiranchel i Adjinaurskogo regionov Azerbaijana]. *Elm*, Baku, 236pp. (in Russian).

Akhmedova, S. Z. 2005. The origin of deserts' phytocoenoses of Djeiranchel-Adjinauri regions [Proiskhozhdenie pustynnykh fitocenozov Djeirantchel-Adjinaurskogo regionov]. *Izvestia NAN Azerbaijana* [*Herald of Nat. Ac. Sci. of Azerb.*], 3-4:81-85, Baku (in Russian).

Akhmetiev, M. A. 2010. Paleocene and Eocene floristic and climatic change in Russia and Northern Kazakhstan. *Bull. Geosci.*, 81,1:77-94.

Akhmetiev, M. A., Beniamovski, V. N., 2009. Paleogene floral assemblages around epicontinental seas and straits in Northern Central Eurasia: proxies for climatic and palaeogeographic evolution. *Geologica Acta* 7 (1–2), 297–309.

Akhundov, G. F., Gogina, E. E., Prilipko, L. I. 1978. Narrowly endemic and rare species of natural flora of Nakhichevan ASSR [Uzkoendemichnye i redkie vidy prirodnoii flory Nakhichevanskoi ASSR]. In: *Bull. Glavn. Bot. Sada*, 107:54-62 (in Russian).

Akjigitova, N. I. 1973. The halophite vegetation – Halophyta [Galofitnye rastenija-Halopyta]. In: *Vegetation cover of Uzbekistan,* 2:211-302. Tashkent (in Russian).

Akjigitova, N. I. 1982. *Halophyte vegetation of Central Asia and its indicative features [Galofitnye rasteniya Tcentral'noi Azii i ikh indikatornye osobennosti]*. Tashkent, FAN Publ. [Publ. House of Ac. Sci Uzbekist], Tashkent, 192pp. (in Russian).

Akopian, J. A. 2007. On the *Pyrus* L. (Rosaceae) species in Armenia. In: Flora, vegetation and plant resources of Armenia [O vidakh roda *Pyrus* L. (Rosaceae) v Armenii] (Gabrielian, E. Ts. et al. – Eds.) (*Inst. Bot. Ac. Sci. Rep. Arm., Arm. Bot. Soc.*). Yerevan, 16:15-26 (in Russian).

Akopian, J. A. 2011. Genus *Salsola* L. sensu lato (Chenopodiaceae) in South Transcaucasia. In: *Takhtajania,* 2011, 1:124-132 (in Russian with English abstract).

Akopian, J. A. 2015. Notes on *Argyromalon* Fed. Section species from the genus *Pyrus* L. of Armenian flora (81-87pp.). In: Vardanyan, Zh. H. (Ed.). Bot. Sci. in the Modern World. *Proc. Int. Conf. dedicated to the 80th anniversary years Bot. Gard.* (Yerevan, 5-9 Oct., 2015), 440pp.

Akymushkin, I. 1973. *The World of animals. Narrative about birds [Mir zhivotnykh. Rasskazy o pticakh]*. Molodaiya Gvardiya Publ., Moscow, 379pp. (in Russian).

Alaback, P. B. 1991. Comparative ecology of temperate rainforests of the Americas along analogues climatic gradients. *Revista Chilena de Historia Natural*, 64:399-412.

Aladin, N., Plotnikov, I. 2004. The Caspian Sea. Lake Basin Management Initiative*, Thematic Paper*: 1-29pp.

Al'bov, N. M. 1899. On the abandoned sites of the Caucasus [K zabroshennykh raijonov Kavkaza] *Zemlevedenie*, I-II:71-84 (in Russian).

Aleksanyan, A. S. 2011a. Biomorphological structure of the flora of arid open woodlands of South Armenia [Biomorphologisheskaiya struktura flory aridnykh redkolesii Juzhnoi Armenii]. *Biol. Zhurn. Armenii*, 2(63): 30-36 (in Russian).

Aleksanyan, A. S. 2011b. The phytogeographical analysis of the flora of the arid woodlands of southern Armenia [Fitogeographicheski analiz flory aridnykh redkolesii Juzhnoi Armenii]. *Biol. Zhurn. Armenii*, 4(63): 83-91 (in Russian).

Aliyev, G. A., Volobuev, V. P., Salaev, M. M. et al. 1953. *Soils of Azerbaijan [Pochvy Azerbaidjana]*. Izd. Acad. Sci. Azer.SSR, 598 pp. (in Russian).

Aliev, R. A., 1954. The salsoletum semi-deserts of Azerbaijan [Sarsazanovye polypustyny Azerbaijana]. *Elm*, Baku, 128 (in Russian).

Aliev, R. A. 1971. Flora of Artemisietum of Azerbaijan. *Uchiyon. Zapiski ATU im. S. M. Kirova. Ser. Biol. Nauk* [*Sci. Papers ATU after S. M. Kirov, ser. Biol. Sc.*]. 14-19 (in Russian).

Aliev, R. A., Gadjiev, V. D., Isaev, J. M., Mailov, A. I., Nabili, D. G., Prilipko, L. I. 1965. *Improvement and rational use of Azerbaijan's winter and summer pastures [Ulucshenie i ratciolal'noe ispol'zovanie zimnykh i letnikh pastbich Azerbaijana].* Izd. AN Az.SSR, Baku: 1-45 (in Russian).

Aliev, S. Y., Gadjiev, E. M. 1986. Materials regarding morphological investigations of the underground part of some vegetation-edificators and plant communities of the Shirvan area [Materialy k morfologicheskomu izucheniju podzemnoi chasti nekotorykh rastenii edifikatorov i rastitel'nykh soobchestv Shirvanskogo raiona]. *Izvest. Akad. Nauk., Ser. Biologich. Nauk*, 3: 19-27 (in Russian).

Alpaidze, V. S., Shengelija, F. K. 1989. Geology. The Colchis Lowland–natural conditions and socio-economic aspects [Kolkhidskaiya nizmennost' – prirodnye usloviya i social'no-ekonomicheskie aspekty] In: Svanidze, G. G. (Ed.). *Gidrometeoizdat*: 9-44. Leningrad (in Russian).

Al-Saghir, M. G., 2006. Phylogenetic analysis of the genus *Pistacia* (Anacardiaceae). *Dissert. Doctor of philosophy in Biol. Sci.* Blacksburg, Virginia 112pp.

Al-Saghir, M. G., Porter, D. M. 2012. Taxonomic revision of the genus *Pistacia* L. (Anacardiaceae) *Amer. J. Pl. Sci.* 3:19.

Al-Saghir, M. G., Porter, D. M., Nilsen, E. T. 2006. Leaf anatomy in *Pistacia* species (Anacardiaceae). *J. Biol. Sci.*, 6: 242-244.

Altman, N. 1994. Sacred trees. *Sierra Club Books*, San Francisco, 120pp.

Amiryan, A. 1999. Fauna of staphylid beetles (Coleoptera, Staphylinidae) in Goravan Sands. In: '*Izvestia 2-go Vsesojuznogo Kongressa po Pochven. Zoologii*' 'Bioraznoobrazie i Zhizn' Pochvenych Sistems' (*Proceedings of the 2nd All-Russian Congr. Soil Zool. 'Biodiversity and Life of the Soil System'*), Moscow, p. 18 (in Russian).

Anderson, R. C. 2006. Evolution and origin of the Central Grassland of North America: climate, fire, and mammalian grazers. *J. Torrey Bot. Soc.,* 133, 4: 626-647.

Anderson, R. C. 2012. Ecology and management of the prairie division. *Cumulative Watershed Effects of Fuel Management in the Eastern United States*. Chapt. 9:175-201.

Andguladze, D. L. 1971. *Apiculture en Géorgie [Beekeeping in Georgia].* Ministère de l'Agriculture RSS de Géorgie. 185 pp.

Apakidze, A. M. 1968. *Towns of ancient Georgia [Goroda drevnei Gruzii].* Metsnieraba, Tbilisi, 296pp. (in Russian).

Arabuli, G., Kvavadze, E., Kikodze, D., Arabuli, T. 2008. The krummholz beech wood of Mt Tavkvetili Javakheti Plateau, (Southern Georgia) – a relict ecosystem. *Proc. Inst. Zool.*, Tbilisi, 23:194-213.

Arakelyan, M. C. 2012. Microevolutionary processes in sympatric populations of some reptiles from the Republic of Armenia and adjacent territories. *Doctoral dissert. (PhD thesis) biol. sci.*, Yerevan, 47pp. (in Russian with English summary).

Archibald, O. W. 1995. *Ecology of world vegetation.* Champan & Hall, London, 510pp.

Arnol'di, K. V. 1957. In relation to the theory of species – an area with respect to ecology and origination of species population [O teorii areala v sviyazi s ekologii i proiskhozhdeniem vidovykh populacii]. *Zool. Zhurn.*, 36,2:1609-1626 (in Russian).

Arnol'di, K. V. 1968. The zonal, zoogeographical and ecological peculiarities of the mirmecofauna and the settling of the plains of Russia by ants [Zonal'nye, zoogeograficheskie i ecologicheskie osobennosti mirmekofauny i naseleniya muraj'ev Russkoi ravniny]. *Zool. Zhurn.*, 7,8: 1155-1178 (in Russian).

Aronson, J. A. 1989. *Halophyte: a data of salt tolerant plants of the world. Office of Arid Lands Studies.* The Tucson, Arizona: University of Arizona Press. 114pp.

Arsdale, A. P., Lordkipanidze, D. 2012. A quantitative assessment of mandibular variation in the Dmanisi Hominins. *Paleoanthropology*: 134-144.

Asadova, K. K. 2008. Features of flora and vegetation in conditions of winter pastures [Osobennosti flory i rastitel'nosti v usloviyakh zimnykh pastbysch]. *Nauchnye Vedomosti Belgorodskogo Gos. Univ-ta, Ser. Estestvennye nauki*, 3(43) 70-76 (in Russian).

Asadova, K. K. 2013. Dynamic of fodder resources of the most important fodder plants of the winter pastures of north-western part of Azerbaidjan. *Sustainable Agriculture Research*, 2,2:29-36.

Asanidze, Z., Akhalkatsi, M., Gvritishvili, M. 2011. Comparative morphometric study and relationships between the Caucasian species of wild pear (*Pyrus* spp.) and local cultivars in Georgia. *Flora*, 206:974-986.

Asanidze, Z., Akhalkatsi, M., Henk, A. D. 2014. Genetic relationships between wild progenitor pear (*Pyrus* L.) species and local cultivars native to Georgia, South Caucasus. *Flora* 209:504-512.

Ashburner, K., McAllister, H. A. 2013. *The genus Betula: a taxonomic revision of birches. Ser.: A Botanical Magazine Monograph*. Royal Botanic Gardens, Kew Publishing, 448pp.

Askin, R., A., Spicer, R. A. 1995. The Late Cretaceous and Cenozoic history of vegetation and climate at northern and southern high latitudes: a comparison. In: *Effects of Past Global Change on Life. National Research Counsil (US) Panel on Effects of past Global Change on Life.* Washington (DC): National Academies Press (USA): 156-173.

Aslanyan, A. V., Petrosyan, R. P., Arakelyan, M. S. 2014. About an extinction alarm situation for two species of lizards in the Ararat Valley of Armenia [K ugrozu ischeznovaniyu dvux vidov iacheritis Araratskoi doliny Armenii]. *Proc. Int. Conf. 'Biol. Divers. Conserv. Problems of the fauna of the Caucasus'*. 21 Sept., 23-26, 2014, Yerevan, Armenia (in Russian).

Atamov, V. V. 2002. *The steppe vegetation of Azerbaijan [Stepnye rasteniya Azerbaidjana*]. Elm. Baku, 208pp. (in Russian).

Atamov, V. V. 2008. Phytosociological characteristics the vegetation of Caspian's shores in Azerbaijan. *Intern. J. of Bot.*, 4(1): 1-13.

Atamov, V. V., Gabbarov, K., Gurbanov, E. 2006. The phytosociological characteristics of ecosystems of the mountains of Talysh region, Azerbaijan. *Asian J. Plant Sci.*, 5 (5):899-904.

Atli, H. S., Arpaci, S., Kaska, N., Ayanoglu, H.1998. Wild *Pistacia* species in Turkey: 35-40pp. In: S. Padulosi and Hadj-Hassan (Eds.) *Report of the IPGRI Workshop,* 14-17 December 1998, Irbid, Jordan. 105pp.

Avanesova, A. A. 2006. The successions of steppes phytocoenosis of European forest-steppe: from the example of V. V. Alekhin Central-Chernozem Biosphere Reserve after M. M. Aliyokhin [Sukksessii stepniykh fitocenosov Evropeiskikh lesostepi: na primere…]. *PhD thesis biol. sci*. Voronezh, 22pp. (in Russian).

Awramik, S. M. 1992. The oldest records of photosynthesis. *Photosyn. Res.*, 33:75-89.

Axelrod, D. I. 1965, A method of determining the altitudes of Tertiary floras: *The Paleobotanist*, 14:144-171.

Axelrod, D. I. 1966. Origin of deciduous and evergreen habitats in temperate forests. *Evolution*, 20:1-I5.

Axelrod, D. I. 1968, Tertiary floras and topographic history of the Snake River Basin, Idaho: *Geol. Soc. America Bull.*, 79: 713-734.

Axelrod, D. I. 1985. Rise of the grassland biome, central North America. *Bot. Rev.* 51:163–201.

Aydin, Z. U., Dönmez, A. A. 2015. Taxonomic and nomenclatural contributions to *Pyrus* L. (Rosaceae) from Turkey. *Turk. J. Bot.*, 39:841-849.

Azimov, V. A. 2005. The complex analysis of biodiversity of the flora of arid, open woodlands of the coastal and foothill area of Dagestan [Kompleksny analiz bioraznoobraziya flory aridnykh redkolesii pribrezhnykh i predgornykh regionov Dagestana] (115pp.). *PhD thesis boil. sci.* Makhatchkala, 1-24 (in Russian).

Azimov, V. A., Teymurov, A. A. 2003. Ecological-coenological analysis of flora of the arid, open woodland of Nagorny (Inter-Mountain) Dagestan [Ecologo-tsenologicheski analiz flory aridnogo redkoles'iya Nagornogo Dagestana]. *Regional'nye Problemy Geografii i Geologii [Regional Problems of Geogr. & Geolog.]:* 2:80-86, Makhatchkala (in Russian).

Babaev, A. G., Zonn, I. S., Drozdov, N. N., Freikin, Z. G. 1986. *Deserts, the nature of the world [Pustyni, priroda mira]*. Moscow, Mysl, 320 pp. (In Russian).

Babayev, M. P., Gerayzade, A. P. 2010. Problems of degradation of soils of dry subtropics of the southern Caucasus. Institute of Soil Science and Agrochemistry, Baku, Azerbaijan. *Proceedings of the 19th World Congress of Soil Science: Soil solutions for a changing world,* Brisbane, Australia, 1-6 August *2010. Symposium 1.3.2 Geochronological techniques and soil formation*, 2010:16-18.

Baeva, V. G. 1985. Psyllids or jumping plant-lice (Homoptera, Psylloidea). In: *Psyllids or jumping plant-lice. Fauna of Tadjik. SSR,* Vol. 8. Acad. Sci. Tadjik. SSR. Donish Publ., Duschanbe (in Russian), 332pp.

Bagdasaryan, G. P., Karapetyan, K. I., Gukasyan, R. Kh. 1973. On the stratigraphical position and age of Manichar lavae [Stratigrafichescheskaija pozitsia i vozrast' Manycharskikh lav]. *Bull. AN ArmSSR, issue Nauki o Zemle*, 3: 49-51 (in Russian).

Bakhshiev, V. 2009. Saltwort small-bush deserts of the central part of the Shirvan Plain [Soliyankovye melkokustarnikovye pustyni tsentral'noi chasti shirvanskoi ravniny]. *Natural Sciences*, Baku State Univ.,113-117 (in Russian).

Bakker, R. T. 1986. *The dinosaur heresies: new theories unlocking the mystery of the dinosaurs and their extinction.* Eilliam Morrow and Company, New York, 481pp.

Ballantyne, A. P., Alden, C. B., Miller, J. B., Tans, P. P., White, J. W. C. 2012. Increase in observed net carbon dioxide uptake by land and oceans during the last 50 years. *Nature,* 488:70-72.

Balyan, G. A. 1972. *Prostrate summer cypress (Kochia prostrate) and its culture in Kirghizia.* Frunze, USSR: Kyrgyzstan Publ. (Translated from Russian; U. S. Department of Commerce, TT77 -59026, 1979). 296pp.

Bamshad, M. J., Olson, S. 2003. Does race exist? *Scientific American*: 289,6: 78-85.

Banks, H. P. 1975. Reclassification of Psilophyta. *Taxon*, 24: 401–413.

Bannikov, A. G. (Ed.). 1969. *Natural Preserves (zapovednyks) of the Soviet Union [Prirodnye zapovednyki Sovetskogo Soyuza]*. Kolos Publ. Moscow, 552pp. (in Russian).

Bannikov, A. G., Golgofskaja, K.Yu., Kotov, B. A. 1969. The Caucasus [Kavkasus] (337-403) In: *Natural reserves (Zapovedniks in the Soviet Union (Bannikov, A. G. (Ed.),* Kolos, Moscow, 459pp. (in Russian).

Barnes, B. V., Zak, D. R., Denton, S. R., Spurr, S. H. 1998. *Forest Ecology.* John Wiley & Sons, New York.

Barseghian, A. M. 1966. Saline and solonchak vegetation of the Ararat plain. Flora and vegetation of Armenia [Solontsevye i solonchakovye rasteniya Araratskoi doliny. Flora i rastitelnost' Armenii]. *Trudy Bot. Inst. AN Arm.SSR* [*Proceed. Bot. Inst. AN Arm.SSR*], 15:6-65, Yerevan (in Russian).

Bashkirov, I. S. 1939. Caucasian bison [kavkazskii zubr]. In: *Caucasian European bison (Moscow: Central Board for Reserves, Forest Parks and Zoological Gardens, Council of the People's Commissars of the RSFSR*): 1–72. (In Russian).

Basinger, J. F., Greenwood, D. R., Sweda, T. 1994. Early Tertarian vegetation of arctic Canada and its relevance to paleoclimatic interpretation. In: *NATO ASI Series, Vol. 127. Cenozoic Plants and Climates of the Arctic (Boulter, M.C.& Fisher, H. C. –Eds.). Springer-Verlag* Berlin Heidelberg, 1994: 175-198p.

Batello, C., Azzu, N. 2010. Gardens of Biodiversity: Conservation of genetic resources and their use in traditional food production systems by small farmers of the Southern Caucasus. Rome, *FAO*, 363pp.

Baudena, M., Dekker, S. C., Van Bodegom van, P. M. et al. 2015. Forests, savannas, and grasslands: bridging the knowledge gap between ecology and dynamic global vegetation models. *Biogeosciences*, 12:1833-1848.

Baum, B. D. 2006. *The rise and fall of the Caucasian race: a political history of racial identity.* New York University Press, New York, 342pp.

Baum, B. R. 1978. *The genus Tamarix*. Israel Academy of Sciences and Humanities, Jerusalem, 209 pp.

Beaman, M. 1994. *Palaearctic Birds: a cheklist of the birds of Europe, North Africa and Asia north of the foothills of the Himalayas.* Harrier Publications. Stony hurst, UK.168 pp.

Beard, J. B. 2012. Origin, biogeographical migrations and diversifications of turfgrasses. *Research Report*. SR,132:1-26.

Bebia, S. M. 1989. Relict forest of chestnut [Reliktovye lesa kashtana]. *Problemy Ekologii Abkhazii*, [*Problems of Ecology of Abkhazia*], 4: 77-82 (in Russian).

Beerling, D. J. 2002. Low atmospheric CO_2 levels during the Permo-Carboniferous glaciation inferred from fossil lycopsids. *PNAS,* 99, 20:12567-12571.

Beerling, D. J. 2006. Leaf evolution: gases, genes and geochemistry. *Ann.Bot.,* 93,3:345-352.

Behrensmeyer, A. K., Damuth, J. D., Dimichele, W. A. et al. 1992. *Terrestrial ecosystems through time: evolutionary palaeoecology of terrestrial plants and animals.* Chicago: University of Chicago Press. Chicago, IL, 551pp.

Bei-Bienko, G.Ya. 1966. Habitat changes by terrestrial organisms as a biological principle [Smena mestoobitanii nazemnykh origanizmami kak biologicheskii printsip]. *Zhurn. Obch. Biol.,* 27,1:5-21 (in Russian).

Belcher, C. M., McElwain, J. C. 2008. Limits for combustion in low O_2 redefine paleoatmospheric predicitons for the Mesozoic. *Science,* 321:1197-1200.

Belcher, C. M., Yearsley, J. M., Hadden, R. M., McElwain, J. C., Rein, G. 2010. Base-line intrinsic flammability of Earth's ecosystems estimated from paleoatmospheric oxigen over the past 350 million years. *Proc. Nat. Acad. Sci.* USA, 107,52:22448-22453.

Bell, R. L., Itai, A. 2011. *Pyrus*. In: *Kole, C. (Ed.), Wild crop relatives: genomic and breeding resources temperate fruits.* Springer-Verlag, Berlin Heidelberg: 147-179.

Belousova, I. P. 2000. Use of methods of genealogical analysis for preservation of the gene pool of European wisent and Przhevalski's horse [Ispol'zovanie metodov genealogitcheskogo analiza dliya sokhraneniya genofonda evropejskogo zubra i loshadi Przheval'skogo*]. PhD thesis biol. sci*, 24pp. S.-Petersburg (in Russian).

Bendrey, R. 2011. Some like it hot: environmental determinism and the pastoral economies of the later prehistoric Eurasian steppe. *Pastoralism*: *Research, policy and practice*, 1,1-16.

Berg, L. S. 1947. *Geographic zones of the Soviet Union [Geograficheskye zony Sovetskogo Soyuza]. Vol. 1*, Geographgiz (State Publ. Geogr. Lit.), Moscow, 397pp. (in Russian).

Berg L. S., 1952. *Geographic zones of the Soviet Union [Geographicheskie zony Sovetskogo Soyuza]. Vol. 2.* Geographgiz (State Publ. Geogr. Lit.), Moscow, 510 pp. (In Russian.).

Beridze, T., Pipia, I., Beck, J. et al. 2011. Plastid DNA sequence diversity in a worldwide set of grapevine cultivars (*Vitis vinifera* L. subsp. *vinifera*), *Bull. Geogr. Nat. Ac. Sci.* 15,1:98-103.

Berna, F., Goldberg, P., Horwitz, L. K. et al. 2012. Microstratigraphic evidence of *in situ* fire in the Acheulean strata of Wonderwerk Cave, Northern Cape Province, South Africa. *PNAS*:1215-1220.

Bernbaum E., 1998. *Sacred mountains of the world.* Berkeley-Los Angeles-London: University of California Press., 291pp.

Berner, R. A., Beerling, D. J., Dudley, R, Robinson, J. M., Wildman, R. A. 2003. Phanerozoic atmospheric oxygen. *Ann. Rev. Earth Planet Sci.* 31:105–134.

Bikov, B. A. 1981. The main peculiarities of halophyte vegetation of Middle Asia and Kazakhstan [Osnovnye osobennosti galofitnykh rastenii Srednei Azii i Kazakhstana]. *Bull. AN Kaz. SSR, Ser. Biol.*,1:1-19 (in Russian).

Billwitz, K. 1997. The pedogenetically undeveloped soil like the formation 'Takyr' [Lehrbuch der Allgemeinen Physischen Geographie]. *Gotha (Justus Perthes Verlag)*, 234-327 (in German).

Bisaz, A., Escher, F., Grosjean, M., Ives, J. D., Messerli, B., Price, M. F. 1997. Mountains of the world: challenges for the twenty-first century. A contribution to Chapter 13, *Agenda 21. Mountain Agenda*. Bern, Paul Haupt.

Björn, L. O., Govindjee, 2009. The evolution of photosynthesis and chloroplasts. *Current Science,* 96, 11:1466-1473.

Blair, J., Nippert, J., Briggs, J. 2014. Grassland ecology, pp. 389–423. In: Monson, R. K. (Ed.). *Ecology and the Environment*. Springer Science+ Business Media, New York., 659pp.

Blondel, L., Aronson, J. 1999. *Biology and wildlife of the Mediterranean Region.* Oxford University Press, Oxford, 328 pp.

Blumenbach, J. F. 1795 (3rded.). *On the natural varieties of mankind [De generis humani varietate nativa]. Edit.3*. Gottingen http://dx.doi.org/10.5962/bhl.title. 35972. 35972.

Blumenbach, J. F. 1865. *The anthropological treatises of Johann Friedrich Blumenbach.* London: Publ. for the Anthtopological Society, by Longman, Green, Longman Roberts & Green, 1865, 382pp.

Bobe, R. 2006. The evolution of arid ecosystems in eastern Africa. *Journal of Arid Environments,* 66,3: 564-584.

Bobek, H.1951. The natural forests and woodlands of Iran]Die natürlichen Wälder und Gehölzfluren Iran]. *Bonner Geographische Abhandlungen*: 8, pp. 1-62.

Bobek, H. 1968. Vegetation. In: *The Cambridge History of Iran.* Cambridge University Press, Cambridge, England, 1:280-293.

Bobrinsky, N. A. 1951. *Animal Geography [Geografiya zhivotnykh]*. Moscow: Prosveshchenie Publ., 384pp.

Bobrov, E. G. 1946. About Asiatic species of Genus *Nitraria* L. (K Aziatskikh vidakh roda *Nitraria* L.) *Sov. Bot*., 14,1:19-30 (in Russian).

Bocharnikov V., Laletin A., Angelstam P. et al. 2012. Russia, Ukraine, the Caucasus, and Central Asia. In: *Traditional Forest-Related Knowledge. Sustaining Communities, Ecosystems and Biocultural 31 Diversity*. Springer Science+Business Media B. V. Dordrecht, Heidelberg, London, New York: 251-280.

Bochud, M. 2011. *Tectonics of the eastern Greater Caucasus in Azerbaijan. Geo Focus, Vol. 30 (Thesis of the academic grade of Doctor rerum naturalium*). Hauterive (Fribourg), Switzerland, 187pp.

Boissier, P. E. 1867. *Flora orientalis [Flora Orientalis...]*. Vol 1. H. Georg, Basel, Geneva, 1017p.

Bond, W. J., Keeley, S. F. 2005a. Fire as a global 'herbivore': the ecology and evolution of flammable ecosystems. *Trends in Ecology and Evolution*, 20,7:387-394.

Bond W. J., Woodward F. I., Midgley G. F. 2005b. The global distribution of ecosystems in a world without fire. *New Phytologist,* 165: 525–537.

Borodin, A. M. et al. (Eds.) 1984. *USSR Red Data Book. [Krasnaya Kniga SSSR]*. Sec. edit. Lesnaya Promyshlennost' Publ. [Forestr. Industr. Publ.], Moscow, Vol., 392pp. (in Russian).

Borowski, D., Munteanu, C. 2011. Biosphere Reserves in European Mountains: an Exploratory Survey. *Biosphere Reserves in the Mountains of the World.* MAB pp 35-40.

Boskhomdzhiev, A. C. 2010. The Caspian coastal greenaway vision plan. *Landscape Architecture & Regional Planning Masters Projects*, 6p.

Bouchenak-Khelladi, Y. Verboom, G. A., Savolainen, V., Hodkinson, T. R. 2010. Biogeography of the grasses (Poaceae): a phylogenetic approach to reveal evolutionary history in geographical space and geological time. *Bot. J. Linn. Soc.,* 162: 543-557.

Bowman, D., Balch, J. K., Artaxo, P. et al. 2009. Fire in the Earth system. *Science*, 324(5926):481-484.

Bowman, D., Balch, J., Artaxo, P. et al. 2011. The human dimension of fire regimes on Earth. *J. of Biogeogr*. 38(12): 2223–2236.

Bowman, D., O'Brein, J. A., Goldammer, J. G. 2013. Pyrogeography and the global quest for sustainable fire management. *Rev. of Environ. & Resourc*., 38,1:57-80.

Box, E. O., Fujiwara, K (Eds.). 2015. Warm-temperate deciduous forests: concept and global overview. In: *Warm-Temperate Deciduous Forests around the Northern Hemisphere* (E. O. Box and K. Fujiwara Eds.). Springer-Verlag:7-26.

Braasch, Th., Peš, T., Michel, S., Jacken, H. 2011. The subspecies of the common pheasant *Phasianus colchicus* in the wild and captivity. *Intern. Journ. Galliformes Conserv.*, 2:6-13.

Brackenbury, J. 1995. *Insect and flowers – A biological partnership*. Blandford Books, Wellington House, 160pp.

Bramwell, M. (Ed.) 1976. *The international book of wood*. Mitchell Beazley Publishers Limited, London, 276pp.

Braund, D. 1994. *Georgia in antiquity: a history of Colchis and Transcaucasian Iberia,* 550 BC-AD 562. Oxford, Clarendon Press, 562p.

Breckle, S.-W., Veste, M., Wucherer, W. 2002. Deserts, land use and desertification (pp. 3-13). In: Breckle, S.-W., Vaste, M., Wucherer, W. 2001 (Eds.). *Sustainable land-use in Deserts'* Springer-Verlag Berlain, Heidelberg, 435 pp.

Bremer, K. 2002. Gondwanan evolution of the grass alliance of families (Poales). *Evolution,* 56:1374-1387.

Brezhnev, D. D., Korovina, O. N. 1981. *Wild relatives of cultivated plants of the flora of the USSR [Dikie rodichi kul'turnykh rastenii vo flore SSSR].* Kolos Publ., Leningrad, 376pp. (in Russian).

Brotherson, J. D., Field, D. 1987. *Tamarix*: impacts of a successful weed. *Rangelands.*, 9(3): 110-112.

Browicz, K. 1989. Chorology of the Euxinian and Hyrcanian element in the woody flora of Asia. *Plant Systematics and Evolution*, 162:305-314.

Browicz, K. 1993. Conspect and chorology of the genus *Pyrus* L. *Arboretum Kornickie*, 38:13-33.

Brown, R. G., Hodkinson, I. D. 1988. Taxonomy and ecology of the jumping plant-lice of Panama (Homoptera. Psylloidea). *Entomograph "9 (Lyneborg, L. –Ed.). E.J."* Brill/Scandinavian Science Press Ltd, 304pp.

Brown, J. H., Lomolino, M. V. 1998. *Biogeography (second edition).* Sunderland (Massachetts): Sinauer Association, USA, 691pp.

Brown, S. A. E., Scott, A. C., Glasspool, I. J., Collinson, M. E. 2012. Cretaceous wildfires and their impact on the Earth system. *Cretaceous Res.*, 36:162-190.

Brundrett, M. C. 2002. Coevolution of roots and mycorrhizas of land plants. *New Phytologists,* 154, 2: 275-304.

Buchmann, S. L., Nabhan, G. P. 1996. *The forgotten pollinators.* Island Press, Washington, D. C. Shearwater Books, Covelo, California, 299pp.

Budantsev, L. Yu. 1992. Early stages of formation and dispersal of the temperate flora in the Boreal Region. *Bot. Rev.* 8,1:1-48.

Budyko, M. L. 1980. *Climate in the past and in the future [Klimat v proshlom i budushchem].* Gidrometeoizdat, Leningrad, 350pp. (in Russian).

Bukhnikashvili, A. 2004. *On the cadaster of small mammals: Insectivora, Chiroptera, Lagomorpha, Rodentia of Georgia [Saqartvelos mcire zomis dzuzumtsovarta kadastrisatvis].* Universal Publ. Tbilisi, 142 (in Georgian).

Bukhsianidze, M. 2015. Catalogue of the holotypes of fossil land mammals in the Georgian National Museum, *Georgian National Museum,* Tbilisi, 53pp. (with 50 plates).

Burchak-Abramovich, N. T. 1955. The ancient cattle of Trialeti Ridge in Tsalka district [Drevnii krupny rogatyi scot Trialetskogo khrebra v Tsalkinskom raijone]. *Proceed. Zardabo Natural History Museum [Trudy est.-ist. muzeiya im. zardabn]* 9:137-142.

Burckhardt, D. 1987. Jumping plant lice (Homoptera, Psylloidea) of the temperate Neotropical region. Part 1: Psyllidea (subfamilies Aphalarinae, Rhinocolinae and Aphalaroidinae). *Zool. J. Lin. Soc.*, 89: 299-392.

Burckhardt, D. 1988a. Information on psyllid fauna of northeastern Turkey (Homoptera: Psylloidea) [Angaben zur Psyllidenfauna der Nordosttürkei (Homoptera: Psylloidea)]. *Mitt Entomol. Ges. Basel* 38, 1:31–44.

Burckhardt, D. 1988b. The jumping plant lice or psyllids (Homoptera, Psylloidea) from Grece. *Biologia Gallo-Helenica* 13:107-110.

Burckhardt, D., 1989. Psyllids (Insecta, Homoptera, Psylloidea) of Algeria [Les psylles (Insecta, Homoptera, Psylloidea) de l'Algérie]. *Archs Sci. Genéve,* 42,2: 367-424.

Burckhardt, D., Basset, Y. 2000. The jumping plant-lice (Hemiptera, Psylloidea) associated with Schinus (Anacardiaceae): systematics, biogeography and host plant relationships. *J. Nat. Hist.*, 34:57-155.

Burckhardt, D., Lauterer, P. 1993. The jumping plant lice of Iran (Homoptera, Psylloidea). *Revue Suisse de Zoologie*, 100 (4): 829-898.

Burckhardt, D., Önuçar, A. 1993. A review of Turkish jumping plant- lice (Homoptera, Psylloidea). *Revue Suisse de Zoologie*, 100(3): 547-574.

Burckhardt, D., Ouvrard, D. 2012. A revised classification of the jumping plant-lice (Hemiptera: Psylloidea). *Zootaxa* 3509:1-34.

Burmester, A. C., 2005. *Socio-economic situation and land use conflicts in Shirvan National Park Regions*, Azerbaijan. Greifswald, 167pp.

Burmistrov, L. 1995. *New crops and wild fruits and nuts.* URL: http://www.newcrops.uq.edu.au/ecotanc/papers/burmist2.htm (Stand:6.10.2011).

Bush, N. A. 1898. A previous account of a second journey through north-western Caucasus in 1898 [predvaritel'nyii otchiot vtorogo puteshestviia v Zapadnom Kavkaze v 1898]. *Izv. Russk.Geogr. Ob-va*, 34: 41-47 (in Russian).

Bush, N. A. 1935. *Botanical and geographical study of the Caucasus [Botaniko-geograficheskii ocherk Kavkaza].* Izd. AN SSSR [USSR Ac. Sci. Publ. House]. Moscow-Leningrad, 135pp. (in Russian).

Butkhuzi, L. 2004. Breaking news: leopard in Georgia. *Caucasus Environment*, 2:48-51.

Butnik, A. A. 1991. *The ecological anatomy of desert's vegetation of Middle Asia. Tashkent [Ecologicheskaiya anatomiia pustynnykh rastenii Sredneii Azii]*, I:74-82 (in Russian).

Buttiker, W. 1979. Fauna of Saudi Arabia. Zoological collections from Saudi Arabia. *Fauna of Saudi Arabia*, 1: 1-22.

Cairney, J. W. C. 2000. Evolution of mycorrhiza systems. *Naturwissenschaften,* 87:467-475.

Calagari, M., Modirrahmati, A. R., Asadi, F. 2006. Morphological variation in leaf traits of *Populus euphratica*. Oliv. Natural populations. *Intern. J. Agric. & Biology*, 8(6):754-758.

Callaway, R. M., Kikvidze, Z., Kikodze, D. 2000. Facilitation by unpalatable weeds may conserve plant diversity in overgrazed meadows in the Caucasus Mountains. *Oikos,* 89,2: 275-282.

Calow, P. (Ed.). 1999. The Blackwell's Concise Encyclopedia of Ecology, 11th Ed. *Wiley-Blackwell*, 168pp.

Cann, R. L., Stoneking, M., Wilson, A. C. 1987. Mitochondrical DNA and human evolution. *Nature,* 1:325(6099):31-36.

Cardinale, B. J., Duffy, J. E., Gonzalez, A., Hooper, D. U. et al. 2012. Biodiversity loss and its impact on humanity. *Nature*, 486, 59–67.

Cardona, T., Murray, J. W., Rutherford, A. W., 2015. Origin and evolution of water oxidation before the last common ancestor of the cyanobacteria. *Molecular Biology and Evolution*, 32,5: 1310-1328.

Carella, R. 2013. First observation of *Quercus pedunculiflora* C. Koch in the Italian Peninsula – short communication. *J. Forest Science*, 59,3:130-135.

Carroll, C., Noss, R. F., Paquet, P. C., Schumaker, N. H. 2004. Extinction debt of protected areas in developing landscapes. *Conserv. Biol.*, 18:1110-1120.

Cavalli-Sforza, L. L., Menozzi, P., Piazza, A. 1994. The history and Abashidze, J. L. 1947. Zelkova in Georgia [Dzelkva v Gruzii]. *Soobch. AN Gr.SSR*, 7(1): 29-32 (in Russian).

Abashidze, J. L. 1952. On the fruiting of sweet chestnut [K plodonosheniyu kashtana posevnogo]. *Proceed. Forest. Inst. Ac. Sci.,* Tbilisi, 4:55-77 (in Russian).

Abashidze, J. L. 1959. *Dendrology [Dendrologia]*. Vol.1. Sas. Sam. Inst. Gamomc. [Publ. House of Agric. Inst.], Tbilisi, 348pp. (in Georgian).

Abashidze, J. 1962. *Dendrology on the basis of plant geography [Dendrologiya mtsenareta geografiis safudzvlebit]. Part II. Pterocarya fraxinicolia* (Lam.). (pp34-35). Georgian Agricult. Inst. Publ., Tbilisi, 42pp. (in Georgian).

Abbasian, H., Kiabi, B. H., Kavousi, K. 2004. Food habits of wild goat, *Capra aegagrus aegagrus*, in the khorramdasht area, Kelardasht, Iran. *Zoology in the Middle East,* 33:119-124.

Abbott, L. B., Bettmann, G. T., Sterling, T.M. 2006. African rue (*Peganum harmala*) seedling response to herbicides applied under water-deficit–stress. *Weed Sci*, 54:198-204.

Abdieva, R. T. 2005. The mosaics of vegetation of solontchak deserts in Gobustan [Mozaichnost' rastitel'nosti solonchakovykh pustyn' v Gobustane]. *Nat. Acad. Sci. Azerb. Inst. Bot.*, Baku: 1-32 (in Russian).

Abramova, T. A. 1972. The reconstruction of paleogeographic conditions of Quarternary's transgression and regression of the Caspian Sea (according to palaeobotanical research) [Rekonstruktciya paleogeograficheskikh uslovii Chetvertichnoi transgresii i regresii Kaspiiskogo moriya (na osnove paleobotanicheskikh issledovanii] (*PhD thesis*) *geogr. sci.* Moscow (1-24pp.) (in Russian).

Abuladze, A. 2013. *Birds of prey of Georgia. Materials towards fauna of Georgia [Khischnye pticy Gruzii. Materialy k faune Gruzii]. issue 6*. Inst. of Zoology, Ilia State University. Lasha Khvichia Publ. Tbilisi, 118p.

Abutalibov, M. 1976. Vegetation of Azerbaijan [Rastitel'nost' Azerbaijana]. *Tshygi*, 256p. Baku (in Russian).

Adamia, Sh., Akhvlediani, K. T., Kilasonia, V. M., Nairn, A. E. M., Papava, D., Patton, D. K. 1992. Geology of the Republic of Georgia: a review: *Inst. Geology Review*, 34,5: 447-476.

Adamia, Sh., Alania, V., Chabukiani, A., Chichua, G., Enukidze, O. Sadradze, N. 2010. Evolution of the Late Cenozoic basins of Georgia (SW Caucasus): a review. In: Sosson, M., Kaymakci, N., Stephenson, R. A., Bergerat, F., Starostenko, V. (Eds.). *Sedimentary Basin Tectonics from the Black Sea and Caucasus to the Arabian Platform.* Geol. Soc., London, Spec. Publ. 340, 239–259.

Adamia, Sh., Alania, V., Chabukiani, A., Kutelia, Z., Sadradze, N. 2011a. Great Caucasus (Kavkasioni): a long-lived North-Tethyan Back-Ark basin. *Turkish Journal of Earth Sciences*, 20:611-628.

Adamia, Sh. A., Chabukiani, A., Kakabadze, M., Kotetishvili, E., Lordkipanidze, M., Tabidze, D. 1991. Geological background of the Caucasus and adjacent area. Geological events of the Cretaceous-Paleogene boundary. In: *Proceed. of Intern. Sympos.*, Tbilisi, 12: 18-5.

Adamia, Sh. A., Chkhotua, T., Kekelia, M., Lordkipanidze, M. B., Shavishvili, I., Zakariadze, G. 1981. Tectonics of the Caucasus and adjoining regions– implications for the evolution of the Tethys Ocean. *J. Struct. Geology*. 3 (4):437-447.

Adamia, Sh., Zakariadze, G., Chkhotua, T., et al. 2011b. Geology of the Caucasus: a review. *Turkish Journal of Earth Sciences*, 20, 489-544.

Adamovich, L. 1899. The Mediterranean elements of the Serbian flora [Die Mediterranen elemente der serbischen Flora]. *Engl. Bot. Jahrb., Bd.* 27, Leipzig, 351-389pp.

Adamovich, L. 1905. The development of the Balkan flora since the Tertiary period [Die Entwicklung der Balkanflora seit der Tertiärzeit]. *Engl. Bot. Jahrb., Bd.* 36, Beibl 81.

Adamovich, L. 1909. The vegetation conditions of the Balkan countries [Die Vegetationsverhältnisse der Balkanländer]. *Vegetation der Erde*. Leipzig, 1909, Bd. 11.

Adams, K. 2013. Genomics. Genomic clues to the ancestral flowering plant. *Science,* 342, 6165:1456-1457.

Adams, R.P. 2014. *Junipers of the world: the genus Juniperus. 4th edition* Trafford Publ. North America & international, 413pp.

Adem, C., Arancli, S., Zeydanli, U. 2007. Camili Biosphere Reserve, Turkey. *GEM-CON-BIO Case Study Report*, 64pp.

Adveev, B., Niemi, N.A. 2011. Rapid Pliocene exhumation of the central Greater Caucasus constrained by low-temperature thermochronomentry. *Tectonics*, 30, TC 2009, doi:10.1029/2010TC00280.

Adzinba, Z.I. 2000. Calcareous endemism of Colchic flora. In: *Mater. Sci. Session dedicated to 90th iubil. Date A.A. Kolakovski*. Sukhum:28-38.

Afanasjeva, I. N., Vasetskii, S. G., Vasina-Popova, E. T. et al. 1979. English-Russian biological dictionary. *Russian Language Publishers*, Moscow, 732pp.

Afkhazava, I. S. 1969. Lakes – the beauty of our nature [Tbebi-Čveni bunebis silamaze]. *Sabchota Saqartvelo*, 167pp. (in Georgian).

Agadjanov, S. D. 1962. The solonchak deserts of the coastal sands of Azerbaijan [Solontchakovye pustiny pribrezhnykh peskov Azerbaijhana]. In: *Sci. Notes Az. State Univ. Ser. Biolog.* 2: 16-221, Baku (in Russian).

Agadjanov, S. D. 1967. *Flora and vegetation of the sea coastal sands of Azerbaijan and its importance for fixing and development of sands. PhD thesis biol. sci.*, Baku: 1-32 (in Russian).

Agaguliev, I. (Ed.). 2000. *Flora and vegetation of south-east Shirvan [Flora i rastitel'nost' jugo-vostochnogo Shirvana].* Baku State Univ., Baku, 147pp. (in Russian).

Aghababyan, K., Khanamiryan, G., Ghukasyan, E., Badalyan, N. 2015. Abundance of narrow-clawed crayfishes (*Astacus leptodactylus* Esch., 1823) and its trends in Lake Sevan, Armenia. *J. Aquacult. Engin. Fisher.* Res., 1,3:116-124.

Aghasyan, L. A. 2014. Contemporary status and perspectives of the Herpetofauna of 'Goravan Sands' Sanctuary. *Proc. Int. Conf. 'Biol. Divers. and Conserv. Problems of the fauna of the Caucasus -2'*. 16-19pp. Sept. 23-26, 2014. Yerevan, Armenia (in Russian).

Agladze, G. D., Zotov A. A., 1987. *Mountainous rangelands and haylands of the Caucasus [Gornye pastbyscha i senokosnye ugodiya Kavkaza].* Soviet Georgia Publ. Tbilisi, 198pp. (in Russian).

Agusti, J., Lordkiphanidze, D. 2011. How 'African' was the early human dispersal out of Africa? *Quat. Sci. Revi.*, 30,11-12:1338-1342.

Ahmadi, A., Kavosi, M. R., Soltanloo, H. 2014. *Zelkova carpinifolia* reservoir from Hyrcanian forests, Northern Iran, a new sacrifice of *Ophiostoma novo-ulmi*. *Biodiversitas*, 15,1:48-52.

Akatov, V. V. 1999. *The insular effect as a factor of high-mountain phytocoenosis formation in the Western Caucasus [Ostrovnoi effect kak factor formirovaniya vysokogornykh fitocenozov Zapadnogo Kavkaza].* MGTI, Maikop: 33-42 (in Russian).

Akatov, V. V., Akatova, T. B., Bebin, A. R. et al. 2009. *Natural complexes of the Imeretian Lowland: bioecological diversity; zoological importance and recommendation on the protection [Prirodnye kompleksy Imeretinskoi nizmennosti: biologicheskoe raznoobrazije, sozologicheskaiya znachimost', rekomendacii po sokhroneniyu]*. Kopi-Print Publ. Sochi-Maikop, 93pp. (in Russian).

Akhalkatsi, M. 2015. *Forest habitat restoration in Georgia, Caucasus Ecoregion,* Orchis Publ. Tbilisi, 101pp.

Akhalkatsi, M., Kimeridze, M. 2012. Implementation of the classification system of forest habitats in accordance with the 'Natura2 000' standarts in the Georgian Legislation. In: *'Legal Aspects of European Forest Sustainable Development'. Proceed. of the 12th Intern. Symp.,* Cypros (Šulek et al. - Eds.).

Akhalkatsi, M., Tarkhnishvili, D. 2012. *Habitats of Georgia*. Natura, 2000.118pp.

Akhani, H. 2006. Biodiversity of halophytic and sabkha ecosystems of Iran:71-88pp. In: *Ajmal Khan, M, Boer, B., Kust, G. S., Banth, H.-J. (Eds.), Sabakha ecosystems.* vol. II: West and Central Asia. Springer, 259pp.

Akhani, H., Djamali, M., Ghorbanalizadeh, A., Ramezani, E. 2010. Plant biodiversity of Hyrcanian relict forests, N.Iran: an overview of the flora, vegetation, palaeocology and conservation. *Pak. J. Bot. Special Issue*, 42:231-258.

Akhani, H., Ghorbanli, M. 1993. A contribution to the halophytic vegetation and flora of Iran. In: Lieth, H. & Al Masoom, A. (Eds.). *Towards the rational use of high salinity tolerant plants*, 1: 35 - 44. Kluwer Academic Publishers. Dordrecht.

Akhmedov, E. G., Yarovenko, Y. A., Nasrullaev, N. I., Babaev, E. A., Akhmedov, S. G. 2009. Conservation of the Bezoar goat in the Eastern Caucasus. 25-31pp. In: Zazanashvili, N. & Mallon (Eds.). *Status and Protection of Globally Threatened Species in the Caucasus.* Tbilisi: CEPF, WWF, Contour Ltd.

Akhmedova, S. Z. 2004. *Biodiversity of the ecosystem of the vegetation cover of Djeiranchel and Adjinauri regions of Azerbaijan [Bioraznoobrazie ekosistem rastitel'nogo pokrova Djeiranchel i Adjinaurskogo regionov Azerbaijana]*. Elm, Baku, 236pp. (in Russian).

Akhmedova, S. Z. 2005. The origin of deserts' phytocoenoses of Djeiranchel-Adjinauri regions [Proiskhozhdenie pustynnykh fitocenozov Djeirantchel-Adjinaurskogo regionov]. *Izvestia NAN Azerbaijana* [*Herald of Nat. Ac. Sci. of Azerb.*], 3-4:81-85, Baku (in Russian).

Akhmetiev, M. A. 2010. Paleocene and Eocene floristic and climatic change in Russia and Northern Kazakhstan. *Bull. Geosci.*, 81,1:77-94.

Akhmetiev, M. A., Beniamovski, V. N., 2009. Paleogene floral assemblages around epicontinental seas and straits in Northern Central Eurasia: proxies for climatic and palaeogeographic evolution. *Geologica Acta* 7 (1–2), 297–309.

Akhundov, G. F., Gogina, E. E., Prilipko, L. I. 1978. Narrowly endemic and rare species of natural flora of Nakhichevan ASSR [Uzkoendemichnye i redkie vidy prirodnoii flory Nakhichevanskoi ASSR]. In: *Bull. Glavn. Bot. Sada*, 107:54-62 (in Russian).

Akjigitova, N. I. 1973. The halophite vegetation – Halophyta [Galofitnye rastenija-Halopyta]. In: *Vegetation cover of Uzbekistan,* 2:211-302. Tashkent (in Russian).

Akjigitova, N. I. 1982. *Halophyte vegetation of Central Asia and its indicative features [Galofitnye rasteniya Tcentral'noi Azii i ikh indikatornye osobennosti]*. Tashkent, FAN Publ. [Publ. House of Ac. Sci Uzbekist], Tashkent, 192pp. (in Russian).

Akopian, J. A. 2007. On the *Pyrus* L. (Rosaceae) species in Armenia. In: Flora, vegetation and plant resources of Armenia [O vidakh roda *Pyrus* L. (Rosaceae) v Armenii] (Gabrielian, E.

Ts. et al. – Eds.) (*Inst. Bot. Ac. Sci. Rep. Arm., Arm. Bot. Soc.*). Yerevan, 16:15-26 (in Russian).

Akopian, J. A. 2011. Genus *Salsola* L. sensu lato (Chenopodiaceae) in South Transcaucasia. In: *Takhtajania,* 2011, 1:124-132 (in Russian with English abstract).

Akopian, J. A. 2015. Notes on *Argyromalon* Fed. Section species from the genus *Pyrus* L. of Armenian flora (81-87pp.). In: Vardanyan, Zh. H. (Ed.). Bot. Sci. in the Modern World. *Proc. Int. Conf. dedicated to the 80th anniversary years Bot. Gard.* (Yerevan, 5-9 Oct., 2015), 440pp.

Akymushkin, I. 1973. *The World of animals. Narrative about birds [Mir zhivotnykh. Rasskazy o pticakh].* Molodaiya Gvardiya Publ., Moscow, 379pp. (in Russian).

Alaback, P. B. 1991. Comparative ecology of temperate rainforests of the Americas along analogues climatic gradients. *Revista Chilena de Historia Natural*, 64:399-412.

Aladin, N., Plotnikov, I. 2004. The Caspian Sea. Lake Basin Management Initiative, *Thematic Paper*: 1-29pp.

Al'bov, N. M. 1899. On the abandoned sites of the Caucasus [K zabroshennykh raijonov Kavkaza] *Zemlevedenie*, I-II:71-84 (in Russian).

Aleksanyan, A. S. 2011a. Biomorphological structure of the flora of arid open woodlands of South Armenia [Biomorphologisheskaiya struktura flory aridnykh redkolesii Juzhnoi Armenii]. *Biol. Zhurn. Armenii*, 2(63): 30-36 (in Russian).

Aleksanyan, A. S. 2011b. The phytogeographical analysis of the flora of the arid woodlands of southern Armenia [Fitogeographicheski analiz flory aridnykh redkolesii Juzhnoi Armenii]. *Biol. Zhurn. Armenii*, 4(63): 83-91 (in Russian).

Aliyev, G. A., Volobuev, V. P., Salaev, M. M. et al. 1953. *Soils of Azerbaijan [Pochvy Azerbaidjana].* Izd. Acad. Sci. Azer.SSR, 598 pp. (in Russian).

Aliev, R. A., 1954. *The salsoletum semi-deserts of Azerbaijan [Sarsazanovye polypustyny Azerbaijana].* Elm, Baku, 128 (in Russian).

Aliev, R. A. 1971. Flora of Artemisietum of Azerbaijan. *Uchiyon. Zapiski ATU im. S. M. Kirova. Ser. Biol. Nauk* [*Sci. Papers ATU after S. M. Kirov, ser. Biol. Sc.*]. 14-19 (in Russian).

Aliev, R. A., Gadjiev, V. D., Isaev, J. M., Mailov, A. I., Nabili, D. G., Prilipko, L. I. 1965. *Improvement and rational use of Azerbaijan's winter and summer pastures [Ulucshenie i ratciolal'noe ispol'zovanie zimnykh i letnikh pastbich Azerbaijana].* Izd. AN Az.SSR, Baku: 1-45 (in Russian).

Aliev, S. Y., Gadjiev, E. M. 1986. Materials regarding morphological investigations of the underground part of some vegetation-edificators and plant communities of the Shirvan area [Materialy k morfologicheskomu izucheniju podzemnoi chasti nekotorykh rastenii edifikatorov i rastitel'nykh soobchestv Shirvanskogo raiona]. *Izvest. Akad. Nauk., Ser. Biologich. Nauk*, 3: 19-27 (in Russian).

Alpaidze, V. S., Shengelija, F. K. 1989. Geology. The Colchis Lowland–natural conditions and socio-economic aspects [Kolkhidskaiya nizmennost' – prirodnye usloviya i social'no-ekonomicheskie aspekty] In: Svanidze, G. G. (Ed.). *Gidrometeoizdat*: 9-44. Leningrad (in Russian).

Al-Saghir, M. G., 2006. Phylogenetic analysis of the genus *Pistacia* (Anacardiaceae). *Dissert. Doctor of philosophy in Biol. Sci.* Blacksburg, Virginia 112pp.

Al-Saghir, M. G., Porter, D. M. 2012. Taxonomic revision of the genus *Pistacia* L. (Anacardiaceae) *Amer. J. Pl. Sci.* 3:19.

Al-Saghir, M. G., Porter, D. M., Nilsen, E. T. 2006. Leaf anatomy in *Pistacia* species (Anacardiaceae). *J. Biol. Sci.*, 6: 242-244.

Altman, N. 1994. Sacred trees. *Sierra Club Books*, San Francisco, 120pp.

Amiryan, A. 1999. Fauna of staphylid beetles (Coleoptera, Staphylinidae) in Goravan Sands. In: '*Izvestia 2-go Vsesojuznogo Kongressa po Pochven. Zoologii*' 'Bioraznoobrazie i Zhizn' Pochvenych Sistems' (*Proceedings of the 2nd All-Russian Congr. Soil Zool. 'Biodiversity and Life of the Soil System'*), Moscow, p. 18 (in Russian).

Anderson, R. C. 2006. Evolution and origin of the Central Grassland of North America: climate, fire, and mammalian grazers. *J. Torrey Bot. Soc.,* 133, 4: 626-647.

Anderson, R. C. 2012. Ecology and management of the prairie division. *Cumulative Watershed Effects of Fuel Management in the Eastern United States*. Chapt. 9:175-201.

Andguladze, D. L. 1971. *Apiculture en Géorgie [Beekeeping in Georgia]*. Ministère de l'Agriculture RSS de Géorgie. 185 pp.

Apakidze, A. M. 1968. *Towns of ancient Georgia [Goroda drevnei Gruzii]*. Metsnieraba, Tbilisi, 296pp. (in Russian).

Arabuli, G., Kvavadze, E., Kikodze, D., Arabuli, T. 2008. The krummholz beech wood of Mt Tavkvetili Javakheti Plateau, (Southern Georgia) – a relict ecosystem. *Proc. Inst. Zool.*, Tbilisi, 23:194-213.

Arakelyan, M. C. 2012. *Microevolutionary processes in sympatric populations of some reptiles from the Republic of Armenia and adjacent territories. Doctoral dissert. (PhD thesis) biol. sci.*, Yerevan, 47pp. (in Russian with English summary).

Archibald, O. W. 1995. *Ecology of world vegetation.* Champan & Hall, London, 510pp.

Arnol'di, K. V. 1957. In relation to the theory of species – an area with respect to ecology and origination of species population [O teorii areala v sviyazi s ekologii i proiskhozhdeniem vidovykh populacii]. *Zool. Zhurn.*, 36,2:1609-1626 (in Russian).

Arnol'di, K. V. 1968. The zonal, zoogeographical and ecological peculiarities of the mirmecofauna and the settling of the plains of Russia by ants [Zonal'nye, zoogeograficheskie i ecologicheskie osobennosti mirmekofauny i naseleniya muraj'ev Russkoi ravniny]. *Zool. Zhurn.*, 7,8: 1155-1178 (in Russian).

Aronson, J. A. 1989. *Halophyte: a data of salt tolerant plants of the world. Office of Arid Lands Studies.* The Tucson, Arizona: University of Arizona Press. 114pp.

Arsdale, A. P., Lordkipanidze, D. 2012. A quantitative assessment of mandibular variation in the Dmanisi Hominins. *Paleoanthropology*: 134-144.

Asadova, K. K. 2008. Features of flora and vegetation in conditions of winter pastures [Osobennosti flory i rastitel'nosti v usloviyakh zimnykh pastbysch]. *Nauchnye Vedomosti Belgorodskogo Gos. Univ-ta, Ser. Estestvennye nauki*, 3(43) 70-76 (in Russian).

Asadova, K. K. 2013. Dynamic of fodder resources of the most important fodder plants of the winter pastures of north-western part of Azerbaidjan. *Sustainable Agriculture Research*, 2,2:29-36.

Asanidze, Z., Akhalkatsi, M., Gvritishvili, M. 2011. Comparative morphometric study and relationships between the Caucasian species of wild pear (*Pyrus* spp.) and local cultivars in Georgia. *Flora*, 206:974-986.

Asanidze, Z., Akhalkatsi, M., Henk, A. D. 2014. Genetic relationships between wild progenitor pear (*Pyrus* L.) species and local cultivars native to Georgia, South Caucasus. *Flora* 209:504-512.

Ashburner, K., McAllister, H. A. 2013. *The genus Betula: a taxonomic revision of birches. Ser.: A Botanical Magazine Monograph*. Royal Botanic Gardens, Kew Publishing, 448pp.

Askin, R., A., Spicer, R. A. 1995. The Late Cretaceous and Cenozoic history of vegetation and climate at northern and southern high latitudes: a comparison. In: *Effects of Past Global Change on Life. National Research Counsil (US) Panel on Effects of past Global Change on Life.* Washington (DC): National Academies Press (USA): 156-173.

Aslanyan, A. V., Petrosyan, R. P., Arakelyan, M. S. 2014. About an extinction alarm situation for two species of lizards in the Ararat Valley of Armenia [K ugrozu ischeznovaniyu dvux vidov iacheritis Araratskoi doliny Armenii]. *Proc. Int. Conf. 'Biol. Divers. Conserv. Problems of the fauna of the Caucasus'*. 21 Sept., 23-26, 2014, Yerevan, Armenia (in Russian).

Atamov, V. V. 2002. *The steppe vegetation of Azerbaijan [Stepnye rasteniya Azerbaidjana*]. Elm. Baku, 208pp. (in Russian).

Atamov, V. V. 2008. Phytosociological characteristics the vegetation of Caspian's shores in Azerbaijan. *Intern. J. of Bot.*, 4(1): 1-13.

Atamov, V. V., Gabbarov, K., Gurbanov, E. 2006. The phytosociological characteristics of ecosystems of the mountains of Talysh region, Azerbaijan. *Asian J. Plant Sci.*, 5 (5):899-904.

Atli, H. S., Arpaci, S., Kaska, N., Ayanoglu, H.1998. Wild *Pistacia* species in Turkey: 35-40pp. In: S. Padulosi and Hadj-Hassan (Eds.) *Report of the IPGRI Workshop,* 14-17 December 1998, Irbid, Jordan. 105pp.

Avanesova, A. A. 2006. *The successions of steppes phytocoenosis of European forest-steppe: from the example of V. V. Alekhin Central-Chernozem Biosphere Reserve after M. M. Aliyokhin [Sukksessii stepniykh fitocenosov Evropeiskikh lesostepi: na primere...]. PhD thesis biol. sci.* Voronezh, 22pp. (in Russian).

Awramik, S. M. 1992. The oldest records of photosynthesis. *Photosyn. Res.*, 33:75-89.

Axelrod, D. I. 1965, A method of determining the altitudes of Tertiary floras: *The Paleobotanist*, 14:144-171.

Axelrod, D. I. 1966. Origin of deciduous and evergreen habitats in temperate forests. *Evolution*, 20:1-I5.

Axelrod, D. I. 1968, Tertiary floras and topographic history of the Snake River Basin, Idaho: *Geol. Soc. America Bull.*, 79: 713-734.

Axelrod, D. I. 1985. Rise of the grassland biome, central North America. *Bot. Rev.* 51:163–201.

Aydin, Z. U., Dönmez, A. A. 2015. Taxonomic and nomenclatural contributions to *Pyrus* L. (Rosaceae) from Turkey. *Turk. J. Bot.*, 39:841-849.

Azimov, V. A. 2005. *The complex analysis of biodiversity of the flora of arid, open woodlands of the coastal and foothill area of Dagestan [Kompleksny analiz bioraznoobraziya flory aridnykh redkolesii pribrezhnykh i predgornykh regionov Dagestana] (115pp.). PhD thesis boil. sci.* Makhatchkala, 1-24 (in Russian).

Azimov, V. A., Teymurov, A. A. 2003. Ecological-coenological analysis of flora of the arid, open woodland of Nagorny (Inter-Mountain) Dagestan [Ecologo-tsenologicheski analiz flory aridnogo redkoles'iya Nagornogo Dagestana]. *Regional'nye Problemy Geografii i Geologii* [*Regional Problems of Geogr. & Geolog.]:* 2:80-86, Makhatchkala (in Russian).

Babaev, A. G., Zonn, I. S., Drozdov, N. N., Freikin, Z. G. 1986. *Deserts, the nature of the world [Pustyni, priroda mira].* Moscow, Mysl, 320 pp. (In Russian).

Babayev, M. P., Gerayzade, A. P. 2010. Problems of degradation of soils of dry subtropics of the southern Caucasus. Institute of Soil Science and Agrochemistry, Baku, Azerbaijan. *Proceedings of the 19th World Congress of Soil Science: Soil solutions for a changing world,* Brisbane, Australia, 1-6 August *2010. Symposium 1.3.2 Geochronological techniques and soil formation*, 2010:16-18.

Baeva, V. G. 1985. Psyllids or jumping plant-lice (Homoptera, Psylloidea). In: *Psyllids or jumping plant-lice. Fauna of Tadjik. SSR, Vol. 8*. Acad. Sci. Tadjik. SSR. Donish Publ., Duschanbe (in Russian), 332pp.

Bagdasaryan, G. P., Karapetyan, K. I., Gukasyan, R. Kh. 1973. On the stratigraphical position and age of Manichar lavae [Stratigrafichescheskaija pozitsia i vozrast' Manycharskikh lav]. *Bull. AN ArmSSR, issue Nauki o Zemle*, 3: 49-51 (in Russian).

Bakhshiev, V. 2009. Saltwort small-bush deserts of the central part of the Shirvan Plain [Soliyankovye melkokustarnikovye pustyni tsentral'noi chasti shirvanskoi ravniny]. *Natural Sciences*, Baku State Univ.,113-117 (in Russian).

Bakker, R. T. 1986. *The dinosaur heresies: new theories unlocking the mystery of the dinosaurs and their extinction.* Eilliam Morrow and Company, New York, 481pp.

Ballantyne, A. P., Alden, C. B., Miller, J. B., Tans, P. P., White, J. W. C. 2012. Increase in observed net carbon dioxide uptake by land and oceans during the last 50 years. *Nature,* 488:70-72.

Balyan, G. A. 1972. *Prostrate summer cypress (Kochia prostrate) and its culture in Kirghizia.* Frunze, USSR: Kyrgyzstan Publ. (Translated from Russian; U. S. Department of Commerce, TT77 -59026, 1979). 296pp.

Bamshad, M. J., Olson, S. 2003. Does race exist? *Scientific American*: 289,6: 78-85.

Banks, H. P. 1975. Reclassification of Psilophyta. *Taxon*, 24: 401–413.

Bannikov, A. G. (Ed.). 1969. *Natural Preserves (zapovednyks) of the Soviet Union [Prirodnye zapovednyki Sovetskogo Soyuza]*. Kolos Publ. Moscow, 552pp. (in Russian).

Bannikov, A. G., Golgofskaja, K.Yu., Kotov, B. A. 1969. The Caucasus [Kavkasus] (337-403) In: *Natural reserves (Zapovedniks in the Soviet Union (Bannikov, A. G. (Ed.),* Kolos, Moscow, 459pp. (in Russian).

Barnes, B. V., Zak, D. R., Denton, S. R., Spurr, S. H. 1998. *Forest Ecology.* John Wiley & Sons, New York.

Barseghian, A. M. 1966. Saline and solonchak vegetation of the Ararat plain. Flora and vegetation of Armenia [Solontsevye i solonchakovye rasteniya Araratskoi doliny. Flora i rastitelnost' Armenii]. *Trudy Bot. Inst. AN Arm.SSR* [*Proceed. Bot. Inst. AN Arm.SSR*], 15:6-65, Yerevan (in Russian).

Bashkirov, I. S. 1939. Caucasian bison [kavkazskii zubr]. In: *Caucasian European bison (Moscow: Central Board for Reserves, Forest Parks and Zoological Gardens, Council of the People's Commissars of the RSFSR*): 1–72. (In Russian).

Basinger, J. F., Greenwood, D. R., Sweda, T. 1994. Early Tertarian vegetation of arctic Canada and its relevance to paleoclimatic interpretation. In: *NATO ASI Series, Vol. 127. Cenozoic Plants and Climates of the Arctic (Boulter, M.C.& Fisher, H. C. –Eds.). Springer-Verlag* Berlin Heidelberg, 1994: 175-198p.

Batello, C., Azzu, N. 2010. Gardens of Biodiversity: Conservation of genetic resources and their use in traditional food production systems by small farmers of the Southern Caucasus. Rome, *FAO*, 363pp.

Baudena, M., Dekker, S. C., Van Bodegom van, P. M. et al. 2015. Forests, savannas, and grasslands: bridging the knowledge gap between ecology and dynamic global vegetation models. *Biogeosciences*, 12:1833-1848.

Baum, B. D. 2006. *The rise and fall of the Caucasian race: a political history of racial identity.* New York University Press, New York, 342pp.

Baum, B. R. 1978. *The genus Tamarix.* Israel Academy of Sciences and Humanities, Jerusalem, 209 pp.

Beaman, M. 1994. *Palaearctic Birds: a cheklist of the birds of Europe, North Africa and Asia north of the foothills of the Himalayas.* Harrier Publications. Stony hurst, UK.168 pp.

Beard, J. B. 2012. Origin, biogeographical migrations and diversifications of turfgrasses. *Research Report*. SR,132:1-26.

Bebia, S. M. 1989. Relict forest of chestnut [Reliktovye lesa kashtana]. *Problemy Ekologii Abkhazii*, [*Problems of Ecology of Abkhazia*], 4: 77-82 (in Russian).

Beerling, D. J. 2002. Low atmospheric CO_2 levels during the Permo-Carboniferous glaciation inferred from fossil lycopsids. *PNAS,* 99, 20:12567-12571.

Beerling, D. J. 2006. Leaf evolution: gases, genes and geochemistry. *Ann.Bot.,* 93,3:345-352.

Behrensmeyer, A. K., Damuth, J. D., Dimichele, W. A. et al. 1992. *Terrestrial ecosystems through time: evolutionary palaeoecology of terrestrial plants and animals.* Chicago: University of Chicago Press. Chicago, IL, 551pp.

Bei-Bienko, G.Ya. 1966. Habitat changes by terrestrial organisms as a biological principle [Smena mestoobitanii nazemnykh origanizmami kak biologicheskii printsip]. *Zhurn. Obch. Biol.,* 27,1:5-21 (in Russian).

Belcher, C. M., McElwain, J. C. 2008. Limits for combustion in low O_2 redefine paleoatmospheric predicitons for the Mesozoic. *Science,* 321:1197-1200.

Belcher, C. M., Yearsley, J. M., Hadden, R. M., McElwain, J. C., Rein, G. 2010. Base-line intrinsic flammability of Earth's ecosystems estimated from paleoatmospheric oxigen over the past 350 million years. *Proc. Nat. Acad. Sci.* USA, 107,52:22448-22453.

Bell, R. L., Itai, A. 2011. *Pyrus*. In: *Kole, C. (Ed.), Wild crop relatives: genomic and breeding resources temperate fruits.* Springer-Verlag, Berlin Heidelberg: 147-179.

Belousova, I. P. 2000. *Use of methods of genealogical analysis for preservation of the gene pool of European wisent and Przhevalski's horse [Ispol'zovanie metodov genealogitcheskogo analiza dliya sokhraneniya genofonda evropejskogo zubra i loshadi Przheval'skogo]. PhD thesis biol. sci*, 24pp. S.-Petersburg (in Russian).

Bendrey, R. 2011. Some like it hot: environmental determinism and the pastoral economies of the later prehistoric Eurasian steppe. *Pastoralism*: *Research, policy and practice*, 1,1-16.

Berg, L. S. 1947. *Geographic zones of the Soviet Union [Geograficheskye zony Sovetskogo Soyuza]. Vol. 1*, Geographgiz (State Publ. Geogr. Lit.), Moscow, 397pp. (in Russian).

Berg L. S., 1952. *Geographic zones of the Soviet Union [Geographicheskie zony Sovetskogo Soyuza]. Vol. 2.* Geographgiz (State Publ. Geogr. Lit.), Moscow, 510 pp. (In Russian.).

Beridze, T., Pipia, I., Beck, J. et al. 2011. Plastid DNA sequence diversity in a worldwide set of grapevine cultivars (*Vitis vinifera* L. subsp. *vinifera*), *Bull. Geogr. Nat. Ac. Sci.* 15,1:98-103.

Berna, F., Goldberg, P., Horwitz, L. K. et al. 2012. Microstratigraphic evidence of *in situ* fire in the Acheulean strata of Wonderwerk Cave, Northern Cape Province, South Africa. *PNAS*:1215-1220.

Bernbaum E., 1998. *Sacred mountains of the world.* Berkeley-Los Angeles-London: University of California Press., 291pp.

Berner, R. A., Beerling, D. J., Dudley, R, Robinson, J. M., Wildman, R. A. 2003. Phanerozoic atmospheric oxygen. *Ann. Rev. Earth Planet Sci.* 31:105–134.

Bikov, B. A. 1981. The main peculiarities of halophyte vegetation of Middle Asia and Kazakhstan [Osnovnye osobennosti galofitnykh rastenii Srednei Azii i Kazakhstana]. *Bull. AN Kaz. SSR, Ser. Biol.*,1:1-19 (in Russian).

Billwitz, K. 1997. The pedogenetically undeveloped soil like the formation '*Takyr*' [Lehrbuch der Allgemeinen Physischen Geographie]. *Gotha (Justus Perthes Verlag)*, 234-327 (in German).

Bisaz, A., Escher, F., Grosjean, M., Ives, J. D., Messerli, B., Price, M. F. 1997. Mountains of the world: challenges for the twenty-first century. A contribution to Chapter 13, *Agenda 21. Mountain Agenda*. Bern, Paul Haupt.

Björn, L. O., Govindjee, 2009. The evolution of photosynthesis and chloroplasts. *Current Science,* 96, 11:1466-1473.

Blair, J., Nippert, J., Briggs, J. 2014. Grassland ecology, pp. 389–423. In*:* Monson, R. K. (Ed.). *Ecology and the Environment*. Springer Science+ Business Media, New York., 659pp.

Blondel, L., Aronson, J. 1999. *Biology and wildlife of the Mediterranean Region.* Oxford University Press, Oxford, 328 pp.

Blumenbach, J. F. 1795 (3rded.). *On the natural varieties of mankind [De generis humani varietate nativa]. Edit.3*. Gottingen http://dx.doi.org/10.5962/bhl.title. 35972. 35972.

Blumenbach, J. F. 1865. *The anthropological treatises of Johann Friedrich Blumenbach.* London: Publ. for the Anthtopological Society, by Longman, Green, Longman Roberts & Green, 1865, 382pp.

Bobe, R. 2006. The evolution of arid ecosystems in eastern Africa. *Journal of Arid Environments,* 66,3: 564-584.

Bobek, H.1951. The natural forests and woodlands of Iran]Die natürlichen Wälder und Gehölzfluren Iran]. *Bonner Geographische Abhandlungen*: 8, pp. 1-62.

Bobek, H. 1968. Vegetation. In: *The Cambridge History of Iran.* Cambridge University Press, Cambridge, England, 1:280-293.

Bobrinsky, N. A. 1951. *Animal Geography [Geografiya zhivotnykh].* Moscow: Prosveshchenie Publ., 384pp.

Bobrov, E. G. 1946. About Asiatic species of Genus *Nitraria* L. (K Aziatskikh vidakh roda *Nitraria* L.) *Sov. Bot.*, 14,1:19-30 (in Russian).

Bocharnikov V., Laletin A., Angelstam P. et al. 2012. Russia, Ukraine, the Caucasus, and Central Asia. In: *Traditional Forest-Related Knowledge. Sustaining Communities, Ecosystems and Biocultural 31 Diversity.* Springer Science+Business Media B. V. Dordrecht, Heidelberg, London, New York: 251-280.

Bochud, M. 2011. *Tectonics of the eastern Greater Caucasus in Azerbaijan. Geo Focus, Vol. 30 (Thesis of the academic grade of Doctor rerum naturalium*). Hauterive (Fribourg), Switzerland, 187pp.

Boissier, P. E. 1867. *Flora orientalis [Flora Orientalis...].* Vol 1. H. Georg, Basel, Geneva, 1017p.

Bond, W. J., Keeley, S. F. 2005a. Fire as a global 'herbivore': the ecology and evolution of flammable ecosystems. *Trends in Ecology and Evolution*, 20,7:387-394.

Bond W. J., Woodward F. I., Midgley G. F. 2005b. The global distribution of ecosystems in a world without fire. *New Phytologist,* 165: 525–537.

Borodin, A. M. et al. (Eds.) 1984. *USSR Red Data Book. [Krasnaya Kniga SSSR]. Sec. edit.* Lesnaya Promyshlennost' Publ. [Forestr. Industr. Publ.], Moscow, Vol., 392pp. (in Russian).

Borowski, D., Munteanu, C. 2011. Biosphere Reserves in European Mountains: an Exploratory Survey. *Biosphere Reserves in the Mountains of the World.* MAB pp 35-40.

Boskhomdzhiev, A. C. 2010. The Caspian coastal greenaway vision plan. *Landscape Architecture & Regional Planning Masters Projects*, 6p.

Bouchenak-Khelladi, Y. Verboom, G. A., Savolainen, V., Hodkinson, T. R. 2010. Biogeography of the grasses (Poaceae): a phylogenetic approach to reveal evolutionary history in geographical space and geological time. *Bot. J. Linn. Soc.,* 162: 543-557.

Bowman, D., Balch, J. K., Artaxo, P. et al. 2009. Fire in the Earth system. *Science,* 324(5926):481-484.

Bowman, D., Balch, J., Artaxo, P. et al. 2011. The human dimension of fire regimes on Earth. *J. of Biogeogr.* 38(12): 2223–2236.

Bowman, D., O'Brein, J. A., Goldammer, J. G. 2013. Pyrogeography and the global quest for sustainable fire management. *Rev. of Environ. & Resourc.*, 38,1:57-80.

Box, E. O., Fujiwara, K (Eds.). 2015. Warm-temperate deciduous forests: concept and global overview. In: *Warm-Temperate Deciduous Forests around the Northern Hemisphere* (E. O. Box and K. Fujiwara Eds.). Springer-Verlag:7-26.

Braasch, Th., Peš, T., Michel, S., Jacken, H. 2011. The subspecies of the common pheasant *Phasianus colchicus* in the wild and captivity. *Intern. Journ. Galliformes Conserv.*, 2:6-13.

Brackenbury, J. 1995. *Insect and flowers – A biological partnership.* Blandford Books, Wellington House, 160pp.

Bramwell, M. (Ed.) 1976. *The international book of wood.* Mitchell Beazley Publishers Limited, London, 276pp.

Braund, D. 1994. *Georgia in antiquity: a history of Colchis and Transcaucasian Iberia,* 550 BC-AD 562. Oxford, Clarendon Press, 562p.

Breckle, S.-W., Veste, M., Wucherer, W. 2002. Deserts, land use and desertification (pp. 3-13). In: Breckle, S.-W., Vaste, M., Wucherer, W. 2001 (Eds.). *Sustainable land-use in Deserts'* Springer-Verlag Berlain, Heidelberg, 435 pp.

Bremer, K. 2002. Gondwanan evolution of the grass alliance of families (Poales). *Evolution,* 56:1374-1387.

Brezhnev, D. D., Korovina, O. N. 1981. *Wild relatives of cultivated plants of the flora of the USSR [Dikie rodichi kul'turnykh rastenii vo flore SSSR].* Kolos Publ., Leningrad, 376pp. (in Russian).

Brotherson, J. D., Field, D. 1987. *Tamarix*: impacts of a successful weed. *Rangelands.*, 9(3): 110-112.

Browicz, K. 1989. Chorology of the Euxinian and Hyrcanian element in the woody flora of Asia. *Plant Systematics and Evolution*, 162:305-314.

Browicz, K. 1993. Conspect and chorology of the genus *Pyrus* L. *Arboretum Kornickie*, 38:13-33.

Brown, R. G., Hodkinson, I. D. 1988. Taxonomy and ecology of the jumping plant-lice of Panama (Homoptera. Psylloidea). *Entomograph "9 (Lyneborg, L. –Ed.). E.J."* Brill/Scandinavian Science Press Ltd, 304pp.

Brown, J. H., Lomolino, M. V. 1998. *Biogeography (second edition).* Sunderland (Massachetts): Sinauer Association, USA, 691pp.

Brown, S. A. E., Scott, A. C., Glasspool, I. J., Collinson, M. E. 2012. Cretaceous wildfires and their impact on the Earth system. *Cretaceous Res.*, 36:162-190.

Brundrett, M. C. 2002. Coevolution of roots and mycorrhizas of land plants. *New Phytologists,* 154, 2: 275-304.

Buchmann, S. L., Nabhan, G. P. 1996. *The forgotten pollinators.* Island Press, Washington, D. C. Shearwater Books, Covelo, California, 299pp.

Budantsev, L. Yu. 1992. Early stages of formation and dispersal of the temperate flora in the Boreal Region. *Bot. Rev.* 8,1:1-48.

Budyko, M. L. 1980. *Climate in the past and in the future [Klimat v proshlom i budushchem].* Gidrometeoizdat, Leningrad, 350pp. (in Russian).

Bukhnikashvili, A. 2004. *On the cadaster of small mammals: Insectivora, Chiroptera, Lagomorpha, Rodentia of Georgia [Saqartvelos mcire zomis dzuzumtsovarta kadastrisatvis].* Universal Publ. Tbilisi, 142 (in Georgian).

Bukhsianidze, M. 2015. Catalogue of the holotypes of fossil land mammals in the Georgian National Museum, *Georgian National Museum,* Tbilisi, 53pp. (with 50 plates).

Burchak-Abramovich, N. T. 1955. The ancient cattle of Trialeti Ridge in Tsalka district [Drevnii krupny rogatyi scot Trialetskogo khrebra v Tsalkinskom raijone]. *Proceed. Zardabo Natural History Museum [Trudy est.-ist. muzeiya im. zardabn]* 9:137-142.

Burckhardt, D. 1987. Jumping plant lice (Homoptera, Psylloidea) of the temperate Neotropical region. Part 1: Psyllidea (subfamilies Aphalarinae, Rhinocolinae and Aphalaroidinae). *Zool. J. Lin. Soc.*, 89: 299-392.

Burckhardt, D. 1988a. Information on psyllid fauna of northeastern Turkey (Homoptera: Psylloidea) [Angaben zur Psyllidenfauna der Nordosttürkei (Homoptera: Psylloidea)]. *Mitt Entomol. Ges. Basel* 38, 1:31–44.

Burckhardt, D. 1988b. The jumping plant lice or psyllids (Homoptera, Psylloidea) from Grece. *Biologia Gallo-Helenica* 13:107-110.

Burckhardt, D., 1989. Psyllids (Insecta, Homoptera, Psylloidea) of Algeria [Les psylles (Insecta, Homoptera, Psylloidea) de l'Algérie]. *Archs Sci. Genéve,* 42,2: 367-424.

Burckhardt, D., Basset, Y. 2000. The jumping plant-lice (Hemiptera, Psylloidea) associated with Schinus (Anacardiaceae): systematics, biogeography and host plant relationships. *J. Nat. Hist.*, 34:57-155.

Burckhardt, D., Lauterer, P. 1993. The jumping plant lice of Iran (Homoptera, Psylloidea). *Revue Suisse de Zoologie*, 100 (4): 829-898.

Burckhardt, D., Önuçar, A. 1993. A review of Turkish jumping plant- lice (Homoptera, Psylloidea). *Revue Suisse de Zoologie*, 100(3): 547-574.

Burckhardt, D., Ouvrard, D. 2012. A revised classification of the jumping plant-lice (Hemiptera: Psylloidea). *Zootaxa* 3509:1-34.

Burmester, A. C., 2005. *Socio-economic situation and land use conflicts in Shirvan National Park Regions*, Azerbaijan. Greifswald, 167pp.

Burmistrov, L. 1995. *New crops and wild fruits and nuts.* URL: http://www.newcrops.uq.edu.au/ecotanc/papers/burmist2.htm (Stand:6.10.2011).

Bush, N. A. 1898. A previous account of a second journey through north-western Caucasus in 1898 [predvaritel'nyii otchiot vtorogo puteshestviia v Zapadnom Kavkaze v 1898]. *Izv. Russk.Geogr. Ob-va*, 34: 41-47 (in Russian).

Bush, N. A. 1935. *Botanical and geographical study of the Caucasus [Botaniko-geograficheskii ocherk Kavkaza]*. Izd. AN SSSR [USSR Ac. Sci. Publ. House]. Moscow-Leningrad, 135pp. (in Russian).

Butkhuzi, L. 2004. Breaking news: leopard in Georgia. *Caucasus Environment*, 2:48-51.

Butnik, A. A. 1991. *The ecological anatomy of desert's vegetation of Middle Asia. Tashkent [Ecologicheskaiya anatomiia pustynnykh rastenii Sredneii Azii]*, I:74-82 (in Russian).

Buttiker, W. 1979. Fauna of Saudi Arabia. Zoological collections from Saudi Arabia. *Fauna of Saudi Arabia*, 1: 1-22.

Cairney, J. W. C. 2000. Evolution of mycorrhiza systems. *Naturwissenschaften,* 87:467-475.

Calagari, M., Modirrahmati, A. R., Asadi, F. 2006. Morphological variation in leaf traits of *Populus euphratica*. Oliv. Natural populations. *Intern. J. Agric. & Biology*, 8(6):754-758.

Callaway, R. M., Kikvidze, Z., Kikodze, D. 2000. Facilitation by unpalatable weeds may conserve plant diversity in overgrazed meadows in the Caucasus Mountains. *Oikos,* 89,2: 275-282.

Calow, P. (Ed.). 1999. The Blackwell's Concise Encyclopedia of Ecology, 11th Ed. *Wiley-Blackwell*, 168pp.

Cann, R. L., Stoneking, M., Wilson, A. C. 1987. Mitochondrical DNA and human evolution. *Nature,* 1:325(6099):31-36.

Cardinale, B. J., Duffy, J. E., Gonzalez, A., Hooper, D. U. et al. 2012. Biodiversity loss and its impact on humanity. *Nature*, 486, 59–67.

Cardona, T., Murray, J. W., Rutherford, A. W., 2015. Origin and evolution of water oxidation before the last common ancestor of the cyanobacteria. *Molecular Biology and Evolution*, 32,5: 1310-1328.

Carella, R. 2013. First observation of *Quercus pedunculiflora* C. Koch in the Italian Peninsula – short communication. *J. Forest Science*, 59,3:130-135.

Carroll, C., Noss, R. F., Paquet, P. C., Schumaker, N. H. 2004. Extinction debt of protected areas in developing landscapes. *Conserv. Biol.*, 18:1110-1120.

Cavalli-Sforza, L. L., Menozzi, P., Piazza, A. 1994. The history and geography of human genes. *Princeton University Press*, Princeton, 1,032pp.

Cerling, T. E., Harris, J. M., MacFadden, B. J. et al. 1997. Global vegetation change through the Miocene/Pliocene boundary. *Nature,* 389:153-158.

Chaloner, W. G., Sheerin, A. 1979. Devonian macrofloras. In: House, M. R., Scrutton, C. T., Bassett, M. G. (Eds.). The Devonian System. *Palaeontological Society Special Paper in Palaeontology*, 23:145-161.

Chaney, R. W. 1947. Tertiary centers and migration routes. *Ecological Monographs*, 17:139-148.

Chape, S., Spalding, M., Jenkins, M. D. 2008. The world's protected areas: status, values and protects in the 21st century. *Prepared by the UNEP World Conservation Monitoring Centre, University of California Press*, Berkley, USA.

Chatterjee, R., Gosh, A. K., Kar, R., Narasimha Rao, G. M. 2014. Dwarfism and Lilliput effect: a study on the *Glossopteris* from the late Permian and early Triassic of India. *Current Science*, 107/10: 1735-1744.

Chen, Y. N., Wang, Q., Ruan, X., Li, W. H., Chen, Y. P. 2004. Physiological response of *Populus euphratica* to artificial water-recharge of the lower reaches of Tarim river. *Acta Bot. Sin*, 46: 1363-1401.

Chernov, Yu. I. 1975. *Natural zonation and the terrestrial animal world [Prirodnaiya zonal'nost'izhivotnyi mir sushi]*. Mysl, Moscow, 222 pp. (in Russian).

Chestin, I. E., Guber, Y. P., Sokolov, V. E., Lobachev, S. V. 1992. The brown bear (*Ursus arctos* L.) in the USSR: numbers, hunting and systematics. *Ann. Zool. Fenniei*, 29:57-68.

Chibilev, A. A. 1990. The face of the steppes [Lik stepeii]. *Gidrometeoizdatel'stvo [Hydrometpublish]* Leningrad, 192pp. (in Russian).

Chincharauli, G. P. 1968. *Vegetation of open woodlands of the Vashlovani reserve: seasonal dynamics and role in the nutrition of some vertebrates [Rastitel'nost'aridnykh redkolesii Vashlovanskogo zapovednika: sezonnaiya dinamica i rol' v pitanii nekotorykh pozvonochnykh]*. Publ. House of Tbilisi St. Univ. PhD thesis biol. sci., Tbilisi, 30pp. (in Russian).

Christenhusz, M. J. M., Byng, J. W. 2016. The number of known plants species in the world and its annual increase. *Phytotaxa*, 261,3:201-217.

Chubukov, L. A. 1966. Climate [Klimat]. In: *The Caucasus [Kavkaz]*. Moscow, Nauka, 85-125 (in Russian).

Clauss, M., Kaiser, T., Hummel, J. 2008. The morpho-physiological adaptations of browsing and grazing mammals. In: *Gordon, I. J. & Prins, H. H. T. (Eds.), The Ecology of Browsing and Grazing*: 47–88. Springer, Heidelberg.

Clements, F. E., Shelford, V. E. 1939. Bio-ecology. New York: *John Wiley*, 425pp.

Climate change and agriculture in Kakheti. 2014. *UNDP Georgia 2014*. Tbilisi, 239pp.

Cochrane, M. A. 2003. Fire science for rainforests. *Nature,* 421:913-919.

Coene, F. 2010. The Caucasus: an introduction. *Routledge*, New York, USA, 238pp.

Coffin, L. 2015. Shota Rustaveli 'The knight in the panther skin'. *Poezia Press*, Tbilisi, Georgia

Conci, C., Rapisarda, C., Tamanini, L., 1996. Annotated catalogue of the Italian Psylloidea. Second part (Insecta, Homoptera). *Atti Acc. Rov. Agiati*, a. 245 (1995), ser. VII. vol. V, B: 5-207.

Connor, S. E. 2011. *A Promethean Legacy: Late Quaternary vegetation history of Southern Georgia, the Caucasus.* Leuven: Peeters, 387pp.

Connor, S. E., Kvavadze, E. V. 2008. Modelling late Quaternary changes in plant distribution, vegetation and climate using pollen data from Georgia, Caucasus. *J. Biogeogr*. 36:529–545.

Cope, M. J., Chaloner, W. G. 1980. Fossil charcoals as evidence of past atmospheric composition. *Nature* 283:647–49.

Corso, M. 2013. *Georgia offers a super bee to help ailing American bees.* www.eurasianet.org/node(66821/). Retrieved 21.11. 2014.

Costanza, R., d'Arge, R., de Groot, R. et al., 1997. The value of the world's ecosystem services and natural capital. *Nature,* 387: 253-260.

Courtillot, V. 2003. *Evolutionary catastrophes. The science of mass extinction.* Cambridge University Press, 173pp.

Cox, C. B. 2001. The biogeographic regions reconsidered. *J. Biogeogr*., 28, 511–523.

Crane, E. 1983. *The archeology of beekiping*. Curnell University Press. Duckworth, London, 360pp.

Crepet, W. L., Niklas, K. J. 2009. Darwin's second 'Abominable Mystery': Why are there so many angiosperm species? *Americ. J. Bot.:* 96 (1):366-381.

Crous, P. W., Groencwald, J. Z., Hill, C. F. 2002. *Cylindrocladium pseudonaviculatum* sp. nov. from New Zealand, and new *Cylindrocladium* records from Vietnam. *Sydowia*, 54(1):23-34.

Cullen, J. 1986. *Yucca.* In: *Walters S. M. et al. (Eds.), The European Garden Flora, vol. 1.* Cambridge University Press, Cambridge: 273-276.

Curty, H. L., Sofletea, N., Toader, A. V., Enescy, M. 2011. Leaf morphological and genetic differentiation between *Quercus robur* and its closet relative, the drought tolerant *Quercus pedunculifolia* k. Koch. *Annals of Forest Sicence*, 68:1163-1172.

Czerepanov, S. K. 2007. *Vascular plants of Russia and adjacent states (the former USSR).* Cambridge University Press, Cambridge, UK. 532pp.

Dadant, C. C. 1984. *The hive and the honeybee,* Dadant and Sons, Hamilton, Illinois, 556pp.

Dai, A. 2011. Drought under global warming: a review. *WIREs climate change.* John Wiley & Sons Ltd., 2:45-65.

Darlington, P. J. Jr. 1957. *Zoogeography, the geographical distribution of animals.* New York: John Wiley and Sons Ltd, 1-688.

Davidson, A. D., Detling, J, K., Brown, J, H. 2012. Ecological roles and conservation challenges of social, herbivorous, burrowing mammals in the world's grasslands. *Frontiers in Ecology and the Environment*, 10,9: 477-486.

Davis, P. H. 1982. *Flora of Turkey and the East Aegean Islands. Vol. 7.* Edinburgh: Edinburgh University Press.

Davis, P. H. (Ed.) 1965- 1988. *Flora of Turkey and the East Aegean Islands.* Edinburgh: Edinburgh University Press. pp. 1–11.

Davis, P., Kenrick, P. 2004. Fossil Plants. *Natural History Museum*, Washington, DC: Smithsonian Books, 216pp.

Davies, T. J., Barraclough, T. G., Chase, M. W., Soltis, P. S., Soltis, D. E., Savolainen, V. 2004. Darwin's abominable mystery: in sights from a surface of the angiosperms. *Proc. Nat. Acad. Sci.*, USA, 101:1904-1909.

De Candolle, A. 1883. *Origin of cultivative plant [Origine des Plantes Cultivées].* Paris: 258-260.

de Waal, Th. 2010. *The Caucasus. An Introduction.* Oxford Press, 272pp.

DeConto, R. M., Pollard, D. 2003. Rapid Cenozoic glaciation of Antarctica induced by declining atmospheric CO2. *Nature,* 421, 245–249.

Delin, D., Kazmer, D., Thompson, D. 2009. Genetic characterization of the *Diorhabda* species complex released for biological control of *Tamarix. Envir. Entomol.*, 38:1373-1378.

Delle, G. V. 1962. On the history of forest vegetation of lake Sevan Basin [K istorii lesnoi rastitel'nosti basseina ozera Sevan]. *Botan. Zhurn.* 47:198-204 (in Russian).

DeLoach, C. J., Carruthers 2004. Biological control programs for integrated invasive plant management. In: *Proceedings of Weed Science Society of America Meeting,* Kansas City, MO. Weed Science of America (SD-Rom), 17pp.

Denevan, W. M. 1992. The pristine myth: the landscape of the Americas in 1942. *Annals of the American Geographers*, 82,3:369-385.

Denk, Th., Frotzler, N., Davitashvili, N. 2001. Vegetational patterns and distribution of relict taxa in humid temperate forests and wetlands of Georgia (Transcaucasia). *Biol. J. Linn. Soc.* 72,2:287-333.

Denk, Th., Grimm, G., Stögerer, K., Langer, M., Hemleben, V. 2002. The evolutionary history of *Fagus* in western Eurasia: evidence from genes, morphology and the fossil record. *Plant Systematics and Evolution*, 232:213-236.

Dhar, A. 2008. *Biodiversity of English yew (Taxus baccata L.) populations in Austria. PhD thesis*. Universität für Bodenkultur, Wien, 245pp.

Dhief, A., Guasmi, F., Triki, T., Mohamed, N., Aschi-Smiti, S. 2011. Natural genetic variation in *Calligonum*. Tunisian genus analyzed by RAPD markers. *African Journal of Biotechnology*, 10(48): 9766-9778.

Didmanidze, E. A. 1979. Butterflies of the arid part of Transcaucasia [Babochki aridnoii chasti Zakavkaz'iya]. In: Gegechkori, A. M. (Ed.). In: *Some animal groups of the arid landscapes of Transcaucasia* [*Nekotorye gruppy aridnoii chasti Zakavkaz'iya*]: 43-114 (in Russian).

Didmanidze, E. 2005. The butterflies of Georgia (Gegechkori A. M. -Ed.). *Studia – 99*, Tbilisi, 87pp. (in Georgian and English).

Dietz, R. S., Holden, J. C. 1970. Reconstruction of Pangaea: breakup and dispersion of continents, Permian to present. *J. Geophys. Res.*, 75(26):4939-4955.

Dimeyeva, L. A. 2010. Psammophyte succession of the northern coast of the Caspian Sea. *Bull. NAN RK. Ser.Biol. i Med.,* 3: 11-15 (in Russian).

Dinnik, N. J. 1882. Caucasian mountain goat [Kavkazskie gornye kozly]. *Trudy S.-Petersb. Obch. Estestvoispytateleii* [*Proceed. S.-Petersb. Nat. Soc.*] 13:16-21 (in Russian).

Dixon, A. P., Faber-Langendoen, C. J., Morrison, J., Loucks, C. J. 2014. Distribution mapping of world grassland types. *Journal of Biogeography:* 1-17.

Djanashvili, A. 1963. *The animal kingdom of Georgia [Saqartvelos tskhovelta samkaro]. Vol. 3, vertebrates*. Metsniereba, Tbilisi, 458pp. (in Georgian).

Djanelidze, Ch. P., Margalitadze, N. A. 1977. To the question of history of Kazbegi region forest vegetations in the Holocene [K voprosu istorii lesnykh rastenii Kazbegskogo regiona v Golocene]. In: *Nakhutsrishvili, G. (Ed.): Kazbegi Alpine Ecosystems*:17-21. Tbilisi-Moscow (in Russian).

Djanelidze, Z., Tatashidze, Z. 2010. Palaeogeographic interpretation of archaeological sites of swamps of the Black Sea coast of Georgia [Paleogeograficheskaya traktovka arkheologicheskikh ob'iektov v bolotakh prichernomoriya Gruzii]. *Bull. Georg. Acad. Sci.*, 4,3: 159-161 (in Russian).

Dokuchaev, V. V. 1948. The study of nature's zones [K izucheniyu o zonakh prirody] (317-329pp.). In: *Selected works*. Gosudarstv. Izd. Sel.-Khoz. Lit. (State Publ. Agricult. Lit.), Moscow, 480pp. (in Russian).

Dolaberidze, R. M. 1984. The pottery with depictions of the turs from Shenako [Metuneoba djikhvis gamosakhulebit Shenaqodan]. *Bull. Georg. State Mus.* 37b: 98-111. Tbilisi (in Georgian).

Dolukhanov, A. G. 1948. Relict stand of yew trees in the Upper Alazani River [Tissovaiya reliktovaiya roscha v verkhnem techenii reki Alazani]. *Trans. Tbil. Bot. Inst.*, 12:91-105 (in Russian).

Dolukhanov, A. G. 1956. Subalpine crook-stem forests of the Caucasus [Subal'piiskie krivoles'iya Kavkaza]. In: *Acad. Sukachev 75th birth:* 158-170. Moscow-Leningrad (in Russian).

Dolukhanov, A. G. 1964. *Dark coniferous forests of Georgia [Temnokhvoijnnye lesa Gruzii]*. Metsniereba [Sci. Publ.], Tbilisi, 126pp. (in Russian).

Dolukhanov, A. G. 1966. Regulations of geographical diversity of forest vegetation and the upper timberline in the Transcaucasian mountains [Reguliyatsia geograficheskoi raznoobrazii lesnoi rastitel'nosti verkhnego predela lesov Zakavkazkikh Gor]. *Probl. Bot.*, 8:196-207, Moskva-Leningrad (in Russian).

Dolukhanov, A. G. 1969. Flora and vegetation of subnival landscapes of the Didi Liakhvi River and the Keli Upland (Greater Caucasus) [Flora i rastitel'nost' subnival'nogo landshafta reki Bol'shogo Liakhvi i Keli Vozvyshennost (Bol'shoii Kavkaz.]. *Botan. Zhurn.*, 54: 1662- 1674 (in Russian).

Dolukhanov, A. G. 1978. The timberline and the subalpine belt in the Caucasus Mountains, USSR. *Arctic and Alpine Research*. Vol. 10,2:404-422.

Dolukhanov, A. G. 1980. *Colchic underforest [Kolkhidskii podlesok]*. Metsniereba, Tbilisi, 261pp. (in Russian).

Dolukhanov, A. G. 1989. Forest vegetation of Georgia [Lesnaiya rastitel'nost' Gruzii]. In: *Vegetation of Georgia*, vol. I. Sci. Publ. House, Tbilisi, 240pp. (in Russian).

Dolukhanov, A. G. 1992. Vegetation of Georgia. Vol. 2. *Forest vegetation of Georgia (unpublished manuscript).*

Dolukhanov, A. G. 2010. *Forest vegetation of Georgia [Lesnaija rastitel'nost' Gruzii].* Universal Publ. Tbilisi, 532pp. (in Russian).

Dolukhanov, A. G., Nakhutsrishvili, G. 2003. The natural vegetation formations of Europe and their subordination. Hygrophilic, thermophytic foliage mixers [Die Natirlichen vegetations formationen Europas und ihre unterglie – derung. Hygrophile thermophytische laubmischuvälder. pp. 384-388]. In: *Karte der Natürlichen Vegetation Europas (Map of the Natural Vegetation of Europe. Mabshtab (Scale) 1:2.500.000. U. Bohn, G. Gollub and C. Hettwer – Eds.* Federal Agency for Nature Conservation. Bonn, Germany.

Dolukhanov, A. G., Sakhokia, M. F., Kharadze, A. L. 1942. On the zonation of high-mountain vegetation of the Caucasus [K zonal'nosti vysokogornykh rastenii Kavkaza]. - *Proceed. Tbilisi Bot. Inst.* 8: 114-130. (in Russian).

Dolukhanov, A. G., Sakhokia, M. Ph., Kharadze, A. L. 1945. About high mountainous vegetation belts of the Caucasus [K voprosu o vysokogornykh rastitel'nykh poiyasakh Kavkaza]. *Proceed. Tbilisi Bot. Inst.* [*Trudy Tbil. Bot. Inst.*] 8:61-72 (in Russian).

Drews, R. 2015. Evolutionary of ice-shelf channels in Antarctic ice shelves. *The Cryosphere*, 9:1169-1181.

Dudley, N., Stolton, S. 2015. An assessment of the role of protected landscapes in conserving biodiversity in Europe. In *Nature Policies and Landscape Policies: Towards an Alliance*; Gambino, R., Peano, A., Eds. Springer Intern. Publ.: Cham, Germany, pp. 315–322.

Dugarjav, C., Gunin, P. D., Bazha, S. N., Saandar, M. 2010. Ecosystem approach for evaluating degradation processes and nature protection in inner Asia. *Geography Environmental Sustainability*, 3(2): 50-66.

Dumitrashko N. V., 1966. *The Caucasus [Kavkaz]*. Nauka, Moscow, 481pp. (in Russian).

Dumitrashko, N. V. 1977. *General characteristics and history of development of the relief of the Caucasus [Obchaiya kharakteristica i istoriya razvitii rel'iefa Kavkaza]*. Nauka Publ., 288pp. Moscow (in Russian).

Ebrahimi, A. M. 2007. Armenian mouflon on Kabudan Island, Iran. *Caprinae*. Newsletter of the IUCN/SSC Caprinae Specialist group: p.4.

ECPC (Environmental Conservation and Protection Center), 2005.

ECPC (Environmental Conservation and Protection Center), 2012.

Eckert, W. G. 1997. *Introduction to forensic science. 2nd edit.* CRC Press. 711 Third Avenue, New York, NY 10017. United States of America. 398pp.

Edwards, E. J., Osborne, C. P., Strömberg, C. A. E., Smith, S. A., and C4 Grasses Consortium, 2010. The origins of C4 grasslands: Integrating evolutionary and ecosystem science. *Science,* 328: 587-591.

Efremov, Y. V. 1988. *The blue necklace of the Caucasus [Goluboe ozherel'ie Kavkaza].* Gydrometeoizdat, Leningrad, 160pp. (in Russian).

Ehleringer, J. R., Cerling, T. E., Dearing, M. D. 2002. Atmospheric CO_2 as a global change driver influencing plant-animal interactions. *Integrative and Comparative Biology*, 42:424-430.

Ehleringer, J. R., Cerling, T. E., Helliker, B. R. 1997. C_4 photosynthesis, atmospheric CO_2, and climate. *Oecologia,* 112:285-299.

Ehleringer, J. R., Sage, R. F., Flanagan, L. B., Pearcy, R. W. 1991. Climate change and the evolution of C_4 photosynthesis. *Trends in Ecololy and Evolution,* 6:95-99.

Eicher, D. L., McAlester, A. L. 1980. *History of the Earth*: Prentice-Hall, Englewood Cliffs, 413 p.

Elagin, I. N. 1953. Oak forests of the extreme western part of the northern slope of the Caucasus [Dubovye lesa zapadnoii chasti severnogo sklona Kavkazskogo khrebta]. In: *Broad-leaved forests of the north-western part of the Caucasus.* Publ. House Ac. Sci. USSR: 142-183 (in Russian).

Elashvili, K. 2008. The language of tree [Khis ena]. *Georgian Folklore*, 4, 20:31-36 (in Georgian).

Eldarov, E. M., Holland, E. C., Aliyev, S. M., Abdulagatov, Z. M., Atayev, Z. V. 2007. Resettlement and migration in post-soviet Dagestan. *Eurasian Geography and Econimics*, 48,2:226-248.

Elenberg, H. 1996. Vegetation of Central Europe with the Alps in an ecological, dynamic and historical perspective. [Vegetation Mitteleuropas mit den Alpen in ökologischer, dynamischer und historischer Sicht]. Stuttgart*: Verlag Eugen Ulmer.* 5. Auflage. 1095 Seiten.

El-Raz, F. 1988. Origin and evolution of the desert. *Interdisciplinary Science Review.* 13,4:331-347.

Elman, E., Salisbury, N. 2009. The state of Seattle's conifer forests. *Seattle Urban Nature,* 14pp.

Emanuelsson, U. 2009. The rural landscapes of Europe – How man has shaped European nature. Stockholm (Sweden): *Swedish Research Council Formas*, 383pp.

Emeljanov, A. F. 1974. Proposals on the classification and nomenclature of ranges [Predlozhenye po klassifikatcii i nomenclature arealov]. *Entomol. Obozr.* [*Entomol. Rev.*], 53(3), 497–522 (in Russian).

Emeljanov, A. F. 1999. On the historical and ecological approaches to biogeographic zonation. Methodological issues in zoological research. *Intern. Conf. dedicated to the 275th anniversary of the Russian Ac. Sci. St.Petersburg,* 26-29 October 1999, Abstract:27-29 (in Russian with English summary).

Engel, M. S., Hinojosa-Diaz, I. A., Rasnitsyn, A. P. 2009. A honey bee from the Miocene of Nevada and the biogeography of Apis (Hymenoptera: Apidae: Apini). *Proc. Califor. Ac. Sci.,* 60, 3: 23 – 38.

Eppelbaum, L. V., Khesin, B. E. 2012. *The origin of the Caucasus, geological evolution and main features.* XVI, Springer, 401pp.

Erschbamer, B., Mallaun, M., Unterluggauer, P. et al. 2010. *Plant diversity along altitudinal gradients in the Central Alps (South Tyrol, Ilaly) and in the Central Greater Caucasus (Kazbegi region, Georgia).* Göttingen, Tuexenia, 30: 11-29.

Esmail-pour, A. 1998. Evaluation, identification and collection of male pistachio cultivars. *Reports for 1998, Pistachio Research Institute*, Rafajan, Iran.

Estes, R. D. 1991. *Behavioral guide to African mammals including hoofed mammals, carnivores and primates.* Russel Friedman Books Publishers. South Africa, 611pp.

Even, R., Odegaard, F., Oug, E., Sandvik, H. 2012. Alien speceis: introduction, establishment and spread (9-56pp.). In: *Gedereas L., Moen, T. L., Skjelseth, S., Larsen, L.-K. (Eds.) 2012.* Alien species in Norway – with the Norwegian Black List. Norwegian Biodiversity Information Centre, 212pp.

Ezcurra, E, et al. 2006. Chapter 1: Natural history and evolution of the world's deserts, p 148. In: Ezcurra E, (Ed.), *Global deserts outlook.* United Nations Environment Programme, Nairobi, Kenya.

Fairbridge, R. W. 1968. Hamada, reg, serir, gibber, sai. *Geomorphology, Encyclopedia of Earth Science:* 520-521.

Fajvush, G. M. 1990. On the autochtonal and allochtonal processes in the development of steppe flora in the Armenian SSR [K avtokhtonnym i allokhtonnym processam razvitii stepnoi flory v Armianskoii SSR]. *Biol. Zhurn. Arm.* 43(3): 220-225 (in Russian).

Fajvush, G. M. 1991. The analysis of the taxonomic structure of Armenian steppe flora [Analiz sistematicheskoi struktury flory steppei Armenii]. *Flora & Rast. Res. Armenii.* [*Flora & Veget. Res. Armenia*], 13: 184-206 (in Russian).

Fajvush, G. M. 1992. *Flora and vegetation of the Armenian steppes [Stepnaiya flora i rastitel'nost' Armenii]. PhD thesis biol. sci.*, 23pp.Yerevan (in Russian).

Fajvush, G. M. 1996. On the origin of mountain steppes of Armenia. *Thaiszia. J. Bot., Košice,* 5 (1995): 153-158.

Fang, J., Lechowicz, M. J. 2006. Climatic limits for the present distribution of beech (*Fagus* L.) species in the world. *J. Biogeogr.*, 33, 1804-1819.

Farhadinia, M, 2004. The last stronghold: cheetah in Iran. *Cat News.* 40:11-14.

Farajova, M. 2017. About specific of rock art of Gobustan and some innovative approaches to its interpretation ('Firuz 2' shelter). *Quaternary International*, http://dx.doi.org./10.1016/ J.Quaint. 2017.07.034.

Farhadinia, M., Hemami, M. R. 2010. Prey selection by the critically endangered Asiatic cheetah in central Iran. *Journ. Nat. Hist.*, 44, 19-20:1239-1249.

Farouk, El-Baz, 1988. Origin and evolution of the desert. *Interdisciplinary Science* Reviews. 13:331–347.

Fedorov, An. A. 1942. Alpine carpets of the Caucasus and and their origin [Al'piiskie kovry Kavkaza i ikh proiskhozhdenie]. *Trudy AN ArmSSR [Proceed. Acad. Sci.Arm.SSR]*, 4:137-155 (in Russian).

Feeny, P. P., Arnold, R. W., Bailey, R. S. 1968. Autumn migration in the south Caspian Region. *Ibis.*110: 35–68.

Ferret, C. 2009. *Role of the horse in civilization [Une civilization du cheval].* Paris, Berlin, 350pp.

Ferring, R., Oms, O., Agusti, J., Berna, F. et al. 2011. Earliest human occupations at Dmanisi (Georgian Caucasus) dated to 1.85-1.78 Ma. *Proceed. Nat. Acad. Sci. of the USA*. 108, 26:10432-10436.

Fert, G. 2001. Georgia, birthplace of the Caucasian bee, *Amer. Bee J.*, 4,11: 789-792.

Fet, V., Polis, G. A., Sissom, W. D. 1998. Life in sandy deserts: The Scorpion model. *Journal of Arid Environments*, 39,4: 609-622.

Fidarova, A. Ch. 2006. *Distribution, bioecological peculiarities and resources of the blueberry of the Caucasus Republic of Severnaja Ossetia – Alania [Rasprostranenie, bioekologicheskie osobennosti i resursy cherniki Kavkaza v Respublike Severnaiya Osetia - Alaniya]. Candidate dissert. PhD thesis biol. sci.*, 22pp. Vladikavkaz (in Russian).

Figurovski, I. V. 1919. The climates of the Caucasus. Predvarit. Ocherki [Klimati Kavkaza. Preliminary Essay]. *Zap. Zakavk. Otd. Russk. Geogr. Obsch-va*) (*Papers of Transc. Branch of Rus. Geogr. Soc.*), 29, 5:27-35. Tiflis (in Russian).

Figurovski, I. V. 1926-1927. The climatical regionalization of Azerbaijan SSR [Klimaticheskoe rajonirovanie Azerbaijanskoi SSR]. *Materialy Po Raionirov. Az. SSR [Materials about Regionaliz. Azerb. SSR]*, I, 1-2:32-41 Baku (in Russian).

Filonov, K. P., Kaletskaya, M. L. 1985. The Influence of wolf predation on wild ungulates [Vliyanie vol'chikh khishnichestv na dikikh kopytnykh]. In: *The wolf.* D. I. Bibikov, Ed.), Science Publ., Moscow: 336-354. (in Russian).

Firsov, G. A. 1998. *Quercus robur* spp. *imeretina*. The IUCN Red List of Threatened Species 1998: *e.T34682A9877851*.

Fischer, A. von, Marshall, P., Camp, A. 2013. Disturbances in deciduous temperate forest ecosystems of the northern hemisphere: Their effects on both recent and future forest development. *Biodivers. Conserv.* Springer; Netherlands; 22: 1863–1893.

Fischer, J. C., Tieszen, L. L., Schimel, D. S., 2008, Climate controls on C3 vs. C4 productivity in North American grasslands from carbon isotope composition of soil organic matter. *Global Change Biology*, 14, 5: 1141-1155.

Flanagan, L. B., Ehleringer, J. R., Pataki, D. E. (Eds.) 2005. *Stable isotopes and biosphere – atmosphere interactions: processes and biological contols*. Elsevier Academic Press. San Diego, US, 318pp.

Fleischer, Th. L. (Ed.). 2011. *The way of Natural History*. Trinity University Press, San Antinio, Texas, 204pp.

Flerov, A. F. 1951. The vegetation of the Colchis Lowland [Rastitel'nost' Kolkhidskoii nizmennosti]. *Bull. VNICNISK*, 1:9-123 (in Russian).

Flerov, A. F., Balandin, V. N. 1931. Steppes of North-Caucasus Krai [Stepy Severo-kavkazskogo Kraiya]. *Rostov na Donu*, 130pp. (in Russian).

Flerov, K. K., Belyaeva, E. I., Yanovskaya, N. M. et al. 1974. *The Paleogene zoogeography of Asia [Paleogenovaia zoogeografiya Azii]*. Nauka Publ., Mocow, 303pp. (in Russian).

Flora of the USSR, 1961. Vol.26,432pp.

Formozov, A. N., 1966. Adaptive modifications of behavior in mammals of the Eurasian steppes. *Journal of Mammalogy,* 47,2:208-223.

Fotjanova, L. I. 1984. Late Eocene preturgaian flora from Old Beringia. *Bot. Zhurn.*, 6(4):425-436 (in Russian with English summary).

Free, J. B. 1970a. Insect Pollination of Crops. *New York Academic*. 544pp.

Freitag, H., Golub, V. B. Yuritsina, N. A. 2001. Halophytic plant communities in the northern Caspian Lowlands: 1, annual halophytic communities. *Phytocoenologia*, 31: 63-108.

Fridland, V. M. 1950. Mountainous meadow-steppe soils of Inner Dagestan [Gornye lugovo-stepnye pochvy vnutrennogo Dagestana]. *Doklady AN SSSR,* 70, 4:695–697 (in Russian).

Fridland, V. M. 1966. Soils ([Pochvy]. In: *Caucasus* [*Kavkaz*]. Nauka, Moscow, 187-222pp.

Fritz, R. S., Simms, E. L. (Eds.) 1992. *Plant resistance to herbivores and pathogens: Ecology, evolution, and genetics.* University of Chicago Press. Chicago, 600pp.

Gabrielian, G. (Ed.) 1986. *Physical geography of Transcaucasia [Fizicheskaiya geografiya Zakavkaz'iya].* Yerevan State Univ. Publ. Yerevan, 155pp. (in Russian).

Gabunia, L., Vekua, A. K. 1963. *The fossil elephant of Taribana [Taribanas namarkhi spilo].* Metsniereba [Science], Tbilisi, 69pp. (in Georgian).

Gabunia, L., Gabashvili, E., Vekua, A., Lordkipanidze, D. 2001. The Late Miocene hominoid from Georgia. In: *L.de Bonis, C. D. Kaufos & P. J. Andrews (Eds.) Hominoid evolution and environmental change in the Neogene of Europe. De Bonis, L. et al. (Eds.).* Vol. 2. Cambridge, UK: Cambridge Univ. Press: 316-325.

Gabunia, L., Vekua, A. 1995. Plio-Pleistocene hominid from Dmanisi, East Georgia, Caucasus. *Nature*, 373:509-512.

Gabunia L., Vekua, A., Lordkipanidze, D., et al. 2000a. Early Pleistocene hominid cranial remains from Dmanisi. Republic of Georgia: taxonomy, geological setting and age. *Science*, 288: 1019-1025.

Gabunia L., Vekua, A., Lordkipanidze, D. 2000b. The environment contexts of early human occupation of Georgia. *J. Hum. Evol.*. 38:785-802.

Gabunia L., Vekua, A., Lordkipanidze, D., Ferring, R., Justus, A., et al. M. 2000c. Current research on hominid site of Dmanisi. In: *Early Humans at the gates of Europe.* (Lordkipanidze, D. et al.-Eds.), 92, 6-23.

Gadjiev, G. A., Ragimov, V. A. 1977. *The climatic characteristic of administrative regions of Azerbaijan SSR [Klimaticheskaya kharakteristika administrativnikh raijonov Azerbaijanskoi SSR].* Elm, Baku, 269pp. (in Russian).

Gagnidze, R. I. 1966. Floristic peculiarities of the subalpine, tall, herbaceous vegetation of the Greater Caucasus [Floristicheskie osobennosti subal'piiskogo vysokotraviya Bol'shogo Kavkaza]. *Probl. Bot.*, 8:30-40. Moscow-Leningrad (in Russian).

Gagnidze, R. I. 1970. On the typology of altitudinal zonation in the Caucasus [Kavkasiis vertikaluri sartklebis tipologiisatvis]. *Bull. Acad. Sci. Georgia*, 59,3:673-676 (in Georgian).

Gagnidze, R. I. 1974. *Botanical and geographical analysis of the subalpine tall-grass florocoenotic complex of the Caucasus [Botaniko-geograficheskii analiz florocenoticheskogo kompleksa subal'piiskogo vysokotrav'iya Kavkaza]*, Metsniereba, Tbilisi, 162p. (in Russian).

Gagnidze, R. I. 1976. Geographical analysis of the florocoenitic complex of subalpine tall herbaceous plants [Geograficheskii analiz florostenoticheskogo komplekssa subal'piiskogo vysokotrav'iya Kavkaza]. In: *The ranges of the plants of the USSR* [*Arealy rastenii SSSR*], 3;27pp., Leningrad (in Russian).

Gagnidze, R. I. 2000. Diversity of Georgia's flora. In: *Biological and landscape diversity of Georgia.* WWF Georgia, Tbilisi, 21-32 (in Georgian and English).

Gagnidze, R. 2005. *Vascular plants of Georgia, a nomenclatural checklist.* Universali Publ.., Tbilisi, Georgia, 247pp. (In Latin with Georgian preface).

Gagnidze, R., Davitadze, M. 2000. *Local Flora (the world of Georgian flora) [Adgilobrivi flora (Saqartvelos floris samkaro)].* Gamomtcemloba Adjara [Adjara Publ.], Batumi, 274pp. (in Georgian).

Gagnidze, R. I., Shetekauri, Sh. K. 1988. The analysis of the high mountain rock and rocky outcrop florocenotic complexes of the southern slopes of the Central Caucasus. In: 'Rastitel'ny mir vysokogornykh ekosistem of the USSR'. Vladivostok: *DVO AN SSSR:* 202-208 (in Russian).

Gailing, O., von Wuehlisch, G. 2004. Nuclear markers (AFLPs) and chloroplast microsatellites differ between *Fagus sylvatica* and *F. orientalis*. *Sylvae Genetica*, 53(3): 105-110.

Galanos, C. J. 2015. The alien flora of terrestrial and marine ecosystems of Rodos island (SE Aegean), Greece. *Willdenowia* 45(2): 261–278.

Galstian, B. J. 1930. The main peculiarities of geography of the soils of Armenian SSR [Osnovnye osobennosti geografii pochv Armiyanskoii SSR]. In: *Zakavkazskii kraevedcheski Sbornik*, 'Zakavkaz'ie', Ser. A., Vol. 1:72-79, Tiflis (in Russian).

Galušhko, A. I. 1978-1980. *Flora of the Northern Caucasus [Flora Severnogo Kavkaza]*. Vol. 1-3. Izd. Rostovskogo Universiteta, Rostov-na-Donu (in Russian).

Gambarova, Y. M. 2011. Rare vegetation monitoring in the Gobustan National Park, Azerbaijan. Final Report, Baku, Azerbaijan. *The Ruffor Small Grant Foundation*: 3-25pp.

Gambarova, Y. M., Gambarov, A. Y., Rustamov, R. B., Zeynalova, M. H. 2010. Rare vegetation classification of remotely sensed images, Gobustan National Park., Azerbaijan. *Intern. Geoinform. Res. and Devel. J.*; 1,2:1-7.

Gamble, C. 1994. *Timewalkers: the prehistory of global colonization.* Harvard University Press. 309pp.

Gamkrelidze, I. P.1977. Main trend of tectonic development of Caucasus and adjacent areas of the Mediterranean belt. *Himalayan geology*, 7:88-96.

Gamkrelidze, I. P. 1986. Geodynamics evolution of the Caucasus and adjacent areas in Alpine time. *Tectonophysics*, 127: 261-277.

Gamkrelidze, I. P. 1997. Georgia. In: Encyclopedia of European and Asian regional geology. London, *Chapmen and Hal*, London, 256- 261.

Gamkrelidze, P. D. 1949. Geological structure of the Adjara-Trialeti folded system [Geologicheskaiya structura skladchatych system Adjara-Trialet]. *Proceed. Geol. Inst. AN Gruz. SSR*, 2:31-34, Tbilisi, (in Russian).

Gamkrelidze, P. D. 1960. The main features of the tectonic composition of the Caucasus. *Geotektonika*, 3:3-14 (in Russian).

Ganbarli, A. A. 1973. *The composition, biological and economic productivity of some friganoid phytocoenosis of the Nakhichevan ASSR [Sostav, biologich. i ekonomich. produktivnost' nekotorykh friganoidnykh fitocenozov Nakhichevanskoi ASSR]. Doctor dissertation (PhD thesis) biol. sci.*, Baku, 31pp. (in Russian).

Garcia, A., Revilla, E. 2013. *The current status of wild grape-vine populations (Vitis vinifera spp. sylvestris) in the Mediterranean Basin.* http://dx.doi.org./10.5772/52933.

Garn, S. M. 1971. *Human races.* Charles C. Thomas, Springfield, Illinois, 196 pp.

Gauger, K., Pietzsch, D. 2005. The National park Lake Ag-Gol – Contributions to the aviafauna of Azerbaijan. *Archiv für Naturschutz und Landschaftsforschung. [Archives of Nature Conservation and Landscape Research]*, 44(4): 57-68.

Gavashelishvili, A., Lukarevskiy, V. 2008. Modelling the habitat requirements of leopard *Panthera pardus* in West and Central Asia. *J. Appl. Ecol.*, 45:579-588.

Gavriljuk, F. S. 1947. Soils of Stavropol district [Pochvy Stavropol'skogo Kraiya]. *Stavropol*, 95pp. (in Russian).

Gazaryan, S., Shevchenko, D. 2014. Sochi 2014: ten years without rights to law. How 'the best world' was being built [Sochi-2014: desiyat let bez prava na zakon. Kak stroili 'luchii mir'] Analytic Report. North Caucasus. *CEE Bankwatch network (bankwatch org/sites/...)*. 1-77pp. (in Russian).

Gederaas, L., Salvesen, I., Viken, A. (Eds.) 2007. Norwegian Black List, 2007 [Norsk Svarteliste 2007]. *Okologiske risikovunderinger av fremmede arten (Ecological risk analysis of alien species)*. Troudheim, 152pp.

Gegechkori, A. M. 1968. Materials for the study of the diet of the psyllids (Homoptera, Psylloidea), narrowly specialized phytophages [Novye materialy k pischevoi specializatcii psyllid, uzko speciailizirovannykh fitofagov]. *Soobch. Acad. Nauk. Gruz. SSR*, 52:509-514 (in Russian).

Gegechkori, A. M. 1969. Some peculiarities of ecological investigations of Psyllids (Homoptera, Psylloidea) inhabiting the drylands of Georgia [Nekotorye osobennosti ekolog. izuchenii psyllid obitaiychikh v aridnykh usloviyakh Gruzii]. *Soobch. AN Gr.SSR*, 54,1:221-224. (in Russian with English summary).

Gegechkori, A. M. 1973. New species of psyllids (Homoptera, Psylloidea) from Georgia [Novye vidy psyllid iz Gruzii]. *Bull. Ac. Sci. Georg.SSR*, 69,1:181-184 (in Russian).

Gegechkori, A. M. 1974. New jumping plant-lice (Homoptera, Psylloidea) from Georgia [Novie listobloshki iz Gruzii]. *Bull. Ac. Sci. Georg. SSR*, 75,5: 702-704 (in Russian).

Gegechkori, A. M. 1975a. *Psyllids (Homoptera, Psylloidea) of Pshav-Khevsureti [Psillidy Pshav Khevsureti]*. Metsniereba, Tbilisi, 92pp. (in Russian).

Gegechkori, A. M. 1975b. Psyllids (Homoptera, Psylloidea) of the Lesser Caucasus, Part 1 (within the border of Meskhet-Djavakheti, South Georgia) [Psyllidy Malogo Kavkaza, Chast 1 (v predelakh Meskhet-Diavaketi, Yuzhnaiya Gruziya). *Vestnik Gos. Muzeia Gruzii*, vol. 28-A:24-35pp. (in Russian).

Gegechkori, A. M. 1976. Psyllids (Homoptera, Psylloidea) of the Lesser Caucasus (within the borders of Zemo Adjara and Guria) [Psyllidy Malogo Kavakza (v granitsakh Zemo Adjarii i Gurii]. *Vestnik Gos. Muzeija Gruzii* [Herald of the State Muz. of Georgia], 29 A: 301-330 (in Russian).

Gegechkori, A. M. 1978. *Psyllids of the middle reaches of the Aras River [Psillidy srednego techeniya reki Araks]*. Metsniereba, Tbilisi, 76pp. (in Russian with English summary).

Gegechkori, A. M. 1979. An ecological and faunistical review of psyllid fauna (Homoptera, Psylloidea) of the Kura-Aras Lowland. In: *'Some Groups of Animals of Arid Regions of Transcaucasia'*, Tbilisi: 19-42 (in Russian).

Gegechkori, A. M. 1980. New data about the psyllid fauna (Homiptera, Psylloidea) of Vashlovani Preserve [Novye Dannye po faune psyllid Vashlovanskogo zapovednika]. *Trudy Zapovedn. Gruzii* (*Bull. Georg. Preserv.*): 5:32-43 (in Russian).

Gegechkori, A. M. 1981. A new genus and species of Psylloidea (Homoptera) from Transcaucasia and Yugoslavia [Novy rod i vidy Psylloidea iz Zakavkaziya i Jugoslavii]. *Zool. Zhurn.*, 50,5: 695-703 (in Russian).

Gegechkori, A. M. 1982. *Vashlovani – smart arid woodland [Vashlovani – tke siketisa]*. Tbilisi, Sabchota Sakartvelo, 106pp. (in Georgian).

Gegechkori, A. M. 1983. *Tusheti is above the clouds [Grublebs zemot Tushetia]*. Sabchota Sakartvelo Publ., Tbilisi, 110pp. (in Georgian).

Gegechkori, A. M. 1984a. *Psyllids (Homoptera, Psylloidea) of the Caucasus [Psyllidy Kavkaza]*. Metsniereba, Tbilisi, 295pp. (in Russian).

Gegechkori, A. M. 1984b. *Species composition and origin of psyllids (Homoptera, Psylloidea) of the Caucasus [Vidovoi sostav i proiskhozhdenie psyllid Kavkaza]. Doctoral dissert. (PhD thesis) boil. sci.*, Tbilisi, 56pp. (in Russian).

Gegechkori, A. M. 1985. *Some aspects of evolution of psyllids [Nekotorye aspekty evolutcii psyllid].* Metsniereba, Tbilisi, 305pp. (in Russian with English summary and contents).

Gegechkori, A. M. 1987a. Psyllids (Homoptera, Psylloidea) of wormwood (clay) deserts of South Caucasus [Psyllidy polynovoi (glinistoi) pustyni Yuzhnogo Kavkaza]. *Soobch. AN GrSSR,* 126(2): 49-57 (in Russian).

Gegechkori, A. M. 1987b. Psyllids (Homoptera, Psylloidea) of various desert biotopes of the Caucasus [Psyllidy raznykh pustynnykh biotopov Kavkaza]. *Soobch. AN GrSSR*, 128, 3: 111-118 (in Russian).

Gegechkori, A. M. 1989. A new species of psyllids (Homoptera, Psylloidea) from Kazakhstan. *Entomol. Obozr.* 62, 2: 68-71. Leningrad (in Russian).

Gegechkori, A. M. 1990. A new species of Psylloidea from Kazakhstan. *Entomol. Soc. of America, Entomol. Obozr.* (in Russian) (translation), USA, 68, 6, 79-81.

Gegechkori, A. M. 1993. The role of Middle Asia in the formation of desert fauna of Psyllids (Homoptera, Psyllinea) of the Caucasus [Rol' Srednei Azii v formirovanii pustynnykh vidov psyllid Kavkaza] (the achievements of entomology in the USSR: ecology and fauna of a small size groups of insects). *Material. 10 S'ezda Vses. Entomol. Obsch.* [In: *Materials of the All Union Entomol. Soc., Leningrad* 1989]. Zool. Inst. RAN Publ.: 114-115 (in Russian).

Gegechkori, A. M. 1997a. *The ecological-geographical peculiarities of psyllid (Homoptera, Psylloidea) fauna of the Caucasus.* Meridiani, Tbilisi (in Russian with English summary).

Gegechkori, A. M. 1997b. *Nature – the creator [Buneba shemokmedia].* Meridiani, Tbilisi, 525pp. (in Georgian with English summary).

Gegechkori, A. M. 1999. Faunal relations between the Caucasus and Anatolian regions using the example of psyllids (Homoptera, Psylloidea). *XXXII congresso Societa di Biogeographia dell' Anatolia*, Roma:16.

Gegechkori, A. M. 2000. Relicts, endemics and Georgia's biodiversity in the background of the Caucasus orogenesis. In: *Biological and Landscapes Biodiversity of Georgia.* WWF Georgia, Tbilisi, 83-96 (in Georgian and English).

Gegechkori, A. M. 2004. *Human evolution, disappearance and tribal civilizations.* Tbilisi University Press (in Georgian with English and Russian prefaces), 325pp.

Gegechkori, A. M. 2007. *The Caucasus – a treasure of a Nature.* CEPF, WWF, Meridiani, 168pp. (in English and Georgian).

Gegechkori, A. M. 2008. *Biogeography, Chapter 1 (Biomes of the world) [Biogeografia, nats. I (Dedamitsis biomebi)*, Tbilisi State University Publ. House (in Georgian with English summary and contents), 527pp.

Gegechkori, A. 2009a. Animal Kingdom in 'The knight in the pather's skin'. In: *monograph: 'About 'The knight in panther's skin',* Vol. N2 (Ghvinjilia, N., Ed-in-chief), 'Sezani', Tbilisi, 230-246 pp. (in Georgian).

Gegechkori, A. 2009b. Countries mentioned in the poem 'The knight in panther's skin'. In *Monograph: About 'The knight in the panther's skin',* Vol. N2 (Ghvinjilia, N., Ed-in-chief), 'Sezani', Tbilisi, pp. 246-255.

Gegechkori, A. M. 2010a. '*The knight in the panther's skin' from biogeographer's point of view*'. Meridiani, Tbilisi (in Georgian with English preface).

Gegechkori, A. M. 2010b. Arid woodlands of the South Caucasus. *Georgian Nat. Mus. Proceedings*, National and Prehistoric Section, 2:137-146.

Gegechkori, A. M. 2010c. Pest psyllids (Hemiptera, Psylloidea) complex of horticulture (near trees plantations) of Georgia and the ways to avoid interspecies competition among some species. *Annals of Agrarian Science*, 8,4:11-15.

Gegechkori, A. M. 2011. The results of biogeographical study of Arcto-Tertiary refugia (Colchis and Talysh) of the southern Caucasus. *Annals of Agrarian Science*, 9,1:16-33.

Gegechkori, A. M. 2012a. The two refugia of the southern Caucasus, Colchis and Talysh: a comperative biogeographical analysis. *International Conference of IGU*, Santiago, Chile (*acreditacion. isa.cl/ugi/.../Gegechkori Arnold_20112324VO3ZVH_f.doc.*).

Gegechkori, A. M. 2012b. A short history of the foundations and development of the zoological department. *Newsletters of the State Museum of Georgia*. Tbilisi, 44:94-107 (In Georgian with Russian summary).

Gegechkori, A. M. 2013. Ecological relationship of distribution patterns of riparian softwood forests and their consumers – species of psyllids (Hemiptera, Psylloidea). *Annals of Agrarian Science,* 12,4:8-17.

Gegechkori, A. 2016. Insular distribution patterns of Goravan Sands (Armenia) using the example of psyllid (Hemiptera, Psylloidea) fauna. *Annual Sci. Conf. of Fac. of Exact and Nat. Sci*. Tbilisi State University, 25-29 January, Tbilisi.

Gegechkori, A. 2018. Patterns of distribution and survival of European yew (*Taxus baccata* L.) in an alpine tree line ecotone in the Greater Caucasus (Georgia). *Annals of Agrarian Science*. V. 16, 2, Pages 170-176.

Gegechkori A. M. 2019. Insular distribution patterns in the Goravan Sands (Armenia) exemplified by the Calligonum-feeding psyllids (Hemiptera: Psylloidea). *Turkish Journal of Zoology*, 34pp. (in press).

Gegechkori, A. M., Chubinidze, M., Popiashvili E. 2013. Fauna of psyllids (Psylloidea) of the riparian forests of the Caucasus and some peculiarities of their intrazonality. *Proc. Georg. Nat. Mus., Nat. Prehist. Sect.* Tbilisi:145-150.

Gegechkori A., M., Didmanidze, E. 2015. Has *Brahmaea ledereri* Rogenhofer, 1873 ever inhabited the Colchis refugium? *Amer. J. Environm. Protection*, *Science PG*, 4 (3-1): 82-92.

Gegechkori, A. M., Djibladze, D. S. 1976. *Psyllids (Homoptera, Psylloidea) of Colchis [Psillidy Kolkhidy]*. Metsniereba, Tbilisi, 110pp. (in Russian with English summary).

Gegechkori, A., Ginturi, Z. 2005. Study of the biology of pear psylloid – *Cacopsylla bidens* (Šulc) (Hemiptera, Psylloidea) in East Georgia (city of Gori environs). *Proc. Georg. Acad. Sci. Biol. Ser. B*: 3(3):73-77.

Gegechkori, A. M., Joosten, H. (Eds.) 2009. The biomes of the Caucasus: *Draft for the IMCG Field Symposium,* Georgia/Armenia, Sept. 2009. Greifswald (pp. 1-169).

Gegechkori, A. M., Loginova, M. M. 1990. *Psyllids (Hemiptera, Psylloidea) of the USSR (an annotated checklist) [Psillidy SSSR (annotirovanny spisok)]*. Metsniereba, Tbilisi, Georgia. 161pp. (in Russian with English summary).

Geideman, T. S., Doroshko, I. 1931. An Essay on the vegetation of salayan steppes [Ocherk o rastitel'nosti sal'yanskikh steppei]. In: *Trudy po Geobot. Obsled. Pastbysch* [*Proceedings Geobotan. Research of Rangelands*]. Izdanie Narkomzema [Narkomzem Publ.], 8,66pp. (in Russian).

Gerasimov, I. P. 1949. Materials of soils geography and cartography of the USSR [Materialy po geografii i kartografii pochv SSSR]. *Trudy Pochven. Inst. after V. V. Dokhuchaev*. 30 (Paraslov, L. I., Gerasimov, I.P.-Eds). Izd. AN SSSR, 323pp. (in Russian).

Gerasimov I. P. (Ed.). 1966. *The Caucasus. Natural conditions and natural resources of the USSR [Kavkaz. Prirodnye usloviya i prirodnye resursy SSSR]*. Nauka, Moscow, 482p. (in Russian).

Gerber, E., Kozlowski, G. 2014. *Relict tree of the Hyrcanian forest in the Talysh Mountains of Azerbaijan*. Natural History Museum. Fribourg, 38pp.

Ghukasyan, E. 2012. Thorny pillow-like vegetation site (11-12pp.). In: *Ecosystem Assessment Report.,* Yerevan, 2012, 46pp.

Gibson D. J. 2009. *Grasses and grassland ecology*. Oxford University Press. Oxford, New York, 305pp.

Gigauri G. N. 2000. Biodiversity of Georgian mountainous forest ecosystems. In Biological and landscape diversity of Georgia. *Proc. First Nat. Conf.*, 28–29 May 1999, Tbilisi, Georgia. WWF Georgia Country Office, Tbilisi: 69–82 (in Georgian and English).

Gigauri, G. N. 2010. *Criptic signs in mountainous eastern Georgia ('pickaxe' script) part 1, Chveni Saunje (Our Treasury),* Tbilisi (saunje.ge/index.php?id+1199&lang=ka) (in Georgian).

Giljarov, M. S. 1965. *A zoological method of diagnostics of soils [Zoologicheski metod dyagnostiki pochv]*. Nauka. Moscow, 276pp. (in Russian with English summary).

Giorgadze, P., Kikalishvili, Sh. 2013. The resources of minerals and fresh waters [Mineraluri da mtknari tsklebis resursebi]. In: *Agriculture, Georgia. Our valuables (Gulua – Ed-in-chief)* Iverioni, Tbilisi: 221-242 (in Georgian).

Gladkova, V. N. 1990a. Review of species of *Pyrus* L. (Rosaceae) from Transcaucasia [Obzor vidov roda *Pyrus* L. iz Zakavkaz'iya]. *Novosti Sistem. Vysh. Rast.* [*News of taxonomy of higher plants*]. Vol. 26:104-113 (in Russian).

Gladkova, V. N. 1990b. New taxa of genus *Pyrus* L. (Rosaceae) from Transcaucasia [Novye taksony roda *Pyrus* L. iz Zakavkaz'iya]. *Novosti Sistem. Vysh. Rast.* [News of taxonomy of higher plants], vol. 27:69-73 (in Russian).

Glasspool, I. J., Scott, A. C., Waltham, D., Pronina, N., Shao, L. 2015. The impact of fire on the Late Paleozoic Earth system. *Front. Plant Sci.* 6:756-771.

Gobejishvili, R. G., Kotlyakov, V. 2006. *Glaciology [Glaciologiya]*. Universali, Tbilisi, 292pp. (in Georgian).

Goetsch, L., Andrew, J. E., Hall, B. D. 2005. The molecular systematics of *Rhododendron* (Ericaceae): A phylogeny based upon *RPB2* Gene sequences. *Syst. Bot.*, 30,3: 616-626.

Goginashvili, N., Tvauri, I. 2013. Beech forest with yew (Fageta Taxceto) in Georgia. *Intern. Caucasian Forestry Symp.* 24-26 October. Artvin, Turkey: 976-981.

Gokhelashvili, R., Javakhishvili, Z., Gavashelishvili, A., Azniashvili, L. 2003a. Caucasian black grouse. *Buneba Print Ltd. GCCW*, Tbilisi, 63pp. (in Georgian with English summary).

Gokhelashvili, R., Kikodze, A. 2006. *Kolketi protected areas. Field Guide.* Buneba Print Publ., 96pp.

Gokhelashvili, R., Reese, K. P., Gavashelishvili, L. 2003b. How much do we know about the Caucasian Black Grouse *Tetrao mlokosiewiczi*? Sandgrouse 25: 33-40.

Ȩolak, A. H., Rotherham, I. D. 2006. A review of the forest vegetation of Turkey: its status, past and present and its future conservation. *Biol.Environ.,* 3:343-354.

Golonka, J. 2004. Plate tectonic evolution of the Southern Margin of Eurasia in the Mesozoic and Cenozoic. *Tectophysics*, 381:235-273.

Goloskokov, V. P. 1969. *Genus Atriplex. Illustration checklist of vegetation of Kazakhstan [Rod Atriplex. Illjustrirovanny opredelitel' rastenii Kazakhstana]. Vol. 1* (Goloskokov, V.P.-Ed.). Nauka, Alma-Ata: 271-320 (in Russian).

Golub, V. B. 1995. *Halophytic desert and semi-desert plant communities on the Tertiary of the former USSR*. Togliatti, 32pp.

Golubev, V. N. 1965. *Ecological-biological peculiarities of herb plants and vegetation communities of the forest-steppe [Ecologo-biologicheskie osobennosti traviyanistikh rastenii i rastitel'nikh soobchestv lesostepi]*. Nauka. Moscow, 287pp. (in Russian).

Gorbrachev, A. N. 1916. *The grey mountain Caucasian bee (Apis mellifera caucasica) and its place among other bees [Seraiya gornaiya Kavkazskaija pchela i eiyo mesto sredi drugikh pchiyol]*. Tiflis (Russian with English summary), 101pp.

Gottsberger, G., Silberbauer-Gottsberger, I. 2006. Life in the cerrado: a South American, tropical seasonal vegetation. In: *Origin, Structure, Dynamics and Plant Use*, Vol. 1. Ulm. Reta Verlag, 277pp.

Gould, J. L., Gould, C. G. 1995. The honeybee. *Scientific American Library*, New York, 239pp.

Gould, S. J. 1994. The geometer of race. *Discover*, 15:64-69 (*http://discovermagazine.com/1994/nov/thegeometerofrace44*).

Grabherr, G., Messerli, B. 2011. An overview of the world's mountain environments (8-14pp.). In: *Biosphere Reserves in the mountains of the world. Excellence in the clouds?* Austrian MAB Committee (Ed.), Austrian, Acad. Sci. Press, Vienna. 2011.

Gracheva, R. G., Belanovskaya, E. A. 2010. Pastoral ecosystems of the North Caucasus under the pressure of modern animal husbandry [Sostoyanie pastbishnykh ecosystem Tsentralnogo Kavkaza pod vozdeistviem sovremennogo gornogo zhivotnovodstva]. *News, RAN, ser. Geogr.*, 1:90-103 (in Russian with English summary).

Graffari, S. M, Saydrasi, L., Ebrahimzadeh, H., Akhani, H. 2000. Chromosome number and karyotype analyses of species of subfamily Salicornoideae (Chenopodiaceae) from Iran. *Iranian J. Bot.*, 12(2): 128-135.

Graham, A. 1999. *Late Cretaceous and Cenozoic history of North American vegetation: north of Mexico.* Oxford University Press, 370pp.

Gratsianskii, A. N. 1971. *Nature of the Mediterranean [Priroda Sredizemnogo moriya]*. Mysl. Publ., Moscow (in Russian), 507pp.

Gray, E. C. 2011. *An evaluation of the invasion potential of Kochia prostrata (forage kochia) in southwestern Idaho, USA (thesis).* Corvallis, OR, USA: Oregon State University, 73pp.

Gray, J. 1993. Major Paleozoic land plant evolutionary bio-events. *Palaegeogr., Palaeclim., Palaeoecol.,* 104:153-169.

Gray, J., Chaloner, W. G., Westoll, T. S. 1985. The microfossil record of early land plants: advances inunderstanding of early terrestrialization, 1970-1984. *Philosophical Transactions of the Royal Society* B, 309 (1138): 167-195.

Grebenshikov, O. S. 1957. The vertical zonation of vegetation in the mountains of the eastern part of western Europe [Vertikal'naiya zonal'nost' rastitel'nosti v gorakh vostochnoi chasti Zapadnoi Evropy]. *Botan. Zhurn.*, 42,6: 192-211 (in Russian).

Grebenshikov, O. S. 1965. *Geobotanical dictionary [Geobotanicheski slovar'] (Russian-English-German- French).* Nauka, Moscow, 226p.

Gregore, M.N. et al. 2014. *Halophytes: an integrative anatomical study*. Springer, New York: 5-25.

Gribova S. A., Isachenko T. I., Lavrenko E. M. 1980.*Vegetation of the European part of the USSR [Rastitel'nost' Evropeijskoi chasti SSSR]*. Nauka, Leningrad, 236 pp.

Griffiths, H. I., Kryštufek, B., Reed, J. M. 2004. *Balkan biodiversity. Pattern and process in the European hotspot.* Kluw. Acad. Publ., Dordrecht, 191 pp.

Grigorjev, A. A. 1960. The modern state of the theory of geographical zonation [Sovremiyonnoe sostoiyanie teorii geograficheskoi zonal'nosti]. In: *Soviet Geography*, Moscow, 289-299pp. (in Russian).

Grigorjev, A. A. 1964. *Regularity of the structure and development of the geographical environment [Zakonomernosti stroeniya i razvitiya geograficheskoi sredy]*. Mysl, Moscow, 382pp. (in Russian).

Grimson, F., Zetter, R., Grimm, G., Pedersen, G. K. Pedersen, A. K., Denk, Th. 2015. Fagaceae pollen from the early Cenozoic of west Greenland: revisiting Engler's and Chaney's Arcto-Tertiary hypotheses. *Plant. Syst. Evol.*, 301:809-832.

Gromova, V. 1935. *On the problems of origin and evolution of domestic animals [K problemam proiskhozhdeniya i evolutsii domashnykh zhivotnykh]*. Nauka Publishing House. Moscow, USSR, 134pp. (in Russian).

Grossheim, A. A. 1926. *The flora of Talysh [Flora Talysha]*. Narkomzem Az. SSR Publ. Tiflis, 273pp. (in Russian).

Grossheim, A. A. 1929. *Introduction into the geobotanical study of winter pastures of the Azerbaijan SSR [Vvedenie v geobotanicheskoi izuchennost zimnykh pastbich Azerbaijanskoi SSR]*. Izdanie Narkomzema, Baku: 1-75 (in Russian).

Grossheim, A. A. 1930. *Analysis of the vegetation of the Mil steppe [Analyz rastenii Mil'koi stepi]*. Izd. Narkomzema. Baku. 6th edition (in Russian).

Grossheim, A. A. 1932a. *The vegetation of winter pastures of Azerbaijan and their forage value [Rastitel'nost' zimnykh pastbych Azerbaidjana i ikh kormovoe znachenie]*. Izdanie Narkomzema, Baku: 1-71 (in Russian).

Grossheim, A. A. 1932b. Analysis of the vegetation of the Kura-Araks lowlands [Analiz rastitel'nosti Kura-Araksinskoi nizmennosti]. In: *Materials about the general utilization of the water resources of the Kura-Aras basin.* Izd. Zakvodproiz, Tiflis, 130pp. (in Russian).

Grossheim, A. A. 1936. *Analysis of the flora of the Caucasus*. Publ. (Azerbaijan Branch) Fil. Vol. I. Baku., 1-257 (in Russian with English summary).

Grossheim, A. A. 1946. The tasks of floristic investigation of the Caucasus [Zadachi floristicheskogo izutcheniya Kavkaza]. *Sovetsk. Botan.*, 2:116-121 (in Russian).

Grossheim, A. A. 1948. *Plant cover of the Caucasus [Rastitel'ny pokrov' Kavkaza]*. MOIP, Moscow, 264pp. (in Russian).

Grossheim, A. A. 1949. *The vegetation type key of the Caucasus [Opredelitel' rasteniy Kavkaza]*. Sovetskaya Nauka, Moscow, 747pp. (in Russian).

Grossheim, A. A. 1952a. *Flora of the Caucasus [Flora Kavkza]. vol. 5, Rosaceae - Leguminosae*, xxi 455pp. - USSR Ac. Sci. Publ. House, Moscow & Leningrad (in Russian).

Grossheim, A. A. 1952b. *Plant richness of the Caucasus [Rastitel'nyi bogatstva kavkaza]*. MOIP Publ., Moscow, 631 pp. (in Russian).

Grossheim, A. A., Sakhokia, M. F. 1929. *Geobotanical investigation of Gobustan [Geobotanicheskoe issledovanie Gobustana]*. Izdanie Narkomzema [Narkomzem Publ.], Baku, 107pp. (in Russian).

Groves, C. P. 1981. Systematic relationships in the Bovini (Artiodactyla, Bovidae). *Zeitschrift für Zoologische Systematik und evolutionsforschung,* 19:264-278.

Grudzinskaja, I. A. 1953. Broad-leaved forests of the foothills of the north-western part of the Caucasus [Shirokolistvennye lesa predgorii severo-zapadnogo Kavkaza]. In: *Broad-leaved Forests of north-east Caucasus.* Publ. House Ac. Sci. USSR., Moscow, 5:5-187. (in Russian).

Gueguechkori, A. M. 1998. Origin of the fauna of the deserts using the example of the Psyllids distributed in the deserts from the old world [Gènese de la faune des dèserts L'exemple des Psyllides et des dèserts. de l'ancien monde]. In: J. F. Richard et N. Beroutchachvili (Eds.). *La Gèographien Gèorgie*. Orstom, Paris. 103-115.

Guliev, S. M., Weinberg, P. J., Askorov, E. 2009. The Daghestan tur (*Capra cylindricornis* Blyth) conservation strategy in Azerbaijan. 52-68pp. In: Zazanashvili, N., Mallon, D. (Eds.). *Status and Protection of Globally threatened Species in the Caucasus.* Tbilisi:CEPF. WWF. Countour, Ltd.

Gulisashvili, V. Z. 1937. Batsara Strict Nature Reserve of the yew tree [Batsarski tissovyii zapovednik]. In: *Nature Reserves and National Monuments of Georgia*, 1:12-20, Tbilisi (in Georgian).

Gulisashvili, V. Z. 1942. *On the forest-steppe zone of the eastern Transcaucasia [K aridnoi redkolesii Vostochnogo Zakavkaz'iya].* Soobch. AN Gr. SSR, [Bull. Ac. Sci. Georg. SSR], 3,4: 345-352 (in Russian).

Gulisashvili, V. Z. 1948. Forest-steppes and steppes of the eastern Transcaucasia and co-action between arboreal and herb vegetations into them [Lesostep'ii stepy Vostochnogo Zakavkaz'iya i vzaimootnoshenie mezhdu drevesnoi i traviyanistoi rastitel'nosty v nikh]. *Bull. All-Union Geogr. Soc.*, 80,2: 127-140 (in Russian).

Gulisashvili, V. Z. 1950. Chestnut (*Castanea sativa* Mill.) and its distribution in Transcaucasia in relation to edaphic and climatic conditions [Kashtan i ego rasprostranenie v Zakavkaz'ije v svijazi edaphicheskikh i klimaticheskikh uslovii]. *Bull. Tbilissk. Bot. Gard.*, 39: 51-63, Tbilisi (in Russian).

Gulisashvili, V. Z. 1961. *Dendroflora of the Caucasus (Ed-in-chief) [Dendroflora Kavkaza]. vol. 2. (Wild and cultivated trees and shrubs).* AN Gruz.SSR. Publ., Tbilisi, 334 pp. (in Russian).

Gulisashvili, V. Z. 1964. *Natural zones and natural-historical regions of the Caucasus [Prirodnye zony i estestvenno-istoricheskye oblasti Kavkaza].* Nauka, Moscow, 326pp. (in Russian).

Gulisashvili, V. Z. 1980. *Savannas–open woodlands [Savanny-aridnye redkoles'iya].* Metsniereba (Sci. Publ.), Tbilisi, 97pp. (in Russian).

Gulisashvili, V. Z., Makhatadze, L. B., Prilipko, L. I. 1975. *Vegetation of the Caucasus [Rastitel'nost' Kavkaza].* Nauka [Sci. Publ. House], Moscow, 232pp. (in Russian).

Gunter, G. 2005. *On the great plains. Agricultural and environment.* Texas AGM University Press, 292pp.

Guo, Y., Olson, R. T. 2015. Effective bioassays for evaluating boxwood blight susceptibility using detached stem inoculations. *Hort. Science*, 50(2):268-271.

Gurbanov, E. M. 2006. The phytosociological characteristics of ecosystems of mountain of Talysh region of Azerbaijan. *Asian J. Plant Sci.*, 5(5):899-904.

Gvasalia, J. 1987. Khevi. *Georgian Soviet Encyclopedia,* vol. 11: 451 (in Georgian).

Gvozdetsky, N. A. 1954. *Caucasus. Lecture Course [Kavkaz. Kurs lektcii]*. Univers. Press. Moscow, Vol. 2, Issue 2. Precaucasus and Transcaucasus, 264pp. (in Russian).

Gvozdetsky, N. A. 1958. *Physiogeography of the Caucasus. Transcaucasia and Ciscausasia [Fizicheskaiya geografiya Kavkaza, Zakavkaz'iya i Severnogo Kavkaza]*. Izdat. Moskovsk. Univ-ta, Moscow, 264pp. (in Russian).

Gvozdetsky, N. A. 1960. The physical-geographical division of the European part of the USSR and the Caucasus [Fiziko-geografitcheskoe rajonirovanie Evropeiskoi chasti SSSR i Kavkaza]. *Izvestiya Vsesoiuzr. Geogr. Obsch-va*, 92, 5: 381-391 (in Russian).

Gvozdetsky, N. A. 1963. Kavkaz. Essays of nature [Kavkaz. Ochereki o prirode]. *Geografizdat*, Moscow, 264pp. (in Russian).

Gvozdetsky, N. A., Golubchikov, Yu. N. 1987. *The mountains [Gory]*. Mysl Publ. Moscow, 403pp. (in Russian).

Halliday, G., Beadle, M. 1983. *Consolidate index of flora Europaea*. Cambridge University Press, 158 pp.

Haltofova, P., Jankovsky, L. 2003. Distribution of sweet chestnut *Castanea sativa* Mill. in the Czech Republic. *J. Forest Sci.* (Prague). 49 (6): 259-272.

Harris, R. B., Reading, R. *Ovis ammon* 2008. *The IUCN Red List Of Threatened Species. Version 2014*, www.iucnredlist.org

Hatch, M. D. 1999 (Sage, R. F., Monson, R. K. (Eds.). C4 photosynthesis: a historical overview: 17-40. In: *Academic Press*, 597pp.

Hedge, I. C., Akhani, H., Feritag, H., Kothe-Heinrich, G., Podlech, D., Rilke, S., Uotila, P. 1997. Chenopodiaceae. In: Rechinger, K. H. (Ed.), *Flora Iranica,* 172. Akademische Druck – and Verlagsanstalt. –Graz.

Hehn, V. 1870. *Crops and domestic animals in their transition from Asia to Greece and Italy as well as to the rest of Europe [Kult urpflanzen und Hausthiere in ihrem Übergang aus Asien nach Griechenland und Italiensowie in das übrige Europa. – Berlin]*.

Heideman, T., Doroshka, I. 1931. Essay on the vegetation of the Salyan steppes [Ocherk rastenii Salyanskikh steppei]. In: *Proceed. Geobot. Res. Pasture*. Baku, issue'Izdanie Narkomzema' [Narkomzem Publ.] 8:66p. (in Russian).

Henze, P. B. 2013. *Tourism in Georgia. An unexploited opportunity for economic development* (www.georgianwinehouse.com/publications/PaulHenzeTourism.doc).

Heptner, V. G., Nasimovich, A. A., Bannikov, A. G. 1961. *Mammals of the Soviet Union: Artiodactyla and Perissodactyla [Mlekopytayuschie Sovetskogo Soiuza. Parnokopytnye i neparnokopytnye]*. Vol. 1, Higher School Publ. [Vyshaija Shkola], Moscow, 776pp. (in Russian),

Heptner, V. G., Nasimovich, A. A., Bannikov, A. G. 1989. *Mammals of the Soviet Union: ungulates*. Brill Academic Publishes. Leiden.

Herbel, C. H., Barnes, R. F., Heady, H. F., Purdy, L. N. 1981. Range research in the Soviet Union. *Rangelands*. 3: 61-63.

Hewitt, B. G. 1994. Peoples of the Caucasus. In: Fernández-Armesto, F. (Ed.). *The Times' Guide to the Peoples of Europe,* 366-384.

Heywood, V. 1999. Trends in agricultural biodiversity. In: *Janick, A., Alexendria, A. (Eds.). Perspectives on new crops and new uses.* ASHS Press, Alexandria, 2-14.

Hirbo, J., Muigai, A., Naqvi, A. N. et al. 2006. The genetic diversity of Transcaucasian native sheep breeds. *Asian-Aust. J. Anim. Sci.*, 19,7: 943-952.

Hodkinson, I. D. 1974. The biology of the Psylloidea (Homoptera): a review. *Bull. Ent., Res.*, 64: 325-339.

Hodkinson, I. D. 1980. Present-day distribution patterns of the Holarctic Psylloidea (Homoptera, Insecta) with particular reference to the origin of the Nearctic fauna. *Journ. Biogeogr.*,7:127-146.

Hodkinson, I. D. 1983. The psyllids (Homoptera, Psylloidea) of the Austro-Oriental, Pacific and Hawaiian zoogeographical realms; an annotated checklist. *J. Nat. Hist.*, 17:341-377.

Hodkinson, I. D. 1986. The psyllids (Homoptera: Psylloidea) of the Oriental Zoogeographical Region: an annotated checklist. *J. Nat. Hist.*, 20(2):299-357.

Hodkinson, I. D. 1988. The Nearctic Psylloidea (Insecta: Homoptera): an annotated checklist. *J. Nat. Hist.*, 22:1179-1243.

Hodkinson, I. D. 2009. Life cycle variation and adaptation in jumping plant lice (Insecta: Hemiptera: Psylloidea): a global synthesis. *Journal of Natural History*, 43(162): 65-179.

Hoetzel, S., Dupont, L., Schefuß, E., Rommerskirchen F., Wefer G. 2013. The role of fire in Miocene to Pliocene C_4 grassland and ecosystem evolution. *Nature Geoscience,* 6:1027-1030.

Hofmann R. R. 1968. Comparison of the rumen and omasum structure in East African game ruminants in relation to their feeding habits. *Symp. Zool. Soc. London* 21: 179–194.

Hofmann, R. R. 1989. Evolutionary steps of ecophysiological adaptation and diversification of ruminants: a comparative view of their digestive system. *Oecologia*, 78:443–457.

Hofmann, R. R., Stewart, D. R. M. 1972. Grazer or browser: a classification based on the stomach-structure and feeding habit of East African ruminants. *Mammalia,* 36:226-240.

Hofstadter, R. 1992. *Social Darwinism in American thought.* Beacon Press, Boston, 288pp.

Hollis, D. 1976. Jumping plant-lice of the tribe Ciriacremini (Homoptera, Psylloidea) in the Ethiopian Region. *Bull. Br. Mus. Nat. Hist.* (*Ent. Ser.*) 34,1:1-83.

Hollis, D. 1984. Afrotropical jumping plant lice of the family Triozidae (Homoptera, Psylloidea). *Bull. Br. Mus. Nat. Hist.* (*Ent. Ser.*) 34:1-102.

Hollis, D. 2004. Australian Psylloidea: Jumping plant-lice and lerp insects. *Goanna Print, Canberra*, 216 pp.

Holtmeier, F. K. 2003. Mountain timberlines – ecology, patchiness, and dynamics. *Advances in Global Changes Research,* 14, 369.. Kluwer Academic Publishers, Dordrecht, Boston, London.

Holubec, V., Krivka P. 2006. *The Caucasus and its flowers.* Loxia Publ. Pardubice, Czech Republic. 390 pp.

Howard, P. 2014. *Flammable planet: wildfires and the social cost of carbon.* Environmental Defence Fund. Institute for Policy Integrity. New York University School of Low, 1-46pp.

Howells, W. 1997. *Getting here: the story of human evolution.* Howells House, 2nd edit. The Compass Press, Washington, 278pp.

Hreidarsson, S. 2011. *Tree-ring studies of chestnut (Castanea sativa) in the Belasitsa Mountain in southern Bulgaria. BS – thesis.* Agricultural University of Iceland: 1-36pp.

Hsü, K. J., Bernoulli, D. 1978. Genesis of the Tethys and the Mediterranean. *Initial Report of the Deep Sea Drilling Project*. 42 :943-949.

Hsü, K. J., Giovanoli, F. 1979. Messinain event in the Black Sea. *Palaegeogr. Palaeoclimatol., Palaeoecol.* 29:75-93. https://www.academia.edu/214960/Power_Racism_and_Superority 1-85pp.

Hsü, K. J., Montadert, L., Bernoulli, D., Cita, M. B. et al. 1978. History of the Mediterranean salinity crisis. In: *Hsü, K. J., Montadert, L., et al. (Eds.), Initial Report of the Deep Sea Drilling Project. Washington (U. S. Govt. Printing Office)*, 42,1 : 1053-1078.

Humboldt, A. von. 1805. An essay on the geography of plants: accompanied by a physical picture of the equinoctial regions, based on measurements carried out, from the tenth degree of northern latitude to the tenth degree of southern latitude, during the years from 1799 to 1803 [Essai sur la géographie des plantes: accompagné d'un tableau physique des régions équinoxiales, fondé sur des mesures executees, depuis le dixieme degré de latitude boréale jusqu'au dixieme degré de latitude australe, pendant les années 1799, 1800, 1801, 1802 et 1803[. A Paris: *Chez Levrault, Schoell et Compagnie*, libraires, XIII-1805. Paris, 1807.

Huseynova, A. D. 2004. About Kura-Aras Lowland plant types, Azerbaijan [K rastitel'nym tipam Kura-Araksinskoi doliny Azerbaijana]. *Nauchn. Trudy AN Bot. Inst.* [*Nat. Sc. Acad. Bot. Inst. Sci. Works*], Baku, 25 :279-282 (in Russian).

Huseynova, A. D. 2008. Perspectives and ways of the rational usage of plants in halophyte deserts of Gobustan (Azerbaidjan) [Perspektivy i puti ratcional'nogo ispol'zovaniya rastenii galofitnykh pustyn' Gobustana, Azerbaijan]. *Int. Symposium Scientific-methodological Conference 'Introduction of non-traditional and rare plants'*. Michurinsk (Russia), III:175-177 (in Russian).

Ibadullayeva, S., Movsumova, N., Gasymov, H., Mamedli, T. 2011. Protection of some rare and endangered vegetable plants in the flora of the Nakhichevan AR. *Intern. J. Biod. Cons.*, 3(6) : 224-229.

Ibragimov, A. 2005. *Vegetation of Nakhichivan AR and its national economic importance [Rastitel'nost' Nakhichevanskoi AR i ego natcional'noe ekonomicheskoe znachenie]*. Elm, 230pp. Baku (in Russian).

Ibragimov, A. Sh., Nabiyeva, F. Kh. 2015. Analysis of endemic plants of the Nakhichevan Autonomous Republic of Azerbaijan. *Int. J. Mult. Res. and Develop*; 2,5.:143-149.

Ibragimov, A. Sh., Nabiyeva, F.Kh. 2016. Cushion plants in Nakhichevan Autonomous Republic of Azerbaijan and their phytocoenosis [Podushechnye rasteniya Nakhichevanskoi Avtonomnoi Respubliki Azerbaijdzhana i ikh fitotsenozy], *Nauka i Mir [Science and World]*, 3:9-12 (In Russian)

Igamberdiev, A. U. Lea P. J. 2006. Land plants equilibrate O_2 and CO_2 concentrations in the atmosphere. *Photosynth. Res.* 87,2:177-194.

Iglesias, A. A., Gonzalez, D. H., Andreo, C. S. 1986. The C_4 pathway of photosynthesis and its regulation. *Biochemical Education,* 14,3 :98-102.

Iglesias, V., Yospin, G. I., Whitlock, C. 2015. Reconstruction of fire regimes through integrated paleoecological proxy data and ecological modeling. *Fronti. Plant Sci.,* 5 :785-792.

Il'jin, M. M. 1936. *Atriplex* (Chenopodiaceae). Flora USSR (Komarov, V. L. (Ed.), Vol. 6:77–116). *Acad. Sci.USSR*, Moscow-Leningrad (in Russian).

Il'insky, A. P. 1937. *The vegetation of the globe [Rastitel'nost' zemnogo shara]*/Ac. Sci. USSR Publ. M.-L., 454pp. (in Russian)

Imat, F. M., Bababeyli, N. 2010. Reasons of desertification in the Middle of Aras River basin (Nakhichivan AR and Turkey) and measures of combating against desertification. *IBSU Sci. J.*, 4(1): 99-105.

Imkhanitskaya, N. N. 1990. Critical notes on the Caucasian species from the genus *Juniperus* L. Section *Juniperus* (Cupressaceae) [Kriticheskie zametki o Kavkazskogo roda *Juniperus*

L. sekcii *Juniperus* (Cupressaceae)]. *Novosti Sist. Vyshikh Rast.* [*News Sist. Higher Veget.*], 27:5-16 (in Russian).

Iniguez, L., Mueller, J. (Eds.). 2008. Characterization of small ruminant breeds in Central Asia and the Caucasus. *International Center for Agricultural Research in the Dry Areas (ICARDA)*, Aleppo, Syria, 411pp.

Isachenko, A. G. 1971. Systems and rhythms of zonality [Systemy i ritmy zonal'nosti]. *Izv. Vsesojuzn. Geogr-go Obschesstva*, 103, 1:10-26 (in Russian).

Isachenko, A. G. 1973. *Principles of landscape science and physical-geographic regionalization*. Massey, J. S. (Ed.). Melbourne Univ. Press., 311pp.

Isachenko, A. G., Shlyapnikov, A. A. 1989. *Nature of the world: Landscapes [Priroda Zemli. Landshafty]*. Mysl, Moscow, 504p. (in Russian).

Isachenko, T. I., Lavrenko, E. M. 1980. Botanical and geographical regionalization. 10-20pp. In: *Gribova et al. (Eds): Vegetation of the European Part of the USSR* [*Botaniko-Geograficheskoe raionirovanie rastitel'nosti Evropeiskoi chasti SSR*], Nauka, Leningrad:10-23 (in Russian).

Isaev, J. 1957. Wormwood (*Artemisia*) semi-deserts in Azerbaijan's winter pastures. *Trudy Botan. Inst. AN Az. SSR*, 20:5-20 (in Russian).

Ivanishvili, M. A. 1969. To the investigations of tragacanth astragals of the Caucasus [k izuchenyu tragakantovykh astragalov Kavkaza]. *Zam. Sist. Geogr. Rast. [Notice Taxon. Geogr. Veget.]*, 27:15-19. Tbilisi (in Russian).

Ivanishvili, M. A. 1973. *Flora of thorn tragacanth formations of the northern slope of the Greater Caucasus [Flora formatcii kolyucheastragalovykh tragakantnikov severnogo sklona Bol'shogo Kavkaza]*. Metsniereba [Science], Tbilisi, 213pp. (in Russian).

Ivanov, A. L. 2002. Flora and florogenesis of *Rhododendron caucasicum* Pall. thickets [Flora i florogenez zaroslei *R. caucasicum*]. *Izvest. Stavropol. State Univ.*: 30-42 (in Russian).

Ivanov, A. N., Bochkarev, Yu. N., Kozlov, D. N., Khaug, A. A. 2000. Basic natural features of the Abrau Peninsula [Osnovnye prirodnye osobennosti poluostrova Abrau]. *In: The Nature of the Abrau Peninsula: Collected Scientific Papers [Priroda poluostrova Abrau: Sb. Nauch. Trud.]*, Moscow: Mosc. State Univ.: 4–14pp. (in Russian).

Ivanov, D., Utescher, T., Mosbrugger, V. et al. 2011. Miocene vegetation and climate dynamics in Eastern and Central Paratethys (Southeastern Europe). *Palaeogeography, Palaeclimatology, Palaeecology,* 304:262-275.

Ivanov, N. N. 1948. *Landscape-climatic zones of the Earth [Landshaftno-klimaticheskie zony zemli]*. Moscow: Izd. Akad. Nauk SSSR [USSR Ac. Sci. Publ. House], 130pp. (in Russian).

James, S. R. 1989. Hominid use of fire in the Lower and Middle Pleistocene: A review of the evidence. *Curr. Anthropol.*, 30,1:1-26.

Javakhishvili, I. 1987. *Economic history of Georgia [Saqartvelos ekonomikuri istoria]. Second ed., vol.5* Metsniereba, Tbilisi, 677pp. (in Georgian).

Joannin S., Ali A. A., Ollivier V., et al. 2014. Vegetation, fire and climate history of the Lesser Caucasus : a new Holocene record from Zarishat fen (Armenia). *Journ. Quarter. Sci.*, 29, 1:70-82.

Jonson, J. 1976. *The international book of wood.* Mitchell Beazley Publishers Limited, 275pp.

Jones, Rh. 2012. Fire-stick farming. *Fire Ecology*, 8,3:3-8.

Joosten, J., Kaffke, A. Machutadze, I. 2003. The mires of the Kolkheti Lowlands (Georgia). *International Mire Conservation Group Newsletter*, 3:19-23.

Jorgensen, S. E. 2009. *Ecosystem ecology.* Academic Press. 521pp.

Kacharava, V. [Eds.-in-chief], Ketskhoveli, N., Kurashvili, B., Maruashvili, L. 1982. Red Data Book of Georgia. Tbilisi, *Sabchota Sakartvelo* (in Georgian), 239pp.

Kafkas, S., Ak B, E., Ozgüven, A. I. 2006. Performances of different *Pistacia* genotypes, *Acta Hortic.* 726:143-145.

Kakulia, A. V. 1952. Main fodder grasses of the Shirak-Eldari winter pastures [Shiraq-Eldaris zamtris sadzovrebis umtavresi sakvebi balakhebi].*Trudy Inst. Bot. AN GrSSR* [*Proceedings Inst. Bot. Sci. Goerg. SSR*], 14:26-34 (in Georgian).

Kaliyakin, V. N., Turubanova, S. A., Smirnova, O. V. 2016. The origin and development of the East European taiga in the Late Ceozoic. *Russ. J. Ecosyst. Ecol.* Vol. 1 (1). DOI:10.21685/2500-0578-2016-1-2.

Kantschaveli, G., Supatashvili, Sh. 1968. *Forest entomology.* Ganatleba Publ., Tbilisi, 371pp. (in Georgian).

Kanwal, H., M., Hameed, T., Nawaz, M. S., Ahmad, A., Younis, A.2012. Structural adaptations for adaptability in some exotic and naturalized species of Agavaceae. *Pakist. J. Bot.* 44: 129-134.

Kapustina, L. A., 2001. Biodiversity, ecology and microelement composition of Kyzylkum desert shrubs (Uzbekistan). In: *USDA Forest Service Proceedings RMRS-P-21.* 2001. 98-103.

Kara, M., Sezgin, E., Kara, A. 2012. Importance of Caucasian honeybee and its characteristics as a gene resource. *J. Agr. Sci. Techn. A* 2:1197-1202.

Kariagin, I. I. 1950-1961. *Flora of Azerbaijan.* Vols. 1-8, Izd. AN Azerb. SSR, Baku (in Russian).

Karpachevsky, L. O. 2006. Biota. In: *Encyclopedia of Soil Science, Second Edition*, Vol. 1 (Lal, R.-Ed.). Taylor & Francis: 174-178.

Kassioumis, K., Papageorgin, K., Glezakos, T., Nogiatzakis, I. N. 2004. Distribution and stand structure of *Taxus baccata* populations in Greece: Results of the first national inventory. *Ecologia Medit.,* 30:159-170.

Kaukhchishvili, K. S. 1941. Discovered Greek script in Armazi [Armazshi agmochenili berdznuli tsartserebi]. *Georg. SSR Sci. Acad Bull.*, 2(1-2): 172-173 (in Georgian).

Kaukhchishvili, K. S. (Ed.), 1955. *History of Georgia [Kartlis tskhovreba].* Vol. 1, Ganatleba Publ., Tbilisi, 463pp. (in Georgian).

Kaya, Z., Raynal, D. J. 2001. Biodiversity and conservation of Turkish forests. *Biological Conservation*, 97:131-141.

Kearey, Ph., Klepeis, K. A., Vine, F. J. 2009. *Global Tectonics, 3rd Edition.* Wiley-Blackwell, West Sussex, UK, 496pp.

Kebadze, Z. G. 1965. The Tertiary relict grow of *Populus euphratica* in Vashlovani Preserve [*Populus euphratica*-s mesameuli relikturi raka vashlovanis nakrzalshi]. *Sakartv. Metcnier. Akad. Moambe* [*Bull. Georg. Ac. Sci.*] 37, 3:14-17 (in Georgian).

Kebadze, Z. 1969. On the biology of xerophilous plants of the arid open woodlands [Nateli tkeebis kserofilur mtcenareta biologiisatvis]. *Bull. Bot. Soc.Georg.*, Tbilisi, 4:26-36 (in Georgian with Russian summary).

Keeley, J. E., Rundel, P. W. 2005. Fire and Miocene expansion of C_4 grasslands. *Ecology Letters,* 8:683-690.

Kefu, Z., Hai, F. Mugar, I. A. 2002. Survey of halophyte species in China. *Plant Science*, 163:495-498.

Keita, S. O. Y., Kittles, R. A., Royal, C. D. M. et al. 2004. Conceptualizing human variation. *Nat. Genet.* 36 (Suppl 1):S17-S20.

Keller, B. A. (Ed.) 1931. *Steppes of the Central-Chernozem region [Stepy Central'no-Chernezemnoi oblasti]:* 319-327, Nauka. Moscow & Leningrad, 327pp. (in Russian).

Kemertelidze, E. P. 2007. Biologically active compounds and original remedies from plants growing in Georgia. *Bull. Georg. Nat. Ac. Sci.*, 175,1:91-96.

Kenrick, P., Crane, P. R. 1997. The origin and early evolution of plants on land. *Nature,* 389,4:33-39.

Kenrick, P., Strullu-Derrien, C. 2014. The origin and early evolution of roots. *Plant physiology,* 166:570-580.

Kendrick, J. 2014. Rethinking modern human origins: getting omt of out of Africa. *Assamblage* 2014. 13pp.

Ketenoglu, O., Tug, G. N., Kurt, L. 2010. An ecological and syntaxonomical overview of *Castanea sativa* and a new association in Turkey. *J. Envir. Biol.*, 31, 81–86.

Ketskhoveli, N. N. 1958. On the dynamics of vegetation of our steppes and open woodlands [Chveni stepebisa da nateli tkeebis dinamikis shestcavlisatvis]. *Proceed, Tbil. Bot.Inst., AN Georg. SSR* 19:45-69 (in Georgian).

Ketskhoveli, N. 1959. Once more regarding the origin of the open arid woodlands [Ertxel kidev nateli tkeebis tsarmoshobis shesakheb]. *Trudy Inst-ta Botan. AN Gruz. SSR.* (*Proceed. Tbil. Bot. Inst. AN Georg. SSR*), Tbilisi, 20:347-362.

Ketskhoveli, N. N. 1960. *Vegetation cover of Georgia [Saqartvelos mtcenareuli safari].* Metsniereba. Tbilisi, 441pp. (in Georgian).

Ketskhoveli, N. N. 1980. Xerophitic (arid) open woodlands [Kserofitnye (aridnye) redkoles'iya]. In: *Vegetation of the European part of the Soviet Union.* Leningrad, 273-276pp. (in Russian).

Ketskhoveli, N. N. 1982. *The nature wounded in the breast [Mkerdshi dachrili buneba].* Metsniereba, Tbilisi, 167pp. (in Georgian).

Kevan, P. G. 1999. Pollinators as bioindicators of the state of environment: species, activity and diversity. *Agricult. Ecosyst. Environm.*, 74 (1-3): 373-393.

Khachidze, M. 1975. On steppe vegetation according to Shiraki example [Stepis mtcenareuloba Shirakis magalitze]. *Bull. Ac. Sci.Georg.* SSR, Tbilisi, 3:102-111 (in Georgian).

Khachidze, M. 1985. *Vegetation cover of Shiraki Plateau and Eldari Lowland (East Georgia) [Rastitel'ny pokrov' Shirakskogo Plato i El'darskoii Nizmennosti (Vostochnaiya Gruzija)] (173pp.), PhD thesis biol. sci.* 20pp.Tbilisi, (in Russian).

Khain, V. E. 1975. Structure and main stages in the tectono-magmatic development of the Caucasus: an attempt at geodynamic interpretation. *Amer. J. Sci.*, 275-A:131-156.

Khain, V. E., Koronovski, N. V. 1997. Caucasus. *In: Moores, E. M., Fairbridge, R. W. (Eds.) Encyclopedia of European and Asian regional geology.* Chapman & Hall, London: 127–136.

Khain, V. E., Koronovski, N. V. 1998. Caucasus. 127–136 In: *Moores, E. M., Fairbridge, R. W. (Eds.) Encyclopedia of European and Asian regional geology.* Springer Science & Business Media. 804pp.

Khaleghi, R. et al. 1998. A profile of Caspian forests. *Iranian Research Institute of Forest and Rangelands Publication*, Tehran, Iran.

Khanjyan, N. 2004. Specially protected nature areas of Armenia [Osobo okhraniyaemye prirodnye territorii Armenii]. *Tigran Mets*, Yerevan, 54 pp. (in Russian).

Kharadze, A. A. 1944. Sketches of flora of the subnival zone of the Upper Svaneti [Ocherk flory subnival'noi zony Verkhnei Svanetii]. *Notes Syst. & Georg. Veget.*12:51-83 (in Russian).

Kharadze, A. L. 1965. On the subnival zone of the Greater Caucasus [K subnival'noi zone Bol'shogo Kavkaza]. *Notes Syst. & Plant Georg. Inst. Bot. Ac. Sci. Georg. SSR,* 25:103-104 (in Russian).

Kharadze, A. L. 1974. *On some florogenetic endemic groups in the Great Caucasus. Problems of Botany. Nauka (Science),* Leningrad, 12:70-76 (in Russian).

Khasanova, L. V. 2003. The ecological peculiarities of biodiversity of Adigea [Ekologicheskie osobennosti bioraznoobraziya Adigei]. *Maikop, Adige Publ*: 7-82 (in Russian).

Khintibidze, L. S. 1990. *Xerophytes floristical complexes of South Georgia [Kserofitnye floristycheskie kompleksy Yuzhnoi Gruzii]. Doct. dissertation (PhD thesis) of biol. sci.),* Tbilisis, 48pp. (in Russian).

Kholodov, V. N. 2001. On the nature of the mud volcano [K prirode vulkanevykh griyazh]. *Priroda* [*Nature*], 11:47-58 (in Russian).

Khorozyan, I. G. Weinberg, P. I., Malkhasyan, A. G. 2009. Conservation strategy for Armenian mouflon (*Ovis orientalis gmelini* Blyth) and bezoar goat (*Capra aegragus* Erxleben) in the Republic of Armenia. 37-45pp. In: *Zazanashvili, N. & Mallon, D. (Eds.), Status and Protection of Globally Threatened Species in the Caucasus. Tbilisi: CEPF, WWF. Contour Ltd.*

Khromova, T., Nosenko, G., Kutuzov, S., Muraviev, A., Chernova, L. 2014. Glacier area changes in Northern Eurasia. *Envir. Res. Lett.*, 9, 015003, doi:10.1088/1748-9326/9/1/015003.

Kialashaki, A., Shabani, S. 2011. Differentiation ecosystem units of Caspian forests in relation to physiographic factor and some of soil characteristics. *Annals Biol. Res.*, 2(3): 276-282.

Kikava, G., Chkhetiani, I., Jugeli, N., Todua, V. 1997. *The wild fruits of Georgia [Saqartvelos veluri khili].* Tbilisi, 163pp. (in Georgian).

Kikodze, E. 1967. *Nitraria schoberi* in Georgia. Nature Syst. Geogr. Plants. *Inst. Bot. of Tbilisi,* 26:63-6 (in Russian with Georgian summary).

Kikodze, A., Gokhelashvili, R. (Eds.) 2007. Protected areas of Georgia [Sakartvelos datsuli teritoriebi]. *Tsignis Sakhelosno*, Tbilisi (in Georgian).

Kikodze, D., Memiadze, N., Kharazishvili, D. Manvelidze, Z., Mueller-Schraerer, H. 2010. *The alien flora of Georgia.* Second edition. Tbilisi, 1-36pp (in Georgian and English).

Kikodze, D., Tavartkiladze, M., Svanidze, T. 2007. Plants of Georgia. Gagnidze, R. (Ed-in-chief) *Field Guide. Department of Protected Areas.* Tbilisi, 224pp. (In Georgian and English).

Kimeridze, K. R. 1966. Distribution pattern of wetland vegetation in the Caucasus [Chaobis mcenareulobis gavrcelebis kanonzomierebebi Kavkasiashi]. *Bull. Georg. Acad. Sci.*, 43,2:234-245 (in Georgian with English summary).

Kimeridze, K. R. 1975. Tufted sedge wetlands of mountainous Georgia [Sakartvelos mtianetis kolbochovani islianebi]. *Bull. Georg. State Museum*, 28-A: 43-56 (in Georgian).

Klimaszewski, S. M. 1973. The jumping plant lice or psyllids (Homoptera, Psyllodea) of the Palaearctic: annotated checklist. *Annales Zoologici, Warszawa* 30(7): 155–286.

Klopotovskaya, N., Kvavadze, E., Lorrkipanidze, D. 1989. Preliminary account of the palaeobotanic. The ancient Paleolithic site Dmanisi in Georgia (Caucasus) [Vorläufige.

Mitteilung zur palaöbotanik. Der altpaläolithische Fundplatz Dmanisi in Georgien (Kaukasus)]. *Jahrbuch des RGZM Mainz.*, 36,9:32-44.

Knapp, H. D. 2005. The global importance of the Caspian forests [Die globale Bedeutung der kaspischen Wälder]. In: *Nosrati, K., Marvie Mohadjer, R., Bode, W. and Knapp, H. D., Schutz der Biologischen Vielfalt und integriertes Management der Kaspischen Wälder (Nordiran). Bundesamt für Naturschutz.* Bonn: 45-70.

Kobakhidze, D. N. 1965. Some results and prospects of the utilization of beneficial entomophagous insects in the control of insect pests in the Georgian SSR (USSR). *Entomophaga*, 10:323-330.

Kodrul, T. M. 2009. Paleocene-Eocene freshwater aquatic wetland ecosystems of the Zeya-Bureya Basin, Russian Far East and their responses to climate change. *Berichte Geol. B.-A.*,78pp. *RECCCE Workshop,* Cams (25.04-28.04. 2009):24.

Kokhodze, T. S. 1991. *Impact of human-economic activity on the game species distribution in Georgia [Vliyanie ekonomicheskoi aktivnosti cheloveka na promyslovykh zhivotnykh v Gruzii].* Tbilisi, Academy of Science Publ. 162pp. (in Russian).

Kolakovski, A. A. 1961. *The flora and vegetation of Colchida [Flora i rastitel'nost' Kolkhidy].* Mosc. Univ. Publ. House, Moscow, 459pp. (in Russian).

Kolakovski, A. A. Ratiani, H. K. 1967. The Pliocene flora of Patara Shiraki [Pliocenovaiya flora Patara Shiraki]. *Trudy Sukhumsk. Bot. Sada [Bull. Sukh. Bot. Gard.]*,16:21-29, Tbilisi, (in Russian).

Komarov, V. L. (Ed.) 1934-1978. Flora of the USSR. *Leningrad (English transl. by IPST).* Vol. 1-24. Jerusalem: Israel.

Kopaliani, J. 1996. *The citadel of Dmanisi [Dmanisis tsikhe].* Tbilisi, 146pp. (in Georgian).

Kopaliani, N., Gurielidze, Z. 2009. Status of turs in Georgia and conservation action plan. Pp. 61-68. In: *Zazanashvili, N. & Mallon, D. (Eds.). Status and protection of Globally Threatened Species in the Caucasus.* Tbilisi: CEPF, WWW, Contur Ltd.

Körner, Ch. 1998. A re-assassment of high elevation treeline positions and their explanation. *Oecologia*, 115:445-459.

Körner, Ch. 2000. The alpine life zone under global change, *Gayana Bot.* (Chile), 57:1-17.

Körner, Ch. 2002. Mountain biodiversity, its causes and function: an overview (3-21pp.). In: Körrner, Ch., Spehn E. M. (Eds.), (2002). *Mountain Biodiversity: A Global Assessment.* Parthenon Publishing Group, London, NY, 348pp.

Körner, Ch. 2003. *Alpine plant life: functional plant ecology of high mountain ecosystems. Second edition.* Berlin: Springer, 344pp.

Körner, Ch. 2012. *Alpine treelines. Funcional ecology of the global high elevation tree limits.* Basel: Springer, 220pp.

Körner, C., Pausen, J. 2004. A world-wide study of high altitude treeline temperatures. *Journal of Biogeography*, 31:713-732.

Körner, Ch., Spehn, E. M. (Eds.). 2002. *Mountain biodiversity. A global assessment.* Parthenon Publ. 350pp.

Kotlyakov, 1966. Modern glaciations. The Caucasus. In: *'Natural conditions and natural resources of the USSR' [Prirodnye usloviya i estestvennye resursy SSSR],* Moscow, Nauka, 127-131 (in Russian).

Kovalev, O. V., Zaitsev, V. F. 1966. A new theoretical approach to the selection of promising agents for biological weed control. In: *Proc. of the IX Int. Sym. on Biological Control of*

Weeds. 19-26, January 1996 (Moran, V. C. & Hoffmann, J. H. (Eds.), Stellenbosch, South Africa. *Univ. of Cape Town*, 283-285.

Kovalev, O. V. 1995. Co-evolution of the tamarisks (Tamaricaceae) and pest arthropods (Insecta; Arachnida; Acarina), with special reference to biological control prospects. In: *Proceedings of the Zoological Institute, Russian Academy of Sciences*. St.Petersburg. Vol. 259. PENSOFT, Sofia-Moscow, 109pp.

Kozhoridze, G., Orlovsky, N., Orlovsky, L. et al. 2015. Geographic distribution and migration pathways of *Pistacia* – present, past and future. *Ecogeography*, 38, 11:1141-1154.

Kozlowski, G., Gibbs, D., Huan, F., Frey, D., Gratzfeld, J. 2012. Conservation of threatened relict trees through living *ex situ* collections: lessons from the global survey on the genus *Zelkova* (Ulmaceae). *Biodiver. Conserv.*, 21: 671-685.

Kozlowski, G., Gratzfeld, J. 2013. Zelkova – an ancient tree: global status and conservation action. *Natural History Museum Eribourg*. Switzerland, 60pp.

Krebs, P., Conedera, M., Pradella, M., Torriani, D., Felber, M., Tinner, W. 2004. Quaternary refugia of the sweet chestnut (*Castanea sativa* Mill.): an extended palynological approach. *Vegetation History and Archaeobotany*, 13:145-160.

Krebs, M., Joosten, H. 2006. The Golden Fleece in trouble - the endangering of the Kolkheti peatlands (Georgia). *International Mire Conservation Group Newsletter*. 1:6-9.

Kreft, H., Jetz, W. 2010. A framework for delineating biogeographical regions based on species distributions. *J. Biogeogr.*, 37:2029-2053.

Krever, V., Zazanashvili, N., Jungius, H., Williams, L., Petelin, D. (Eds.) 2001. Biodiversity of the Caucasus Ecoregion: an analysis of biodiversity and current threats and initial investment portfolio. *Signar and WWF*, Baku, Erevan, Gland, Moscow, Tbilisi, 184p.

Krivokhatsky, V. A., Emeljanov, A. F. 2000. Use of zoogeographical subdivisions in particular zoogeographical researches for the example of the Palaearctic antlion fauna (Neuroptera, Myrmeleontidae) [Ispol'zovanie vydelov obcheij biogeografii dlija chastnikh zoogeograficheskikh issledovanii na primere Palearkticheskoi fauny]. *Entomolog. Rev*, 80(9): 1042–1056 (in Russian).

Krivonogova, M. B. 1960. Cushion and thorn-cushion vegetation, their geographic distribution and main peculiarities [Podushechniki i kolyuchie-podushechniki, ikh geograficheskoe rasprostranenie i osnovnye osobennosti]. In: *Probl. Bot.* USSR Ac. Sci. Publ. House, 5, 226-232, M.-L. (in Russian).

Kryshtofovich, A. N. 1955. Development of the phytogeographical regions the Northern Hemisphere during the Early Tertiary period [Razvitie botaniko-geograficheskikh oblasteii severnogo polusharija s nachala tretychnogo perioda]. *Vopr. Geob. Azii* (*Questions of Geology of Asia), AN SSSR*, Moscow, 2:824-844 (in Russian).

Kryzhanovsky, O. L. 1965. *Composition and origin of terrestrial fauna of Middle Asia [Sostav i proiskhozhdenie nazemnoi fauniy Srednei Azii]*. Nauka, Moscow-Leningrad, 419pp. (in Russian).

Kryzhanovsky, O. L. 2002. *Composition and distribution of entomofaunas of the Earth [Sostav i rasprostranenie entomofaun zemnogo shara]*. KMK Press, Moscow, 237pp. (in Russian).

Kudaktin, A. H. 1982. Interrelationships of wolves and ungulates in the Caucasus Reserve [Vzaimootnoshenie volkov i kopytnykh v Kavkazskom zapovednike]. 65-70pp. In: *Collect. Sci. Trans. Ecology, protection and use of carnivorous mammals in the RSFSR*. Moscow, *Science* (in Russian).

Kurdoğlu, O., ÇokÇaliskan, B. A. 2011. Assessing the effectiveness of protected area management in the Turkish Caucasus. *African Journal of Biotechnology*, 10 (75): 208-222.

Kurtubadze, M. 2008. *Cattle in the Caucasus ecoregion. Caucasus ecoregion – environment and human development issues.* www.grida.no(graphicslib/Search?q=(?caucasus-ecoregion).

Kutateladze, Sh. 1947. Wild growing pears of Georgia [Dikorastiyuschie grushi Gruzii]. *Bull. Tbilisk. Bot. Inst.*, 11:205-240 (in Russian).

Kutateladze, Sh. 1980. *Pyrus* L. In: Ketskhoveli, N (Ed.). *Flora of Georgia [Saqartvelos flora].* Metsniereba (Science), 6:171-187 (in Georgian).

Kuyéma, N. A. 2003. *Ecological assessment of riparian forests in Benin: Phytodiversity, phytosociology and spatial distribution of tree species.* Wageningen University Plant Science, Benin, West Africa.

Kuznetsov, B. A. 1949. The experience of the zoogeographic region partitioning of the Caucasus and Transcaucasia [Opyt zoogeograficheskogo rajonirovaniya Kavkaza i Zakavkaz'iya]. *Trudy Mosk. Pushno-Mekh. Inst.* 2:109-143 (in Russian).

Kuznetsov, I. I. 1891. The elements of the Mediterranean region in Western Transcacuasia. *Letters of Russian Geograf. Soc.* [Zapiski Russk. Geogr. Obchestva], 23,3, S.-Petersb., 190 pp. (in Russian).

Kuznetsov, N. I. 1901. *Ericaceae B. Materials of the flora of the Caucasus (Flora caucasica critica) [Materialy k flore Kavkaza].* 4, 1-49pp. (in Russian).

Kvavadze, E. 1997. The perspective of the coprolitical study of mammals by pollen analysis for solution of palaecological and anthrological problems. *Proceedings Actes du Congres BIOCHROM 97,* Montpellier: 121-128.

Kvavadze, E., Gambashidze, I., Mindiashvili, G., Gogochuri, G. 2006. The first finding in southern Georgia of fossil honey from the Bronze Age, based on palaeontological data. *Vegetation History and Archaeobotany*, 16, 5:399-404.

Lachashvili, I., Khachidze, M. 2010. *Desert flora and vegetation of Georgia.* Universali Publ., Tbilisi, Georgia, 319pp. (in Georgian and English).

Lachashvili, I., Lachashvili, N., Khachidze, M. 2004. *Synopsis of flora of the Vashlovani: State Reserve.* Universaly Publ., Tbilisi, Georgia, 157pp. (in Georgian and English).

Lachashvili, I., Lachashvili, N., Khachidze, M. 2007. *Conspectus of the flora of Khiziki (East Georgia).* Universal, Tbilisi, 378pp. (in Georgian with English summary).

Lachashvili, N. I. 1989. *Structural analysis of Eldari vegetation (East Georgia) [Strukturny analiz rastitel'nosti Eldari, Vost. Gruziya]. PhD thesis biol. sci.*, 23pp. (in Russian).

Lachashvili, N. I., Khachidze, M. N. 2005. The typology of vegetation of the Vashlovani State Reserve [Vashlovanis saxelmtcifo nakrdzalis mtcenareulobis tipologia]. *Metsniereba da Tekhnologiebi*, 1-3: 143-146 (in Georgian).

Lancaster, N. 2009. Aeolian features and process. *The geological Society of America*: 1-26.

Landolt, E. 1969. Unsere Alpenflora. *Verl. Schweizer Alpen-club.* Zurich, 223pp.

Lang, D. M. 1962. *A modern history of Georgia.* Grove Press, New Jork, 340pp.

Larcher, W. 1980. Climate stress in the mountains - selection filter for plants [Klimastress im Gebirge: Adaptationstraining und Selektionsfilter für Pflanzen]. *Rhein. Westf. Akad. Wiss. Vortr*, 291:49-79.

Latchininsky, A., Sword, G., Sergeev, M. et al. 2011. Locusts and grasshoppers: behavior, ecology, ecology and biogeography. *Psyche*: *A Journal of Entomology*, 1-4pp.

Laurance, W. F. 2010.Habitat destruction: death by a thousand cuts. In: Sodhi N.S, Ehrlich P.R, (Eds.). *Conservation Biology for All.* Oxford: Oxford University Press, 73–87pp.

Laurin, M. 2004. The evolution of body size, Cope's rule and the origin of amniotes. *Syst. Biol.*, 53,4:594-622.

Lavrenko, E. M. 1940. The steppes of the USSR [Stepy SSSR]. In: *The Vegetation of the USSR, Vol. 2.* [Izdatel'stvo AN SSSR] [The USSR Acad. Sci. Publ. House], Moscow-Leningrad. 265pp. (in Russian).

Lavrenko, E. M. 1951. The age of the botanical region of nontropical Eurasia [Vozrast botanicheskikh oblastey vnetropicheskoii Evrazii]. *Izv, AN SSSR, Ser. Geogr.* [*Bulletin of Ac. Sci. of the USSR, Ser Geogr.*], 2:17-28 (in Russian).

Lavrenko, E. M. 1954. The steppes of the Eurasian steppe region, their geography, dynamics and history [Stepy Evraziatskoi stepnoii oblasti, ikh geographiya, dinamika i istoriya]. In collection: *Questions of Botany.* Moscow-Leningrad, 1:155-191 (in Russian).

Lavrenko, E. M. 1960. On the Sahara-Gobian desert biogeographic region and its subdivision [K Sakhara- Gobiiskomu biograficheskomu regionu u ego podrazdelenyu]. *Dokl. Akad. Nauk SSSR* [*Rep. Ac. Sci. USSR*], 134, 1:149–152 (in Russian).

Lavrenko, E. M. 1962. Basic features of botanical geography of the deserts of Eurasia and North Africa [Osnovnye cherty botanicheskoi geografii pustyn' Evrazii i Severnoi Afriki] In: *'Komarovskiye Chteniya' [Komarov Memorial Lectures],* Izd. AN SSSR [Ac. Sci. Publ. of the USSR], 15, 169pp, Moscow-Leningrad (in Russian).

Lavrenko, E. M. 1964. Types of vertical belt structure of vegetation cover in the mountains of the USSR [Tipy vertikal'noi poiyasnosti rastitel'nogo pokrova v gorakh SSSR]. In: *Gerasimov, I. P. et al. (Eds.). Sovremen. Probl. Geografii*, 189-195. Moscow (in Russian).

Lavrenko, E. M. 1965. The provincial delineation of the Central Asian and Irano-Turanian subregions of the Afro-Asian Desert Region [Provincial'noe razdelenie tsentral'noii Azii i Irano-Turanskoii podoblasteii Afro-Aziatskoii pustynnoi oblasti]. *Botan. Zhurn.* [*Bot. J.*] Leningrad. 50(1), 3 – 15 (in Russian).

Lavrenko, E. M., Karamysheva, Z. V., Nikulina, R. I. 1991. *Steppes of Eurasia [Stepy Evrazii].* Nauka, Leningrad, 146pp. (in Russian).

Le Pichon, X. 1968, Sea-floor spreading and continental drift. *J. Geophys. Res.*, 73(2): 3661-3697.

Lenoir, J., Gegout, J. C., Marquet, P. A., De Ruffay, P. A., Brisse, H. 2008. A significant upward shift in plant species optimum elevation during the 20th century. *Science*, 320:1768-1771.

Leppik, E. E. 1957. Evolutionary relationships between entomophilous plants and anthophilous insects. *Evolution,* 11:466-481.

Levantsova, Y. V. 2009. *Genus Stipa L. in flora of the North Caucasus and North-Western Transcaucasia [Rod Stipa L. vo flore Severnogo Kavkaza i Severo-Zapadnogo Zakavkaz'iya] (206pp.). PhD thesis bio. sci.*, 23p. Stavropol (in Russian).

Li, F. 2011. Psyllidomorpha of China (Insecta: Hemiptera). *Huayu Nature Book Trade Co. Ltd,* Beijing, China. 1–2: 1–1976pp.

Li, Z., Zheng, C. 2005. Structural characteristics and eco-adaptability of the heteromorphic leaves of *Populus euphratica. Forest Stu.* China, 7:11-15.

Liamine, N. (Ed.) 1961. Biogeographical regions in Europe. The Black Sea region. *European Environmental Agency*, 3-16pp.

Lieberman, P. 2002. On the nature and evolution of the neural bases of human language. *Yearbook of Phys. Anthr.* 45: 36–62.

Likhonos, E. D., Tuz, A. S., Lobachev, A. Y. (Eds.) 1983. Cultivated flora of the USSR. Pome fruit trees (an apple, pear and quince). Vol. 14, [Kul'turnaiya flora SSR. 14. Semechkovye (yabloniya, grusha, aijiva], Moscow, *Kolos*, 163-164.

Lilienberg, D. A., Shirinov, N. Sh. 1977. The modern plate tectonic movement of the Caucasus [Sovremyionnye tektonicheskie dvizhenie Kavkaza]. In: *Dumitrashko, N. V. (Ed.). General characteristics and the history of development of the Caucasus:* 48-59. Nauka, Moscow (in Russian).

Lindenmayer, D. B., Franklin, J. F. & Fischer, J. 2006. General management principles and a checklist of strategies to guide forest biodiversity conservation. *Biol. Conserv.* 131: 433–445.

Liniger, H. P., Weingartner, R., Grosjean, M. Et al. 1998. Mountains of the world: water towers for the twenty-first century. A contribution to global freshwater management. *Mountain Agenda*. Bern, Paul Haupt.

Linnaeus 1753. *Species Plantarum,* vol. 1, London, 392p.

Lioubimtseva, E., Henebry, G. M. 2009. Climate and environmental change in arid Central Asia: impacts, vulnerability and adaptations. *Journ. Arid. Environm.* 73(11): 963-977.

Lipka, O. N. 2006a. The botanical diversity and conditions nowadays of the vegetation of Markotkh Ridge (North-West Caucasus) [Botanicheskoe raznoobrazie i sovremyonnoe sostoiyanie rastitel'nosti khrebta Markotkh]. *PhD biol. sci.*, 1-19pp., Moscow (in Russian).

Lipka, O. N. 2006b. Floristical and coenotical richness of junipers' forests and open woodlands of the Markotkh Ridge [Floristicheskoe i tsenoticheskoe bogatstvo mozhevel'nikovikh lesov i aridnykh redkolesii khrebta Markotkh]. *Mater. Mosk-go Geogr. Obch-va – Biogeographia* (*Materials Mosk. Geogr. Soc. – Biogeography*): 13:36-42. Moscow (in Russian).

Lipsky, V. I. 1899. Flora of the Caucasus [Flora kavkaza]. *Trudy Tiflisk. Bot. Sada. SPb.,* 4,585pp. (in Russian)

Lisitskaya, Y. V. 2005. On the analysis of genus *Stipa* L. in the flora of Ciscaucasica [K analizu roda *Stipa* L. vo flore Predkavkaz'iya]. In: *Problemy razvitiya biologii i ekologii na Severnom Kavkaze (In: Problems of development biology and ecology in North Caucasus). Stavropol, SGU [St.Gos.Un.]*: 177-179 (in Russian).

Lockwood, J. A., Latchininsky, A. V., Sergeev, M. (Eds.). 2000. *Grasshoppers and grassland health.* Springer Netherlands, 215pp.

Loginova, M. M. 1960. New and little known jumping plant-lice (Homoptera, Psylloidea) of Middle Asia and Kazakhstan. *Trudy Vsesojuzn. Entomol. Obschestva* [*Proceedings of the All-Union Entomological Society*], 47: 53-93 (in Russian).

Loginova, M. M. 1968. New data on the fauna and biology of the Caucasian Psylloidea (Homoptera) [Novye dannye po faune i biologii psyllid Kavkaza]. *Trudy Vsesoyuzn. Entomolog*. Obshch. Akad. Nauk SSSR, 52: 275-328 (in Russian).

Loginova, M. M. 1969a. Psyllids (Homoptera, Psylloidea). In: *Plant communities and animal populations of the steppes and deserts of Central Kazakhstan (Biocomplex studies in Kazakhstan, part1)* [*'Rastitelnie soobchestva i zhivotnoe naselenie stepei i pustin centralnogo Kazachstana' (Biokompleksnoe issledovanie v Kazachstane, chast 1*), Nauka Publ., Leningrad: 381-392 (in Russian).

Loginova, M. M. 1969b. New species of psyllids (Homoptera, Psylloidea) from the Mongolian People's Republic [Novye vidy psyllid iz Mong. Nar. Resp.] Part I. *Entomol. Obozr.*, 48: 162-170 (in Russian).

Loginova, M. M. 1970. New species of psyllids (Homoptera, Psylloidea) from Middle Asia [Novye vidy psyllid iz srednei Azii]. *Entomol. Obozr.*, 49,3:601-623 (in Russian).

Loginova, M. M. 1971. On the taxonomy of the Palaearctic jumping plant-lice (Homoptera, Psylloidea) [O taxonomii listobloshek Palearktiki]. *Entomol.Obozr.*50: 628-631 (in Russian).

Loginova, M. M. 1974a. Psyllids (Psylloidea, Homoptera) of the Mongolian People's Republic [Psyllidy Mongol. Nar. Resp.] In: '*Nasekomie Mongolii*' (*Insects of Mongolia*), 2, Leningrad, 2: 51-66 (in Russian).

Loginova, M. M. 1974b. The adaptive peculiarities of psyllids of the tribe Pachypsylloidini (Homoptera, Aphalaridae) inhabit deserts [Adaptivnye osobennosti psyllid triby Pachypsylloidini obytayuschie v pustyniyakh]. In: *Mater. VII S'ezda Vsesojuzn. Entomol. Obchestva [Proceedings of the VII All-Union Entomol. Society]* I: 75-76 (in Russian).

Loginova, M. M. 1976. Psyllids of the tribe Pachypsylloidini (Psylloidea, Homoptera) [Psyllidy tribe Pachipsylloidini]. *Zool. Zhurn.*, 55: 612-614 (in Russian with English summary).

Longuet –Marx, F. (Ed.), 1998. The Caucasus: old axes, new networks [Caucase: Axes anciens, nouveaux enjeux]. Geneva, Switzerland: *CRES, Nauveaux Mondes*, 214p.

Lordkipanidze, B. L. 2003. The Brown bear (*Ursus arctos* L.) in arid and semi-arid ecosystems of the South Caucasus [Bury Medved' v aridnych i semi-aridnych ekosistemakh Yuzhnoi Gruzii], (35-40pp.) In: *Conservation of arid and semi-arid ecosystems of the South Cauacsus. Proc. of NACRES – Centre for Biodiversity Conservation and Research.* Tbilisi, Georgia (in Russian).

Lordkipanidze, D., Jashashvili, T., Vekua, A., et al. 2007. Postcranial evidence from early *Homo* from Dmanisi, Georgia. *Nature*, 449:305-310.

Lordkipanidze, D., Vekua, A. Ferring, R., et al. 2005. The earliest toothless hominin skull. *Nature*, 434:717-718.

Lordkipanidze, O. 2000. Phasis: the river and city in Colchis. *Geographica Historica 15.* Franz Steiner Verlag, Stuttgard, 146pp.

Lubchenco, J., Olson, A. M., Brubaker, L. B., Carpenter, S. R. et al. 1991. The Sustainable Biosphere Initiative: an ecological research agenda. *Ecology*:72(2):371–412.

L'vov, P. L. 1964. Forests of Dagestan [Lesa Dagestana]. *Dagestanskoe Knizhnoe Izdatel'stvo*, Makhachkala (in Russian), 293pp.

Lysenko, T., Muchina, L., Lakushenko, D. 2011. Nomenclatural notes on saline vegetation of Ukraine, southern Russia and Kazakhstan. *Lazaroa*, 32:187-189.

Lyubenova, M., Sokolovska, M., Bratanova-Doncheva, S., Radonova, M. 2008. Content of macro- and trace elements in sweet chestnut phytomass in Belasitza Mountain, Bulgaria. *Silva Balcanica*, 9(1):1-18.

MacFadden, B. J. 1997. Origin and evolution of the grazing guild in New World terrestrial mammals. *Trends Ecol. Evol.*, 12(5):182-187.

Magakian, A. K. 1941. *Vegetation of the Armenian SSR [Rastitel'nost' Armiyanskoi SSR]*. Izd. Ak. Nauk SSSR [The USSR Ac. Sci. Publ.], Moscow-Leningrad, 274pp (in Russian).

Magin Ch. 2005. *World Heritage Thematic Study for Central Asia: a regional overview*. Paris: UNESCO World Heritage Centre, (2009-11-17). http://whcunesco.org.

Magomedov, M. R., Yarovenko, Y. 2009. Current status of Dagestanian tur (*Capra cylindricornis*) in the eastern Caucasus (Dagestan) [Tekuschii status Dagestanskogo tura v vostochnom Kavkaze (Dagestan]. 68-73pp. In*: Zazanashvili, N & Mallon, D. (Eds.). Status and protection of Globally Threatened Species in the Caucasus.* Tbilisi: CEPF, WWF. Countour Ltd.

Maharramova, E. 2015. Genetic diversity and population structure of the relict forest trees *Zelkova carpinifolia* (Ulmaceae) and *Pterocarya fraxinicolia* (Juglandaceae) in the South Caucasus. *Dissert. Doctor rerum naturalium (Dr.rer. nat)*, Freie Universitat, Berlin, 1-123pp.

Maharramova, E. H., Safarov, H. M., Kozlowscki, G., Borsch, T., Muller, L. A. 2015. Analysis of nuclear microsatellites reveals limited differentiation between Colchic and Hyrcanian populations of the wind-pollinated relict tree *Zelkova carpinifolia* (Ulmaceae). *Amer. J. Bot.,* 102,0:119-128.

Mailov, A. I. 1989. National resources in Azerbaijan deserts [National'niye resursy pustyn' Azerbaijana]. *J, Problems Desert. Develop.* Ashkhabad, 5:63-65 (in Russian).

Makalatia, S. 1933. *Tusheti.* Tfilisi Publ. Tfilisi, 225pp. (in Georgian).

Makashvili, A. 1995. *The trees and shrubs of Georgia [Dereviya i kustarniki Gruzii].* WWF, Tbilisi (in Georgian with Georgian-English-German-French-Russian dictionary of plant names), 66p.

Makhatadze, L. B. 1957. *Oak forests of Armenia [Dubravy Armenii].* Publ. House Ac. Sci. Arm.SSR, Yerevan, 326pp. (in Russian).

Makhatadze, L. B. 1966. *Dark coniferous forests of the Caucasus [Temnokhvoijnnye lesa Kavkaza]* Lesopromyshlennost [Forest Industry Publ]. Moscow, 159pp. (in Russian).

Makhatadze, L. B., Urushadze, T. F. 1972. *Subalpine forests of the Caucasus [Subal'piiskie lesa Kavkaza].* Lesnaya Promyshlennost [Forest Industry Publ.], Moscow, 112pp. (in Russian).

Maleev, V. P. 1935. A review of the Caucasian oaks and their taxonomic and geographic relations due to evolution of the genus *Quercus* [Obzor dubov Kavkaza i ikh sistematicheskikh i geograficheskikh otnosheniyakh v sviyazi s evolutsiei roda *Quercus*] *J. Botan.* XX, 2:153-177 (in Russian).

Maleev, V. P. 1939. Pear *Pyrus* L. [Grushi *Pyrus* L.]. In: *Flora USSR,* vol. 9, Moscow-Leningrad, 352pp. (in Russian).

Maleev, V. P. 1940. Vegetation of the Black Sea coastal countries in the Euxinean Province of the Mediterranean region, its origin and connections [Rastitel'nost' prichernomorskikh stran (Evksinskoyi provincii Sredizemnomor'ija, eijo proiskhozhdenie i svijazi]. *Trudy Botan. Inst. AN SSSR, Ser. Geobotanica,* 3(4):135-252 Leningrad (in Russian).

Maleev, V. P. 1941. *Tertiary relicts in the flora of the Western Caucasus and major stages of the Quaternary History of its flora and vegetation [Tretychnye relikty vo flore Zapadnogo Kavkaza i osnovnye etapy Chetvertichnoi istorii ego flory i rastitel'nosti].* Moscow, 1: 61-144 (in Russian).

Malhi, Y., Roberts, J. T., Betts, R. A., Killeen, T. J., Li, W., Nobre, C. A. 2008. Climate change, deforestation, and the fate of the Amazon. *Science*, 319(5860), 169-172.

Mallon, D. P. 2008. *Saiga tatarica.* In: IUCN 2012. *IUCN Red List of Threatened Species.*

Mamedaliev, J. G., 1963. Atlas of the Azerbaijan SSR. *Gosudarstv. Geologich. Komitet SSR* [*State Geolog. Org. USSR*], Baku-Moscow (in Russian).

Mamedov, E.Yu., Esenov, P. E., Durikov, M. Kh., Zverev, N. E., Tsukanova, S. K. (Redjepbaev-Ed.). 2009. Experience of halophyte cultivation on saline soils [Opyt Viraschivaniya galofitov na zasoliyonnykh zemliyakh]. Ministry of Nature Protection of Turkmenistan, National Institute of Deserts, *Flora and Fauna*, Ashgabad (1-44pp.) (in Russian).

Manafzadeh, S., Staedler, Y. M., Conti, E. 2017. Visions of the past and dreams of the future in the Orient: the Irano-Turanian region from classical botany to evolutionary studies. *Biol. Rew. Camb. Philos. Soc*, 92(3):1365-1388.

Manjavidze, D. V. 1982. *Relict forests of Adjara [Reliktovye lesa Adjarii]*. Metsniereba, Tbilisi, 262pp (in Russian).

Manthey, M., Peper, J. 2010. Estimation of grazing intensity along grazing gradiets – the bias of nonlinearity. *J. Arid Environ.*, 74:1351-1354.

Manuel, C., Molles, Jr. 1999. *Ecology: Concepts and Applications*. McGraw-Hill Inc. [Reprinting by Science Press], McGraw Hill, Boston, 342 pp.

Mao, K., Hao, G., Liu, J-Q., Adams, R. P., Milne, R. I. 2010. Diversification and biogeography of *Juniperus* (Cypressaceae): variable diversification rates and multiple intercontinental dispersals. *New Phytologist*, 188:254-272.

Marlon, J. R., Bartlein, P. J., Daniau, A. L. et al. 2013. Global biomass burning: a synthesis and review of Holocene paleofire records and their controls. *Quaternary Science Reviews,* 65:5-25.

Marques, E. S. D., Price, P. W., Cobb, N. S. 2000. Resource abundance and insect herbivore diversity on woody fabaceous desert plants. *Environmental Entomology*, 29(4):696–703.

Martin, P. S. 2005. *Twilight of the mammoths: ice age extinctions and the rewilding of America*. University of Coliformin Press, 270pp.

Martirossian, A. A. 1982. *The rock carvings of the Chegham Mountain Range [Naskal'nye risunki Chegamskogo gornogo khrebta]*. Ac. Sci. Arm.SSR Publ., Yerevan, 124pp. (in Russian).

Maruashvili, L. I. 1964. The physical geography of Georgia [Saqartvelos fizikuri geografia]. *Tsodna*, Tbilisi, 343 pp. (in Georgian).

Maruashvili, L. I. 1970. *The physical geography of Georgia [Saqartvelos fizikuri geografia]*. Part II. Tbilisi State Univ. Publ., Tbilisi, 215pp. (in Georgian).

Maruashvili, L. I. 1975. The Pleistocenic climatic deterioration in Georgia. In: *Geography in the Georg*. SSR Symposium [Geografiya v Gruzinskoi SSR], Metsniereba, 2:135-141, Tbilisi (in Russian).

Maruashvili, L. I. 1981. Features of mountainous karst of the southern part of the USSR. Georgia as an example. In: *Sharma, H. Sh. (Ed.), Perspectives in Geomorphology*, vol. 1:270p., Concept Publishing Company, New Dehli.

Maruashvili, L. I. 1985. *Palaegeographical dictionary [Paleogeograficheski slovar]*. Mysl, Moscow, 366pp. (in Russian).

Maruashvili, L. I. 1987. Khvamli. *Georgian Soviet Encyclopedia*, Vol.11:474 (in Georgian).

Matikashvili, V. I. 1953. Boxwood stand in Kakheti [Samshitovaiya roscha v Kakheti], *Vestnik Tbilisk Bot. Sada*. [*Proc. Tbilisi. Bot. Gord.*], 15:206-209 (in Russian).

Matsiakh, I. 2016. Assessment of forest pests and diseases in native boxwood forests of Georgia. *Final report*. IUCN, WWF (www.enpi.Reg.org.).

Mattia, F., Imazio, S., Grassi, F., Doulati, H., Scienza, A., Labra, M. 2008. Study of genetic relationships between wild and domesticated grapevine distributed from Middle East Regions to European Countries. *Rendiconti Lincei*, 19:223-240.

Mayer, C. A. 1831. List of plants found and collected during the journey undertaken in the Caucasus and the provinces on the western shore of the Caspian Sea during 1829 and 1830. St. Petersburg. [Verzeichnis der pflanzen, welche während der auf allerhöchsten Befehl in der Jahren 1829 und 1830 unternommenen Reise im Caucasus und in den provinzen am westlichen Ufer des Caspischen Meeres gefunden und eigesammelt worden sind. St. Petersburg]. *Gustav Fischer Verlag, Stuttgart*, 386 pp.

Mayr, E. 2002. The biology of race and the concept of equality. *Daedalus*, 131,1:89-94.

McAllister, H. A. 1993. Cytology and the conservation of rare birches. In: *Hunt D. (Ed.): Betula*:61-66.

McGovern, P. E. 2003. *Ancient wine: the scientific search for the origin of viniculture.* Princeton University, Princeton. New Jersey

McInerney, F. A., Wing, S. L. 2011. The Paleocene-Eocene thermal maximum: a perturbation of carbon cycle, climate, and biosphere with implications for the future. *Ann. Rev. Earth Planet. Sci.* 39: 489-516.

McLoughin, S. 2001. The breakup history of Gondwana and its impact on pre-Cenozoic floristic provincialism. *Aust. J. Bot.*, 49:271-300.

McNaughton, S. J. 1984. Grazing lawn: animals in herds, plant form, and co-evolution. *The American Naturalist,* 124,6:863-886.

McNaughton, S. 1985. Ecology of a grazing ecosystem: the Serengeti. *Ecol. Monogr.*, 55,3:259-294.

McNeely, J. A., Mooney, H. A., Neville, L. E., Schei, P., Waage, J. K. (Eds.). 2001. *A global strategy on invasive alien species.* Gland, Switzerland and Cambridge, UK, World Conservation Union (IUCN).

Medvedev, J. S. 1915. Vegetation of the Caucasus. An experience of botanical geography of the Caucasus Isthmus [Rastitel'nost' Kavkaza. Opyt botanicheskoi geografii Kavkazskogo peresheika]. In: *Trudy Tiflisk Bot. Sada* [*Proceed. Tiflis. Bot. Gard.*], 18,1-2, Tiflis, 108pp. (in Russian).

Medvedev, J. S. 1919. *Trees and shrubs of the Caucasus [Dereviya i kustarniki Kavkaza]*, 3-rd ed., Tiflis (in Russian).

Meigs, P. 1953. World distribution of arid and semi-arid bioclimate. *IUNESCO Arid Zone Res. Ser.,* 1:203-209.

Melikov, R. K., Hatemov, V. V. 2002. The formation of Artemisietum of lowland of Kur-Araz in Azerbaijan. In: *VIth plant of life Southeast Asia Symposium*. Van. Turkey: 91-92.

Menabde, Ts. 1983. *Concise English-Russian-Georgian biological dictionary.* Metsniereba, Tbilisi, 267pp.

Menitski, Y. L. 1981. *Oaks of the Caucasus: a review of the Caucasian representatives of section Quercus [Duby Kavkaza: obzor Kavkazskikh predstaviteleii sektsii Quercus]* Leningrad Sci. Publ., 196pp. (in Russian).

Menzbir, M. A. 1934. *An essay of the history of fauna of the European part of the USSR (from the Early Tertiary) [Ocherk po istorii fauny Evropeiiskoi chasti SSSR [ot nachala Tretichnoi ery]*. Moscow-Leningrad, 223pp. (in Russian).

Merev, N., Yavuz, H. 2000. Ecological wood anatomy of Turkish *Rhododendron* L. (Ericaceae). Interspecific variation. *Turk. J. Bot.*, 24:227-237.

Merino, F., Donat, P. 2011. Invasive plants in the coastal vegetal communities in Valencia (Spain). *Not. Bot. Horti Agrobot. Cluj-Napoca,* 39(1), 9-17.

Messerly, B., Ives, J. D. (Eds). 1997. Mountains of the world: a global priority. Carnforth, UK and New York, USA, Parthenon Publishing Group and Ltd, 495pp.

Metera, E. Sakowski, T. Sloniewiki, K., Pomonowicz, B. 2010. Grazing as a tool to maintain biodiversity of grassland - a review. *Animal Science Papers and Reports,* 28,4:315-334.

Metreveli, R. (Ed-in-Chief). 2008. *History of Georgia [Kartlis tskhovreba].* Meridiani, & Artanuji Publ., Tbilisi, 828pp. (in Georgian).

Metreveli, R. (ed-in-chief). 2008. *History of Georgia [Kartlis tskhovreba].* Meridiani, & Artanuji Publ., Tbilisi, 828pp. (in Georgian).

Meusel, H., Jäger, E., Weinert, E. 1965. *Comparative chorology of Central European flora. [Vergleichende Chorologie der Zentraleuropäischen flora].* Jena: Gustav Fischer Verlag, Jena, 583 pp.

Meyen, S. V. 1981. *Traces of herbs [Sledy trav indeiiskikh].* Mysl Publ., Moscow, 159 pp. (in Russian).

Michel, S. 2008. Conservation and use of wild ungulates in Central Asia - potentials and challenges. In: R. D. Baldus, Damn, G. R., Wollscheid, K. (Eds.). *Best practices in sustainable hunting. A guide to best practices from around the world,* 32-40pp.

Michener, C. D. 2000. *The bees of the world. 2nd edit.* Johns Hopkins University Press, Baltimore, 953pp.

Mikeladze, R. G. 1960. On the study of alpine carpet-like meadows of South Ossetia [K izucheniyu al'piiskikh kovreobraznykh lugov Yuzhnoi Ossetii]. *Probl. Bot.*, 5:174-181. M.-L. (in Russian).

Mikhailovskiy, V. B. 1972. *Genetical features of takyr like soils of south-eastern Shirvan and the Siazan-Sumgait massif [Geneticheskie osobennosti takyrnykh pochv Yugo-Vostochnoi Shirvan i Siazan-Sumgaitskogo massiva]. PhD thesis biol. sci.*, Baku, 21pp. (in Russian).

Milanovsky, E. E. 1968. Neotectonics of the Caucasus. *Nedra*, 483pp. (in Russian).

Milanovsky, E. E. 1977. Neotectonic and modern volcanism of the Caucasus. In*: Geology of the Quaternary Period* [*Geologia Chetvertichnogo Perioda*]: 35-43, Yerevan (in Russian).

Milanovsky, E. E. 2008. Origin and development on Pliocene and Quaternary Glaciations in northern and eastern Europe, Iseland, Caucasus and Siberia. *Geological Soc., London, Special Publications*, 301:87-115.

Milanovsky, E. E., Khain V. E. 1963. Geological structure of the Caucasus. Moscow, *Moscow State University*, 671p. (in Russian).

Milchunas, D. G., Sala, O. E., Lauenroth, W. K. 1988. A generalized model of the effects of grazing by large herbivores on grassland community structure. *Am. Nat.*, 132:87–106.

Mil'kov, F. N. 1950. *Russian forest-steppe plains. Experience landscape characteristics [Lesostep' Russkoi ravnyny. Opyt Landshaftnoi charakteristiki].* Izd-vo AN SSSR Publ. (USSR Ac. Sci. House), Moscow, 296p (in Russian).

Mil'kov, F. N. 1977. *Natural zones of the USSR [Prirodnye zony SSSR].* Mysl, Moscow, 294pp. (in Russian).

Mil'kov, F. N., Gvozdetsky, N. A. 1976. *Physical geography of the USSR [Fizicheskaiya geografiya SSR].* Mysl. Moscow, 450pp. (in Russian).

Miller, J. S. McCue, K., Consiglio, T., Stone, J., Eristavi, M. et al. 2005. Endemic medicinal plants of Georgia (Caucasus). *Missouri Botanical Garden Press* (in English and Georgian), 41pp.

Milne, R. I., Abbott, R. 2002. The origin and evolution of Tertiary relict flora. *Advances in botanical Research*, 38:281-314.

Milne, L. J., Milne, M. 1976. The mountains. Life Nature Library. *Time-Life International* (Netherland) B. V. (135pp.).

Milner-Gulland, E. J., Kholodova, M. V., Bekenov, A. et al. 2001. Dramatic declines in saiga antelope populations, *Oryx,* 35,4:340-345.

Minahi, K., Goudriaan, J., Latinga, E. A., Kimura, T. 1993. Significance of grasslands in emission and absorbtion of greenhouse gases. In: *Grasslands for our World. Barker, M. J. (Ed.). SIR Publishing*, Wellington, NZ.

Mirimanian, Kh. P. 1959. The origin of mountainous steppes after logging of forests [Proiskhozhdenie gornykh stepei posle vyrubki lesov]. *Botan. Zhurn.*, 44(5):116-122 (in Russian).

Mirzaei, M., Nouri, N., Kiya, A. K., Alizadeh, N. 2016. Analysis on rock art in Qarasu River Basin in Meshgin Shahr (Azerbaijan). *Int. J. Archaeol*, 4(5): 54-60.

Mitichishvili, R., Pirtskhalaishvili, T., Begeluri, G. 2013. The Georgian traditional products of Stockbreeding [Metskhoveleobis qartuli tradiciuli produqtebi]. In: *Agriculture, Georgia. Our valuables (Gulua, B. – Ed.- in-chief). Iverioni*, Tbilisi: 165-176 (in Georgian).

Mittermeier, R. 2012. *Language diversity is highest in biodiversity hotspots.* www.Twitter.com/PrimatelWatcher.

Molchanov, G. I., Molchanova, L. P., Gulko, N. M., Molchanova, A. G., Suchkov, I. C. 1989. Edible medicinal plants of the Caucasus [S'edobnye lekarstvennye rasteniya Kavkaza]. Reference book. Rostov-on-Don. Rostov. Univers. Publ., 464pp. (in Russian).

Molles, M. C., 1999. *Ecology.* Concepts and Applications. WCB/McGraw-Hill, Dubuque, IA. 509 pp.

Monroe, J. S., Wicander, R., 1998, Physical geology: Exploring the Earth (3rd edition). Belmont, California, Wadsworth Publishing Company, 646 p.

Moon, D. 2008. The Russian steppe: an environmental history, 1700–1914. Center to Russian, Central and East European Studies, *Working Paper Series*: 1-11pp.

Moore, J. C., Berlow, E. L., Coleman, D. C., et al. 2004. Detritus, trophic dynamics and biodiversity. *Ecology Letters,* 7:584-600.

Moore, T., Stevens, R., McArthur, E.1982. Preliminary study of some insects associated with rangeland shrubs with emphasis on *Kochia prostrata*. *J. Range Manag.* 35(1): 128-130.

Moores, E. M., Fairbridge, R. W. (Eds.) 1998. Encyclopedia of European and Asian regional geology. *Encyclop. Earth Sci. Ser.*, London, 825 pp.

Morales, R. 1997. Synopsis of the genus *Stachys* L. in the Mediterranean area. *Lagascalia*, 19 (1-2):249-266.

Mordkovich V. G. 2005. *The principles of biogeography [Osnovy biogeografii*]. KMK Publ. Moscow, 236pp. (in Russian).

Mordkovich, V. G. 2014. Steppe ecosystems [Stepnye ekosistemy] (2nd ed.). Novosibirsk Geo Publ. (in Russian), 170pp.

Mordkovich, V. G., Giljarov, A. M., Tishkov, A. A. Balandyn, S. A. 1997. The fate of steppes [Sud'ba stepei]. *Mangazeja*, Novosybirsk, 208 pp. (in Russian).

Mordukhai-Boltovskoy, P. D. 1979. Composition and distribution of Caspian fauna in the light of modern data. *Int. Rev. Gesamten Hydrobiol.*, 64:383-392.

Morozov, Yu. L. 2009. *Life mode and economical use of the Caucasian turs [Biologija i ekonomicheskoe ispol'zovanie Kavkazskogo tura]. PhD thesis biol. sci.* (23pp.), Stavropol (in Russian).

Mosar, J., Kangarli, T., Bochud, M., Glasmacher, U. A., Rast, A., Brunet, M.-F., Sosson, M. 2010. Cenozoic-recent tectonics and uplift in the Greater Caucasus: a perspective from Azerbaijan. Geol. Soc., London. Special Publications, 340:261-280. In: *Sedimentary Basin Tectonics from the Black Sea and Caucasus to the Arabian Platform* (*Sosson, M. et al. - Eds).*

Mosner, E., Schneider, S., Lehmann, B., Leyer, I., 2011. Hydrological prerequisites for optimum habitats of riparian *Salix* communities – identifying suitable reforestation sites. *Applied Vegetation Science* 14, 367-377.

Mousavi, R. 2012. Non-wood forest products, the utilization, and harvesting methods in Sardasht, North West Iran. *Intern. J. Forest, Soil and Erosion* (UFSE), 2(3):133-136.

Mukhashavria, A. 1984. Great spruce bark beetle (*Dendroctonus micans*) in Georgia [Nadzvis didi lafnichamia Saqartveloshi). *Georgian Great Encyclopedia*, vol. 7, 342 (in Russian).

Müller, P. 1986. *Biogeography*. Harper & Row Publishers, New York, 377pp.

Mumladze, L. D., Tarkhnishvili, D., Pokryszko, B. M. 2008. A new species of the genus *Helix* from the Lesser Caucasus (SW Georgia). *J. Conch.*, 39:483-485.

Munkhbaatar, M. 2012. Hypothesis of mutualism between of Gobi racerunning (*Eremias przewalskii*) and Siberian nitrebush *(Nitraria sibirica). Proceedings of the 5th Congress of the Alexander M. Nikolsky Herpetological Society.* 22-27 September 2012, Minsk. Belarus:203-205.

Murtskhvaladze, M., gavashelishvili, A., Tarkhnishvili, D. 2010. Geographic and genetic boundaries of brown bear (*Ursus arctos*) population in the Caucasus. *Molecular Ecology*, 19,9:1829-1841.

Musayev, A., Akparov, Z., Huseynova, T. 2014. Importance of biodiversity fruit crops and their wild relatives for food and nutritional security in Azerbaijan. *J. Crop and weed*, 10,2:142-146.

Mushketov, I. V. 1896. Geological essay of the glacial area of Teberda and Chkhalta in the Caucasus [Geologicheskji otcherk lednikovoi oblasti Teberdy i Tchkalty na Kavkaze]. *SPB, Trudy Geol. Komiteta, [Proceed. Geol. Committee]*, 17, 4:22-29 (in Russian).

Myers, N., Mittermeier, R. A., Mittermeier, C., G., B. Da Fonseca, G. A. B., Kent, J. 2000. Biodiversity hotspots for conservation priorities. *Nature,* 403:853-858.

Myers, R. 2006. Living with fire-sustaining ecosystems and livelihoods through integrated fire management. The Nature Conservancy Global Fire Initiative, Tallahassee, F. L. 32pp. *http://tncfire.org/documents/Integrated-Fire-Management_Myers - 2006 pdf.*

Nadiradze, T. 2014. Bio-ecological peculiarities of *Taxus baccata* L. in decorative gardening of Eastern Georgia. *European Researcher,* 71,3-2:560-563.

Nagy, L., Grabherr, G. 2009. *The biology of alpine habitats.* New York: Oxford University Press, 376pp.

Nagy, L., Grabherr, G., Körner, Ch., Thompson, D. B. A. (Eds.). 2003. *Alpine biodiversity in Europe. Ecological studies*, vol 167. Springer-Verlag Berlin Heidelberg, 477pp.

Nakaidze, E. K., Sokhadze, E. V. 1979. *Typological peculiarities of vegetations and soils of subtropical steppes of the eastern Transcaucasia [Tipologicheskie osobennosti rastitel'nosti i pochvy subtropicheskikh stepei vostochnogo Zakavkaz'iya].* Metsniereba 26-46 (in Russian).

Nakhutsrishvili, G. 1998. The vegetation of the subnival belt of the Caucasus Mountains. *Arctic and Alpine Research*, 30,3:222-226.

Nakhutsrishvili, G. 1999. The vegetation of Georgia (Caucasus). Camerino, Italy. *Braun-Blanquetia*, vol. 15:69pp.

Nakhutshishvili, G. 2003. High mountain vegetation of the Caucasus region. In: Nagy L., Grabherr G., Körner Ch., Thompson D. B. A. (Eds). *Alpine biodiversity in Europe.* Springer Verlag Berlin Heidelberg: 93-103.

Nakhutsrishvili, G. 2013. The vegetation of Georgia (South Caucasus). *Geobotany Studies.* Springer-Verlag Berlin Heidelberg, 235pp.

Nakhutsrishvili, G., Abdaladze, O., Akhalkatsi, M. 2006. Biotopes types of the treeline of the Central Greater Caucasus. In: Gatta, D., Akeroyd, J. (Eds.). *Nature Conservation: Concepts and Practice. Springer*, Berlin, New York:211-225.

Nakhutsrishvili, G., Akhalkatsi, M., Abdaladze, O. 2009. Main threats to mountain biodiversity in Georgia (the Caucasus) 2009. *Mountain Forum Bulletin* 9 (2):18–19.

Nakhutsrishvili, G., Zazanashvili, N., Batsatsashvili, K. 2011. Regional profile: Colchis and Hyrcanic temperate rainforests of the western Eurasian Caucasus: 214-221. In: *Temperate and Boreal Rainforests of the World. Ecology and Conservation* (DellaSala, D. A. (Ed.). Island Press.

Nakhutsrishvili, Sh. G. 1966. The main features of vegetational cover of the southern upland of Georgia [Osnovnye osobennosti rastitel'nogo pokrova Yuzhnogo Nagor'iya Gruzii]. In: *Rastitel'nost' Vysokogorii i Voprosy Khos. Ispol'z-a* (*Vegetation of highland and questions of economic use*). Sci. Publ. House. Moscow-Leningrad (in Russian).

Narynyan, S. G. 1962. Alpine carpets of the Caucasus as a particular type of vegetation (essays of their genesis and classification [Al'piiskie kovry Kavkaza kak osoby tip rastitel'nosti (ocherk ikh genesisa i klassifikatcii]. *Trans. Bot. Inst. Acad. Sci. Arm. SSR*, 13:5-28 (in Russian).

Naseka, A. M. 2010. Zoogeographical freshwater divisions of the Caucasus as a part of the West Asian transition region [Zoogeograficheskie presnevodnye regiony Kavkaza kak chast' zapodnoaziatskoi perekhodnoi oblasti]. *Proc. Zool. Inst. RAS*, 314, 4:469-492.

Nasy, R., Wunder, S., Campos, J. 2002. Forest ecosystem services: can they pay our way out of deforestation? *Discussion paper prepared for the GEF for the Forestry Roundtable*, Costa Rica, March 11. 29 pp.

Nasidze, I. S. 1995. Genetic polymorphisms of the Caucasus ethnic groups: distribution of some serum protein and red cell enzyme genetic markers. *Gene Georgr.*, 9(2):91-116.

Nechaeva N. T., Vasilevskaya B. K., Antonova K. G., 1973. *Plant Life Forms in Karakum Desert. [Zhiznennye formy rastenii pustyni Karakum],* Nauka. Moscow, 243 p.

Nekrutenko, Y. P. 1990. *Diurnal butterflies of the Caucasus [Dnevnye babochki Kavkaza]. A key of families Papilionidae, Pieridae, Satyridae and Danaidae [Opredelitel' semistv...].* Naukova Dumka Publ., Kiev, 213pp. (in Russian).

Nesterov, A. I., Fedotov, V. I. 2005. Toward the questions about the northern border of the forest-steppe zone in mid-Russia elevation [K voprosu o severnoi granitse lesostepnoi zony na Srednerusskoi vozvyshennosti]. *Vestnik VGU, Ser. Geografia, Geoekologia*, 2:151-154.

Neudert, R., Allahverdiyeva, N. 2009. The economic performance of transhumant sheep farming in Azerbaijan and prospects for its development. *Annals of Agrarian Science*, 7(4):153-158.

Newlands, G. 1972. Notes on psammophilic scorpions and a description of a new species (Arachnida: Scorpionides). *Annals of the Transvaal Museum*, 27:241-253.

Nikol'skaya, N. I. 1972. Species of the genus *Kalidium* Moq., their distribution and role in plant communities [Vidi roda *Kalydium* Moq., ikh rasprostranenie i rol' v rastitel'nom soobchestve]. *Bot. Zhurn.*, 57:299-312 (in Russian).

Nomade, S., Scao, V., Guillou, H., Messager, E. et al. 2013. Upper Pliocene to Lower Pleistocene volcanism in Javakheti (Republic of Georgia). Impact on the landscape morphology and human occupation. Environments and societies in the southern Caucasus during the Holocene. *Quaternary International*, 395, 45–59.

Novruzova, Z. A., Orlova, S. Y., Sadikhov, I. A. 1978. The peculiarities of foliar anatomy of some species of the feather grasses of the Nakhichevan ASSR [Osobennosti anatomii lista nekotorykh vidov kovyleyii Nakhichevanskoi ASSR]. *Izv. AN Azerb.SSR, ser. Biol.* (*Bul. Ac. Sci. Azerb. SSR, Ser. Biol. Sci.*), 2:3-7 (in Russian).

Ochiauri, D. 1965. *Materials of the flora of Pshav-Khevsureti [Masalebi Pshav-Khevsuretis floristvis].* Tbilisi, Metsniereba, 97pp. (in Georgian).

O'Loughlin, J., Kolossov, V., Radvanyi, J. 2007. The Caucasus in a time of conflict, demographic transition, and economic change. *Eurasian Geography and Economics*, 42,2:135-156.

Onipchenko, V. G. 2002a. Alpine vegetation of the Teberda Reserve, the Northwestern Caucasus. *Veröff. Geobot. Inst. ETH, Stiftung Rübel*, Zürich, 130:1-168.

Oreskes, N. (Ed.). 2003. *Plate Tectonics-an insider's history of the modern theory of the Earth. Cambridge*, MA: Westview Press. ISBN 0-8133-4132-9, 452pp.

Organ, C., Nunn, C. L., Machanda, Z., Wrangham, R. W. 2011. Phylogenetic rate shifts in feeding time during the evolution of Homo. *Proc. Natl. Acad. Sci. USA*: 108: 14555-14559.

Osborne, C. P., 2008. Atmosphere, ecology and evolution: what drove the Miocene expansion of C_4 grasslands? *Journal of Ecology*, 96:35-45.

Osborne, C. P., Beerling, D. J. 2006. Nature's green revolution: the remarkable evolutionary rise of C_4 plants. Philosophical Transactions of the Royal Society B: *Biological Sciences*, 361 (1465): 173-194.

Osborne, C. P., Chaloner, W. G., Beerling, D. J. 2004. Falling atmosphere CO_2 – the key to megaphyll leaf origins. In: *Poole, I., Hemsley, A. R.-Eds. The evolution of plant physiology. Kew, UK: Elsevier Academic Press*: 197-215.

Ospanov, M. O., Stamkulova, K. Zh. (Eds.) 2012. Biodiversity of wetlands in the Syrdarya River delta front [Bioraznoobrazie bolot peredovoii linii del'ti reki Sirdarii] (Chapt.3: *Flora and Vegetation*, 18-27). Almaty, 62pp. (in Russian)

Oxford dictionaries 2012. Oxford University Press.

Paffengolts, K. N. 1959. *Geological overview of the Caucasus [Geologicheski ocherk Kavkaza].* Acad. Sci. Tbilisi, 1-49 (in Russian).

Pagani, M., Huber, M., Liu, Z., et al. 2011. The role of carbon dioxide during the onset of Antarctic glaciation. *Science*, 334, 6060:1261-1264.

Painter, K., Ralphs, M., Pfister, J., Gardner, et al. 2003. Why white people are called 'Caucasian'? *Proceedings of the Fifth Annual Gilder Lehrman Center International Conference at Yale University.* Collective degradation: Slavery and the construction of race. Yale University, Nov. 7-8:1-37.

Painter, N. I. 2011. *The history of white people.* W. W. Norton & Company, USA, 448pp.

Palagin, I. D. 1988. *Apiculture by the laws of hollow trees [Pchelovodstvo po zakonam duplistogo dereva].* Mordovskoe Knizh. Izd. [Mordovian Book Publ], 131pp. (in Russian).

Palone, R. S., Todd, A. H. (Eds.) 1998. Chesapeake Bay riparian handbook: a guide for establishing and maintaining riparian forest buffers. *USDA Forest Service*. NA-TP-02-97.

Penov, V. P., Il'ychev, Y. G., Salpegarov, A. D. 2008. *Fluctuations of the North Caucasus Glaciers during 19-20th century*. Severokavkazskoe Izd. MIL [North Caucasus MIL Publ.] Pyatigorsk (in Russian), 183pp.

Papuashvili, R., Papuashvili, N. 2005. *Ergeta (Former settlement of Kolkheti) [Ergeta (Kolkhetis adreuli dasakhleba],* Metsniereba, Tbilisi, 79pp. (in Georgian).

Parfitt, D, Badens, M. L. 1997. Phylogeny of the genus *Pistacia* as determined from analysis of the chloroplast genome. *Proc. Nat.Acad. Sci*. USA, 94:7987–7992.

Parida, A. K., Das, A. B. 2005. Salt tolerance and salinity effect on plants: a review. *Ecotoxicol. Envir. Soc.*, 60:324-249.

Parolo G, Abeli T, Gusmeroli F, Rossi G. 2011. Large-scale heterogeneous cattle grazing affects plant diversity and forage value of alpine species-rich *Nardus* pastures. *Grass Forage Sci*. 66(4): 541–550.

Patarkalishvili, T. 2017. Forest biodiversity of Georgia and endangered plant species. *Annals of Agrarian Science*, 15,3:349-351.

Paton, A. J., Brumitt, N., Govaerts, R. et al. 2008. Towards Target 1 of the global strategy for plant conservation: a working list of all known plant species - progress and prospects. *Taxon,* 57:602-611.

Patrikeev, M. V. 1991. On the birds of the Bejuk-Dash (Gobustan, eastern Azerbaijan). Fauna, population and ecology of birds of the North Caucasus [O ptitcakh gory Beijuk-Dash (Gobustan, Vostochnyi Azerbaijan). Fauna, naselenie i ecologyiy aptitc Sev. Kavkaza]. *In: Materialy Nauchn.-Praktich. Konf. [Materials for the Sci. Appl. Conf.]*, Stavropol: 56-59.

Pausas, J. G. Keeley, J. E. 2009. A burning story: the role of fire in the history of life. *BioScience:* 59,7:593-601.

Pearce, D. W., Moran, D., Wendelaar, A. 1996. Global biodiversity priorities: a cost-effectiveness index for investments, *Global Environmental Change* 6 (2), pp. 103-119.

Pearson, P. N., Foster, G. L. Wade, B. S. 2009. Atmospheric carbon dioxide through the Eocene-Olygocene climate transition. *Nature,* 461:1110-1113.

Pearson, P. N., Palmer, M. R. 2000. Atmospheric carbon dioxide concentrations over the past 60 million years. *Nature,* 406,17:695-699.

Peer, T., Gruber, J. H., Millinger, A., Hussain, F. 2007. Phytosociology, structure and diversity of the steppe vegetation in the mountains of Northern Pakistan. *Phytocoenologia*, 37: 1-65.

Peper, J., Pietzsch, D., Manthey, M. 2010. Semi-arid rangeland vegetation of the Greater Caucasus foothillsin Azerbaijan and its driving environmental conditions. *Phytocoenologia*, 2-3: 73-90.

Perkins DF. 1968. Ecology of *Nardus stricta* L. annual growth in relation to tiller phenology. *J. Ecol*. 56,3:633–646.

Petrov, M. P. 1973. *Deserts of the world [Pustyni Mira]*. Nauka, Leningrad, 435 pp. (in Russian).

Philip, K., Klepeis, A., Frederick, J. V. 2009. *Global tectonics*. Blackwell Publishers., 496pp.

Piermattei, S. 2013. Local farmers vs environmental universalism: conflicts over nature conservation in the Parco Nazionale dei Monti Sibillini, Italy. In: *Koensler, A. and Papa, C. (Eds.), 2013. After anthropocentrism? Environmental conflicts, social movements and power. J. Polit. Ecol.*, 20:255-341.

Pietzsch, D. 2008. *Vegetation response along grazing gradients in Artemisia-semi-desert of Gobustan (Azerbaijan). Diploma thesis in the study Program of Biology*. Greifswald, 152pp.

Pimental, D., McNair, M., Buck, L., Pimental, M., Kamil, J. 1997. The value of forests to world food security. *Human Ecology,* 25,1:91-120.

Piperno, D. R., Sues, H.-D. 2005. Dinosaurs dined on grass. *Science,* 310:1126-1128.

Pipia, I., Gogniashvili, M., Tabidze, V. et al. 2012. Plastid DNA sequence diversity in wild grapevine samples (*Vitis vinifera* subsp. *sylvestris*) from the Caucasus region. *Vitis,* 51,3:119-124.

Poinar, G., Alderman, S., Wunderlich, J. 2015. One hundred million year old ergot: psychotropic compounds in the Cretaceous? *Palaeodivesrsity*, 8:13-19.

Pollard, D., DeConto, R. M.2015. Hysteresis in Cenozoic Antarctic ice-sheet variations. *Glob. Planet. Change,* 45: 9–21.

Ponomarenko, L. I. 1986. Coenopopulations of wormwoods (*Artemisia fragrans* Willd. and *Artemisia szowitsiana* (Bess.) Grossh.) In: *The desert phytocoenosis of the Shirvan. AN Tadj. SSR, Dushanbe*:1-21p. (in Russian).

Ponting, C. 1991. *A green history of the world: the environment and the collapse of great civilizations.* New York: Penguin. 432p.

Pontzer, H., Scott, J. R., Lordkipanidze, D., Unger, P. S. 2011. Dental microwear texture analysis and diet in the Dmanisi hominins. *J. Hum. Evol.* 61:683-687.

Popović, R.., Kojić, M., Karadžić B. 1997. Ecological characteristics of six important submediterranean tree species in Serbia. *Bocconea* 5(2): 431–438.

Prasad, V., Strömberg, C. A. E., Alimohammadian, H., Sahni, A. 2005. Dinosaur coprolites and the early evolution of grasses and grazers. *Science,* 310 (5751):1177-1180.

Prasad, V., Strömberg, C. A. E., Leaché, A. D. et al. 2011. Late Cretaceous origin of the rice tribe provides evidence for early diversification in Poaceae. *Nat. Commun.,* 2,480:1-8.

Price M. F. 1995. Mountain Research in Europe: An Overview of MAB Research from the Pyrenees to Siberia, *Paris and Carnforth: UNESCO and Parthenon*, 230pp.

Prilipko, L. I. 1945. The vegetation of Azerbaijan SSR. *Physical geography of Azerb. SSR [Rastitel'nost' Azerbaidjanskoi SSR. Fizicheskaiya geografiya AzSSR]*. Izd. AN Az. SSR [Ac. Sci. Az.SSR], Baku.162pp. (in Russian).

Prilipko, L. I. 1949. Results of experiments to improve the pasture systems in the Shirvan-steppe. *Trudy Bot. Inst. AN Az. SSR*, 11:3-43 (in Russian).

Prilipko, L. I. 1970. *The vegetation cover of Azerbaijan [Rastitel'ny pokrov' Azerbaijana]*. Elm, Baku, 170pp. (in Russian).

Prilipko, L. A. 1980. Deserts of the Kura-Aras (East Transcaucasia) [Pustyny Kura-Araksinskoi nizmennosti, Vostochnoe Zakavkaz'ie]. In: *Vegetation of European part of the USSR* [*Rastitel'nost Evropeiskoi chasti SSSR*], Leningrad, pp 295-298 (in Russian).

Prilipko, L. I., Agadjanov, S. D. 1972. *The vegetation of the Caspian Sea shore of Azerbaijan and its changing according to dynamics of water level [Rastitel'nost' poberezhii Kaspiiskogo Moriya Azerbaijana ego kolebanie v sviyazy s dinamikoi ego urovniya]*. Elm. Publ. Baku, 147-149, (in Russian).

Prima, V. 1974: Subnival flora of the East Caucasus, its composition, ecological-biological and geographical analysis [Subnival'naiya flora Vostochnogo Kavkaza, ejo sostav, ekologo-biologicheskii i geograficheski analiz] (46-69pp.). In: *Galushko, A. J. (Ed.). Flora and vegetation of the East Caucasus*. Ordjonikidze (in Russian).

Proctor, M., Yeo, P., Lack, A. 1996. *The natural history of pollination.* Herper Collins, London, 487pp.

Proctor, M., Yeo, P., Lack, A. 2003. *The natural history of pollination.* Timber Press, London, 577pp.

Prokaev, V. I. 1967. *Basic methods of physiographic regionalization [Osnovnye metody fiziko-geograficheskogo rajonirovaniya].* Nauka, Leningrad, 168pp. (in Russian).

Prokhorov, A. M. (Ed.-in-chief) 1973. *The Great Soviet Encyclopedia [Bol'shaiya Sovetskaiya Entciklopedia].* vol. 11 (p. 113). Moscow, Sovetskaia Entsiklopediya (in Russian).

Prydnia, M. 2002. *Taxus baccata* in the Caucasus region. *Journal of Yew Protection.*9:146-166.

Pucek, Z. 1986. *Bison bonasus* (Linnaeus, 1758) – Wisent. In: *Handbuch der Säugetiere Europes.* J. Niethammer & F. Krapp, Eds.). Aula verlag Wiesbaden: 2/II. Parhufer*: 278-315.*

Pucek, Z., Belousova, I. P., Krasinska, M., Krasinski, Z. A., Olech, W. (Pucek, Z.-Ed.) 2002. European Bison *bison bonasus*: current state of the species and an action plan for its conservation. *Mammal Research Institute. Polish Ac. Sci.,* Bialowieza. http/bison.zbs.biolowieza.pl/pdf/bis-2003. Pdf. (1-51pp.).

Pyne, S. J. 2001. *Fire: a brief history. Seattle*: University of Washington Press, 204pp.

Pysek, P. 1998. Alien and native species in Central European urban floras: a quantitative comparison. *J. Biogeogr.* 25:155–163.

Qoranashvili, G. 1995. Questions of ethnic identity according to Leonti Mroveli's Historical Chronicles. *Studies,* Vol.1, 32-51pp. Tbilisi (in Georgian).

Radde, G. I. 1874. Four lectures on the Caucasus [Vier Vorträge über den Kaukasus]. *Petermanns Geographische Mitteilungen*, 8(36): 1-74.

Radde, G. I. 1899. *Collection of the Caucasian Museum [Kolektcii Muzeiya Gruzii].* I. Tiflis, 521pp. (in Russian).

Ranson, S. L. Thomas, L. 1960. Crassulacean acid metabolism. *Annual Rev. Plant Physiol.,* 11,1:81-110.

Raup, D. M. 1986. Biological extinction in earth history: *Science*, 231, (№4745): 1528-1533.

Raupach, M. R., Marland, G., Ciais, Ph., Le Quere, C., Canadell, I, G., Klepper, G., Field, C. B. 2007. Global and regional drivers of accelerating CO_2 emissions. *Proc. Nat. Ac. Sci.* USA, 104,24:10288-10293.

Rautain, G. S., Kalabushkin, B. A., Nemtsev, A. S. 2000. A new subspecies of the European bison [Novyi podvid Evropeijskogo bizona], *Bison bonasus montanus*, ssp. *nova* (Bovidae, Artiodactyla). *Proceed. Biol. Sci.,* [*Trudy Biol. Nauk]* 375,4:563-567 (in Russian).

Ravanbakhsh, M., Amini, T. 2013. Study on floristic composition, chorology snd ecological structure of Forest Reserve, Telesh, Iran. *World Appl. Sci. J.* 28(11): 1782-1788.

Raven, J. A., Edwards, D. 2001. Roots: evolutionary origins and biochemical significance. *Journal of Environlemtal Botany,* 52: 381-401.

Rawat, G, S., Uniyal, V. K. 1993. Pastoralism and plant conservation: the valley of flowers dilemma. *Environ. Conserv.,* 20:164-167.

Rayfield, D., (Ed.) 2006. *A comprehensive Georgian-English dictionary.* Vol. I, II. Cornett Press, London.

Raymo, M. E., Ruddiman, W. F. 1992. Tectonic forcing of Late Cenozoic climate. *Nature,* 359:117-122.

Rcheulishvili, M. D. 1957. *Sheep farming in Georgia [Ovcevodstvo v Gruzii*]. Gozdat Publ. House, Tbilisi, Georg. SSR, 143pp. (in Russian).

Reich, P. B., Frelich, L. 2002. Temperate deciduous forests, vol.2. The Earth system: biological and ecological dimensions of global environmental change, pp. 565-569. Mooney, H. A., Canedell, J. G. (Eds.). In: *Encyclopedia of global environmental change [Munn, T (Ed-in-chief).* Sohi Wilee & Sons ltd, Chichester, 2002].

Reid, R. E. H. 1997. Dinosaurian physiology: the case for 'intermediate dinosaurs', In: Farlow, I. O., Brett-Surman, M. K. (Eds.). *The Complete Dinosaur.* Bloomington: Indiana University Press: 449-473.

Reid, W. V. Mooney, H. A., Cropper, A. et al. 2005. *Ecosystems and human well-being synthesis. Myllenium Ecosystem Assessment.* Island Press, Washington, 137pp.

Reimers, N. E. 1990. *Environmental management (The ecological-socio-economic glossary).* Mysl, Moscow, 639pp. (in Russian).

Reisner, O. 2009. Travelling between two worlds – the Tergdaleulebi, their identity conflict and national life. *Identity Studies*: 1:36-50.

Rhopal R. 2007. The beautiful skull and Blumenbach's errors: the birth of the scientific concept of race. *British Med. J.*, 335 (7633) 1308-1309.

Richardson, A. D., Friedland, A. J. 2009. A review of the theories to explain arctic and alpine treelines around the world. *J. Sustain. Forestry*, 28: 218-242.

Rightmire, G., Lordkipanidze, D. 2011. Comparisons of early Pleistocene skulls from East Africa and the Georgia, Caucasus: Evidence bearing on the origin and systematics of genus *Homo*. In*: Grine, F. E., Fleage, S. G., Leaky, R.E.-Eds. The first humans, origins and early evolution of the genus Homo. Vertebrate Paleobiology and Palaeoanthropology.* Springer. Netherlands, Dordrecht:39-48.

Rightmire, G. P., Lordkipanidze, D.., Vekua, A. 2006. Anatomical descriptions, comparative studies and evolutionary significance of the hominin skulls from Dmanisi, Republic of Georgia. *J. Hum. Evol.*, 50,2:115-141.

Rikhter, G. D., Mil'kov, F. N. (Eds.) 1956. *Forest-steppe and steppe of the Russian plains [Leso-step i step Russkoi ravnyni].* Inst. Geografii AN SSSR, Izd. AN SSR (Inst. Geogr. AN USSR, Publ. House Ac. Sci. USSR), 296pp. (in Russian).

Rikli, M. 1943-1948. The plants extended across to the Mediterranean countries [Das Pflanzenkleid der Mittelmeerländer]. *Zweite Auflage*. Bd. 1-3. Verlag Hans Huber, Bern

Risser, P. G. 1995. The status of the science examining ecotones. *BioScience*, 45:318-325.

Roberts, W. G., J. G. Howe, and J. Major. 1977. A survey of riparian forest flora and fauna in California. p. 3–19. In: *A. Sands (Ed.). Riparian forests in California: their ecology and conservation. Institute of Ecology Publication No. 15.,122pp.* University of California, Davis.

Robinson, J. M. 1989. Phanerozoic O_2 variation, fire, and terrestrial ecology. *Palaeogeogr., palaeclimat., palaecology*: 75,3:223-240.

Roebroeks, W., Villa, P. 2011. On the earliest evidence for habitual use of fire in Europe. *Proc. Nat. Acad. Sci.*. USA, 108, 3: 5209-5214.

Rögl, F. 1999. Mediterranean and Paratethys. Facts and hypotheses of an Oligocene to Miocene paleogeography, short overview. *Geologica Carpathica*, 50,4:339-349.

Rois-Diaz, M., Mosquera-Losada, M. R., Rigueiro-Rodriguez, A. 2006. Biodiversity indicators on silvopastoralism across Europe. *FFI, Technical Report*, 21,66pp.

Romer A. 1966. *Vertebrate paleontology (3rd edit).* Univ. of Chicago Press, New York, 446pp.

Rosanov, A. N. 1954. The zonal soils of lowlands and foothills of the Kura-Aras Lowland [Zonal'nye pochvy nizmennostei i predgor'iya Kura-Araksinskoi nizmennosti]. In: *To the study of origination of saline soils and their melioration.* Izd. AN SSSR, Trudy Pochven. Inst. AN SSSR [Ac. Sci. USSR Publ., Proceed. Inst. of Soil Ac. Sci. USSR], 44:312-331 Moscow (in Russian).

Roschupkin, G. A. 1958. *Tushetian sheep [Tushetinskiaya ovtsa]. PhD thesis agricult. sci.* Georg. Agric. Inst. Publ., Tbilisi, 37pp. (in Russian).

Royer, D. L. 2006. CO_2-forced climate tresholds during the Phanerozoic. *Geochimic. Cosmochim. Acta,* 70/23:5665-5675.

Royer, D. L., Wing, S. L., Beerling, D. J., Jolley, D. W. et al. 2001. Paleobotanical evidence for near present-day levels of atmospheric CO_2 during part of the Tertiary. *Science,* 292:2310-2313.

Ruban, D. A. 2006. The palaeogeographic outlines of the Caucasus in the Jurassic: the Caucasus Sea and the Neotethys Ocean. *Annales Geologiques de la Peninsule Balkanique,* 67: 1-11.

Ruben, J. A., Jones, T. D. 2000. Selected factors associated with the origin of fur and feathers. *American Zoologist,* 40:585-596.

Rubtsov, G. A. 1930. *Pears [Grushi]* VASKHNIL Publ. [Izd. VASKHNIL] Leningrad, 142pp. (in Russian).

Rudomakha, A. 2005. The world natural heritage site 'West Caucasus'. In: *Russia is in danger SEU Times. The newsletter of the Socio-Ecological Union,* Moscow, Jan. 17.

Ruhlen, M. 1994. *The origin of language: tracing the evolution of the mother tongue.* John Wiley & Sons, New York, ISBN 0-471-58426-6, 526pp.

Ruttner, F. 1988. *Biogeography and taxonomy of honeybees.* Springer-Verlag, Berlin Heidelberg, 284 p.

Ryabushkina, N., Gemedjieva, N., Kobaisy, M., Cantrell, Ch. L. 2008. Brief review of Kazakhstan flora and use of its wild species. *The Asian and Australias. J. Plant Sci & Biotechn.,* 2(2):64-71.

Sabashvili, M. N. 1936. *Soils of the sub-humid, subtropical zone of Georgian SSR [Pochvy sub-gumidnoi subtropicheskoi zony Gruz.SSR].* Metsniereba, Tbilisi, 204pp. (in Russian).

Sabashvili, M. N. 1948. *Soils of Georgia [Pochvy Gruzii].* Metsniereba, Tbilisi, 396pp. (in Russian).

Sabashvili, M. N. 1965. *Soils of Georgian SSR [Sakartvelos SSR niadagebi].* Metsniereba, Tbilisi, 552pp. (in Georgian).

Sadikhov, I. A. 1981. *Taxonomy and bio-ecological peculiarities of species of the genus Stipa L. of the Nakhichevan ASSR [Systematika i bio-ekologicheskie osobennosti vidov roda Stipa L. Nakhichevanskoi ASSR]. PhD thesis biol. sci.* Baku, 1-25pp. (in Russian).

Safarov, I. S. 1962. *The most important Tertiary arboreal relicts of Azerbaijan [Vazhneishie drevesnye tretichnye relikty Azerbaidzhana].* Izd. A N Az.SSR. Baku, 312 pp (in Russian).

Safarov, I. S. 1966. *Dzelkva, its biological and ecological peculiarities and industrial significance [Dzel'kva, eiyo biologicheskoe i ekologicheskoe osobennosti i promyslennaiya znachimost'*]. Izd. AN Az. SSR [AN Az. SSR Publ.], 1:15-17 (in Russian).

Safarov, I. S. 1979. The subtropical forests of Talysh [Subtropicheskie lesa Talysha]. *AN Az. SSR, Inst. Bot.* im V. L. Komarova. Elm, Baku, 157 pp. (in Russian).

Safarov, I. S., Olisaiev, V. A. 1991. *Caucasus forests [Lesa Kavkaza]*. Izdanie IR, Vladikavkaz, 271pp. (in Russian).

Safriel, U., Adeel, Z. 2005. Dryland systems. In: *Hassan, R., Scholes, R., Ash, N (Eds.). Ecosystems and human well-being: Current state and trends.* Island Press. Washington,: 625-658pp.

Safronova, I. N. 1980. The northern Turanian (the Caspian Sea coastal) deserts [Severoturanskie (prikaspijskie) pustny]. In: *Vegetation of the European part of the USSR.* Nauka, Leningrad, 291-295 (in Russian).

Safronova, I. N. 2008. Studies on the halophyte desert vegetation in the northern Caspian Region (Caspian Lowland and Mangyshlak). In: *Biosaline Agriculture and High Salinity Tolerance*, section II (C. Abdelly, M. Oztürk & M. Ashraf-Eds), pp. 221–232. Birkhaüser Verlag AG, Basel.

Sage, R. F. 2001. Environmental and evolutionary preconditions for the origin and diversification of the C_4 photosynthetic syndrome. *Plant Biology,* 3:202-213.

Sage, R. F. 2004. The evolution of C_4 photosythesis. *New Phytol.,* 161:341-370.

Sage, R. F., Christian, P.-A., Edwards, E. J. 2011. The C_4 plant lineages of planet Earth. *J. Exp. Bot.,* 62,9:3155-3169.

Sage, R. F., Li, M., Monson, R. K. (Eds.). 1999. *C_4 plant biology.* San Diego, CA: Academic Press.

Saghatelyan, A. 2006. *Flora of Armenia: its composition, analysis and relationships.* Turczaninowia, 9,3:5-47.

Sagheb-Talebi, K., Sajedi, T., Pourhashemi, M. 2014. *Forests of Iran – a treasure from the past, a hope for the future.* Springer. Dordrecht, Heidelberg, New York, London, 149pp.

Saintot, A., Brunet, M. F., Yakovlev, F., Sébrier, M., Stephenson, R., Ershov, A., Chalot-Prat, F., McCann, T.2006. The Mesozoic-Cenozoic tectonic evolution of the Greater Caucasus. In: *Gee, D. & Stephenson, R. A. (Eds.). European Lithosphere Dynamics. Geological Society*, London, Memoirs, 32:277-289.

Sakai, A. 1971. Freezing resistance of relicts from the Arcto-Tertiary flora. *New Phytol.*, 70:1199-1205.

Sakhokia, M. F. 1958. *Botanical excursions over Georgia [Botanikuri ekskursiebi saqartveloshi]*. Gruz. SSR Ac. Sci. Press, Tbilisi, 108pp. (in Georgian).

Sakhsuvarov, R. T. 1994. *The psammophyte vegetation of samur-divichi (alluvial-sea) plain Caspian shores [Psammofitnye rastenija Samur-Divichiskoii nizmennosti...]*. Elm, Baku, 134 pp (in Russian).

Sala, O. E., Austin, A. T., Vivanco, L. 2001. Temperate grassland and shrubland ecosystems. In: Levin SA (Ed.) *Encyclopedia of Biodiversity*. Academic Press, San Diego, 5:627-635.

Sala, O. E., Chapin, I. F. S., Armesto, J. J. et al. 2000. Global biodiversity scenarios for the year 2100. *Science*, 287(5459):1770-1774.

Salaev, M. E. 1966. *Soils of the Lesser Caucasus (in the border of Az. SSR) [Pochvy Malogo Kavkaza (v granitcakh Az. SSR]*. Elm, Baku, 328pp. (in Russian).

Sales, F., Hedge, I. C. 1996. Biogeographical aspects of selected SW Asiatic woody taxa. *Ann. Naturhist. Mus.Wien*. 98B Supp. 149-161.

Satunin, K. A. 1910. Some viewpoints on the origin of the Caucasian fauna [Nekotorye soobrozheniya o proiskhozhdenii fauny Kavkazskogo Kraiya]. *Izv. Kavk. Otd. Russk. Geogr. Obsch. [Memoirs of the Caucasian Department of the Imperial Russian Geographical Society]*. Tiflis, 20(2):1-12 (in Russian).

Satunin, K. A. 1912. On zoogeographical districts of the Caucasus territory [O zoogeograficheskikh okrugakh Kavkazskogo Kraja]. *Nov. Kavkaz. Muz. (The Caucasus Museum News)*, Tiflis, 7(1), 1-106 (in Russian).

Satunin, K. A. 1920. The mammals of the Caucasian region [Mlekopytaiyuchie Kavkazskogo regiona]. *Trudy Muzeiya Gruzii [Proceed. Muz. Gruzii]*, 2, II, Tiflis (in Russian).

Scheiter, S., Higgins, S. I., Osborne, C. P. et al. 2012. Fire and fire-adapted vegetation promoted C_4 expansion in the Late Miocene. *New Phytologist*:195:653-666.

Schiebinger, L. L. 2004. *Nature's body: gender in the making of modern science*. Rutgers University Press, 314pp.

Schild, A. S. 2008. Foreward, In: *International Mountain Biodiversity Conference,* Katmandu, Nepal, 16-18 November, 2008.

Schmidt, P. A. 2005a. Conservation in the Caucasus. Europe and non-European countries; Caucasus [Naturschutz in Kaukasien. Europa und außereuropäische Länder, Kaukasien], 14-6. In: Konold W., Böcker R., Hampicke U. *Handbuch Naturschutz und Landschaftspflege*, XIV-6. 15. ErgLfg. 4/05: 1-26. Landsberg: ecomed.

Schmidt, P. A. 2005b. Biodiversity and its protection in the Caucasus region [Biologische Vielfalt und ihr Schutz in der Kaukasusregion]. In: Erdmann K.-H. & Bork H. R. (Hrsg.): *Zukunftsfaktor Natur – Blick –Punkt Berge und Gebirge. Bundesamt für Naturschutz*, Bonn-Bad Godesberg. S. 159-177.

Schmidt, S. 2007. Vegetation and soil formation of Shirvan National Park, Azerbaijan. *Archives of Nature Conservation and Landscape Research*, 46: 29–69.

Schmidt, S., Pietzsch, D. 2007. Contributions to the number of a *Gazella subgutturosa subgutturosa* population in Shirvan National Park, Azerbaijan. *Archives of Nature Conservation and Landscape Research*, 46(2):71-78.

Schmidt, S., Uppenbrink, M. 2009. Potential analysis for further nature conservation in Azerbaijan. *Geozon Science Media*, Greifswald, 164pp.

Scott, A. C. 2000. Pre-Quaternary history of fire. *Palaeogegraphy, Palaeoclimatology, Palaeoecology*, 164: 297-345.

Schröter, C. 1926. The plant life of the Alps [Das Pflanzenleben der Alpen], 2nd ed. Verlag A. Raustein, Zürich, 1283pp. *Science Media*, Greifswald, 164p.

Schuerholz, G. 2004. Transboundary ecosystem restoration in politically high conflict areas. Present to the 16th International Conference, Soc. Ecol. Rest. August 24-26, 2004. Victoria, Canada. *Soc. Ecol. Rest.,* Washington, D.C., USA, 1-9.

Sclater, P. L. 1858. On the general geographical distribution of the members of the class. *Aves. Journal of the Proceedings of the Linnean Society (Zoology),* 2: 130-145.

Scott, A. C. 2009. Forest fire in the fossil record (1-37pp.) In: *Fire effects on soil and restoration* strategies (Cerda, A. & Robichaud, P.R.-Eds). Science Publ., USA, 535pp.

Scott, A. C., Bowman, D. M. J. S., M., S., Bond, W. J., Pyne, S. J., Alexander, M. E. 2014. *Fire on Earth: an introduction*. Chichester: John Wiley and Sons, 434pp.

Scott, A. C., Cripps, J., Nichols, G., Collinson, M. E. 2000. The taphonomy of charcoal following a recent heathland fire and some implications for the interpretation of fossil charcoal deposits. *Palaeogeog*raphy *Palaeoclimato*logy *Palaeoeco*logy, 164:1–31.

Scott, R. A., Barghoorn, E. S., Leopold, E. B. 1960. How old are the angiosperms? *Amer. Journ. Sci.,* 258-A:284-299.

Searle, K. R., Shipley, L. A. 2008. The comparative feeding behaviour of large browsing and grazing herbivores. In: *Gordon, I. J., Prins, H. H. T. (Eds.): The Ecology of Browsing and Grazing: 117–148. Springer Series, Ecological Studies* 195, Berlin, Heidelberg.

Senft, A. R. 2009. *Species diversity patterns at ecotones. MSc Thesis.* University of North Carolina at Chapel Hill, Department of Biology. 55 pp.

Sereno, P. C., Martinez, R. N. Wilson, J. A. et al. 2008. Evidence for avian intrathoracic air sacs in a new predatory dinosaur from Argentina. *PloS one*, 3,9:e3303. doi:10.1371/ journal.pone.0003303. Service. Agr. Inf. Bull., 415, 120 pp.

Sessa, E. B., Zimmer, E. A., Givnish, Th. J., 2012. Unraveling reticulate evolution in North America *dryopteris* (Dryopteridaceae). *BMC Evolutionar Biology*, 12:104. doi: 10.1186/ 1471-2148-12-104.

Severtsov, N. A. 1877. *On the zoological (mainly ornithological) regions of nontropical parts of our mainland [O zoologicheskikh (Preimuschestvenno ornitologicheskikh) oblast'yakh vnetropicheskikh chastei nashego materica].* Izv. Imp. Russk. Geogr. Obshch. 13,3: 123-155 (in Russian).

Shackleton, D. M. 1997. Wild sheep and goats and their relatives. Status survey and conservation action plan for Caprinae. *Gland IUCN*, 390pp.

Shahid, S. A., Abdelfattah, M. A. 2009. Gypsum polymorphism in the desert environment of Abu Dhabi Emirate. *Europ. J. Sci. Res.*, 29,2.237-248.

Shaneen, H., Sarwar, R., Firdous, S. S. et al. 2015. Distribution and structure of conifers with special emphasis on *Taxus baccata* in moist temperate forests of Kashmir Himalayas. *Pak. J. Bot.*, 47:71-76.

Shanina, A. A. 1987. The rare species of plants in pistachio-juniper communities of the North-West Caucasus [Redkie rasteniya v fistashnikovo-mozhevelikovye soobchestva Severo-Zapadnogo Kavkaza]. *Sbornik Nauchn. Trudov [Collect. Sci. Works]*: 19-24, Moscow (in Russian).

Sharashidze, R. V. 1967. *Dzelkova woods of Georgia [Dzel'kvovye lesa Gruzii]. PhD thesis biol. nauk.* Metsniereba, Tbilisi, 21pp. (in Russian).

Sharitz, R. R., Mitsch, W. J. 1993. Southern floodplain forests. In: *Martin, W. H., Boyce, S. G., Echternacht, A. C. (Eds.). Biodiversity of the Southeastern United States: lowland terrestrial communities. 1st ed.* New York: John Wiley and Sons: 311–372.

Sharkov, E., Lebedev, V., Chugaev, A. et al. 2015. The Caucasian-Arabian segment of the Alpine-Himalayan collisional belt: geology, volcanism and neotectonics. *Geoscience Frontiers,* 6,4:513-522.

Shaw, K., Stritch, L., Rivers, M. et al. 2014. The red list of Betulaceae. *BGGI, USDA Forest Service, FFI, Global Trees Campaign, IUCN/SSC,* 66pp.

Shcherbakova, E. M. 1973. The Ancient Glaciations of the Great Caucasus [Drevnee Oledenenie Bol'shogo Kavkaza]. University Press, Moscow, 272 pp. (in Russian).

Shetekauri, Sh., Chelidze, D. 2016. *High mountain flora of Meskheti and Javakheti (Lesser Caucasus).* Saari Publ. Tbilisi, 511pp. (in Georgian with English summary).

Shetekauri, Sh., Tsiskarauli, L., Zangurashvili, T. 2006. High mountain flora of Pirikiti Khevsureti and Tusheti (NE Greater Caucasus). *Fl. Medit.*, 16:355-378.

Shiffers, E. V. 1953. *The vegetation of the North Caucasus and its natural meadowlands [Rastitel'nost' Severnogo Kavkaza i ego prirodnye kormovye ugod'iya].* The USSR Acad. Sci. Publ. House [Izdatel'stvo AN SSSR], 399pp. (in Russian).

Shiji, W., Binghao, C., Hugun, L. 1996. *Euphrates poplar forest*. China Environ. Sci. Press, 117p.

Shipley, L. A. 1999. Grazers and browsers: how digestive morphology affects diet selection. In: *Launchbaugh, K. L., Mosley, J. C., Sander, K. D. (Eds.). Grazing behavior of livestock and wildlife*. University of Idaho. 70:20-27.

Schuerholz, G. 2004. Trans-boundary Ecosystem Restoration in Politically High Conflict Areas. *16th International Conference on Restoration on the Edge, Victoria, Canada, Society for Ecological Restoration,* 24-26 August 2004, p. 9. http://www.globalrestorationnetwork.org/uploads/files/LiteratureAttachments/349_transboundary-ecosystem-restoration-in-politically-high-conflict-areas.pdf, 12 May 2015.

Silva, G. J., Souza, T. M., Barbieri, R. L., Costa de Oliveira, A. 2014. Origin, domestication, and dispersing of pear (*Pyrus* spp.). *Advances in Agriculture* 2014: 1–8.

Silvestro, D., Cascales-Miñana, B., Bacon, C. D., Antonelli, A. 2015. Revisiting the origin and diversification of vascular plants through a comprehensive Bayesian analysis of the fossil record. *New Phytol.* doi: 10.1111/nph.13247.

Simpson, G. G.. 1966. The biological nature of man. *Science*, 152(3721):472-478.

Sinitsin, V. M. 1965. Ancient climate of Eurasia. Part I. *Paleogene and Neogene [Drevnye Klimaty Evrazii. Chast' I. Paleogen i Neogen]*. Izd. LGU (Leningrad State University), Leningrad, 166 pp. (in Russian).

Skhirtladze, N. I., Koronovski, N. V., Shirinian, K. G. 1977. The late orographic (Neogene-Anthropogene) volcanism of the Caucasus. *Himalayan Geology*, 7:150-158.

Smith, C. H. 1983. A system of world mammal faunal regions. I. Logical and statistical derivation of the regions. *J. Biogeogr.,* 10:455-466.

Smith, F., M., 1973. Abiotic functions in grassland ecosystems. Analysis of sturcture, function, and utilization of grassland ecosystems. Vol. 2. *A Progress Report (Principal investigator, G. M. Van Dyne), US/IBP Grassland Biome*. Colorado State University, Fort Collins Colorado: 29-96.

Smith, H. 2002. Concepts of race in culture and biology: From Carolus Linnaeus to Colaude Levi-Strauss. https://www.academia.edu/4041722/_Race_and_Biology_A_Historical_Perspective_From_Carolus_Linnaeus_to_Claude_Levi_Strauss_2002.

Snisarenko, T. A. 2008. Adaptation genesis of xerophytes of Ciscaucasia. *Vestnyk Chuvashskogo Gos. Pedag. Univ.* [*Herald of Chuvashia. St. Pedog. Univ.*], Cheboksaria, 280-288pp. (in Russian).

Sochava, V. B. 1949. *On the origin of beech forest of the Caucasus [K proiskhozhdeniyu bukovikh lesov Kavkaza]*. Publ. House Ac. Sci. USSR, Biol. Ser. 2:112-121 (in Russian).

Sokhadze, E. V. 1982. *Limestone and vegetation (Botanical-Geographical analysis on the example of Eurasia) [Izvestniyaki i rastitel'nost' (botaniko-geograficheski analiz na primere Evrazii]*. Metsniereba Publ., Tbilisi, 162pp. (in Russian).

Sokhadze, E. V., Sokhadze, M. E. 1952. About the biological peculiarities of Botriochloetum and structure of Botriochloetum communities of the Caucasus [K biologicheskoi osobennosti Botriochloetum i struktura Botriochloetum soobchestva]. *Bull. MOIP-Moscow Soc. of Nat. Invest. Biol. Dep.*, 57(3): 82-91 (in Russian).

Sokhadze E. V., Sokhadze, M. E. 1991. *Steppe and semi-desert vegetation of East Georgia [Stepnaya i polupustynnaiya rastitel'nost' Gruzii]* (unpublished manuscript) (in Russian).

Sokhadze, M. E. 1970. The rhythm of seasonal development of the vegetation of beard grass steppes and semi-deserts of East Georgia [Ritmy sezonnogo razvitiya rastitel'nosti kovil'nykh stepei i polupustyn' Vostochnoi Gruzii]. *Bull. MOSH, Odt. Biol. (Sect. Biol.)*, 75,17:29-38 (in Russian).

Sokhadze, M. E. 1977. *The ecological-biological and coenitical peculiarities of vegetation of beard grass steppes of East Georgia [Ekologo-biologicheskie i cenoticheskie osobennosti rastenii borodachevoy stepy Vostocnoi Gruzii].* Metsniereba., 177pp. (in Russian).

Sokolov, I. I. 1959. *The USSR fauna: Mammals [Fauna SSSR: Mlekopytaiyschie]* (orders Artiodactyla and Perissodactyla). Leningrad: USSR Acad. Sci. Press (in Russian).

Sokolov, V. 1974. *Saiga tatarica.* Mammalian species. *The American Society of Mammalogists*, 38:1-4.

Sokolov, V. E, Syroechkovski, E. (Eds.) 1990. Nature reserves of the Caucasus [Zapovedniki Kavkaza]. In: *Zapovedniki SSSR* (*Strict Nature Reserves of the USSR*), Mysl, Moscow, 365pp. (in Russian).

Sokos, Ch., Birtses, P. 2014. The east indigenous black-necked pheasant population of Europe. *Newsletter of the Galliformes Specialist Group*: 13-22.

Solomon, J., Shulkina, T., Schotz, G. E. (Eds.) 2014. *Red list of the endemic plants of the Caucasus: Armenia, Azerbaijan, Georgia, Iran, Russia, and Turkey.* St. Louis, US: Missouri Botanical Garden Press, 451p.

Solomonia, O., Bushueva, I., Dolgova, E. et al. 2016. Glacier variations in the northern Caucasus compared to climatic reconstructions over the past millennium. *Global and Planetary Change*, 140: 28-58.

Soltanmuradova, Z. I., Teymurov, A. A. 2010. Taxonomic composition of flora of the coastal lowland of the Republic of Dagestan [Taksonomicheskaiya structura flory primorskoi nizmennosti Respubliki Dagestan]. *Yug Rossii*: *Ecologiya, razvitie*: 3:32-38. Makhachkala (in Russian).

Soskov, Yu. D. 1988. *The genus Calligonum L. (systematics, geography, evolution). Doctor dissert. (PhD thesis) biol. sci.* Leningrad, 31pp. (in Russian).

Sosnovski, D. I. 1928. The modern conditions and near future of Eldarian pine in the South Caucasus [Sovremiyonnoe sostoiyanie i blizhaichee buduchee El'darskoi sosny v Iuzhnom Kavkaze]. In: *Izv. Azerb. Gos. Univers.*, I-VII, I-24 (in Russian).

Sosson, M., Rolland, Y., Danelian, T., et al. 2010. Subductions, obduction and collision in the Lesser Caucasus (Armenia Azerbaijan, Georgia), new insights. In: *Sosson, M. et al. Sedimentary Basin Tectonics from the Black Sea and Caucasus to the Arabian Platform. Geol. Soc.*, London, Spec. Publ., 340:329–352.

Spengenberg, V., Arakelyan, M., Galoyam, E., Matveevsky, S. et al. 2017. *Reticulate evolution of the rock lizards: meiotic chromosome dynamics and spermatogenesis in diploid and triploid males of the genus Darevskia. Genes (basel)* 2017 Jun: 8960: 149. Published online 2017 May 24.doi:10.3390/genes8060149.

Spiegelberger, P. Th., Matthies, A. D., Müller-Schärer, H., Schaffner, Ch. U. 2006. Scale-dependent effects of land use on plant species richness of mountain grassland in the European Alps. *Ecography,* 29:541-548.

Spodek M, Burckhardt D, Freidberg A (2017). The Psylloidea (Hemiptera) of Israel. *Zootaxa,* 4276: 301-345.

Stanjukovitch, K. V. 1955. General types of zonation of the mountains of the USSR [Osnovnye Tipy zonal'nosti gor SSSR]. *Izvestia Vsesoiuzn. Geogr. Obschestva*. 87, 3:412-436 (in Russian).

Stanjukovitch, K. V., Stanjukovitch, M. B. 1972. The basic principles of division of the vegetation into zonal types in the mountains of the USSR [Osnovnye principy vydeleniya tipov poiyasnosti rastitel'nogo pokrova v gorakh SSSR]. *In: Izv. VGO, 104, 3:174-182* (in Russian).

Stanton, M. L. 1983. Spatial patterns in the plant community and their effects upon insect search. In: *(Ahmad, S (Ed.). Herbivorous insects: host-seeking behavior and mechanisms,* Academic Press, NY:125-157.

State soil map legend of Azerbaijan 2003 [Legenda gosudarstvennoi karty pochv Azerbaijana 2003] (in Russian).

Staver, A. C., Levin, S. A. 2012. Integrating theoretical climate and fire effects on savanna and forest systems. *The American Naturalists,* 180,2:211-214.

Stebbins, G. L. 1971. Adaptive radiation of reproductive characteristics in angiosperms, II: seeds and seedlings. *Annual Review of Ecology and Systematics*, 2:237-260.

Stebbins, G. L. 1981. Coevolution of grasses and herbivores. *Annals of the Missouri Botanical Garden*, 68,1:75-86.

Stepanov, N. A. 1952. *Oak forests of the North Caucasus [Dubovye lesa Severnogo Kavkaza].* Goslesbumiz (State Forest Paper Publ.), 31:73-266, Moscow (in Russian).

Still, C. J., Berry, J. A., Collatz, G. J., DeFries, R. S. 2003.The global distribution of C3 and C4 vegetation: carbon cycle implications. *Global Biogeochem. Cycl.*, vol. 17,1006, doi:10.1029/2001 GB001807.

Stockwell, C. E. 2002. *Race and biology a historical perspective from Carolus Linnaeus to Claude Levi Strauss. Mast. of liberal Arts,* The University of Chikago, 1-15pp.

Strauss, A. 2005. Terrestrial vegetation and soil conditions of Ag-Gel National Park in Azerbaijan as a basis for a possible reintroduction of the Goitered gazelle (*Gazella subgutturosa*) (*Archive of Nature Conservation and Landscape Resaerch*), 44,4:37-57.

Stringer, C. B., Adrews, P. 1988. Genetic and fossil evidence for the origin of modern humans. *Science*, 239:1263-1268.

Strömberg, C. A. E. 2011. Evolution of grasses and grassland ecosystems. *Ann. Rev. Earth Planet Sci.:* 39:517-544.

Strömberg, C. A. E., McInerney, F. A. 2011. The Neogene transition from C_3 to C_4 grasslands in North America: assemblage analysis of fossil phytoliths, *Poleobiology,* 37,1:50-71.

Succow, M. 1995. Semi-desert and desert zones (biome) [Halbwüsten und wüstenzone]. In: Fukarek et al. *Urania Planzenreich-Vegetation. Urania Verlag*: Leipzig, Jena, Berlin: 150-177.

Suleimanov, M., Iniguez, L., Musayeva, M. 2006. Policy reforms and livestock development in Central Asia. In: *(Babu, S. C., Djalalov, S., (Eds.). Policy reforms and agriculture development in Central Asia: Springer*, 277-309.

Suttie, J. M., Reynolds, S. G., Batello, C. (Eds.). 2005. Grasslands of the world: Introduction, 1-18pp. (part.1). *Food and Agriculture Organization of the United Nations (FAO)*, Rome, Italy, 512pp.

Swift, M. J., Heal, O. W., Anderson, J. M. 1979. *Decomposition in terrestrial ecosystems.* vol.5. University of California Press. 372pp.

Tadevosyan, T. L., 2001. *Ex Situ Conservation of rare and endangered psammophilic species of the flora and plant associations of Ararat valley. Ph.D thesis boil. sci.*, Yerevan, 27pp. (in Russian).

Tadevosyan, T. L. 2002. On the syntopical distribution of the flora and vegetation in the 'Goravan Sands ecosystem'. *Flora, Vegetation and plant Resources of Armenia*, 14: 96-99 (in Russian).

Tadevosyan, T. 2006a. Habitat suitability for reptiles in the Goravan Sands Sanctuary, Armenia. *Herpetological Conservation and Biology*, 1: 39-44.

Tadevosyan, T. L. 2006b. New data on the abundance and distribution of the Persian toad-headed lizard in the Goravan Sands sanctuary, Armenia. *El. Journ. Nat. Sci.Nat. Ac. Rep. Armen.* 2(7): 45-50.

Tadevosyan, T. 2007. The role of vegetation in microhabitat selection of syntopic lizards, *Phrynocephalus persicus, Eremias pleskei* and *Eremias strauchi* from Armenia. *Amphibia-Reptilia,* 28: 444-448.

Taiz, L, Zeiger, E. 2010. Plant physiology (5th edition). Sunderland M.A: *Sinauer Associates Inc.*, 782pp.

Takhtajan, A. L. 1937. The xerophile vegetation of the skeleton mountains of Armenia [Kserofitnye rasteniya skeletnykh gor Armenii]. *Proc. Arm. Fil. Ac. Sci. USSR, Biol. Ser.*, 2: 44-53 (in Russian).

Takhtajan, A. L, 1941. Botanical-geographical essay of Armenia [Botaniko-geograficheski ocherk Armenii]. *Bull. BIN AN Arm. SSR*, 4:49-61 (in Russian).

Takhtajan, A. L. 1969. *Flowering plants: origin and dispersal.* Edinburg, Oliver and Boyd, 310pp.

Takhtajan, A. L. 1978. Floristic regions of the Earth [Floriticheskie region zemli]. *AN SSSR. Komarov's Bot. Inst.* Nauka Publ. Leningrad, 247pp. (in Russian).

Takhtajan, A. 1986. *Floristic regions of the world.* University of Califoria Press, 522pp.

Takhtajan, A. L. 1988. *Floristic regions of the world.* Dehra Dun: Bishen Singh, Mahendra Pal Singh (India).

Takhtajan, A. L. 1997. *Diversity and classification of flowering plants.* Columbia University Press, New York, 620pp.

Takhtajan, A. L., Fyodorov, An. A. 1972. *Flora of Yerevan [Flora Erevana].* Nauka, Leningrad, 396pp. (in Russian).

Talakhadze, G. M. 1964. *Main soil types of Georgia [Saqartvelos niadagta umtavresi tipebi].* Metsniereba, Tbilisi, 348pp.(in Georgian).

Talibov, T. H., Ibrahimov, A. M. 2009. Wild species of pears (*Pyrus* L.) in the flora of the Nakhichevan Autonomous Republic of Azerbaijan [Dikorastiyuschie grushi (*Pyrus* L.) flory Nakhichevanskoii Avtonomnoi Respubliki Azerbaidzhana]. *Turczaninowia*, 12(3-4): 82-87 (in Russian).

Talibov, T. H., Weinberg, P. I., Mammadov, I. B., Mammadov, E. N., Talibov, S. T. 2009. Conservation strategy of the Asiatic mouflon (*Ovis orientalis gmelini* Blyth) and the bezoar goat (*Capra aegagrus* Erxleben) in Azerbaijan. 46-52pp.. In: Zazanashvili, N. & Mallon, D. (Eds.). *Status and Protection of Globally Threatened Species in the Caucasus.* Tbilisi: CEPF, WWF. Contour Ltd.

Tanfiljev, V. G. 1971a. Essays of steppes vegetation of Stavropol Krai [Ocherk stepnoi rastitel'nosti Stavropol'skogo Kraiya]. *Mater. Po Izuchen. Stavropol'skogo Kraija* (*Mater. of Invest. of Stavropol Krai)*, Stavropol, 12-13: 79-96 (in Russian).

Tanfiljev, V. G. 1971b. The virgin steppes of Stavropol Krai [Tselinnye stepy Stavropol'skogo Kraiya]. *Botan. Zhurn.*, 56,5:692-701 (in Russian).

Tanfiljev, V. G. 1977. The dry and forb-bunchgrass steppes [Sukhie i dernovinnozlakovye stepy]. *Trudy SNIISK* (*Bull. SNIISK*), 5:7-76 (in Russian).

Tanfiljev, V. G. 1980. The meadow and forb steppes of Stavropol Upland [Lugovye i raznotravnye stepy Stavropol'skoii vozvychennosti]. *Trudy SNIISX* (*Bull. SNIISK*):16-23 (in Russian).

Tanfiljev, V. G. 1987. About the flora of virgin steppes of Stavropol [K flore tselinnykh stepeii Stavropol'iya]. In: *Flora Severnogo Kavkaza i voprosy eiyo istorii* (*Flora of North Caucasus and questions of its history*), 4(1-116pp.) (in Russian).

Tanfiljev, V. G., Kononov, V. I. 1972. On the distribution of feather grasses in Stavropol Krai [O rasprostranenii kovyleii v Stavropol'skom Krae]. *Botan. Zhurn.* 57(8): 112-119 (in Russian).

Tarkhnisvili, D. 1996. *The distribution and ecology of the amphibians of the Caucasus. Doctoral dissertation (PhD thesis) Biol. Sci.,* Tbilisi, 48pp. (in Russian and English).

Tarkhnishvili, D., Gokhelashvili, R. K. 1999. The amphibians of the Caucasus. *Advances in Amphibian Research in the former Soviet Union.* Vol. 4 Pensoft, 239pp.

Tarkhnishvili, D., Kaya, U., Gavashelishvili, A., Serbinova, I. 2008. Ecological divergence between two evolutionary lineages of the Caucasian salamander: evidence from GIS analysis. *Herpetol. J.*, 18:155-163.

Tarkhnishvili, D., Kikodze, D. (Eds.). 1996. *Principal characteristics of Georgian Biodiversity.* Natura Caucasica (Tbilisi), 1 (2), 1-46.

Tatashidze, Z. 2000. Abkhazeti. In: *Geography of Georgia, Part I, Physical Geography.* Metsniereba, Tbilisi, 284-293 (in Georgian).

Taylor, J. 1979. Tree worship. *The Mankind Quarterly:* 79-142.

Taylor, K. L. 1990. The tribe Ctenarytainini (Hemiptera: Psylloidea): a key to known Australian genera, with new species and two new genera. *Invertebrate Taxonomy*, 4:95–121.

Taylor, D. W., Hickey, L. J. (Eds.). 1996. Flowering plant origin, evolution and phylogeny. New York. Chapman and Hall. *Introduction: the challenge of flowering plant history* (Taylor & Hickey, 1-7pp.), Chapman & Hall, 403pp.

Tegetmeyer, C., Peper, J., Etzold, J., Schmidt, S. 2007. Conservation value of the shifting sand dune complex and lagoons in Samur Divichi Lowlands (Northern Azerbaijan). *Archiv für Naturschutz und Landschaftsforschung* (*Archives of Nature Conservation and Landscape Research*), 46:79-93.

Teo, Th. 2009. Psychology without Caucasus. *Canadian Psychology*, 50, 2:91-97.

This, P., Lacombe, T., Thomas, M. R. 2006. Historical origin and genetic diversity of vine grapes. *Trends in Genetics*, 22:511-519.

The Great Soviet Enciclopedia, 1978. Vol. 25, 544pp.

Tielidze, L. G. 2016. Glacier change over the last century, Caucasus Mountains, Georgia, observed from old topographical maps, Landsat and ASTER satellite imagery. *The Cryosphere*, 10:713-725.

Tilba, P. A., Mnatsakanov, R. A., Krutolapov, V. A. 2006. Wetland-flooded areas of the Northern Caucasus. Imereti Lowland [Zabolochennye mesta Severnogo Kavkaza. Imeretskaiya nizmennost']. *Wetland-flooded areas of Russia*, vol. 6:98-108 (in Russian).

Tockner, K., Standford, J. A. 2002. Riverine flood plains: present state and future trends. *Environ. Conserv.*, 29:308-330.

Toderich K. N., Tsukatani, T. 2007. New approaches for biosaline agriculture development, management and conservation of Central Asian degraded dry lands. *Discussion paper 638. Kyoto Institute of Economic Research*, Kyoto, Japan, 19pp.

Togonidze, P. 2011. Typology of East Georgian open juniper woodlands. *Bull. Georg. Nat. Acad. Sci.*, 5(3): 103-106.

Tomaškin, J. 2007. Role of non-productional functions of grassland in soil protection and environment. *Carpath. J. of Earth and Environ. Sci.,* 2,1:33-38.

Troitsky, N. A. 1928. Boxwood on Saguramo range [Samshit na Saguramskom khrebte]. *Works Tiflisk. State Polytechnical Inst. [Trudy Tifl. Gos. Polyt. Inst.]*. 3:29-34 (in Russian).

Troitsky, N. A. 1933. 'The Sacred grove' of Zemo (Upper) Aragvi ['Sviyachenaiya roscha' Zemo Aragvi]. *Transc. Tbil. Bot. Inst.*, 1:47-49 (in Russian with German summary].

Troll, C. 1973. The upper timberlines in different climatic zones. *Arctic and alpine Research*, 5:13-18.

Tsereteli, G. V. 1942. The Armazian bilingua [Armazis bilingva]. *Enimkis Moambe*, 13:34-42 (in Georgian).

Tsereteli, G. V. 1949. The Armazian script and problem of the origination of the Georgian alphabet [Armaziiskoe pismennost' i problemy proiskhozhdeniya Gruzinskogo al'favita]. In: *Epigraphika Vostoka.* Vol. 3,59-71, Leningrad (in Russian).

Tsereteli, K. 1991. Armazian script (Semitological investigations) [Armazuli damtcerloba]. *Semitologiuri Dziebani*, 5:60-70 (in Georgian).

Tshikolovets, V. V., Nekrutenko, Y. P. 2012. *The butterflies of Caucasus and Transcaucasia (Armenia, Azerbaijan, Georgia and Russia).* Two volumes. Tshikolevets Publ., Kiev, 424pp. (48 color plates).

Tsitlidze, B., Stephanishvili, V., Kikalashvili, Sh. 2013. The Georgian bee [*Qartuli futkari*] 177-185pp. In: *(Gulua, B. (Ed.-in-chief), Georgia, our treasury. Agriculture*. Tbilisi, Iverioni Publisher, 33-39 (in Georgian).

Tugushi, K. H. 1972. The box tree groves of Abkhazeti. In: *Forests of Abkhazeti* [*Samshitovaija roscha Abkhazii*]. Alashara Publ. Sukhumi: 98-118 (in Russian).

Tuite, K. 1994. *Violet on the Mountain: an anthology of Georgian folk poetry (translated and edited by Kevin Tuite).* Amirani, 141pp. Tbilisi.

Tumadjanov, I. I. 1937. The geobotanical review of pine forests of the Ateni Gorge [Geobotanicheski ocherk sosnovykh lesov Atenskogo ychel'iya]. *Trudy Tbilissk. Bot. Inst., [Proceed. Tbilisk. Bot. Inst.]*, 2:92-99 (in Russian).

Tumadjanov, I. I. 1938. Forests of mountainous Tusheti [Lesa gornoi Tushetii]. *Trudy Tbilissk. Bot. Inst.*, 5:32-43.

Tumadjanov, I. I. 1940. Birch and beech forests in Inner Dagestan. *Trudy Tbilisskogo Bot. Inst-ta*. 7:44-52 (in Russian).

Tumadjanov, I. I. 1960. On the typology of subalpine crook-forest of the Teberdian Nature Reserve [K tipologii subal'piiskikh kryvolesii teberdinskogo zapovednika]. *Trudy Teberd. Zapovedn.* [*Proceed. Teberd. Zapovedn.*] 2, 1:17-23 (in Russian).

Tumadjanov, I. I. 1966. The ancient desert in Mountainous (Inner) Dagestan [Drevnye pustyni Vnutrennogo Dagestana]. *Botan. Zhurn.* 51,6:784-791 (in Russian).

Tunijev, B. S. 1990. On the independence of the Colchis center of amphibian and reptile speciation. *Asiatic Herpetological Research.*, 3:67-84.

Tunijev, B. S. 1995. On the Mediterranean influence on the formation of herpetofauna of the Caucasus Isthmus and its main xerophilous refugia. *Russian J. Herpetol.*, 2,2: 95-119.

Turna, I. 2004. Variation of morphological characters of Oriental spruce (*Picea orientalis*) in Turkey. *Biologia,* Bratislava, 59,4:519-526.

Turner, N. J., Luczaj, L. J., Miglioriani, P. et al. 2011. Edible and tended wild plants, traditional ecological knowledge and agroecology. *Cr. Rev. Plant Sci.*, 30:198-225.

Tushinsky G. K. (Ed.), 1968. *Glaciation of Elbrus Mt [Oledenenie El'brusa].* Proc. Univ. Moscow: 345 pp. (in Russian).

Tutberidze, B. 2012. Cenozoic volcanism of the Caucasian mobile belt in Georgia: its geological–petrological peculiarities and geodynamic conditions. *Turk. J. Earth Sci.,* 21:799-815.

Tutin, T. G., Heywood, V. H., Burges, N. A., Valentine, D. H., Walters, S. M., Webb, D. A. 1964. *Flora Europaea*, vol. 1, Cambridge University Press, Cambridge.

Tuz, A. S. 1972. To the questions of taxonomy of the domestic pear [K voprosu klassifikatsii grushi domashnei]. *Nauchn. Trudy* [*Sci. Collect*], *Moykop experiment. Station VIR*, 5, 165pp. (in Russian).

Tvaradze, M. S. 1984. Rhizophagus grandis in integrated control systems of forest protection against *Dendroctonus micans*. *Proceed. Int. Congr. of Entomol.* Hamburg, 17:610.

Tzvelev, N. N. 1989. The system of grasses (Poaceae) and their evolution. *Bot. Rev*. 55: 141–204.

Tzvelev, N. N. 2006. Brief synopsis of grasses (Poaceae) of Eastern Europe: the beginning of systems (trybe Bambuseae – Bromeae). [Kratkiy Konspekt Zlakov (Poaceae) Vostochnoi Evropy: nachalo systemy (triby Bombuseae-Bromeae). *News of Systematics of Higher Plants. [Nov. Syst. Vish. Rast]*, 38:100-112 (in Russian).

Ucler, A. O., Yucesan, Z., Demirci, A., Yavuz, H., Oktan, E. 2007. Natural tree collectives of pure Oriental spruce (*Picea orientalis* (L.) Link) on mountain forests in Turkey. *Journ. Environ. Biol.*, 28(2):295-302.

Uerpmann, H.-P. 1996. Animal domestication – accident or intention? In: *Harris, D. R. (Ed.). The origins and spread of agriculture and pastoralism in Eurasia.* London, UCL Press: 227-232.

Upchurch, G. R., Wolfe, J. A. 1987. Mid-Cretaceous to Early Tertiary vegetation and climate: evidence from fossil leaves and woods. In*: The origins of angiosperms and their biological consequences (Friis, E. M., Chaloner, W. G., Crane, P.R.-Eds).* Cambridge University Press, Cambridge: 75-105pp.

Urushadze, T. F. 1987. *Soils of the mountain forests of Georgia [Pochvy Gornykh lesov Gruzii].* Metsniereba, Tbilisi, 242pp. (in Russian).

Urushadze, T. F., Blum, W. E. H. 2014. Soils of Georgia*. Nova Sciences Publishers*. 252pp.

Urushadze, T., Manvelidze, Z., Tvauri, I., Dolidze, L. 2010. Oriental beech in Georgia – present state and conservation priorities. Communicationes Instituti Forestalis Bohemicae, Forestry and Game Management Research Institute, *Strnady,* 25:98-112.

Vachnadze, G., Makhatadze, L. 1989. Lowland oak forests of East Georgia and the necessity for their afforestation [Nizinnye lesa Vostochnoi Gruzii i neobkhodimost' ikh oblesenii]. *Transcauc. Tbil. For. Inst., Probl. Mount For*. 53-68. Tbilisi (in Russian).

Villani F, Pigliucci M, Cherubini M (1994) Evolution of *Castanea sativa* Mill. in Turkey and Europe. *Genet Res Cambridge* 63: 109–116.

Valentine, J. F. 1990. *Grazing management.* Academic Press, Inc., San Diego, California:130-136.

Valles, J., Torrell, M., Garnatje, T., Garcia-Jacas, N., Vilatersana, R., Susanna, A. 2003. Genus *Artemisia* and its allies, phylogeny of the subtribe Artemisiinae (Asteraceae, Anthemadea) based on nucleotide sequences of nuclear ribosomal DNA internal transcribed spacers (ITS). *Plant Biol.* 5: 274-284.

Van Klink, R., Van der Plas, F., Van Noordwijk, C. G. E., et al. 2015. Effects of large herbivores on grassland arthropod diversity. *Biol. Rev.* 90,2: 347–366.

Van Thienen, P., Vlaar, N. J., van den Berg, A. P., 2004, Plate tectonics on the terrestrial planets: Physics of the Earth and Planetary Interiors, *Elsevier*, 142:61–74.

Vavilov, N. I. 1926. Center of origin of cultivated plants [Tsentry proiskhozhdenija kul'turnykh rastenii]. *Bull. Appl. Bot. Genet Plant Breed*, 16, 248pp. (in Russian).

Vavilov, N. I. 1931. The role of Central Asia in the origin of cultivated plants. *Bull. Applied Bot. Genetics & Plant Breading* (USSR) 26,3:3-44 (in Russian and English).

Vavilov, N. I. 1992. *Origin and geography of cultivated plants*. Cambridge University Press., 198pp.

Vekua, A., Lordkipanidze, D. 2008. The history of vertebrate fauna in eastern Georgia. *Bull. Georg. Nat. Ac. Sci.*, 2,3:149-155 (in Georgian with English summary).

Vekua, A., Lordkipanidze, D. 2010. Dmanisi (Georgia) – site of discovery of the oldest hominid in Eurasia. *Bull. of the Georgian National Academy of Science*, 4,2:158-164.

Vekua, A., Maisuradze, G., Nadirashvili, R. 2012. Geomorphological, geological and palaeontological description of Gereji Desert. *Proceed. Georg. Nat. Mus., Nat. Sci. and Prehist. Sect.,* 4:85-94 (in Georgian with English summary).

Vereshchagin, N. K. 1947. *Game and commercial animals of the Caucasus [Okhotnitchie i promyslovye zhivotnye Kavkaza]*. Publ. House Ac. Sci. Az. SSR [Izdatel'. AN Az. SSR], Baku, 144pp. (in Russian).

Vereshchagin, N. K. 1958a. History of the terrestrial fauna formation of the Caucasian Isthmus [Istorya formirovaniya nazemnoi fauny Kavkazskogo peresheika], 486-505. In: *Verschagin N. K., Animal World of the USSR, [Zhivotny mir SSSR]*, Moscow-Leningrad, 656pp. (in Russian).

Vereshchagin, N. K. 1958b. Zoogeographical regionalization of the Caucasian Isthmus [Zoogeograficheskoe rajonirovanie Kavkazskogo peresheika], 506-514. In: *Vereshchagin, N. K., Animal World of the USSR [Zhivotniy Mir SSSR]*, Vol. 5:506-514 (in Russian).

Vereshchagin, N. K. 1959. *The mammals of the Caucasus (A history of the formation of the evolution of the terrestrial fauna [Mlekopytayuschie Kavkaza (Istoriya formarovaniya fauny)]*. Izdatel'stvo AN SSSR [The USSR Ac. Sc. Publ. House], 703pp. Moscow-Leningrad (in Russian).

Vickers, M., Kakhidze, A. 2004. *Pichvnari 1: Greeks and Colchians on the East coast of the Black Sea: Results of Excevation conducted by the joint British-Georgian expedition (1998-2002)*, 458pp.

Vigne, J. D. 2011. The origins of animal domestication and husbandry: a major change in the history of humanity and the biosphere. *C. R. Biol.,* 334: 171- 181.

Villani, F., Pigliucci, M., Cherubini, M. 1994. Evolution of *Castanea sativa* Mill. in Turkey and Europe. *Genet. Res.* 63: 109–116.

Vincent, S. J., Morton, A. C., Carter, A., Gibbs, S., Barabadze, T. G. 2007. Oligocene uplift of the Western Greater Caucasus: an effect of initial Arabia-Eurasia collision. *Terra Nova*, 19,2:160-166.

Vinogradov, M. S. 1969. *The Krasnopoliansk bee rearing nursery*. Apiacta, 1:113-114.

Volk, G. M., Richards, C. M., Henk, A. D., Reilley, A. A. 2006. Diversity of wild *Pyrus communis* based on microsatellite analyses. *J. Amer. Soc. Hort. Sci.*, 131,3:408-417.

Volodicheva, N. 2003. The Caucasus. Chapt. 15,350-377 In: *Shahgedanova, M. (Ed.). The physical geography of Northern Eurasia.* Oxford, Oxford University Press, 529pp.

Voronov A. G., Drozdov N. N., Myalo E. G., 1985. Biogeography of the world [Biogeografiya mira]. *Vyshaya Shkola [Higher School]*, Moscow, 272 pp. (in Russian).

Voropaev, G. V. 1986 (Ed.-in-Chief). *The Caspian Sea: hydrology and hydrochemistry.* Baydin, S. S., Kosarev, A. N. (Eds.). Nauka [Science] Publ., Moscow, 261p. (in Russian).

Voskanyan, V. E. 1976. Flora and vegetation of the upper part of the alpine and subnival belts of the Mt Aragats [Flora i rastitel'nost' verkhnei chasti al'piiskogo i subnival'nogo poiyasakh gory Aragats]. *Armen. Bot. Zhurn.*, 29,8:19-26.

Vtorov, P. P., Drozdov, N. N. 2001. *Biogeography [Biogeografia].* Vlados Press. Moscow, 304pp. (in Russian).

Wallace, A. R., 1876. *The geographical distribution of animals: with a study of the relations of living and extinct faunas as elucidating the past changes of the Earth's surface,* Vol. 1, 503 pp., Vol. 2, 607 pp., Macmillan, London.

Walter, H. 1968. Vegetation of the Earth: Ecological-physiological characteristics; Vol. 1. Tropical and subtropical zones (Tropicheskie i subtropicheskie zony). *Progress* Moscow, 552pp. (transl. from Germany in Russian)

Walter, H. 1974. Vegetation of the Earth: ecological-physiological characteristics. Vol.2: Forests of temperate zones [Lesa umerennoi zony]. *Progress.* Moscow. 424pp. (transl. from Germany in Russian).

Walter, H. 1975. Vegetation of the Earth: tundras, meadows, steppes, and non-tropical deserts [tundry, luga, stepy i vne-tropicheskie pustyni]. Vol. 3. *Progress*, Moscow, 428pp. (transl. from Germany in Russian).

Walter H, Breckle S-W. 1991. Ecology of the Earth. Geobiosphere [Ökologie der Erde. Geo-Biosphäre]. Band 4: Spezielle Ökologie der gemäßigten und arktischen Zonen außerhalb EuroNordasiens: *Zonobiom IV–IX, 2nd edn.* Stuttgart, Germany: Fischer,586pp.

Walther, G.-R., Post, E., Convey, P. et al. 2002. Ecological responses to recent climate change. *Nature* (6879):389-395.

Wang, N., H., A., McAllister, Bartlett, P. R., Buggs, J. A. 2016. Molecular phylogeny and genom size evolution of the genus *Betula* (Betulaceae). *Ann. Bot.*, 117(6):1023-1035.

Wang, Yu-Fei, Ferguson, D., K., Zetter, R. et al. 2001. Leaf architecture and epidermal characters in *Zelkova*, Ulmaceae. *J. Linn. Soc.,* 136:225-265.

Wanpo, H., Ciochon, R., Yumin, G. et al. 1995. Early *Homo* and associated artefacts from China. *Nature*, 378:275-278.

Ward, J. V., Tockner, K., Arscott, D. B., Claret, C. 2002. Riverine landscape diversity. *Freshwater Biology,* 47:807-819.

Ward, P. D. 2006. Out of thin air. Dinosaurs, birds, and earth's ancient atmosphere. *Joseph Henry Press*, Washington, D. C., 278pp.

Wardle, D. A. 2002. Communities and ecosystems: linking the aboveground and belowground components. *Monographs in Population Biology,* 34. Princeton University Press, NJ, 392 pp.

Wardle D. A., Bardgett, R. D., Klironomos, J. N., Setala H., van der Putten, W. H., Wall D. H. 2004. Ecological Linkages between aboveground and belowground biota. *Science*, 304: 1629-1633.

Wardle, P. 1974. Alpine timberlines. In: *Ives, J. D., Barry, R. G. (Eds.). Arctic and alpine environements* (371-402pp.). London: Methuen, Publ.

Warldon, B. L. 2011. Strategic use of forage kochia (*Kochia prostrata*) to revegetate wildlife habitat. *Natural Resources and Environmental Issues,* vol. 17, 17:1-4.

Warmuth, V. 2011. On the origin and spread of horse domestication. *A dissertation submitted to the University of Cambridge for the degree of Doctor of Philosophy.* Corpus Christi College, University of Cambridge: 1-138.

Wasson, K., Woolfolk, A., Fresquez, C. 2013. Ecotones as indicators of changing environmental conditions: rapid migration of salt marsh-upland boundaries. *Estuaries and Coasts*, doi 10.1007/s 12237-013-9601-8.

Wataru, Y., Hikosaka, K., Danielle, A. 2013. Temperature response of photosynthesis in C3, C 4, and CAM plants:temperature acclimation and temperature adaptation. *Photosynth. Res.* 119(1-2), 101-117.

Weigard, J. F., Krankina, O. N., Brooks, D. J., Petukhov, Y. P., Shurupova, M. F. 1995. Adaptive learning for scientific management of non-timber forest products in the Pacific Northwest: looking to Russia. *Version 4.3-12/15/95:49pp.*

Weinberg, P. J. 2008. *Capra caucasica.* In: *IUCN 2008, 2008 Red List of Threatened Species.* www. iucnredlist.org.

Weinberg, P. J., Fedosenko, A. K., Arabuli, A. B., et al. 1997. The commonwealth of independent states (former USSR). In: *Shackleton, D. M. (Ed.). Wild sheep and Goat and their relatives. Status survey and conservation action plan for Caprinae.* IUCN, Gland, Switzerland and Cambridge, UK.

Wertsch, J. 2013. Five things worth knowing about the Caucasus. *Pulitzer Center on Crisis Reporting.* Washington University, St. Louis. (https://global.?wustlaedu/five-things-caucasus).

White, I. M., Hodkinson, I. D. 1985. Nymphal taxonomy and systematics of the psylloidea (Homoptera). *Bull. Br. Mus. Nat. Hist.* (*Ent.*), 50(2):153-301.

Whitlock, C., Higuera, P. E., McWethy, D. B., Briles, C. E. 2010. Paleoecological perspectives on fire ecology: revisiting the fire-regime concept. *The Open Ecology Journal,* 3:6-23.

Whittaker, R. H. 1961. Vegetation history of the Pacific coast states and the "central" significance of the Klamath region. *Madroño* 16:5-23.

Williams, D. G., Scott, R. L., Huxman, T. E., Goodrich, D. C., Liv, G. 2006. Sensitivity of riparian ecosystems in arid and semi-arid environments to moisture pulses. *Hydrological Processes*, 20:3191-3205.

Williams, L., Zazanashvili, N., Sanadiradze, G., Kandaurov, A. (Eds.) 2006. *Ecoregional Conservation Plan for the Caucasus.* WWF Caucasus Programme Office. Contour Ltd, Tbilisi, 220 p.

Willis, K. J. 2009. Evolution and function of Earth's biomes: temperate forests. In: *Earth System: History and Natural Variability.* Vol. III. EOLoS. ILtd, Isle of Man, UK (1-12pp.).

Wilson, B. 2004. *The hive: the story of the honeybee and us.* London, John Murray, 311pp.

Wilson, D. E., Reeder, D. M. (Eds.). 2005. *Mammal species of the world: a taxonomic and geographic reference.* John Hopkins University Press, Third edition, Vol. 1-2, xxxv + 2142 pp. Baltimore.

Wing, S. L., Hickey, L. J., Swisher, C. C. 1993. Implications of an exceptional fossil flora for Late Cretaceous vegetation. *Nature,* 363:342-344.

Winter, K. 1981. C_4 plants of high biomass in arid regions of Asia – occurance of C_4 photosynthesis in Chenopodiaceae and Polygonaceae from the Middle East and USSR. *Oecologia*, 48:100-106.

Wolfe, J. A. 1977. *Paleogene floras from the gulf of Alaska region. Geological Survey Professional paper 997*. US Government Printing Office, Washington: 1977, 97pp. (Plates 1-30). [Lower Ravenian: 26-31pp.].

Wood, B. A. 2011. Did early Homo migrate 'out of' or 'into' Africa? *Proc. Nat. Ac. Sci.*, 108(26): 10375-10337.

Wood, F. 2004. *The silk road: two thousand years in the heart of Asia*. University of California Press, 270pp.

Woodward, S. L. 2003. *Biomes of Earth: terrestrial, aquatic, and human-dominated.* Greenwood Press, Westport, CT, 456pp.

Wrangham, R. W., Rachel, N. C. 2010. Human adaptation to the control of fire. *Evolutionary Anthropology*, 19,5:187-199.

Wulf, E. M., Maleeva, O. F. 1969. *The world resources of useful plants [Mirovye resursy poleznykh rastenii]*. Nauka [Science] Leningrad: 374-376.

Wulff, Y. V. 1944. *Historical Geography of Plants [Istoricheskaija geografija rastitel'nosti]*. Izdatel'stvo AN SSSR. [The USSR Ac. Sc. Publishing House], Moscow-Leningrad, 545p (in Russian).

Xiang, Q. Y., Soltis, D. E., Soltis, P. S. 1998. The eastern Asian, eastern and western-North American floristic disjunction: Congruent phylogenetic patterns in seven diverse genera. *Mol. Phylogenet. Evol.*, 10: 178–190.

Xiang, Q.-Y., Zhang, W.-H., Ricklefs, R. E., Qian, H., Chen., Z.-D., Wen, J., Li, J.-H. 2004. Regional differences in rates of plant speciation and molecular evolution: a comparison between Asia and Western North America, *Evolution*, 58:2175-2184.

Xie, L., Yang, Z. Y., Wen, J., Li, D.-Z., Yi, T.-S. 2014. Biogeographic history of *Pistacia* (Anacardiaceae), emphasizing the evolution of the Madrean–Tethyan and the eastern Asian-Tethyan disjunctions. *Mol. Phylogenet. Evol.*, 77:136-146.

Yaroshenko, P. D. 1956. Alternation of Transcaucasian vegetation cover in relation to soil-climatic changes and human impact [Smeny rastitel'nogo pokrova Zakavkaz'iya v ikh sviyazi s pochvenno–klimaticheskimi izmeneniyami i deyatel'nostiyu cheloveka]. *Acad. Sci. USSR: Far East Branch and Botanical Inst. after V. L. Komarova*, Moscow-Leningrad, 242pp. (in Russian).

Yi, T., Wen, J., Golan-Goldhirsh, A., Partiff, D. E. 2008. Phylogenetic and reticulate evolution in *Pistacia* (Anacardiaceae). *Am. J. Bot.*, 95,2:241-251.

Yuritsina, N. A. 2000. The geographical peculiarities of vegetation of the salinizate ecotypes of the northern edge of the northern Caspian Region [Geograficheskoe svoetbrazie rastitel'nosti zasolionnykh ekotipov v severnoi okraine severnogo kaspiiskogo regiona]. In: '*Actual Problems of Ecology*', 3, Samara, 219-220 (in Russian).

Yurtsev, B. A. 1966: *The Hypoarctic botanical-geographical belt and the origin of its flora [Gypoarkticheski botaniko-geograficheski poiyas i proiskhozhdenie ego flory]*. Nauka, Moscow-Leningrad, 94 pp. (in Russian).

Zabelina, N. M., Isaeva-Petrova, L. S., Kuleshova, L. V. 1998. Zapovedniks and National Parks of Russia. *Logata*. Moscow (in Russian and English), 160pp.

Zablotskaya, M. M. 2004. Current problems of the restoration of Caucasian wisent in Russia [Sovremiyonnye problemy vosstanovleniya zubra v Rossii]. In: *Problems of preservation and restoration of wisent. Collect. Sci. Works. Danki*: 43-47 (in Russian).

Zachos, J. C., Pagani, M., Sloan, L., Thomas, E., Billups, K. 2001. Trends, rhythms, and aberrations in global climate 65 Ma to present. *Science,* v. 292, no. 5517:292:686-693.

Zakharov, S. A. 1914. Characterization of the Caucasian high-mountain soils [Kharakteristika vysokogornykh pochv Kavkaza]. In: *Proceedings of the Konf. Instit. of Land Survey* [*Izvest. Zemled. Inst-ta*], Moscow, 1914), 5:62-64 (in Russian).

Zakharov, S. A. 1939. Soils of Ciscaucasia [Pochvy Severnogo Kavkaza]. In: *Pochvy SSSR (Soils of the USSR). Vol. III.* The USSR Acad. of Sci. Publ. House:236-253 (in Russian).

Zaletaev, V. S. 1976. *Life in the desert (geographic-biogeocenotic and ecology problems) [Zhizn' v pustyne (geographo-biocenoticheskie i ecologicheskie problemy].* Moscow, Mysl, 271 p. (in Russian).

Zamani, A., Attar, F., Civeyrel, L. 2015. Leaf epidermis characters of Iranian *Pyrus L.* (Rosaceae) and their taxonomic implications. *Genet. Resour. Crop Evol.* doi:10.1007/s10722-015-0341-4.

Zamani, A., Attar, F., Ghahreman, A., Maroofi, H. 2008. Anatomical studies of the genus *Pyrus* L. (Rosaceae) in Iran and its taxonomical implications. *Ir. J. Bot.* 14:132–142.

Zamtaradze, G. 1985. *Poisonous and weedy plants of hay meadows and pastures of Georgia [Saqartvelos mdelo-sadzovrebis shkhamiani da sarevela mcenareebi].* Metsniereba Press, Tbilisi, 51pp. (in Georgian).

Zazanashvili, N., Gagnidze, R., Nakhutsrishvili, G. 2000. Main types of vegetation zonation on the mountains of the Caucasus. *Acta phytogeogr. Suercica,*85:7-16.

Zazanashvili, N., Garfoth, M., Jungius, H., Gamkrelidze, I. 2012 (Eds.). *Ecoregion conservation plan for the Caucasus. 2012 revised and updated edition.* Edit. WWF-Caucasus Programme Office, Tbilisis, 66pp.

Zazanashvili, N., Gavashelishvili, L., Montalvo, C. et al. (Eds.) 2011. *Strategic guidelines for responding to impacts of global climate change on forests in the southern Caucasus (Armenia, Azerbaijan, Georgia),* WWF. KFW Entwicklungsbank. Available at: http://awsassets.panda.org/downloads/forest_strategy_for_south_caucasus_1.pdf (accessed: 26/03/2013).

Zazanashvili, N., Mallon, D. (Eds.) 2009. Status and protection of globally threatened species in the Caucasus. *Investments in the Caucasus Hotspot 2004-2009.* Tbilisi: CEPF, WWF, Contour Ltd, (232pp).

Zedelmeyer, O. M. 1933. Geobotanical review of vegetation of the western coastal site of lake Sevan [Geobotanicheski ocherk rastitel'nosti zapadnoi pribrezhnoi chasti ozera Sevan]. *Sbornik Basseina Ozera Sevan (Collection of Sevan Lake Basin)*, 3,3:32-38. Arm. Fil. AN SSR Publ. (in Russian).

Zeder, M. A. 2008. Domestication and early agriculture in the Mediterranean Basin: origins, diffusion, and impact. *Proc. Nat. Ac. Sci., USA*, 105,33:11597-11604.

Zelenko, O. V. 1969. *The types of riparian forests and forests communities of the eastern Ciscaucasia [Tipy pribrezhnykh lesov i lesnykh soobchestv vostochnogo Predkavkaz'iya]. Candidate dissert. PhD thesisbiol. sci.* Dnepropetrovsk, 23pp. (in Russian).

Zemlinski, S. E. 1958. *Medicinal plants of the USSR (3rd ed) [Lekarstvennye rasteniya SSSR].* Medgiz Publ., Moscow, 524pp. (in Russian).

Zendedel, A., Burckhardt, D., Fekrat, L. 2016. An annotated checklist of the jumping plant-lice (Hemiptera: Psylloidea) of Iran. *J. Entomol.Res. Soc.*, 18: 37-55.

Zeng, Yu, 2010. Modeling complex dynamics at alpine treeline ecotones. *PhD dissert biol. sci.* 147pp.

Zenkevich, L. A. 1963. *Biology of the seas of the USSR*. Interscience Publishers, New York,. 740pp.

Zerbe, S., Thevs, N., Kühnel, E. 2010. Vegetation, ecosystem dynamics, and restoration of floodplains in Central Asia – the Tarim River (Xinjiang, NW China) as an example. *Forest Ecology, Landscape Research and Nature Conservation*, 10: 85-89.

Zhang, Z., Wang, H., Guo, Z., Jiang, D. 2007. What triggers the transition of palaeoenvironmental patterns in China, the Tibetian Plateau uplift or the Paratethys Sea retreat? *Palaeogeography, Palaeoclimatology, Palaeoecology*., 245:317-331.

Zheng, X., Cai, D., Potter, D., Postmen, J., Liu, J., Teng, Y. 2014. Phylogeny and evolutionary histories of *Pyrus* L. revealed by phylogenetic trees and network based on data from multiple DNA sequences. *Molec. Phylogen. Evol.*, 80:54-65.

Zhisheng, A., Kutzbach, J. E., Prell, W. L., Porter, S. 2001. Evolution of Asian monsoons and phased uplift of the Himalayan Tibetian plateau since Late Miocene times. *Nature,* 411:62-66.

Zhordania, R., Boeme, B., Kuznetsov, A. 1999. Birds of Georgia. A Field Guide. WWF Country Office Georgia, *World Wide Fund for Nature*. Tbilisi 1999, 253pp.

Zhu, J.-T., Li, X.-Y., Zhang, Z.-M., Manzari, S., Namaghi, H., 2010. Ecophysiological adaptation of *Calligonum roborovskii* to decreasing soil water content along an altitudinal gradient in the Kunlin Mountain, Central Asia. *Russian Journal of Plant Physiology*, 57: 826.

Zhukov, V. S. 2004. Chorological analysis of ornithofauna of the Northern Eurasia: landscape and ecological aspects [Khronologichesk analiz ornitofauny Severnoi Evrazii: landshaftnye i ekologicheskii aspekty]. *Novosibirsk*, 182pp. (in Russian).

Zhukovsky, P. M. 1964. *Cultivated vegetations and their wild relatives [kul'turnye rasteniya i ikh dikie sorodichi]*. Kolos Publ. Leningrad, 791pp. (in Russian).

Zocchi, R. 1971. Contribution to the knowledge of the entomofauna associated with tamarisk in Italy [Contributo alla conoscenza dell'entomofauna delle tamerici in Italia]. *Redia,* 52:31-129.

Zoghi, Z. Azadfar, D., Kooch, Y. 2011. Influence of physiographic factors on vegetative and morphological characters of Beech plus trees – A case study in Hyrcanian forest. *Int. J. Environ. Sci.*, 1,7:839-846.

Zohari, M. 1952. A monographic study of the genus *Pistacia*. *Palestine J. Bot.*.15:187-228.

Zohary, D. 1995.Domestication of the grapevine *Vitis vinifera* L. in the Near East. In: McGovern PE, Fleming SJ, Katz SH, Eds. *The origins and ancient history of wine.* Gordon and Breach, New York:pp.23–30.

Zonenshain, L. P., Kuzmin, M. I., Natapov, L. M. 1990. Geology of the USSR: A plate tectonic synthesis. American Geophysical Union. *Geodyn. Ser. Monogr.*, 21, 242p.

Zonn, I., Kostianoy, A., Kosarev, A., Glantz, M. 2010. *The Caspian Sea encyclopedia.* Springer, 525pp.

Zonn, S. V. 1950. *Mountain forest soils of the northwest Caucasus [Gornye pochvy severo-zapadnogo Kavkaza]*. Izd. AN SSSR. Moscow, 333pp. (in Russian).

Zwinger, A., Willard, B. E. 1996. *Land above the trees: A guide to American alpine tundra.* Johnson Books, Boulder, CO, 425pp.

ABOUT THE AUTHOR

Dr. Arnold Gegechkori is a famous zoologist and biogeographer, a Full Professor of Tbilisi State University, Faculty of Exact and Natural Sciences (Georgia, Transcaucasia), Head of the Chair of Zoology (1990-2006), Head of the Chair of Biodiversity (2006-present), Head of the Bachelor's program in 'Ecology'.Visiting Professor of biogeography at Monash University (Melbourne, Australia, 2000), at the UNSW (Sydney, Australia, 2000-2003), at the Greifswald University (Germany, 2008). Head of the Department of Zoology at the Georgian National Museum (1970-2005). Received his PhD (Candidate's) degree from Tbilisi Agricultural Institute (1966) and his PhD (Doctor's) degree from the Institute of Zoology, St. Petersburg (1985). The President of the International Society for 'The Caucasus Environment' (ISCE; 2001-2003); member of the International Geographical Union (IGU), National Geographic Society (USA), International Mire Conservation Group (IMCG) and the Federation of Journalists of Georgia; a State Prize Laureate of Georgia (in fiction literature), P. Melikishvili Prize Laureate (the highest scientific honour of Tbilisi State University), I. Gogebashvili I Prize Laureate (the highest educational honour of Georgia). Awarded the order of Honour (the highest award of the Republic of Georgia, 1985, 2012), Ivane Javakhishvili Medal (the highest award of Tbilisi State University), International man of the Year for 2000/2001 (IBC, Cambridge). Research interests: psylloids (Hemiptera, Psylloidea) - fauna, biology, ecology, biogeography, evolution. Courses delivered at the University (alma mater) include: zoology, historical biogeography, biomes of the world, introduction to biology, fundamentals of ecology. As a guest lecturer and guest researcher had a sabbatical leave in the countries of South-East Asia (1986), South America (1988) and on 5 continent's (2000). The scientific field expeditions involve all biomes of the world, almost all 'Biodiversity Hotspots' (CEPF, 2000), particularly the Caucasus (1962-1992), including all (56) the strict nature reserves and National Parks of the eco-region. Author or co-author of eight monographs and numerous scientific articles in European languages, as well as 11 books of fiction (novels, poetry, etc.). Presented a Georgian TV programe - "Window Through Nature" (1969-2000). Twice awarded a Gold Medal - "Nature Defender" (1985,1994). Hobbies: fine-art photography, sport.

Index of Latin Terms

A

B

C

F

G

H

I

J

K

L

M

Q

R

S

T

U

V

W

X

Y

Z

Related Nova Publications

Armenia: A Year at Erzeroom, and on the Frontiers of Russia, Turkey, and Persia

Author: Robert Curzon

Series: Caucasus and Central Asia Political, Social and Economic Issues

Book Description: The governments of Turkey and Persia requested the assistance of England and Russia to draw up a treaty of peace, and to come to a distinct understanding as to where the line of border ran between the two empires. This book originally published in 1854 describes the Curzon's travels through this region.

Softcover ISBN: 978-1-53615-068-1
Retail Price: $95

Drylands: Biodiversity, Management and Conservation

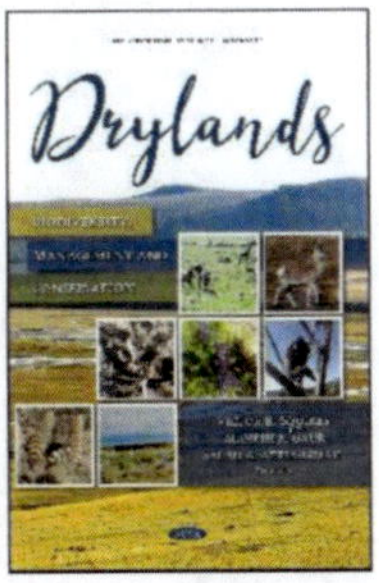

Editors: Victor Roy Squires, Mahesh Gaur, and Salah Attia-Ismail

Series: Environmental Research Advances

Book Description: This book opens up discussion on a range of issues relating to the important matters of biodiversity, land management and efforts at conservation (wise use) of natural resources that support an area's biodiversity.

Hardcover ISBN: 978-1-53615-896-0
Retail Price: $230

To see a complete list of Nova publications, please visit our website at www.novapublishers.com

Related Nova Publications

Microbiological Environmental Hygiene

Editor: Eino Elias Hakalehto, Ph.D.

Series: Microbiological Hygiene

Book Description: The book *Microbiological Environmental Hygiene* contains descriptions of the microbes and natural ecosystems from rain forests to arctic areas, as well as strains in the cleaning processes of human epithelia and food or water sources.

Hardcover ISBN: 978-1-53613-178-9
Retail Price: $395

Central Asia: Perspectives and Present Challenges

Editor: Bárbara Beyer

Series: Asian Political, Economic and Social Issues

Book Description: The opening chapter of *Central Asia: Perspectives and Present Challenges* aims to identify and address the unique features of conical and pointed domes, which topped a majority of distinct buildings in the cities of Merv, Samarkand, Bukhara, and Khiva from the Parthian to the Shaybanids era.

Hardcover ISBN: 978-1-53614-635-6
Retail Price: $160

To see a complete list of Nova publications, please visit our website at www.novapublishers.com